Bioaerosols:
Assessment and Control

Editor:

Janet Macher, Sc.D., M.P.H.

Assistant Editors:

Harriet A. Ammann, Ph.D., D.A.B.T.

Donald K. Milton, M.D., Dr.P.H.,M.O.H.

Harriet A. Burge, Ph.D.

Philip R. Morey, Ph.D., C.I.H.

ACGIH
1330 Kemper Meadow Drive
Cincinnati, OH 45240-1634
www.acgih.org

ISBN: 882417-29-1

Published in the United States of America by:

ACGIH
Kemper Woods Center
1330 Kemper Meadow Drive
Cincinnati, OH 45240-1634
Telephone: 513-742-6163 Fax: 513-742-3355
E-mail: comm@acgih.org
http://www.acgih.org

CONTENTS

FOREWORD
    Robert F. Herrick, Sc.D., C.I.H.

CONTRIBUTORS

ACKNOWLEDGMENTS

LISTS OF ABBREVIATIONS, ACRONYMS, UNITS, AND LATIN NAMES

PART I. BASIC INFORMATION AND GUIDANCE
    Chapter 1.   Introduction

    Chapter 2.   Developing an Investigation Strategy

    Chapter 3.   Health Effects of Bioaerosols

    Chapter 4.   The Building Walkthrough

    Chapter 5.   Developing a Sampling Plan

    Chapter 6.   Sample Analysis

    Chapter 7.   Data Interpretation

PART II. BACKGROUND INFORMATION
    Chapter 8.   Medical Roles and Recommendations
        Cecile S. Rose, M.D., M.P.H., with Kathleen Kreiss, M.D.; Donald K. Milton, M.D.,
        Dr.P.H., M.O.H.; and Edward A. Nardell, M.D.

    Chapter 9.   Respiratory Infections — Transmission and Environmental Control
        Edward A. Nardell, M.D. and Janet M. Macher, M.P.H., Sc.D.

    Chapter 10.   Prevention And Control Of Microbial Contamination
        Richard Shaughnessy, Ph.D. and Philip R. Morey, Ph.D., C.I.H., with Eugene C. Cole,
        Dr.P.H.

    Chapter 11.   Air Sampling
        Klaus Willeke, Ph.D., C.I.H. and Janet M. Macher, M.P.H., Sc.D.

    Chapter 12.   Source Sampling
        John Martyny, Ph.D., C.I.H.; Kenneth F. Martinez, M.S.E.E., C.I.H., with Philip R. Morey,
        Ph.D., C.I.H.

    Chapter 13.   Data Analysis
        Janet M. Macher, M.P.H, Sc.D.

Chapter 14.   Data Evaluation
Harriet A. Burge, Ph.D. with Janet M. Macher, M.P.H, Sc.D.; Donald K. Milton, M.D.,
Dr.P.H., M.O.H.; and Harriet A. Ammann, Ph.D., D.A.B.T.

Chapter 15.   Remediation Of Microbial Contamination
Richard Shaughnessy, Ph.D. and Philip R. Morey, Ph.D., C.I.H.

Chapter 16.   Biocides And Antimicrobial Agents
Eugene C. Cole, Dr.P.H. and Karin K. Foarde, M.S.

PART III. SPECIFIC AGENTS

Chapter 17.   Source Organisms — An Overview
Harriet A. Burge, Ph.D. and Janet M. Macher, M.P.H, Sc.D.

Chapter 18.   Bacteria
James A. Otten, S.M. and Harriet A. Burge, Ph.D.

Chapter 19.   Fungi
Harriet A. Burge, Ph.D. and James A. Otten, S.M.

Chapter 20.   Amebae
James A. Otten, S.M. and Harriet A. Burge, Ph.D.

Chapter 21.   Viruses
James A. Otten, S.M. and Harriet A. Burge, Ph.D.

Chapter 22.   House Dust Mites
Larry G. Arlian, Ph.D.

Chapter 23.   Endotoxin And Other Bacterial Cell-Wall Components
Donald K. Milton, M.D., Dr.P.H., M.O.H.

Chapter 24.   Fungal Toxins And β-(1$\rightarrow$3)-D-Glucans
Harriet A. Burge, Ph.D. with Harriet A. Ammann, Ph.D., D.A.B.T.

Chapter 25.   Antigens
Cecile S. Rose, M.D., M.P.H.

Chapter 26.   Microbial Volatile Organic Compounds
Harriet A. Ammann, Ph.D., D.A.B.T.

FOREWORD

When Theodore Hatch reviewed the advances made in preventing occupational disease in the first half of this century, he observed that progress had been made as a result of interdisciplinary collaboration "... a true melding, even to the point where the different individuals making up the group lost sight of their respective fields and functioned together as a unit, each making his own peculiar contribution, but always as part of the whole" (Hatch, 1964). This book, produced by the ACGIH Bioaerosols Committee in 1998, is a perfect example of the continued progress we can make by following Hatch's guidance. In this case, the product is a comprehensive guide to the assessment and control of bioaerosols. A glance at the list of editors and chapter authors reveals the remarkable diversity in professional backgrounds and specialties of the contributors. This group has converged upon the need to prevent occupational disease by applying the industrial hygiene paradigm of recognition, evaluation, and control to the set of hazards loosely termed *bioaerosols*. The result is *Bioaerosols: Assessment and Control*, which will help the industrial hygienist, indoor environmental specialist, and other occupational health professionals learn from the wide range of people who contributed their expertise to advance the state of knowledge of biologically derived airborne contaminants. Hatch would have been proud.

There is a small measure of irony in the publication of this book by an organization in a profession which still identifies itself by title as "industrial." The ACGIH Bioaerosols Committee has prepared guidance which will be useful in the full range of workplaces, including those which do not fit the traditional image of "industrial." These indoor workplaces, are in fact the environments in which most people spend their working time. The ubiquitous nature of biologically derived contaminants, and their importance as causal agents of work-related disease, are actually among the factors which forced our profession to broaden its scope to the non-industrial workplace. As workers in these apparently benign environments began to demonstrate symptoms which they attributed to their workplace, industrial hygienists first applied the tools we had available for evaluating the manufacturing environment. In many cases, the measurements of chemicals as gases, vapors, and aerosols revealed exposures far below the levels we had found in the manufacturing world. There was a tendency to dismiss the workers' complaints, because the exposures we measured in their environments did not approach the levels we had become accustomed to finding in the workplaces where industrial hygienists had traditionally practiced. In retrospect, it is clear that we were wrong to dismiss the workers for at least two reasons. First, we tended to discount evidence of a causal association between symptoms and the workplace because the exposure levels were below the limits. This was done despite the admonitions of the TLV Committee and others that the exposure limits are not meant to be interpreted as hard lines between safe and unsafe conditions. Second, we were not sufficiently aware of the limitations of our measurement methods, especially the fact that we usually made no measurements at all of the biologically derived contaminants which are the subject of this book. As the ACGIH guide clearly documents, an investigation of work-related disease must include some assessment of possible exposure to biologically derived contaminants, and this book will put this important tool into the hands of future investigators.

Finally, the book illustrates the application of the scientific method to the study of occupational hazards and disease. This is remarkable only by comparison to the conventional approach to industrial hygiene, which is frequently compliance-driven. But how does one design and conduct an investigation in a setting where there are no relevant exposure limits? The approach the Bioaerosols Committee has presented in this book is founded upon the scientific method of investigating potential causal relationships between exposure and disease. It is a thoughtful application of the scientific approach to workplace investigations which can and should be used over the full range of exposures and hazards, not just the biologically derived contaminants which are the subject of this book.

So *Bioaerosols: Assessment and Control* presents in a single volume a comprehensive guide to the recognition, evaluation, and control of biologically derived contaminants. It is a remarkable product of volunteer effort, and it represents a significant step toward the improvement of workplace conditions.

Robert F. Herrick, Sc.D., C.I.H.

References
Hatch, T.: Major Accomplishments in Occupational Health in the Past Fifty Years. *Ind. Hyg. J.* 25:108-113 (1964).

CONTRIBUTORS

Harriet M. Ammann, Ph.D., D.A.B.T.
Office of Environmental Health
    Assessment Services
Department of Health
Olympia, Washington

Larry G. Arlian, Ph.D.
Department of Biological Sciences
Wright State University
Dayton, Ohio

Harriet A. Burge, Ph.D.
Department of Environmental Health
Harvard School of Public Health
Boston, Massachusetts

Eugene C. Cole, Dr.P.H.
Dyncorp
Durham, North Carolina

Karin K. Foarde, M.S.
Research Triangle Institute
Research Triangle Park,
    North Carolinia

Kathleen Kreiss, M.D.
NIOSH
Morgantown, West Virginia

Janet M. Macher, Sc.D., M.P.H.
Environmental Health Laboratory
California Department of Health
    Services
Berkeley, California

Kenneth F. Martinez, M.S.E.E., C.I.H.
NIOSH
Cincinnati, Ohio

John W. Martyny, Ph.D., C.I.H.
Tri-County Health Department
Commerce City, Colorado

Donald K. Milton, M.D., Dr.P.H., M.O.H.
Occupational Health Program
Harvard School of Public Health
Boston, Massachusetts

Philip R. Morey, Ph.D., C.I.H.
Microbiology, AQS Services
Gettysburg, Pennsylvania

Edward A. Nardell, M.D.
The Cambridge Hospital
Cambridge, Massachusetts

James A. Otten, S.M.
Camber Corporation
Oak Ridge, Tennesee

Cecile S. Rose, M.D., M.P.H.
Occupational and Pulmonary Medicine
National Jewish Medical and
    Research Center
Denver, Colorado

Richard Shaughnessy, Ph.D.
University of Tulsa, CERT
Tulsa, Oklahoma

Klaus Willeke, Ph.D., C.I.H.
Department of Environmental Health
University of Cincinnati
Cincinnati, Ohio

ACKNOWLEDGEMENTS

This book was produced through the cooperative efforts of the members of the Bioaerosols Committee of ACGIH, with the invaluable assistance of others. We wish to recognize the significant contributions of authors from outside the committee: Larry G. Arlian, Ph.D.; Eugene C. Cole, Dr.P.H.; and Karin K. Foarde, M.S. Miriam K. Lonon, Ph.D., also contributed significantly to the writing of the book and participated in committee discussions about the contents.

The authors owe a great debt to the outside reviewers who provided helpful comments and criticisms on a first draft of the book: Brian Crook, Ph.D., M.R.I.P.H.H.; Brian Flannigan, Ph.D.; Michael Hodgson, M.D.; J. David Miller, Ph.D.; David G. Taylor, Ph.D., C.I.H.; and John A. Tiffany. The committee also wishes to recognize the valuable input we received from the participants in an interactive course on the book that was held in Orlando, Fla., on February 8–9, 1999. The following persons also reviewed one or more chapters of the book or helped in other ways, for which we are grateful: Leonidas Alevantis, Robert Castellan, Terry Gordon, Michael Hendry, Mark Hernandez, Nitin Kapadia, Dan Lewis, Kai-Shen Liu, Joseph W. Lstiburek, Shelly Miller, David Schnurr, Dennis Shusterman, Daniel Smith, Miriam Valesco, Herbert Venable, Lurance Webber, William Wehrmeister, Alan Weinrich, Lauren Wohl, and JoeyZhou.

We acknowledge the unnamed institutions and agencies that supported the authors while they worked on this book and the many unnamed colleagues and family members who lent their services, suggestions, and encouragement in its production. Without the patience, efficiency, and good humor of the ACGIH staff, this book could not have been produced.

Bioaerosols Committee Members

Harriet M. Ammann, Ph.D., D.A.B.T.
Harriet A. Burge, Ph.D.
Kathleen Kreiss, M.D. (retired)
Janet M. Macher, Sc.D., M.P.H.
Kenneth F. Martinez, M.S.E.E., C.I.H.
John W. Martyny, Ph.D., C.I.H.
Donald K. Milton, M.D., Dr.P.H.

Philip R. Morey, Ph.D., C.I.H.
Edward A. Nardell, M.D.
James A. Otten, S.M.
Tiina Reponen, Ph.D. (new appointment)
Stephen J. Reynolds, Ph.D., C.I.H. (new appointment)
Cecile S. Rose, M.D., M.P.H. (retired)
Richard Shaughnessy, Ph.D.
Klaus Willeke, Ph.D., C.I.H. (retired)

ABBREVIATIONS/ACRONYMS/UNITS/LATIN NAMES

AGENCIES AND ASSOCIATIONS

AAAAI	American Academy of Allergy, Asthma, and Immunology
AATCC	American Association of Textile Chemists and Colorists
ACGIH	American Conference of Governmental Industrial Hygienists
AIHA	American Industrial Hygiene Association
AOAC	Association of Official Analytical Chemists
ARC	American Red Cross
ASHRAE	American Society of Heating, Refrigerating and Air-Conditioning Engineers
ASTM	American Society for Testing and Materials
CDC	Centers for Disease Control and Prevention
CEN	European Standardization Committee
CMHC	Canada Mortgage and Housing Corporation
FEMA	Federal Emergency Management Agency
IAACI	International Association of Allergology and Clinical Immunology
ICRP	International Commission on Radiological Protection
IICRC	Institute of Inspection, Cleaning and Restoration Certification
IOM	Institute of Medicine
ISIAQ	International Society of Indoor Air Quality and Climate
ISO	International Standards Organization
IUATLD	International Union Against Tuberculosis and Lung Disease
NADCA	National Air Duct Cleaners Association
NAIN	National Antimicrobial Information Network
NIOSH	National Institute for Occupational Safety and Health
NIST	National Institute of Standards and Technology
USEPA	U.S. Environmental Protection Agency
USP	U.S. Pharmacopeia

ABBREVIATED AGENT NAMES — Amebae

N. fowleri	*Naegleria fowleri*

ABBREVIATED AGENT NAMES — Arthropods and Arthropodic Antigens

A. siro	*Acarus siro* (storage mite)
B. tropicalis	*Blomia tropicalis* (house dust mite)
Bla g	*Blattella germanica* (German cockroach) allergen
D. farinae	*Dermatophagoides farinae* (house dust mite)
Der f	*Dermatophagoides farinae* (allergen)
D. microceras	*Dermatophagoides microceras* (house dust mite)
D. pteronyssinus	*Dermatophagoides pteronyssinus* (house dust mite)
Der p	*Dermatophagoides pteronyssinus* (allergen)
E. maynei	*Euroglyphus maynei* (house dust mite)
L. destructor	*Lepidoglyphus destructor* (storage mite)
Per a	*Periplaneta americana* (American cockroach) (allergen)

ABBREVIATED AGENT NAMES — Bacteria

E. coli	Escherichia coli
L. bozemanii	Legionella bozemanii
L. jordanis	Legionella jordanis
L. micdadei	Legionella micdadei
L. pneumophila	Legionella pneumophila
M. tuberculosis	Mycobacterium tuberculosis
S. aureus	Staphylococcus aureus
S. pneumoniae	Streptococcus pneumoniae

ABBREVIATED AGENT NAMES — Fungi

A. obclavatum	Acremonium obclavatum
A. flavus	Aspergillus flavus
A. fumigatus	Aspergillus fumigatus
A. niger	Aspergillus niger
A. parasiticus	Aspergillus parasiticus
A. versicolor	Aspergillus versicolor
B. dermatitidis	Blastomyces dermatitidis
B. cineria	Botrytis cineria
C. albicans	Candida albicans
C. herbarum	Cladosporium herbarum
C. sphaerospermum	Cladosporium sphaerospermum
C. immitis	Coccidioides immitis
C. sativus	Cochliobolus sativus
C. neoformans	Cryptococcus neoformans
E. nidulans	Emericella nidulans
E. nigrum	Epicoccum nigrum
F. moniliforme	Fusarium moniliforme
H. capsulatum	Histoplasma capsulatum
P. aurantiogriseum	Penicillium aurantiogriseum
P. chrysogenum	Penicillium chrysogenum
P. commune	Penicillium commune
P. fastigata	Phialophora fastigata
P. roqueforti	Penicillium roqueforti
S. atra	Stachybotrys atra
S. chartarum	Stachybotrys chartarum

ABBREVIATED AGENT NAMES — Mammals and Mammalian Antigens

Can f	Canis familiaris (domestic dog) allergen
Fel d	Felis domesticus (domestic cat) allergen

ABBREVIATED AGENT NAMES — Viruses

FCV	Four Corners virus
HIV	human immunodeficiency virus
SNV	Sin Nombre virus

CHEMICALS AND CHEMICAL FORMULAS

$Ca(OCl)_2$	calcium hypochlorite
Cl	chlorine
ClO_2	chlorine dioxide
CO_2	carbon dioxide

DAPI	4',6-diamidino-2-phenyl-indole
ETO	ethylene oxide
FITC	fluorescein isothiocyanate
FRC	free residual chlorine
H_2O	water
H_2O_2	hydrogen peroxide
HCHO	formaldehyde
K_2HPO_4	dipotassium hydrogen phosphate
$MgSO_4\ 7\ H_2O$	magnesium sulfate heptahydrate
NaOCl	sodium hypochlorite
O_3	ozone

CONDITION AND DISEASE NAMES

ABPA	allergic bronchopulmonary aspergillosis
ABPM	allergic bronchopulmonary mycosis
AIDS	acquired immunodeficiency syndrome
ATA	alimentary toxic aleukia
BRI	building-related illness
BRS	building-related symptom
GAE	granulomatous amebic encephalitis
HP	hypersensitivity pneumonitis
HPS	Hantavirus pulmonary syndrome
MCS	multiple-chemical sensitivity
MDR-TB	multiple-drug resistant tuberculosis
NSBRS	non-specific building-related symptom
ODTS	organic dust toxic syndrome
PAM	primary amebic meningoencephalitis
RADS	reactive airway disease syndrome
RUDS	reactive upper airway disease syndrome
SBS	sick building syndrome
SBRI	specific building-related illness
TB	tuberculosis

OTHER ABBREVIATIONS IN TEXT

AGI	all-glass impinger
AND-I	Andersen impactor, stage 1
AND-VI	Andersen impactor, stage 6
AO	acridine orange
BASE	Building Assessment Survey and Evaluation
BCE	before current era
BRI	building-related illness
BRS	building-related symptom
BURK	Burkard spore trap
CA	cellulose acetate
CBF	ciliary beat frequency
CE	cellulose mixed ester
CFR	Code of Federal Regulations
CFU	colony-forming unit
CNS	central nervous system
CT	computerized tomography
CT	cooling tower
CV	coefficient of variation
DDI	dilution-dependent inhibition
DG	dichloran glycerol

DII	dilution-independent inhibition
DNA	deoxyribonucleic acid
EC	evaporative cooler
EHP	environmental health professional
ELISA	enzyme-linked immunosorbent assay
EMPAT	Environmental Microbiology Proficiency Analytical Testing
ERH	equilibrium relative humidity
ESP	electrostatic precipitator
FID	flame ionization detector
FIFRA	Federal Insecticide, Fungicide, and Rodenticide Act
GC	gas chromatography
GC-MS	gas chromatography-mass spectrometry
GF	glass fiber
GNB	Gram-negative bacterium
GPB	Gram-positive bacterium
Gr.	Greek
HEPA	high-efficiency particulate air
HPLC	high performance liquid chromatography
HVAC	heating, ventilating, and air conditioning
IEQ	indoor environmental quality
IgA	immunoglobulin A
IgE	immunoglobulin E
IgG	immunoglobulin G
IgM	immunoglobulin M
IH	industrial hygienist
LAL	*Limulus* amebocyte lysate
LC	liquid chromatography
LDL	lower detection limit
LPS	lipopolysaccharide
LTA	lipoteichoic acid
MC	moisture content
MEA	malt extract agar
MK-II	New Brunswick Slit Sampler
MS	mass spectrometry
MS-MS	dual or tandem mass spectrometry
MSDS	material safety data sheet
MVOC	microbial volatile organic compound
MWF	metal-working fluid
NSBRS	Non-specific building-related symptoms
OAI	outdoor air intake
PC	polycarbonate
PCR	polymerase chain reaction
PG	peptidoglycan
pH	hydrogen-ion activity
PID	photoionization detector
PM10	particulate matter <10 m
PMN	polymorphonuclear neutrophil
PPE	personal protective equipment
PVC	polyvinyl chloride
R	agreement ratio
RAST	radio-allergosorbent test
RCS	Reuter centrifugal sampler
RH	relative humidity
RLV	relative limit value
RNA	ribonucleic acid

SAS	surface air system
SBS	sick building syndrome
SBRI	Specific building-related illness
SENSOR	Sentinel Event Notification System for Occupational Risks
sp.	species (singular)
spp.	species (plural)
TLC	thin layer chromatography
TLV	Threshold Limit Value
TST	tuberculin skin test
UDL	upper detection limit
UV	ultraviolet
UVGI	ultraviolet germicidal irradiation
UVR	ultraviolet radiation
VOC	volatile organic compound
X	times (magnification)

UNITS OF MEASURE

µg	microgram (10^{-6} gram)
µg/g	microgram per gram
µg/m^2	microgram per square meter
µg/m^3	microgram per cubic meter
µg/mL	microgram per milliliter
µm	micrometer (10^{-6} meter)
°C	degree Celsius
°F	degree Fahrenheit
ACH	air change per hour
atm	atmosphere
a_w A_w	water activity
CFU	colony-forming unit
CFU/cm^2	colony-forming unit per square centimeter
CFU/g	colony-forming unit per gram
CFU/m^3	colony-forming unit per cubic meter
CFU/mg	colony-forming unit per milligram
CFU/mL	colony-forming unit per milliliter
cm	centimeter (10^{-2} meter)
cm/s	centimeter per second
EU	endotoxin unit
EU/g	endotoxin unit per gram
EU/m^3	endotoxin unit per cubic meter
FEV$_1$	forced expiratory volume in one minute
ft	foot
ft^2/gal	square foot per gallon
ft^3/min	cubic foot per minute
g	gram
g/cm^3	gram per cubic centimeter
g/L	gram per liter
g/m^2	gram per square meter
g/m^3	gram per cubic meter
gal	gallon
in	inch
in/ft	inch per foot
in w.g.	inch water gage
kd	kilodalton (10^3 dalton)
kg	kilogram (10^3 gram)
kg/m^3	kilogram per cubic meter

L	liter
L/min	liter per minute
L/s	liter per second
lb	pound
lb/ft^3	pound per cubic foot
m	meter
m/s	meter per second
m^3/day	cubic meter per day
m^3/hour	cubic meter per hour
m^3/min	cubic meter per minute
mg	milligram (10^{-3} gram)
mg/g	milligram per gram
mg/kg	milligram per kilogram
mg/L	milligram per liter
mg/m^3	milligram per cubic meter
mil	0.001 inch
min	minute
mL	milliliter (10^{-3} liter)
mm	millimeter (10^{-3} meter)
ng	nanogram (10^{-9} gram)
ng/g	nanogram per gram
ng/m^3	nanogram per cubic meter
ng/mg	nanogram per milligram
nm	nanometer (10^{-9} meter)
Pa	Pascal
pg	picogram (10^{-12} gram)
pg/m^3	picogram per cubic meter
pH	hydrogen-ion activity
ppm	part per million
RH	relative humidity
s	second

VARIABLES IN TEXT AND EQUATIONS

α	significance level
Δ	change
δ	surface density
η	fluid viscosity
ρ_p	particle density
Σ	summation
τ	particle relaxation time
χ^2	chi-square
A	deposit area
A	total number of species in Sample 1
B	total number of species in Sample 2
C	number of cases
c_a	average air concentration
C_c	Cunningham correction factor
d	particle diameter
d_a	aerodynamic particle diameter
d_{50}	50% cut-off diameter
d_i	rank difference in Spearman rank correlation
E_e	cumulative error
E_i	individual error
f	degrees of freedom

GM	geometric mean
GSD	geometric standard deviation
H	test statistic in Kruskal-Wallis procedure
H_0	null hypothesis
H_1	alternative hypothesis
I	number of infectious persons
ID_{50}	50% infectious dose
k	number of columns in two-way analysis of variance
k	number of populations in Kruskal-Wallis procedure
L	length, slit impactor
LD_{50}	50% lethal dose
m	number of items in first sample in two-sample Wilcoxon test
n	sample size
n	number of items in second sample in two-sample Wilcoxon test
n	number of isolates in Spearman rank correlation
n	number of rows in two-way analysis of variance
n_i	number of items in each sample in Kruskal-Wallis procedure
N	sum of n_i in Kruskal-Wallis procedure
p	breathing rate
P	significance level
q	generation rate of infectious agents
Q	airflow rate, ventilation rate
R	agreement ratio
R_i	mean ranks in Kruskal-Wallis procedure
R_j	sum of ranks in two-way analysis of variance
r_s	Spearman rank correlation
S	number of susceptible persons
S	particle stopping distance
S_{50}	50% stopping distance
SD	standard deviation
Stk	Stokes number
Stk_{50}	Stokes number 50% of particles
t	time
T	correction for ties in Kruskal-Wallis procedure
U_o	initial particle velocity
W	width (impactor nozzle)
W	number of species two samples have in common
x_i	individual measurement
x	sample mean
y_1	zone height above impaction surface within which incoming streamlines are deflected from original paths
y_2	impactor jet-to-plate distance
y_3	greatest distance from an impaction plate within which air moves laterally

LIST OF FIGURES

FIGURE 2.1. Fundamental Components Of Bioaerosol-Related Illnesses.

FIGURE 2.2. Paradigm For Understanding And Controlling Occupational Bioaerosol Exposures.

FIGURE 2.3. Fundamental Steps In An Investigation.

FIGURE 2.4. Outline Of An Investigation Strategy.

FIGURE 4.1. Typical HVAC System Components For A Large Building.

FIGURE 6.1. Enzyme-Linked Immunosorbent Assay (ELISA).

FIGURE 6.2. Short Section Of A DNA Sequence Showing Paired Nucleotide Bases And DNA Replication.

FIGURE 6.3. Illustration Of Hybridization Of A DNA Probe To A Complementary Section Of A Nucleic Acid Sequence.

FIGURE 7.1. Steps Connecting A Biological Agent And A Host Response.

FIGURE 9.1. Schematic Representative Of The Major Regions Of The Respiratory Tract.

FIGURE 9.2. Effect Of Ventilation Rate On TST Conversion Rate.

FIGURE 10.1. Sorption Isotherms For (A) Pine Wood And (B) Concrete; A = Adsorption Curve, ° = Desorption Curve.

FIGURE 10.2. Means Of Water Entry Into A Building.

FIGURE 10.3. Location Of Air And Vapor Barriers* In A Cold Climate.

FIGURE 10.4. Location Of Air And Vapor Barriers* In An Air-Conditioned Building In A Warm, Humid Climate.

FIGURE 10.5. Recommended Pressure Relationship Between Indoors And Outdoors In A Hot, Humid Climate.

FIGURE 10.6. Potential For Condensation When Cooling Exceeds Outdoor Dew-Point Temperature.

FIGURE 10.7. Detail Of Water Trap At Condensate Drain Line. Drain Pan Slopes ~0.2 Cm For Every 10 Cm Of Pan Length (0.25 In/Ft).

FIGURE 11.1. Mechanisms Of Particle Removal From Air — Solid-Plate Impaction.

FIGURE 11.2. Single-Stage, Multiple-Hole Impactor.

FIGURE 11.3. Slit-To-Agar Impactor.

FIGURE 11.4. Pollen And Spore Trap.

FIGURE 11.5. Mechanisms Of Particle Removal From Air — Centrifugal Impaction (A) Agar-Strip Impactor.

FIGURE 11.5. (B) Cyclone Sampler.

FIGURE 11.6. Schematic Representation Of The Collection Process In An Impinger.

FIGURE 11.7. Mechanisms Of Particle Removal From Air – Filtration.

FIGURE 11.8. Cross Section Of Solid-Plate Impactor (A) Critical Parameters Of The Impaction Zone, (B) Stopping Distance And Its Importance In Particle Removal.

FIGURE 11.9. Stopping Distance As A Function Of Aerodynamic Particle Diameter For Selected Bioaerosol Samplers; Cut-Off Diameters.

FIGURE 11.10. Soap-Bubble-Meter (Flowmeter) Calibration.

FIGURE 11.11. Calibration Chamber.

FIGURE 11.12. Equal-Pressure Sampling Train.

FIGURE 13.1. Example: Frequency Distribution Of Repeated Measurements Of The Concentration Of Total Culturable Bacteria From A Single Dust Sample.

FIGURE 13.2. Example: Frequency Distribution Of Mite Allergen Concentration In Dust Samples, Illustrating Skewing To The Right.

FIGURE 13.3. Example: Histogram Of Data From Figure 13.2 After Logarithmic Transformation.

FIGURE 13.4. Frequency Distributions Of Biological Agent Concentrations In Hypothetical Non-Problem And Problem Environments.

FIGURE 13.5. Frequency Distributions Of Log-Transformed Biological Agent Concentrations In The Hypothetical Non-Problem And Problem Environments In Figure 13.4.

FIGURE 13.6. Similarity Triangle Depicting Relatedness Among Sampling Sites.

FIGURE 14.1. Logarithmic Plot Of Bioaerosol Concentration At Paired Sites.

FIGURE 14.2. Frequency Distributions Of Log-Transformed Biological Agent Concentrations In The Hypothetical Non-Problem And Problem Environments In Figure 13.5.

FIGURE 14.3. Logarithmic Plot Of Indoor And Outdoor Bioaerosol Concentrations At Three Sites.

FIGURE 16.1. Sample Biocide Label.

FIGURE 17.1. The Biological Kingdoms.

FIGURE 17.2. Classification Of Life Forms With Examples.

FIGURE 17.3. Lifestyle Adaptations.

FIGURE 18.1. Cell Shapes And Arrangements For Common Bacterial Forms.

FIGURE 18.2. Sample Flowchart For Identification Of Aerobic, Coccoid Bacteria.

FIGURE 19.1. Fungal Growth Showing Aerial And Vegetative Hyphae And Reproductive Structures.

FIGURE 20.1. *Naegleria Spp.* (A) Ameboid Stage, (B) Flagellated Stage, (C) Cyst Stage.

FIGURE 20.2. Acanthamoeba Spp. (A) Ameboid Stage, (C) Cyst Stage.

FIGURE 21.1. Drawings Of Viruses In The Common Virus Families Of Vertebrates.

FIGURE 22.1. Adult Female *D. Farinae.*

FIGURE 22.2. Life Cycle Of *D. Pteronyssinus* At 23°C, 75% RH.

FIGURE 23.1. Group Mean Percentage Change In Forced Expiratory Volume In One Second (Rfev1%) Versus Airborne Dust Concentration (Panel A), Airborne Endotoxin Concentration (B), And Log Airborne Endotoxin Concentration (C).

FIGURE 23.2. Predicted Association Between Log-transformed Endotoxin Exposure And Annual Rate Of Decline Of FEV1 (With Standard Error); Corrected For Age, Baseline FEV1, And Pack-Years Of Smoking.

FIGURE 23.3. Dose-response Curves For Three House Dust Extracts Showing (A) DDI With Both Dilution-Dependent Inhibition And Enhancement, (B) DII, And (C) No Evidence For Interference.

FIGURE 25.1 Primary And Secondary Antibody Responses.

LIST OF TABLES

TABLE 2.1. Outline Of An Investigation Strategy And Related Chapters In This Book.

TABLE 3.1. Terms Used To Describe Building-Associated Medical Conditions.

TABLE 3.2. Alphabetical Listing Of Conditions Related To Exposure To Biological Agents And Chapters In Which The Conditions And Agents Are Discussed.

TABLE 3.3. Components Of An Epidemiological Study.

TABLE 4.1. Potential Sources Of Biological Agents Or Bioaerosol Entry Routes Into Buildings And Factors Related To Microbial Growth Or Bioaerosol Dissemination.

TABLE 5.1. Example: Bioaerosol Sampling Plans.

TABLE 5.2. Fundamental Steps When Collecting Environmental Samples To Test Hypotheses.

TABLE 5.3. Points To Consider When Designing A Sampling Plan.

TABLE 5.4. Deciding Where To Collect Bioaerosol Samples To Compare Anticipated High And Low Exposures.

TABLE 5.5. Time-Related Patterns For Collecting Bioaerosol Samples.

TABLE 5.6. Example: Sampling Plan To Assess The Contribution Of Fungal Growth In A Ventilation System To The Presence Of Fungal Spores In Indoor Air.

TABLE 5.7. Selecting An Air Sampler For Bioaerosol Collection.

TABLE 5.8. Calculated Air Concentration Ranges Suitable For Direct-Agar Impactors And Fungal Culture For Selected Sample Collection Times.

TABLE 5.9. Suggested Sample Numbers.

TABLE 5.10. Sample Information.

TABLE 5.11. Steps In Bioaerosol Sample Collection.

TABLE 5.A.1. Example: The Randomization Process.

TABLE 6.1. Collection And Analysis Of Environmental Samples For Biological Agents.

TABLE 7.1. Example: The Interpretation Process For Several Kinds Of Data.

TABLE 8.1. Components Of An Occupational And Environmental History For Patients With BRI Possibly Due To Bioaerosol Exposures.

TABLE 9.1. Regions And Functions Of The Respiratory Tract.

TABLE 9.2. Inhalable, Thoracic, And Respirable Dust Criteria Of ACGIH-ISO-CEN.

TABLE 10.1. Terms Used To Describe Water In Air And Materials.

TABLE 11.1. Widely Used, Commercially Available Samplers For Collecting Bioaerosols.

TABLE 11.2. Sampling Parameters And Cut-Off Diameters For Selected Bioaerosol Impactors.

TABLE 11.3. Advice On Airflow Calibration.

TABLE 13.1. Samples Used To Assess Errors In Bioaerosol Sample Collection And Analysis.

TABLE 13.2 Airborne Fungi Collected On Cellulose Agar Indoors And Outdoors.

TABLE 13.3 Percentage Of Air Samples Yielding Viable Fungi During A Twelve-Month Period.
TABLE 13.4. Example: Similarity Index For An Unknown GNB From An Air Sample.
TABLE 13.5. Rank-Order Comparison: Concentrations And Ranks.
TABLE 19.1. Classification Of Some Fungal Genera Found In Indoor And Outdoor Air.
TABLE 19.2. How Some Common Asexual Fungi Are Classified By Their Method Of Spore Production.
TABLE 19.3. Moisture Requirements For Some Common Fungi.
TABLE 19.4. Relative Usefulness Of Available Methods For Analysis Of Fungal Samples.
TABLE 19.5. Culture Media For Fungi (Ingredients Per Liter Of Distilled Water).
TABLE 18.1. Classification Of Bacteria That Are Common In Indoor Air Or Are Agents Of Airborne Disease.
TABLE 21.1 Health Effects Of Selected Airborne And Droplet-Borne Viruses.
TABLE 22.1. Common House Dust Mites.
TABLE 22.2. Common Storage (Domestic) Mites.
TABLE 22.3. Length Of Life Cycle (Egg To Adult) For *Dermatophagoides Spp.* At Different Temperatures (75% RH).
TABLE 22.4. Reproductive Statistics For *Dermatophagoides Spp.* Reared At 23° Or 35°C (75% RH).
TABLE 22.5. Rhs For Selected Conditions Of Absolute Water Content Of Air And Temperature.
TABLE 22.6. Environmental Control Measures To Minimize Dust-Mite Populations And Allergen Production.
TABLE 24.1. Some Common Fungi, Mycotoxins, And Health Effects From Ingestion, Dermal, Or Inhalation Exposure.
TABLE 24.2. Toxins From Spores Of Several Fungal Species Cultivated On Laboratory Media; Quantification By High-Pressure Liquid Chromatography Or Thin-Layer Chromatography.
TABLE 24.3. Principal Cell Wall Polysaccharides Of Some Common Environmental Fungi.
TABLE 25.1. Terms Used With Antigen-Mediated Diseases.
TABLE 25.2. Biological Sources Of Some Common Airborne Allergens.
TABLE 25.4. Primary Indoor Aeroallergens.
TABLE 26.1. Selected Studies Of Vocs From Stock Fungi And Bacteria In Laboratory Culture.
TABLE 26.2. VOCS From Fungi Grown In Laboratory Culture And From Mixed Culture On Building Material Substrates.
TABLE 26.3. VOCS Found In Problem Buildings And Fungi And Bacteria Cultured From Air Samples From These Buildings.
TABLE 26.4. VOCS Found In Problem Buildings And Fungi And Bacteria Cultured From Materials From These Buildings.

LIST OF APPENDICES

APPENDIX 4.A. Building Walkthrough Checklist.
APPENDIX 11.A. Manufacturers And Suppliers Of Bioaerosol Samplers
APPENDIX 11.B(I). Positive-Hole Correction Table To Adjust Colony Counts From A 100-Hole Impactor For The Possibility Of Collecting Multiple Particles Through A Hole.
APPENDIX 11.B(Ii). Positive-Hole Correction Table To Adjust Colony Counts From A 200-Hole Impactor For The Possibility Of Collecting Multiple Particles Through A Hole.
APPENDIX 11.B(Iii). Positive-Hole Correction Table To Adjust Colony Counts From A 219-Hole Impactor For The Possibility Of Collecting Multiple Particles Through A Hole.
APPENDIX 11.B(Iv). Positive-Hole Correction Table To Adjust Colony Counts From A 400-Hole Impactor For The Possibility Of Collecting Multiple Particles Through A Hole.
APPENDIX TABLE 11.C(I). Correction Factors For Particle Collection With Multiple-Hole Impactors.

Chapter 1

Introduction

1.1 Bioaerosols and Biologically Derived Airborne Contaminants
1.2 Why No TLVs® for Bioaerosols
    1.2.1 *Sampling for Airborne Biological Agents*
    1.2.2 *Total Culturable or Countable Biological Agents*
    1.2.3 *Specific Culturable or Countable Bioaerosols Other than Infectious Agents*
    1.2.4 *Infectious Agents*
    1.2.5 *Assayable Biological Contaminants*
    1.2.6 *Criteria on Which to Base Exposure Limits for Biologically Derived Airborne Contaminants*
1.3 Overview
1.4 Approaches
1.5 References

1.1 Bioaerosols and Biologically Derived Airborne Contaminants

Bioaerosols are those airborne particles that are living or originate from living organisms. Bioaerosols include microorganisms (i.e., culturable, nonculturable, and dead microorganisms) and fragments, toxins, and particulate waste products from all varieties of living things. Bioaerosols are ubiquitous in nature and may be modified by human activities. All persons are repeatedly exposed, day after day, to a wide variety of such materials. Individual bioaerosols range in size from submicroscopic particles (<0.01 μm) to particles greater than 100 μm in diameter.

ACGIH uses the term *biologically derived airborne contaminants* to describe bioaerosols, gases, and vapors that living organisms produce. Biologically derived materials are natural components of indoor and outdoor environments but, under some circumstances, biological agents may be considered contaminants when found indoors. ACGIH has defined *biological contamination* in buildings as the presence of (a) biologically derived aerosols, gases, and vapors of a kind and concentration likely to cause disease or predispose persons to adverse health effects, (b) inappropriate concentrations of outdoor bioaerosols, especially in buildings designed to prevent their entry, or (c) indoor biological growth and remnants of growth that may become airborne and to which people may be exposed (ACGIH, 1998a). The term *biological agent* is used here to refer to a substance of biological origin that is capable of producing an effect, for example, an infection or a hypersensitivity, irritant, inflammatory, or other response.

1.2 Why No TLVs® for Bioaerosols

Threshold Limit Values (TLVs) refer to air concentrations of substances and represent conditions under which it is believed that nearly all workers may be repeatedly exposed day after day without adverse health effects (ACGIH, 1998b). Standards to prevent harmful exposures to air contaminants have five primary components: (a) the criterion or scientific basis for the standard, (b) a sampling method, (c) an analytical method, (d) a sampling strategy, and (e) a limit value. ACGIH has considered the possibility of recommending TLVs for bioaerosols and concluded that sufficient information is not yet available on these five components for many of the biologically derived airborne contaminants to which workers are exposed in non-manufacturing environments (ACGIH, 1998a). The statement on why there are so few TLVs for bioaerosols is repeated and expanded here.

TLVs exist for certain substances of biological origin, including cellulose; some wood, cotton, and grain dusts; nicotine; pyrethrum; starch; subtilisins (proteolytic enzymes); sucrose; vegetable oil mist; and volatile compounds produced by living organisms (e.g., ammonia, carbon dioxide, ethanol, and hydrogen sulfide). However, there are no mandatory numerical limits against which investigators can compare measurements of air or source concentrations for the majority of substances of biological origin that are associated with building-related exposures. Thus, in the U.S., sampling for biologically derived airborne contaminants is not conducted for the purpose of complying with any federal or state regulations other than for the agents for which existing TLVs have been adopted as standards. Readers should be aware of any national or local standards for exposure to biological agents that have been set elsewhere than in the U.S. Where such standards exist, investigators should use appropriate procedures to measure exposures to specific biological agents and follow applicable guidelines for interpreting exposure measurements.

Given the general absence of compliance monitoring for biological agents, data on the range of inhalation exposures to specific biological agents are limited and the methods that investigators use to collect and analyze these agents vary widely. Even if limits were set, at present they would be arbitrary standards because the available environmental and health data on which to base exposure criteria are few and of inconsistent quality, as described in more detail below. Further, there are more reliable methods to identify environments in need of intervention than by comparing environmental bioaerosol measurements with numerical standards. This book (developed by a consensus of researchers and practitioners under the auspices of ACGIH) defines methods for assessing and controlling exposures to biologically derived airborne contaminants. These methods rely on visually inspecting buildings, assessing occupant symptoms, evaluating building performance, testing potential environmental sources, and applying professional judgement.

1.2.1 Sampling for Airborne Biological Agents

This book provides background information on the major groups of bioaerosols including their sources and health effects. Also described here are methods to collect, analyze, and interpret samples for biological agents from potential environmental sources. Occasionally, environmental sampling detects a single or predominating biological agent. More commonly, environmental sampling reveals a mixture of many biologically derived materials, reflecting the diverse and interactive nature of indoor microenvironments. Some of these biological agents are clearly harmful and their confirmed presence is a cause for concern. Other biological materials are normal components of indoor and outdoor environments. These latter agents typically cause little, if any, harm but may elicit responses in sensitive persons, even at ambient concentrations, or in all persons, when such agents are present in sufficient quantity. Environmental sampling for bioaerosols should only be conducted following the careful formulation of testable hypotheses about potential sources of biological agents and mechanisms by which workers may be exposed to bioaerosols from these sources. Even when investigators work from testable hypotheses and well-formulated sampling plans, results from environmental bioaerosol studies may be inconclusive and occasionally even misleading.

For the reasons identified below, there are no TLVs for interpreting environmental measurements of (a) total culturable or countable bioaerosols (e.g., total bacteria or fungi), (b) specific culturable or countable bioaerosols (e.g., *Aspergillus fumigatus*), (c) infectious agents (e.g., *Legionella pneumophila*), or (d) assayable biological contaminants (e.g., endotoxin, mycotoxins, allergens, or microbial volatile compounds).

1.2.2 Total Culturable or Countable Biological Agents

Culturable biological agents are those bacteria and fungi that can be grown in laboratory culture. The results of such cultures are reported as the number of colony-forming units per sample volume, mass, or area. Countable bioaerosols are those pollen grains, fungal spores, bacterial cells, and other particles that can be identified and counted under a microscope. The results of such samples are reported as the number of particles per sample volume, mass, or area. A general exposure limit for concentrations of culturable or countable biological agents is not scientifically supportable because of the following:

1. Culturable microorganisms and countable biological particles do not comprise a single entity. Bioaerosols in occupational settings are generally complex mixtures of many different microorganisms as well as non-living particles released from fungi, bacteria, animals, and plants.

2. Human responses to bioaerosols range from innocuous effects to serious, even fatal, diseases, depending on the specific material involved and workers' susceptibility to it. Therefore, an appropriate exposure limit for one bioaerosol may be entirely inappropriate for another. Exposure limits for total culturable or countable biological agents would not take these differences into consideration and, therefore, no such recommendations are given in this book.

3. It is not possible to collect and evaluate all bioaerosol components using a single sampling method. Many reliable air sampling methods are available to collect bioaerosols and many sensitive and specific analytical methods are available to assay biological agents. However, different methods of sample collection and analysis may result in different estimates of the relative concentrations of culturable or countable bioaerosols. Therefore, any exposure limits that may be proposed for total culturable or countable biological agents must specify the methods of bioaerosol collection and analysis.

4. At present, information relating culturable or countable bioaerosol concentrations to health effects is generally insufficient to describe exposure–response relationships from which TLVs could be derived.

To overcome the obstacles listed above, information would be needed on the concentrations of the diverse culturable and countable biological agents that may be found in work, residential, commercial, and recreational environments. These data would have to be compiled consistently using carefully chosen methods of sample collection and analysis. In addition to environmental measurements, matching data on health responses for the occupants of the tested environments would have to be collected to determine if there is any value in measuring total culturable or countable biological agents to predict environmental air quality. It would also be valuable to determine where and when air sampling for total culturable or countable bioaerosols may identify environmental contamination that could not otherwise be detected by careful visual examination, questioning of facility operators and occupants, or other means of assessing the potential presence of harmful or undesirable biological agents.

1.2.3 Specific Culturable or Countable Bioaerosols Other than Infectious Agents

Exposure limits for individual culturable or countable bioaerosols have not been established to prevent hypersensitivity, irritant, or toxic responses. At present, information relating health effects with exposures to specific culturable or countable bioaerosols consists largely of case reports and qualitative exposure assessments. Future information on exposure–response relationships may lead to the establishment of exposure limits. However, currently available data are generally insufficient for this purpose. Some obstacles to collecting good epidemiological data on dose–response relationships for specific culturable or countable biological agents include the following:

1. Most data on concentrations of specific bioaerosols are derived from indicator measurements where the relationship of the indicator to the concentration and biological activity of the actual effector agent is unknown. For example, investigators often use the air concentration of culturable fungi to represent potential exposure to airborne fungal allergens and mycotoxins. The accuracy of substituting exposure to airborne fungi for exposure to fungal allergens or toxins has not been determined.

2. Replicate sampling is uncommon in bioaerosol assessments despite the fact that bioaerosol components and relative concentrations vary widely within and among different occupational and environmental settings. Further, the most commonly used air sampling devices for bioaerosol testing are designed to collect "grab" samples over relatively short time intervals. Measurements from single, short-term grab samples are poor predictors of either short-term or long-term air concentrations. Long-term average air concentrations are good predictors for some bioaerosol-related hazards, but grab samples may be orders of magnitude higher or lower than average concentrations and are unlikely to represent workplace exposures accurately. Some organisms and sources release aerosols as "concentration bursts," which may only rarely be detected by limited grab sampling and may be masked in measurements of long-term average concentrations. Nevertheless, such episodic bioaerosol releases may produce significant health effects.

3. In studies of single workplaces, the number of persons affected by exposure to biological agents may be small if contamination is localized and affects only a fraction of building occupants. However, data from different studies can seldom be combined to reach meaningful numbers of test subjects because the specific types of biological agents responsible for bioaerosol-related illnesses are diverse and often differ from study to study. These factors contribute to the low statistical power common in evaluations of cause–effect relationships between exposures to specific biological agents and building-related health complaints.

To overcome the obstacles listed above, better sampling and analytical methods for specific biological agents are needed to estimate exposures to actual effector agents. Also needed are detailed and extensive experimental or epidemiological data on the magnitudes and patterns of exposures to specific biological agents as well as careful and objective measurements of health effects.

1.2.4 Infectious Agents

Human dose–response data are available for only a few infectious bioaerosols. At present, air sampling protocols for infectious agents (other than some opportunistic pathogens) are limited and suitable primarily for research endeavors. Therefore, environmental sampling for infectious agents is not a routine approach to assessing their presence and worker exposure thereto. Rather, infectious diseases are controlled through awareness of the potential for infection and surveillance of workers at risk of exposure to infectious agents. In most routine exposure settings, public health measures (e.g., immunization, postexposure prophylaxis, active case finding, and medical treatment) remain the primary defenses against infectious bioaerosols. In addition to these measures, facilities associated with increased risks for transmission of airborne infectious diseases (e.g., microbiology laboratories, animal handling facilities, and health-care settings) should employ work practices and engineering controls to minimize air concentrations of infectious agents. Further, such facilities should consider the need for administrative controls and, where contact cannot otherwise be avoided, the use of personal protective equipment to prevent the exposure of workers to infectious aerosols. This book also discusses the use of protective equipment in remediation work that may involve skin or inhalation exposure to infectious agents in contaminated building materials or organic debris.

1.2.5 Assayable Biological Contaminants

Assayable, biologically derived contaminants (e.g., endotoxin, mycotoxins, allergens, and microbial volatile compounds) are fungal, bacterial, animal, or plant substances that can be detected using chemical, immunological, or biological assays. Evidence does not yet support exposure limits for any of these substances. However, collection and assay methods for certain common airborne allergens and endotoxin are steadily improving, and field validation of these methods is also progressing. Dose–response relationships for some assayable bioaerosols have been observed in experimental studies and occasionally in epidemiological surveys. Therefore, a relative exposure limit for endotoxin is proposed in this book and exposure limits for other substances may be appropriate in the future. Innovative molecular techniques are becoming available for specific bioaerosols currently detectable only by culture or counting. Use of these techniques may provide more information from which to understand exposure–response relationships for assayable biological agents.

1.2.6 Criteria on Which to Base Exposure Limits for Biologically Derived Airborne Contaminants

Many of the difficulties faced in attempts to establish TLVs for biologically derived airborne contaminants were previously encountered with other substances. Established limits for chemical substances are not fine lines between safe and dangerous concentrations nor are they a relative index of toxicity. A similar caveat would apply to TLVs that may eventually be considered for workplace exposures to biological agents. Because of wide variation in individual susceptibility to chemical and biological agents, a small percentage of workers may experience discomfort from some substances at concentrations at or below a threshold limit. A smaller percentage of workers may be affected more seriously by aggravation of a pre-existing condition or by development of an occupational illness. Among the parameters that may affect workers' reactions to substances are genetic factors, age, personal habits, medication, or previous exposures.

The bases on which TLVs are established may differ from substance to substance because of the diversity of chemical and biological agents to which workers may be exposed and the ranges of health responses workers may have as a consequence of these exposures. Protection against impairment of health may be a guiding factor for some exposure limits, whereas reasonable freedom from irritation, narcosis, nuisance, or other forms of stress may form the basis for others (ACGIH, 1998b). Health impairments that have been considered in the establishment of TLVs include those adverse responses that shorten life expectancy, compromise physiological function, impair a worker's ability to resist other toxic substances or disease processes, or adversely affect reproductive function or developmental processes. These and other criteria (e.g., prevention of infection) may need to be considered in the eventual establishment of exposure limits for biologically derived airborne contaminants.

In defining exposure limits for biological agents, a problem encountered with greater frequency than with many chemical substances is that biological exposures are often to complex mixtures of variable composition. Consequently, qualitative and quantitative information about exposure to biologically derived materials is often imprecise because agents other than those identified and measured may also be present and responsible for some of the health responses experienced by exposed persons. Given the constraints outlined above, at present, the usual approach of sampling workplace atmospheres and comparing measurements with TLVs cannot be applied to bioaerosols. In place of compliance monitoring, investigators can use the approaches outlined in this book and other references to assess and control exposures to biologically derived airborne contaminants.

1.3 Overview

This book is the fourth publication of the ACGIH Bioaerosols Committee to provide guidance on assessing and controlling bioaerosols in indoor environments (ACGIH, 1986, 1987, 1989). The book is divided into three parts: The Basics, Background Information, and Specific Agents. Following this introduction, Part I begins with a chapter on how to develop an investigation strategy. The subsequent chapters describe the primary health effects and symptoms that have been associated with bioaerosol exposure and outline a building walk-through inspection, the development of a sampling plan, sample analysis, and interpretation of environmental sampling data. Part II presents background information on the roles of medical professionals in investigations of bioaerosol-related illnesses, the transmission and control of respiratory infections, prevention and control of microbial contamination, collection of air and source samples for biological agents, data analysis, data evaluation, remediation of microbial contamination, and biocides and antimicrobial agents. Part III, on specific agents, begins with a chapter on the biology of the source organisms that may become airborne as bioaerosols or that may produce biologically derived airborne contaminants. The remaining chapters cover specific biological agents [i.e., bacteria, fungi, amebae, viruses, house dust mites, endotoxin and other bacterial cell-wall components, fungal toxins and β-(1→3)-D-glucans, antigens and allergens, and microbial volatile organic compounds]. Authors are not identified for the chapters in Part I because many committee members and others contributed to these chapters. Primary and secondary authorship is recognized for other chapters.

Within chapters, individual sections are titled and numbered, the section numbers beginning with the chapter number. Section numbers are used throughout the book to direct readers to related discussions and explanations that appear elsewhere. These cross-references are set off in brackets. For example, the notation "[see 1.2]" directs readers to Section 2 in Chapter 1.

Many terms commonly shortened to acronyms and abbreviations are introduced once in the text, with later chapters using only the shortened form. Readers can find the full terms in the listing of acronyms and abbreviations beginning on page xi. The full terms for units of measurement that are abbreviated in the text are also listed for readers who may be unfamiliar with this notation. The Latin genus and species names of organisms are given on first use. Later references within a chapter adopt the standard convention of abbreviating the genus name to a first initial. A listing of genus and species names is provided on pages xi-xii for reference.

Occasionally specific products or tradenames are used in this book. Such mention does not imply endorsement on the part of an author or the publisher.

1.4 Approaches

We encourage users of this book to read Part I before beginning an investigation. We suggest that readers start with the chapters on investigation strategies and health effects. In addition to the information investigators can obtain through building inspections and environmental sampling for biological agents, the committee continues to feel that a health assessment is important for the successful handling of building-related illnesses that are due to bioaerosol exposures. A health assessment may be formal or informal, depending on the situation and circumstances, and does not always require the direct participation of medical professionals as members of an investigation team. Readers should note that the discussion of sample analysis precedes the discussion of sample collection. By presenting the information in this order, the importance is emphasized of considering what biological agents are sought, choosing an appropriate analytical method to identify and perhaps quantify them, and then identifying suitable collection methods.

The chapters in Parts II and III provide technical information that will aid industrial hygienists (IHs), environmental health professionals (EHPs), and indoor environmental quality (IEQ) consultants who conduct building evaluations. These chapters summarize the information needed to (a) understand when and why bioaerosols are important, (b) sample the environment to qualitatively or quantitatively determine if particular biological agents are present, and (c) prevent or mitigate problems related to bioaerosol exposure. ACGIH strives to provide accurate, complete, and useful information and the authors of this book have attempted to provide advice consistent with current knowledge on bioaerosol assessment and control. However, this is an area of active research and investigation and recommendations may change as newer information becomes available. Neither ACGIH nor any persons contributing to or assisting in the preparation of this information—nor any persons acting on behalf of these parties—makes any warranty, guarantee, or representation (express or implied) with respect to the usefulness or effectiveness of any information, method, or process disclosed in this material, nor do these parties assume any liability for the use of, or for damages arising from the use of, any information, method, or process disclosed herein.

The Bioaerosols Committee welcomes comments and suggestions on this book. Please communicate with the ACGIH Bioaerosols Committee at: ACGIH; Attn. Communications Dept.; Bioaerosols Committee; 1330 Kemper Meadow Drive, Suite 600; Cincinnati, OH 45240-1634; phone: 513-742-2020; fax: 513-742-3355; e-mail: comm@acgih.org.

1.5 References

ACGIH: Bioaerosols — Airborne Viable Microorganisms in Office Environments: Sampling Protocol and Analytical Procedures. American Conference of Governmental Industrial Hygienists, Appl. Ind. Hyg. 1(1):R19-R23 (1986).

ACGIH: Guidelines for Assessment and Sampling of Saprophytic Bioaerosols in the Indoor Environment. American Conference of Governmental Industrial Hygienists, Appl. Ind. Hyg. 2(5):R10-R16 (1987).

ACGIH: Guidelines for the Assessment of Bioaerosols in the Indoor Environment. American Conference of Governmental Industrial Hygienists, Cincinnati, OH (1989).

ACGIH: Biologically Derived Airborne Contaminants. In: 1998 TLVs and BEIs. Threshold Limit Values for Chemical Substances and Physical Agents, Biological Exposure Indices, pp. 11-14. American Conference of Governmental Industrial Hygienists, Cincinnati, OH (1998a).

ACGIH: Introduction to Chemical Substances.In: 1998 TLVs and BEIs. Threshold Limit Values for Chemical Substances and Physical Agents, Biological Exposure Indices, pp. 3-4. American Conference of Governmental Industrial Hygienists, Cincinnati, OH (1998b).

Chapter 2

Developing An Investigation Strategy

2.1   Introduction
    2.1.1   *Focus of This Book*
    2.1.2   *Components Contributing to Bioaerosol-Related Illnesses*
    2.1.3   *Steps in an Investigation*
    2.1.4   *Participants in Building Investigations*
2.2   Gathering Preliminary Information
    2.2.1   *Health Assessments — the Host*
    2.2.2   *Bioaerosol Assessments — the Agent*
    2.2.3   *Building Assessments — the Environment*
2.3   Hypothesis Development
2.4   Hypothesis Testing
    2.4.1   *Environmental Data*
    2.4.2   *Bioaerosol Data*
        2.4.2.1   Sample Analysis
        2.4.2.2   Sample Collection
    2.4.3   *Medical and Epidemiological Data*
    2.4.4   *Example of Hypothesis Checking*
2.5   Data Assessments
2.6   Recommendations and Remediation
    2.6.1   *Sources*
    2.6.2   *Pathways*
    2.6.3   *Receivers*
2.7   Risk Communication
2.8   Summary

2.1 Introduction

2.1.1 Focus of This Book

The primary focus of this book is the identification and control of bioaerosol exposures in non-manufacturing workplaces, with an emphasis on the evaluation of actual or potential bioaerosol exposures in office environments. The approaches discussed here may also apply in institutional, commercial, and residential settings as well as some manufacturing and recreational environments. Of secondary concern in this book is the evaluation of biological contamination in the absence of human exposure, in which case investigations are conducted to assess and prevent material or structural damage due to biological contamination rather than to protect human health. The following chapters discuss the interpretation of air and source samples for various biological agents and the identification of bioaerosol exposures that have been associated with health complaints or disease. As discussed in Section 1.2, TLVs exist for certain substances of biological origin, however, ACGIH currently does not support any specific numerical guidelines for interpreting environmental measurements of (a) total culturable or countable bioaerosols (e.g., total bacteria or fungi), (b) specific culturable or countable bioaerosols (e.g., individual genera or species of microorganisms or classes of biological agents), (c) infectious agents, or (d) assayable biological contaminants (e.g., endotoxin (see 1.2.1 and 1.2.5), mycotoxins, allergens, or microbial volatile compounds).

The lack of exposure criteria for many bioaerosols precludes identifying excessive exposures solely by measuring air concentrations of biological agents. Information has been accumulated about the adverse health effects of some bioaerosol exposures from case studies of affected workers, epidemiological studies of groups of workers, and laboratory and clinical evaluations of humans as well as model *in-vivo* and *in-vitro* systems. However, the institution of workplace or ambient exposure limits for bioaerosols has been hampered by (a) the diversity of biological agents and their effects on individuals, (b) the often multiple agents to which workers are exposed, and (c) the lack of epidemiological or toxicological data that establish dose–response relationships for carefully measured exposures to biological agents. In some cases, the lack of exposure data is due to the inadequacies of available sampling or analytical methods for the relevant biological agents. As a result, investigators identify associations between health effects and bioaerosol exposures on a case-by-case basis by combining health assessments with environmental observations and measurements.

2.1.2 Components Contributing to Bioaerosol-Related Illnesses

The strategies investigators use to study bioaerosol-related problems can play a role in how completely they understand what is happening in a work environment. Investigators strive to understand the interactions between a *host* (a worker) and an *agent* (a particular biological agent) that lead to contact and possibly an effect (Figure 2.1). In addition, investigators seek to explain how both of these factors influence and are influenced by the *environment* in which they meet (the workplace or building). Epidemiological studies often focus on identifying the time, place, and persons involved when assessing the effects of exposures to biological agents.

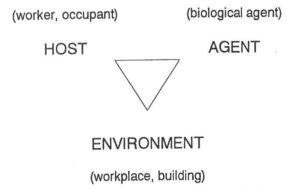

FIGURE 2.1. Fundamental components of bioaerosol-related illnesses.

For the discussion here, a *source* of a biological agent is the object, substance, animal, or person from which the agent was passed to a host. Typical indoor sources of biological agents are (a) people, who shed bacteria and viruses, (b) building materials, furnishings, and ventilation system components that provide a suitable environment for organism survival and growth, (c) accumulations of materials of biological origin on indoor surfaces, and (d) animals that shed allergens. Some texts use the term reservoir to describe indoor sites of microbial multiplication or accumulation. However, in this book, the term *reservoir* is used to designate the site in which an organism normally lives and multiplies and on which it primarily depends for survival. In some cases, the source and the reservoir of a biological agent are the same.

Typical reservoirs for biological agents are people, animals, plants, soil, water, and combinations of these. For example, a cooling tower may be the source of *Legionella pneumophila* that causes an outbreak of Legionnaires' disease in a building, but lakes, rivers, and streams are the natural reservoirs for this bacterium. Similarly, *Cladosporium herbarum* growing on damp walls in a building may be the source for worker exposure to fungal allergens, but the natural reservoir for this fungus is dead plant materials found outdoors. Distinguishing sources and reservoirs may be necessary when interpreting environmental sampling data and deciding how to control microbial growth. For example, *Legionella* spp. are frequently found in municipal water supplies because the bacteria may survive water treatment. Therefore, a one-time cleaning and disinfection of a warm-water storage tank will only temporarily rid it of the bacteria if they are likely to be reintroduced in the water supplied to a building. Conditions in the water storage tank (e.g., water temperature) must be maintained to minimize multiplication of these bacteria should they return.

Another familiar paradigm for understanding exposures to occupational hazards is to identify *sources, pathways,* and *receivers* (Figure 2.2). Separating the components of contaminant transmission in this way fits well with the scheme of identifying *hosts, agents,* and *environments* (Figure 2.1). Air is the common pathway through which the biological agents discussed in this book travel from sources to receivers. The respiratory tract is the primary target organ or site at which the effects of bioaerosol exposure are manifested, but systemic effects may also be seen as a result of the inhalation of some biological agents. Skin contact and ingestion are other potential routes of exposure to biological agents, but these exposure routes are addressed secondarily in this book.

2.1.3 Steps in an Investigation

Through training and experience, IHs, EHPs, and IEQ

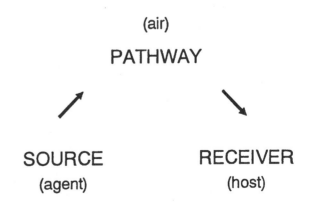

FIGURE 2.2. Paradigm for understanding and controlling occupational bioaerosol exposures.

consultants learn to gather and evaluate information about hosts, agents, and environments as well as sources and pathways. Experienced investigators use such information to reach supportable explanations for problems related to biological agents in indoor environments and to formulate reasonable recommendations to resolve problems. Figure 2.3 shows the main steps of this process, and Figure 2.4 explains in more detail how information about workers, biological agents, and buildings is combined in an investigation strategy. The following chapters of this book explain the implementation of these four steps (Table 2.1).

At each step in a workplace investigation, the participants should define their objectives and state their goals and expectations for the next activity. The overall objective of an investigation may change as further information becomes available, but the reasons for conducting an investigation should be defined as clearly and specifically as possible. The overall and specific purposes of a study should be agreed upon and communicated to the necessary parties as early as possible in the process.

Investigators can view these steps for assessing the presence and significance of biological and other agents in workplaces as similar to how physicians examine, diagnose, and treat patients. Some patients visit a doctor for a general checkup, others because they have particular concerns or complaints, and still others for serious problems that have developed. Likewise, workplaces may be evaluated on a routine basis, because certain conditions have arisen, or when clearly evident and potentially serious problems have become apparent. As for healthcare, measures to avoid bioaerosol-related problems may be categorized as primary, secondary, and tertiary prevention. First is prevention of worker exposure to potentially harmful biological agents. Good building design, operation, and maintenance could also be viewed as primary prevention in workplaces to avoid conditions

GATHER INFORMATION

↓

FORMULATE HYPOTHESES

↓

TEST HYPOTHESES

↓

MAKE RECOMMENDATIONS

FIGURE 2.3. Fundamental steps in an investigation.

GATHER INFORMATION ↓	HEALTH ASSESSMENT. Determine if the symptoms occupants report appear to be building-related and if symptoms and building conditions could be associated with biological agents.
	BIOAEROSOL ASSESSMENT. Determine if the building's history suggests a potential for a bioaerosol problem. Identify biological agents that could be found in the workplace, as described, or that could cause reported symptoms.
	BUILDING ASSESSMENT. Inspect the worksite. Note damp areas, signs of water damage, potential sources of biological agents, and possible mechanisms for bioaerosol generation and transport.
FORMULATE HYPOTHESES ↓	THEORIES. Construct plausible hypotheses using available information about (a) occupant complaints and potential causes, (b) possible sources of biological agents, and (c) the building environment. Seek additional information from appropriate experts and published literature.
TEST HYPOTHESES COLLECT DATA ↓	ENVIRONMENTAL DATA. Collect and analyze relevant environmental information. Identify the fundamental factors responsible for the presence of biological agents.
	BIOAEROSOL DATA. As needed, consult bacteriologists, mycologists, entomologists, public health specialists, medical professionals, and toxicologists. Develop a sampling plan, and collect and analyze relevant air and source samples for specific biological agents.
	MEDICAL AND EPIDEMIOLOGICAL DATA. As needed, consult medical professionals, toxicologists, and epidemiologists. Collect and analyze relevant medical and epidemiological data.
INTERPRET DATA ↓	DATA ASSESSMENT. If evidence supports a hypothesis, proceed to identifying solutions. Otherwise, discard disproven hypotheses, formulate new ones, and collect data to test these.
MAKE RECOMMENDATIONS	REPAIR and PREVENTION. Repair causes of excess moisture and conditions contributing to the presence of biological agents. Clean contaminated materials and remove materials that cannot be reclaimed. Implement measures to prevent a recurrence of bioaerosol-related problems. Revisit site to confirm that control measures are effective.

FIGURE 2.4. Outline of an investigation strategy.

that allow bioaerosol entry or growth of biological agents in buildings. Second is intervention post-exposure, post-injury, or when workers have developed signs or symptoms of bioaerosol-related reactions. For building maintenance, prompt responses to signs of water intrusion or microbial contamination are common secondary prevention measures that limit adverse effects of such events. Last are measures designed to limit the consequences of evident problems (conditions either in workers or buildings) that have occurred through natural causes, accident, neglect, or unforeseen circumstances. Managers, supervisors, and health and safety professionals are responsible for early recognition of potential health problems in workers. Likewise, building operators and managers are responsible for primary and some secondary prevention in facilities under their supervision. Occasionally,

these parties need outside advice to assess and determine how to address more serious problems related to microbial contamination or bioaerosol exposures or situations in which cause–effect relationships are difficult to establish.

2.1.4 Participants in Building Investigations

Investigators who study bioaerosol-related issues generally specialize in one of the three components outlined in Figure 2.1: (a) the health effects that the occupants (human hosts) experience, (b) the biological agents themselves and their detection, identification, and quantification, or (c) the environmental conditions that lead to the indoor presence of these agents and to workplace exposures. A doctor can better examine individual workers with building-related health problems if the treating physician also has information about workplace conditions.

TABLE 2.1. Outline of an Investigation Strategy and Related Chapters in this Book.

Investigation Step		Related Chapters
Gather Information		
Health assessment	Ch. 3	Health Effects Of Bioaerosols
	Ch. 8	Medical Roles And Recommendations
	Ch. 9	Respiratory Infections — Transmission And Environmental Control
Bioaerosol assessment	Ch. 18	Bacteria
	Ch. 19	Fungi
	Ch. 20	Amebae
	Ch. 21	Viruses
	Ch. 22	House Dust Mites
	Ch. 23	Endotoxin And Other Bacterial Cell-Wall Components
	Ch. 24	Fungal Toxins And β-$(1 \rightarrow 3)$-D-Glucans
	Ch. 25	Antigens
	Ch. 26	Microbial Volatile Organic Compounds
Building assessment	Ch. 4	The Building Walkthrough
	Ch. 10	Prevention And Control Of Microbial Contamination
Formulate and Test Hypotheses		
Collection of environmental data	Ch. 4	The Building Walkthrough
	Ch. 10	Prevention And Control Of Microbial Contamination
Collection of bioaerosol data	Ch. 5	Developing A Sampling Plan
	Ch. 6	Sample Analysis (also Ch. 17 to 26)
	Ch. 11	Air Sampling
	Ch. 12	Source Sampling
Collection of medical and epidemiological data	Ch. 3	Health Effects Of Bioaerosols
	Ch. 8	Medical Roles and Recommendations
	Ch. 9	Respiratory Infections — Transmission And Environmental Control
Data assessment	Ch. 7	Data Interpretation
	Ch. 13	Data Analysis
	Ch. 14	Data Evaluation
Make Recommendations		
Repair and prevention	Ch. 10	Prevention And Control Of Microbial Contamination
	Ch. 15	Remediation Of Microbial Contamination
	Ch. 16	Biocides And Antimicrobial Agents

On their part, microbiologists can better advise investigators on what environmental samples to collect and can interpret results more correctly if they know something about the environment under study and what, if any, health problems building occupants have experienced. Similarly, engineers and field investigators inspecting buildings should know what clinical diagnoses have been made for individual workers and what environmental samples have revealed.

Ideally, investigators studying different aspects of a building-related problem would consult one another and share information. Therefore, practitioners in different fields should be familiar with and appreciate the information and qualifications that other disciplines can contribute. Occasionally, investigators from the same or different disciplines may be asked to give their independent assessments of a situation. Rather than conduct an entire new investigation, consultants may also be asked to review a report prepared by others. An employer or building manager may give more weight to findings if multiple investigators reach the same conclusions and give similar advice on what needs to be done.

Investigators can be expected to assess buildings and building-associated complaints in accordance with the standard of care that applies to persons who are experts in conducting such evaluations. Investigators should act only within their areas of competence, as qualified by their education, training, experience, or demonstration of competency. Investigators should follow the standards of practice and comply with any restrictions pertaining to the practice of their profession, occupation, trade, or business. The lack of a title or degree does not mean a person is not qualified to conduct an investigation or render an opinion. However, possession of a title or degree does not automatically qualify a person as an expert. Only if they possess necessary permits, licenses, certifications, or registrations should investigators engage in activities that require such credentials.

Investigations of possible bioaerosol exposures often fall to IHs and EHPs because they are trained to identify work-related exposures and assess environmental conditions. This is logical because bioaerosol dissemination and the environmental conditions that contribute to the presence of biological agents are often the controlling factors in the host–agent–environment triangle (Figure 2.1). Membership in a professional organization indicates active participation in a field, but not all IHs or EHPs are trained or have experience in the investigation of bioaerosol-related hazards. IHs and EHPs can assist in such investigations by conducting site inspections, assessing exposures, and evaluating ventilation systems as well as by advising on the need for and proper selection and use of personal protective equipment.

Finally, investigators should undertake only those inspections, measurements, or tests for which they have or can obtain any necessary inspection, testing, or measurement equipment. Investigators should follow correct scientific procedures when conducting tests or making measurements and should only use methods that have been shown to be reliable. Investigators should avoid situations where they may encounter a conflict of interest that could impair or appear to impair their ability to provide accurate and unbiased assessments of health and safety conditions.

2.2 Gathering Preliminary Information
2.2.1 Health Assessments — the Host
The biological agents found in buildings can be diverse, arising from plants, animals, or microorganisms. However, different biological agents may produce similar health effects. For example, some fungi, bacteria, and amebae cause infections often manifested by common initial symptoms of fever, chills, and malaise. Likewise, the symptoms of hypersensitivity diseases are not specific to single causative agents even though the responsible antigen–antibody interactions are very specific. Inhalation fevers also have similar manifestations but many causes. Some reactions (e.g., asthma and rhinitis) may be caused by non-biological as well as biological agents. Therefore, although symptoms seldom identify the bioaerosols to which building occupants are exposed, the types of health effects observed may identify what category of agents investigators should seek (e.g., infectious agents, allergens, or causes of inhalation fevers).

Health assessments focus on potential hosts. However, it is not always possible to interview building occupants; indeed, some investigations are conducted to identify potential problems rather than to study building occupants who already have health or comfort complaints. Therefore, investigators may begin a study with a walkthrough site inspection and collection of data other than health data. Even in the latter case, a health assessment is not entirely missing from the evaluation process. The investigations discussed in this book are those that address the health and safety of current or future building occupants and how exposures to biological agents could affect these workers. Therefore, a starting point for any investigation is identification of the populations that might occupy a building or be exposed to potential bioaerosols, especially persons who may be highly susceptible to biological agents (e.g., very young or elderly persons, immunocompromised individuals, pregnant women, and persons with pre-existing medical conditions).

Health assessments determine whether conditions in a building could pose a hazard for occupants and if the symptoms that workers report are consistent with bioaerosol exposures. Chapter 3 describes the primary categories of reactions to bioaerosol exposures. Health assessments and occupant interviews need not be elaborate or time consuming activities but are often extremely valuable for developing investigation strategies. Health assessments can help investigators narrow the focus of an investigation and identify the types of environmental information they need to collect. Investigators should consult a physician, other medical professional, or toxicologist—preferably one with experience in occupational and environmental medicine—to evaluate serious health complaints (e.g., asthma, possible hypersensitivity pneumonitis, and infectious diseases) [see 8.5.2].

Epidemiologists occasionally play important roles in investigations of bioaerosol-related health effects. Epidemiological studies establish case definitions for specific health complaints to identify affected and unaffected workers and compare their exposures [see 3.7 and 8.6]. For example, a study may find that previously well persons who recently began experiencing building-related allergy symptoms share a ventilation system and that complaints began after a humidifier was installed in the system. Epidemiological studies may also compare rates of particular symptoms among workers known to be exposed to a biological agent and matched control workers known not to be exposed. A conclusion that workers' symptoms are consistent with bioaerosol exposure warrants follow up with an on-site inspection to identify potential sources of biological agents and possible dissemination routes.

2.2.2 Bioaerosol Assessments — the Agent

Readers are encouraged to study the chapters in Part III, which provide information on the characteristics, ecology, reservoirs, sources, and health effects of the biological agents most commonly encountered in indoor environments. These include (a) living infectious microorganisms (e.g., bacteria, fungi, amebae, and viruses), (b) allergens from plants, microorganisms, arthropods, birds, and mammals, (c) endotoxin and other bacterial cell-wall components, (d) mycotoxins and β-$(1\rightarrow3)$-D-glucans from fungi, and (e) microbial volatile organic compounds. Familiarity with the information in these chapters can help narrow the focus of a study, save valuable time and resources, and increase the likelihood that an investigation will accurately determine if biological agents are a potential source for concern. The fundamental questions regarding biological agents in indoor environments generally are not "Are biological agents present"—they almost certainly are, but "Is any biological agent present in an excessive amount," "Can building occupants be exposed to this material," "Would exposure to this agent cause adverse health effects," and "Could exposure to this agent account for building-associated health effects that building occupants have reported." Each chapter in Part III includes information on sample collection and analysis as well as data interpretation. Some of the chapters address remediation and control issues not covered in Chapter 10 or Chapter 15.

2.2.3 Building Assessments — the Environment

Chapter 4 describes how investigators use a site visit to survey building design and operation, look for potential sources of biological agents, and observe occupant activities. Inspectors should be accompanied by the building manager, maintenance personnel, and an employee representative, if available. Investigators may find potential sources of biological agents outdoors, in ventilation systems, or within the occupied spaces of buildings. Conclusions from an initial walkthrough inspection of a building generally fall into one of three categories. Investigators may (a) find no conditions that appear to contribute to the presence of unusual types or proportions of biological agents and conclude that excessive bioaerosol exposures are unlikely, (b) identify obvious problems and recommend solutions, or (c) note suggestive situations that warrant further evaluation.

2.3 Hypothesis Development

Investigators combine available environmental, epidemiological, medical, and toxicological evidence to develop hypotheses (i.e., carefully formulated logical answers or explanations). Many investigators make assumptions or guesses about problem situations without formally identifying these as hypotheses. Nevertheless, it is the pursuit of such theories that drives the investigative process. New evidence may fit neatly into a proposed scenario and strengthen a premise, or information may refute an explanation and convince an investigator to modify or discard a theory. Besides gathering information and formulating their own explanations, investigators should also listen to building managers and occupants. Ventilation system operators, maintenance personnel, workers, and supervisors often have theories about why their facility does or does not have problems. An investigator may not agree with the explanations these individuals offer, but often must address their concerns or include their theories with the hypotheses the investigator develops independently.

Following is an example of the type of hypothesis an investigator might use to guide a study. After conducting a preliminary site inspection and interviewing various parties, an investigator concludes that there appears to be fungal growth on ceiling panels in an office occupied by workers suffering from apparently building-related rhinitis. Specific subhypotheses in this theory are that (a) the discoloration and surface material on the panels is fungal growth, (b) sufficient moisture has reached this area of the ceiling to support microbial growth, (c) fungal spores from this growth are entering the breathing zone of the rooms' occupants, and (d) the symptomatic workers have developed sensitivity to allergens carried by spores of this specific fungus. Consideration of this example will be continued in Section 2.4.4 to illustrate how investigators could proceed to test the validity of these assumptions as explained in the next section.

2.4 Hypothesis Testing

Reasons for the presence of biological agents and worker exposure thereto may be obvious in some situations, and investigators may proceed to formulate recommendations without confirming their hypotheses. Other situations may be less clear, more than one explanation may be plausible (requiring different resolutions), or substantiating evidence may be needed to warrant repairs or remediation. For these and other reasons, investigators often must test their hypotheses. Hypotheses are generally based on a series of assumptions that can be challenged. Investigators check their assumptions by devising tests that could disprove their theories. A hypothesis that survives challenge may be true [see 14.1.1].

2.4.1 Environmental Data

The proliferation of bacteria, fungi, protozoa, and arthropods in buildings depends on a number of interrelated variables. Moisture is recognized as the primary factor that limits biological growth. Chapters 4 and 10 describe what investigators look for when examining buildings for sources of biological agents. In particular, Section 10.3 describes different measurements of water in air and materials. Section 10.4 describes places investigators should look for indoor and outdoor moisture sources and potential sites of biological growth. Visual inspections and environmental assessments should be conducted by experienced persons. Examples of the tests they may perform are measurement of the relative humidity of indoor air; determination of mechanisms for moisture entry and vapor migration; measurement of pressure differences between indoor and outdoor zones or above-ground and subsurface spaces; conduction of tracer experiments to illustrate patterns of air movement and measure ventilation rates; examination of wall construction (e.g., the types of interior and exterior materials and the location of air and vapor retarders); and tests for rain entry, water drainage, and water wicking.

2.4.2 Bioaerosol Data

The lack of health-based exposure criteria for most types of biological agents precludes environmental sampling and simple comparison of measurements with established air concentrations and dose–response relationships [see 7.3.2 and 14.2.6]. Therefore, environmental samples are primarily collected to test suspected sources of biological agents, identify and quantify the agents present, and demonstrate bioaerosol release from environmental sources. A sampling plan includes specification of (a) the biological agents that are sought, (b) typical sources of the suspected agents, (c) the analytical methods that will be used to detect, iden-

tify, and quantify the agents along with constraints the methods may place on sample collection, (d) the sampling methods to be used, (e) the operating parameters of the sampling instruments, and (f) the locations and times at which samples will be collected [see 5.2].

Used properly, environmental sampling for biological agents can be a valuable investigative and research tool. However, inexperienced investigators are cautioned that such samples are unlikely to provide useful answers unless collected using a carefully considered sampling plan designed to address clearly stated, testable hypotheses. Scientifically sound quantitative sampling may be time consuming and costly. Therefore, investigators should carefully consider whether they need to sample for biological agents to understand the host–agent–environment and source–pathway–receiver interactions that are occurring in a particular setting in order to interrupt one or more of the links. Investigators on limited budgets can begin by designing an optimal sampling plan [see Chapter 5] and can then determine how much testing must be eliminated to stay within budget. At that point, the investigators must decide if the sampling they can afford to perform will be useful and defensible. If not, the investigators may need to adopt another approach to evaluate the situation or seek sufficient support to conduct a reasonable investigation.

2.4.2.1 Sample Analysis Although sample analysis follows sample collection, investigators should consider the kinds of analyses that will be performed before selecting a sampling method and collecting environmental samples. IHs, EHPs, and IEQ consultants who have little background in biological analyses are frequently unclear about what they can realistically learn from such testing. Participation of the analytical personnel in the hypothesis-generation and sample-planning phases of an investigation is strongly encouraged until investigators gain experience in conducting environmental assessments for biological agents.

Data interpretation problems (i.e., being unable to draw clear conclusions from environmental data) are more often the result of unclear objectives or poor planning in the study-design and sample-collection phases of an investigation than in the sample analysis portion. Analyses may be sophisticated, precise, and accurate without being useful if the reason for collecting samples was unclear, the sampling locations or times were poorly chosen, too few samples were collected to be representative, or the samples were collected improperly. For example, a laboratory may spend several weeks to carefully and correctly identify the many fungal and bacterial isolates that may be cultured from air samples. However, culture-based analyses identify only a small fraction of all microorganisms in environmental samples and critical bioaerosol exposures may go undetected if investigators rely solely on culture-based analytical methods. Chapter 6 provides an overview of what investigators can learn from standard methods for analyzing various samples for biological agents. The sample analysis sections of the individual chapters in Part III can also guide investigators.

2.4.2.2 Sample Collection An investigator selects a sampling method by considering (a) the biological agents (or indicators thereof) of interest, (b) the analytical methods by which the laboratory will identify and perhaps quantify the material, and (c) the sites and times at which samples will be collected. Respectively, Chapters 11 and 12 describe air and source sampling. The chapters in Part III describe sampling for particular biological agents and their indicators or carrier particles.

The locations and times at which samples are collected depend on the hypotheses an investigation is pursuing. Chapter 5 discusses collection of representative samples. To identify building-related bioaerosol exposures, many investigators collect air samples using both spore traps and agar impactors, sampling outdoors and indoors at suspected problem and control locations. Settle plates do not collect airborne particles in a representative manner and do not reliably measure bioaerosol concentrations [see 11.3.1]. Investigators should bear in mind that samples provide information about a site as it existed at the time tested. However, the findings may not represent conditions at a time in the past or future, even the relatively recent past or near future. Changes in the kinds, concentrations, and proportions of biological agents in the air can be rapid and substantial. Such fluctuations in undisturbed bulk and surface materials likely occur more slowly.

2.4.3 Medical and Epidemiological Data

The initial health-assessment phase of an investigation may identify a need for physical examinations or clinical testing of workers to diagnose their conditions, evaluate the severity of any illnesses, determine the work-relatedness of their problems, and recommend treatment. Preliminary interviews with workers may also indicate the value of a systematic and thorough collection of epidemiological data. Chapters 3, 8, and 9 describe these activities. Medical and epidemiological data are evaluated to identify supportable associations between bioaerosol exposures and health outcomes.

2.4.4 Example of Hypothesis Checking

This section returns to the example hypothesis presented in Section 2.3 to illustrate how to apply the concepts briefly outlined above. The investigators test the validity of their overall explanation by challenging each of their subhypotheses.

First, the investigators proposed that dark material on ceiling panels was fungal growth. They could check this assumption by inspecting the material more closely. Collection of adhesive tape samples could confirm that the

discoloration on the ceiling was fungal growth and might even allow a presumptive identification of the fungus.

The second subhypothesis was that the ceiling was damp in the area of presumed fungal growth. Learning that the building's roof leaks during heavy rains would support this assumption. The observation of condensation on cold-water pipes above the affected ceiling panels would also explain how moisture could reach the ceiling. Use of a moisture meter or other device may provide an objective indication of the amount of moisture in materials. Measurement of dewpoint and surface temperature may shed light on the moisture dynamics of the space. Ultimately, the presence of microbial growth, which is only possible if water is available, provides evidence that sufficient moisture was present.

The third subhypothesis was that spores from the fungal growth became airborne for the office occupants to inhale. The investigators could collect air samples to see if spores of the fungus identified in surface samples were present at greater concentrations or as a greater fraction of the total air concentration in the room with visible growth than in other locations. The biological agents of interest are fungal allergens, but detection of fungal spores is often used to indicate allergen presence.

The final subhypothesis was that symptomatic occupants were immunologically sensitive to the fungus on the ceiling panels in their work area. Investigators could ask an allergist for advice on how to check this assumption and how helpful it would be to test building occupants for evidence of sensitivity or exposure to allergens from the fungus identified in surface and air samples. The evidence outlined in support of the first three subhypotheses should be sufficient to conclude that the source of excessive moisture in the ceiling should be stopped and the water-damaged ceiling panels should be replaced. However, for completeness, a medical evaluation of affected workers was also considered. In this example, a preliminary walkthrough inspection that included consideration of occupant symptoms led to (a) collection of environmental data and information about the building's history, (b) development of a surface and air sampling plan, followed by collection and analysis of environmental samples, and (c) possible medical and epidemiological evaluations of workers.

2.5 Data Assessments

Data assessment is the step where investigators make decisions on the relevance to human exposure of environmental observations and measurements, the strength of associations between exposures and existing or eventual disease, and the probability of current or future risks. Investigators frequently collect important information simply by observing physical evidence. An obvious interpretation of visible fungal growth in an area of chronic plumbing leaks is that undesirable conditions exist. Even if exposure to airborne fungal spores or other materials has not yet occurred, it may in the future. In addition to observational data, investigators may also need to interpret data from environmental samples and epidemiological surveys.

Chapter 14 explains how investigators weigh evidence to reach decisions. Only investigators confident in their knowledge of biological agents, the methods used to collect and analyze samples for them, and the epidemiology of bioaerosol-related diseases should undertake interpretation of environmental sampling data. Occasionally, investigators need the support of medical professionals or toxicologists to conclude that measured exposures could produce the health effects observed or could pose a hazard for building occupants. Given the state of knowledge on some biological agents, such determinations may require a considerable research effort and be beyond the scope of an investigation. Such uncertainties may be especially problematic in legal deliberations [see 14.2.7]. A goal of data interpretation is to explain clearly how a host and an agent could come into sufficient contact in an environment to produce a particular effect. Other goals are explaining how biological agents enter buildings, identifying what environmental conditions have contributed to organism survival and growth indoors, understanding the pathways by which biological agents reach hosts, estimating the magnitude of exposures to biological agents, and justifying recommendations for correcting potentially or clearly harmful conditions.

2.6 Recommendations and Remediation

Investigators combine the information they have gathered from visual examinations, interviews, and testing to identify the hosts, agents, agent sources, and exposure pathways involved in current or potential problems (Figures 2.1 and 2.2). An understanding of these key features and their interactions can help investigators assess potential health hazards and identify actions that will minimize or prevent bioaerosol exposures that may lead to adverse health effects. Investigators base their conclusions and recommendations on their experience and professional judgement and should clearly communicate their findings, interpretations, and recommendations to the appropriate parties.

2.6.1 Sources

Not surprisingly, control of bioaerosol exposures focuses primarily on identifying environmental sources and interrupting biological processes. A concurrent focus is prevention of particle release from sources of biological agents. Section 10.2 discusses the factors that lead to biological contamination in buildings and how to control such problems. Chapter 15 outlines procedures to remove biologically contaminated building materials. Chapter 16 discusses disinfectants used to treat micro-

bial growth and prevent its recurrence and cautions investigators against inappropriate use of biocides to treat microbial growth in building environments.

Other familiar forms of source control are water treatment to minimize microbial multiplication (e.g., amebae and *Legionella* spp.) as well as measures to minimize generation of water droplets that may carry infectious or allergenic agents [see 10.5 and 10.6]. Humidity control and good housekeeping—as measures to control mite and cockroach populations—are described in Sections 22.6 and 25.6. For infections spread from person to person (e.g., tuberculosis, colds, and influenza), early identification of affected workers and their removal from the workplace until appropriately treated or recovered are routine control measures [see 9.8.1].

2.6.2 Pathways

The exposure route from an identified source of a biological agent to a building occupant is often readily apparent. Still, investigators may be requested to demonstrate that a proposed exposure pathway exists by showing that the biological agent can be detected in indoor air. Occasionally, a means for worker exposure is unclear. For example, a building owner reluctant to undertake expensive repairs may argue that the hazard of microbial growth within a wall space is small if there is no obvious means for bioaerosols to penetrate the interior wall and reach building occupants. Likewise, for the earlier example of fungal contamination on ceiling panels, a building owner could argue that the contamination does not pose a hazard if there is no evidence that the fungal growth releases particles or volatile compounds. In such situations, an investigator's function may be to establish that a plausible or demonstrable exposure pathway exists. Investigators may be able to demonstrate that a proposed exposure pathway exists by showing that the biological agent can be detected in indoor air or by following particle transport or air movement using smoketubes, pressure differentials, or other means.

In some situations, biological agents may be known to be airborne, but their sources may not have been identified. Similarly, a known source may be in the process of remediation, and the investigators' responsibility is to minimize bioaerosol dissemination during the repair process. For such cases, investigators may limit exposures for building occupants and remediation workers by intercepting biological particles on their way from sources to receivers. Examples of means to interrupt agent transmission are the air filters used in large buildings to remove particulate matter from outdoor and return air. Section 15.2 describes the use of negative pressure and air filtration to contain bioaerosols during remediation operations. Section 9.8.6.4 mentions the use of filters to capture airborne pathogens and the use of ultraviolet germicidal irradiation to inactivate airborne infectious agents. Drift eliminators on cooling towers are also designed to trap aerosols that might allow droplet-borne bacteria to escape in exhaust air [see 10.6].

2.6.3 Receivers

Controls that focus on the worker (i.e., the host or receiver) are a last resort in preventing bioaerosol exposures. Therefore, the use of personal protection (e.g., eye, skin, and respiratory protection) is only discussed as a precaution against bioaerosol exposure to be used during site inspections and clean up operations [see 4.6.2 and 15.2]. Some of these precautions are similar to control measures routinely used in microbiology laboratories as well as clinical and veterinary-care settings. For infectious agents, immunization of workers is an established means of protecting them from some airborne infectious diseases [see 8.2.3.4]. For hypersensitivity diseases, prevention of allergen exposure is preferred, but drug treatment may allow workers to tolerate some degree of workplace exposure, at least on a temporary basis [see 8.2.1.4]. Relocation or reassignment or workers may be necessary when exposure is unavoidable, drug treatment is unavailable or unacceptable, and use of personal protective equipment is impractical or unacceptable.

2.7 Risk Communication

Risk communication is too large a topic to cover adequately in this book, other than to emphasize its importance. Risk communication can be viewed as the formulation by experts of accurate, clear messages about the nature, magnitude, significance, and control of risk and the dissemination of this information to nonexperts. It might better be viewed as an interactive process through which individuals and groups exchange information and opinions. The primary factors that have been found to determine people's perception of risks are the fairness, familiarity, and voluntariness of their exposure. Deciding what level of risk is acceptable is not a technical question but a social, political, economic, and value-based one [see also 14.1.4.5].

Participants in building investigations asked to communicate with workers or building occupants can prepare by anticipating common questions and finding answers (or explanations for why the information is not available). Typical questions and concerns are (a) what agencies are involved in the investigation and which individual or group is in charge, (b) what hazardous agents are present and what are the potential consequences of exposure to these agents, (c) what kinds of samples have been collected and what did they show, (d) when will the investigation be completed and when will the audience learn the findings, and (e) in the interim, what is being done to ensure the workers' health and safety. Good communication can improve understanding, allay fears, lead to acceptance of information offered, and influence personal and organizational decisions, but poor communication can lead to confusion and distrust. Key to good communication of workplace risks are the credibility of the source of the information and the trust the audience has in this person or group.

2.8 Summary

This chapter provides an overview of the four steps involved in an investigation of actual or potential bioaerosol exposure (Figure 2.3). These steps represent an idealized approach and actual investigations may not proceed as systematically or be as complete. Therefore, investigators should be flexible and modify these steps on a case-by-case or as-needed basis. The time required for the various steps can vary greatly, depending on the conditions at the study site and the objectives of an investigation. However, an investigation should not be more time consuming or costly than necessary. When allocating resources, a balance should be sought between supporting a thorough investigation and reserving time and funds for remediation and prevention efforts that may be needed.

Investigations of bioaerosol-related hazards should aim to describe (a) what biological or other agents may be present, (b) the environmental conditions that may contribute to their presence, (c) how the agents may affect building occupants, (d) the means by which bioaerosols may reach humans, and (e) what can be done to prevent exposure. Unwary investigators should recognize that even experienced medical professionals, toxicologists, microbiologists, engineers, IHs, EHPs, and IEQ consultants often find it difficult to answer apparently straight-forward questions about biological agents, their effects on people, and the environments in which biological agents may be found. Therefore, investigators should draw on their own knowledge, experience, professional judgement and, above all, common sense during bioaerosol investigations and should recognize when they need assistance from other specialists. In addition, investigators should recognize the importance of effective communication with building designers, owners, operators, and occupants and the necessity and value of educating these parties about biological agents as well as their sources, potential health effects, and control.

Chapter 3

Health Effects Of Bioaerosols

3.1   Introduction
3.2   Building-Related Symptoms
    3.2.1   *Epidemiological Studies of Building-Related Symptoms*
    3.2.2   *Bioaerosols and Building-Related Symptoms*
3.3   Hypersensitivity Diseases
    3.3.1   *Rhinitis and Sinusitis*
    3.3.2   *Asthma*
    3.3.3   *Hypersensitivity Pneumonitis*
3.4   Inhalation Fevers
    3.4.1   *Humidifier Fever*
    3.4.2   *Pontiac Fever*
    3.4.3   *Other Diseases from Irritant or Toxic Exposures*
3.5   Infectious Diseases
    3.5.1   *Legionnaires' Disease*
    3.5.2   *Other Infectious Diseases*
3.6   Other Building-Related Symptoms
3.7   Epidemiological Investigations
3.8   Summary
3.9   References

3.1 Introduction

Building-associated complaints arise from diverse symptoms that workers experience as a result of exposures to various physical, chemical, and biological agents in buildings. The majority of health complaints in problem buildings are related to mucous membrane discomfort (i.e., eye, nose, and throat irritation), headache, and fatigue from unknown causes. The term sick building syndrome (SBS; or tight building syndrome) has been used to describe non-specific building-related symptoms that cannot be associated with an identifiable cause. Although poorly defined, this term is still commonly used and widely understood as a distinction from specific building-related illnesses (BRIs), which are diagnosable diseases with known etiologies (Menzies et al., 1994; Hedge, 1995; Hodgson, 1995; Menzies and Bourbeau, 1997). To avoid possible misunderstandings associated with the term SBS, a distinction is made in this book between non-specific building-related symptoms (NSBRSs) and specific building-related illnesses (SBRIs). For simplicity, these terms have been shortened to BRSs and BRIs (Table 3.1). BRSs can be uncomfortable, even disabling, but permanent sequelae are rare (Redlich et al., 1997). Objective physiological abnormalities generally are not found, although there are several physiological markers for eye and mucosal effects [see 8.2]. Specific BRIs occur less often than BRSs but some are more serious and all are accompanied by physical signs and laboratory findings.

Investigators who are familiar with the range of medical conditions associated with exposures to chemical, biological, and physical agents in indoor environments can design more efficient studies to identify the causes of workers' complaints. Investigators can also use this knowledge to predict problems that may arise as a result of exposure to indoor contaminants for which potential sources were observed during building inspections conducted in the absence of complaints. When appropriate, investigators not trained in medicine, epidemiology, or toxicology involve specialists in their studies. Thus, a health assessment, by a building investigator of any discipline, is part of the information gathering phase of the investigation process described in Chapter 2. As discussed in Section 8.1, medical expertise should be sought for (a) diagnosis and management of an individual worker with a possible BRI, typically building-related hypersensitivity disease, inhalation fever, or infection, and (b) investigation of populations of workers with BRSs. Table 3.2 is an alphabetical listing of the potentially bioaerosol-related health effects discussed in this book. This list covers those bioaerosol-related conditions seen in office, commercial, or recreational environments. Other health effects may be seen as a result of exposures to biological agents in health-care or laboratory settings, agricultural or animal-handling facilities, or manufacturing environments. This table lists known causative agents and the respective chapters that discuss the conditions and the biological agents. Some of these conditions may also be seen as a consequence of exposure to chemical or physical agents, but Table 3.2 identifies only the primary biological agents that may be responsible for the listed conditions.

3.2 Building-Related Symptoms

3.2.1 Epidemiological Studies of Building-Related Symptoms

Work-related irritation of the mucous membranes of the eyes, nose, and throat are common in office workers, even in buildings not identified as having IEQ complaints (Nelson et al., 1995). Eye symptoms may include itching, mild redness, and irritation, and some building occupants may be unable to wear contact lenses in implicated buildings. Nasal symptoms include dryness, stuffiness, congestion, itching, and runny nose. Throat symptoms include feelings of dryness and irritation. Skin symptoms are less common, but workers may report dryness or skin irritation.

Specific causes of BRSs remain unclear, but exposure to biological and chemical agents have been implicated [see 3.2.2 and 26.3]. It is often assumed that BRSs re-

sult from an insufficient supply of fresh (outdoor) air to an enclosed space. However, BRS complaints have usually not been found to correlate with ventilation rates (Robertson et al., 1985; Harrison et al., 1987; Skov and Valbjørn, 1987; Fanger et al., 1988; Menzies et al., 1993). An association has been noted for ventilation rates below 10 L/s (approx. 20 ft^3/min) of outdoor air per person (Mendell 1993), but this is a fairly generous outdoor air supply rate and may not be achieved at all times in buildings operated by energy-conscious managers. European studies have found BRS symptoms to be associated with air-conditioned buildings with and without humidification (Finnegan et al., 1984; Burge et al., 1987; Skov and Valbjørn, 1987; Mendell and Smith, 1990; Mendell, 1993; Jaakkola and Miettinen, 1995).

Job category, job satisfaction, and gender also influence workers' perceptions of job-related symptoms. Further, occupant activities and building furnishings can affect IEQ and complaint rates. Specifically, the surface area of "fleecy" materials, exposed paper materials, and the amount and allergenicity of floor dust have been related to complaint rates (Gravesen et al., 1986; Skov and Valbjørn, 1987; Gyntelberg, et al., 1994). Buildings may have different reasons for poor IEQ, and a contaminated ventilation sys-

TABLE 3.1. Terms Used to Describe Building-Associated Medical Conditions

Sick building syndrome (SBS) (see Building-related symptoms)	
Building-related illness (BRI) or Specific building-related illness (SBRI)	Diagnosable illness whose cause can be directly attributed to exposure to an indoor chemical, biological, or physical agent. Medical condition of known etiology frequently accompanied by documentable physical signs and laboratory findings: • Infection (e.g., acute viral infection, legionellosis, or tuberculosis), • Syndrome associated with exposure to a chemical or physical agent (e.g., carbon monoxide poisoning, dermatitis from glass fibers, or irritant-induced or exacerbated asthma or rhinitis), • Immunologically mediated disease (e.g., allergic rhinitis, sinusitis, asthma, or hypersensitivity pneumonitis), • Inhalation fever associated with exposure to a biological agent (e.g., humidifier fever, Pontiac fever, or other febrile, flu-like illness).
Building-related symptom (BRS) or Nonspecific building-related symptom (NSBRS)	Nonspecific symptom (e.g., eye, nose, or throat irritation, headache, fatigue, or other discomfort) that usually cannot be associated with a well-defined cause but that appears to be linked with time spent in a building.

TABLE 3.2. Alphabetical Listing of Conditions Related to Exposure to Biological Agents and Chapters in which the Conditions and Agents are Discussed

Condition	Description	Ch.	Known or Suspected Causative Agents	Ch.
Allergic bronchopulmonary mycosis (ABPM)	A condition seen in asthmatics resulting from immunological reactivity to colonization of the airways with a fungus (see Allergic broncho-pulmonary aspergillosis).	25	Fungi (e.g., *Penicillium* spp.)	19
Allergic bronchopulmonary aspergillosis (ABPA)	The most common form of allergic broncho-pulmonary mycosis in which the colonizing fungus is an *Aspergillus* sp.	25	*Aspergillus flavus, Aspergillus fumigatus, Aspergillus terreus*	19
Allergy	An immune-mediated hypersensitivity to a foreign material (see Hypersensitivity diseases).	3, 8, 25	–	–
Asthma	A chronic inflammatory condition of the airways; usually characterized by intermittent episodes of wheezing, coughing, and difficulty breathing.	3, 8, 25	Bacterial antigens Fungal allergens Amebic allergens Arthropodal allergens Bird, pollen, or mammalian allergens Endotoxin	18, 25 19, 25 20 22, 25 25 18, 23
Bronchitis	Inflammation of the mucous membranes of the large airways, characterized by excessive mucus production and cough.	8	–	–
Building-related illness (BRI)	Diagnosable illness accompanied by documentable physical signs and laboratory findings that is associated with indoor exposure (see also Infectious diseases, Hypersensitivity diseases, and Inhalation fevers).	3, 8, 9, 25	(see specific health effect)	–
Building-related symptom (BRS)	Nonspecific symptom (e.g., eye, nose, or throat irritation, headache, fatigue, or other discomfort) that cannot be associated with an identifiable cause but that appears to be linked to time spent in a building.	3, 8, 25	Endotoxin Mycotoxins β-(1→3)-*D*-glucans MVOCs	18, 23 19, 24 19, 24 26
Common cold	Head cold, acute viral rhinitis, acute coryza; characterized by rhinitis, sneezing, lacrimation (tearing), irritated nasopharynx, chills, and malaise (see also Influenza).	8, 9	Rhinoviruses or coronaviruses	21
Conjunctivitis	Inflammation of the mucosal surface of the eye surrounding the cornea; characterized by lacrimation (tearing), irritation, and redness; seen with some hypersensitivity and infectious diseases and in response to certain biological and chemical irritants.	25	–	–
Cryptococcosis	Systemic fungal infection usually presenting as a respiratory infection and occasionally as meningitis (e.g., in HIV-infected persons).	3, 9, 15	*Cryptococcus neoformans*	19
Dermatitis	Inflammatory skin condition, skin reaction, skin rash; seen with some hypersensitivity and infectious diseases and in response to contact with certain biological and chemical agents (see also Toxic effects).	25	–	–

TABLE 3.2. Alphabetical Listing of Conditions Related to Exposure to Biological Agents and Chapters in which the Conditions and Agents are Discussed (cont.)

Condition	Description	Ch.	Known or Suspected Causative Agents	Ch.
Granulomatous amebic encephalitis (GAE)	Rare inflammation of the brain due to infection with an ameba	20	*Acanthamoeba* or *Balamuthia* spp.	20
Hantavirus pulmonary syndrome (HPS)	Often fatal acute respiratory infection preceded by a flu-like illness.	3, 9	Sin Nombre virus (SNV) or Four Corners virus (FCV)	21
Histoplasmosis	Systemic fungal infection of varying severity, with the primary lesion usually in the lungs.	3, 9, 15	*Histoplasma capsulatum*	19
Humidifier fever	A form of inhalation fever; characterized by fever, chills, muscle aches, malaise, and chest symptoms.	3, 8	Bacteria (GNB) Amebae Endotoxin	18 20 18, 23
Hypersensitivity disease (allergy)	Disease for which a subsequent exposure to an antigen produces a greater effect than that produced on initial exposure (see Allergy, Allergic bronchopulmonary aspergillosis, Allergic bronchopulmonary mycosis, Asthma, Hypersensitivity pneumonitis, Rhinitis, and Sinusitis).	3, 8, 25	Bacterial antigens Fungal allergens Amebic allergens Arthropodal allergens Bird, pollen, or mammalian allergens	18, 25 19, 25 20, 25 22, 25 25
Hypersensitivity pneumonitis (HP) (extrinsic allergic alveolitis)	An inflammatory disease of the lung due to cell-mediated immunological reactions in pulmonary tissues.	3, 8, 25	Bacterial antigens Bacterial proteases Fungal allergens	18, 25 18 19, 25
Infection	Entry and multiplication of an infectious agent in the body, the result of which may be inapparent or result in an infectious disease (see Infectious disease).	–	–	–
Infectious disease	Clinically manifest disease resulting from an infection (see Common cold, contagious disease, Cryptococcosis, Granulomatous amebic encephalitis, Hantavirus pulmonary syndrome, Histoplasmosis, Influenza, Legionnaires' disease, Primary amebic meningoencephalitis, Pneumonia, and Tuberculosis).	3, 8, 9	(see specific infection)	–
Influenza	Acute infectious disease of the respiratory tract; characterized by fever, headache, myalgia, prostration, rhinitis, sore throat, and cough.	8, 9	Influenza virus	21
Inhalation fever	Febrile, flu-like illness following heavy exposure to certain biological or chemical agents from environmental sources (see Humidifier fever and Pontiac fever).	3, 8	Bacteria, legionellae Fungi Amebae Endotoxin	18 19 20 18, 23
Legionnaires' disease	Bacterial pneumonia; preceded by anorexia, malaise, myalgia, and headache; accompanied by fever, chills, and (occasionally) cough.	3, 8, 9	*Legionella* spp., most often *Legionella pneumophila*	18, 15
Opportunistic fungal infection	Fungal respiratory infection primarily seen in immunocompromised persons.	3, 8, 9, 15	Fungi	19

TABLE 3.2. Alphabetical Listing of Conditions Related to Exposure to Biological Agents and Chapters in which the Conditions and Agents are Discussed (cont.)

Condition	Description	Ch.	Known or Suspected Causative Agents	Ch.
Organic dust toxic syndrome (ODTS)	Poorly characterized condition similar to humidifier fever; short-lived, febrile reaction following exposure to high organic dust levels (see Inhalation fever).	3, 8	–	–
Pneumonia	Lung infection; characterized by filling of the alveoli with secretions, cough, chest pain, shortness of breath, and fever.	3, 8	Bacteria Viruses	18 21
Pontiac fever	A form of inhalation fever; characterized by fever, chills, headache, and myalgia.	3, 8	*Legionella* spp.	18, 15
Primary amebic meningo-encephalitis (PAM)	Rare inflammation of the brain and its lining due to infection with *Naegleria fowleri*.	20	*Naegleria fowleri*	20
Rhinitis (coryza)	Inflammation of the lining of the nose causing rhinorrhea (runny nose), congestion, itching, and sneezing due to either allergy or direct irritation.	3, 8, 25	(see specific health effect)	–
Sick building syndrome (see Building-related symptom)				
Sinusitis	Inflammation of the mucosal lining of the sinuses leading to purulent nasal and pharyngeal drainage and cough (see Rhinitis).	3, 8, 25	–	–
Sore throat	Inflammation of the mucous membranes and underlying parts of the nasopharynx, pharynx, and larynx; associated with pain or discomfort on swallowing (see Common cold, Influenza, and Building-Related Symptom).	3, 8	(see specific health effect)	–
Toxic effect	Reaction to a biological toxin; may involve various organ systems (e.g., inhalation fever, mucous membrane or skin irritation, dizziness or tremors, nausea, immune suppression.	3, 18, 19, 23, 24, 26	Bacterial toxins Mycotoxins MVOCs	18, 23 19, 24 26
Tuberculosis (TB)	Bacterial infection that may affect many organs but most commonly the lungs.	3, 8, 9	*Mycobacterium tuberculosis*	18

tem is sometimes a source for complaints (Fanger et al., 1988). Thermal discomfort and perception of humidity or dryness may also be associated with symptoms (Menzies et al., 1993; Sundell and Lindvall, 1993; Jaakkola and Miettinen, 1995; Nelson et al., 1995). Mendell (1993) summarized the epidemiological literature pertinent to the effects of building, workspace, job, and personal factors. These data showed consistent associations between increased symptoms and the following building characteristics: air conditioning, carpets, occupant density, and video display terminal use. The following personal characteristics have also consistently been associated with increased symptoms: female gender, job stress or dissatisfaction, and a history of allergies or asthma. NIOSH looked for associations between environmental factors and work-related health conditions in 80 complaint office buildings and found increased relative risks associated with presence of debris inside the air intake, poor or no drainage from drain pans, presence of suspended ceiling panels, and recent renovation with new gypsum board (Sieber et al., 1996). It is unclear from cross-sectional studies whether the association between BRSs and allergies/asthma represents an overlap of symptoms or whether either of these conditions predisposes persons to the other.

3.2.2 Bioaerosols and Building-Related Symptoms

Bioaerosols have not been conclusively associated with BRSs, although biological agents are strong contenders in etiologic hypotheses. Burge et al. (1987) and Finnegan et al. (1984) suggested that correlations of BRSs with humidification and cooling were due to microbial contamination. BRSs have been associated with fungal multiplication and the presence of unusual types of fungi (Morey et al., 1984; Morey, 1988). The chief candidates for relationships between BRSs and bioaerosols are endotoxins (from Gram-negative bacterial cell walls), mycotoxins (from certain fungi), and other microbial components or products [e.g., β-(1→3)-D-glucans and microbial volatile organic compounds (MVOCs)], which later chapters discuss in detail. Investigators have targeted these bioaerosols because their effects, theoretically, would not depend on immunological sensitization and might be sufficiently brief to account for complaints that resolve when occupants leave a building. In this regard, reports associating BRSs with endotoxin and perhaps β-(1→3)-D-glucan exposures offer promising leads requiring confirmation and further study (Rylander et al., 1992; Gyntelberg et al., 1994; Teeuw et al., 1994).

Many hypotheses about hosts, agents, and their environments must be considered when investigating particular problem buildings. Potential contributors to BRSs include unsuitable ventilation system design, inadequate ventilation rates, improper ventilation system maintenance, and offgassing of volatile compounds from building materials and furnishings as well as emission of a multitude of airborne substances from occupant activities, microbial growth, and the accumulation of dust containing biological agents. Section 8.3 discusses physiological tests physicians use to assess building-related health effects and medical management of affected workers.

3.3 Hypersensitivity Diseases

Hypersensitivity diseases result from exposure to antigens, which stimulate a specific immunological response [see Chapter 25]. Hypersensitivity pneumonitis (HP) is associated with antigen-specific immunoglobulin G (IgG), elevated T lymphocytes in bronchoalveolar lavage fluid (Fink, 1983), and granuloma on lung biopsy. Other antigen-mediated diseases [e.g., allergic rhinitis (or "hay fever")] and most allergic asthma occur in persons with a genetic makeup that allows production of antigen-specific IgE.

Attack rates of hypersensitivity diseases among exposed building occupants are usually low, because an immunological reaction is required for these diseases to occur. However, even one clearly building-related HP case constitutes a significant public health event because the disease can be so devastating. Such a diagnosis should trigger a search for additional cases and initiation of corrective measures to reduce or eliminate bioaerosol exposures. HP and elevated asthma risk can co-exist in problem buildings (Hoffman et al., 1993). Most building-related antigens are assumed to be of fungal or bacterial origin, but protozoa have also been implicated. Dust-mite and animal allergens are common causes of allergic asthma in the residential environment and, to some degree, in workplaces (Lundblad, 1991; Janko et al., 1995).

3.3.1 Rhinitis and Sinusitis

Allergic rhinitis and sinusitis are diagnosed by patient history, physical examination, eosinophils in nasal smears, immediate skin-prick tests to aeroallergens, elevated total or specific IgE antibody levels, and occasionally by specific nasal challenge [see 8.2.1.1 and 25.2.2.1]. Building-related allergic rhinitis may be common and may overlap with BRS complaints but has been poorly documented (Robertson and Burge, 1985). Building-related allergic rhinitis is often accompanied by conjunctivitis. Future investigation of building-related allergic rhinitis or sinusitis would likely be most successful among office workers at risk for HP or asthma because antigens that cause these diseases might also elicit allergic nasal and sinus responses.

3.3.2 Asthma

Building-related asthma is characterized by complaints of chest tightness, wheezing, cough, and shortness of breath that are worse on work days and improve on weekends or vacations. Symptoms may occur within an hour of exposure, onset may be delayed for 4 to 12 hours, or both. To diagnose asthma, physicians rely on patient history, wheezing, and either reversible air flow limitation or elicitation of air flow limitation on challenge with low doses of pharmacological agents (methacholine or histamine) [see 8.2.1.2 and 25.2.2.1]. A work-related pattern of reduced peak air flows often indicates that asthma is due to a building-related exposure. Patients diagnosed with building-related asthma should be restricted from problem environments to prevent worsening of the condition, although remission may not occur in cases of long-standing exposure. In addition to exposure cessation, patients can be treated with asthma medications such as bronchodilators, corticosteroids, cromolyn, or theophylline.

Building-related asthma in office workers is not commonly recognized. Some case reports have shown asthma to be associated with humidifier use. Biocides used in humidification systems have been suspected as a cause of office-associated asthma (Finnegan and Pickering, 1986). Exacerbation of asthma from bioaerosols generated from a home cool-mist humidifier has been established using quantitative sampling and bronchial provocation testing (Solomon, 1974). Epidemic asthma has also occurred in a printing plant in association with a contaminated humidifier (Burge et al., 1985). A five-fold increased risk of endemic asthma was observed for occupants of an office building with moisture incursion from a below-grade wall (Hoffman et al., 1993). Asthma has also been seen in the industrial setting in high prevalence among workers with elevated daily endotoxin exposures from contaminated water sprays (Milton et al., 1995). In this industrial setting, it is likely that much of the asthma was due to irritant rather than allergic mechanisms.

3.3.3 Hypersensitivity Pneumonitis

HP is characterized by (a) acute, recurrent pneumonia with fever, cough, chest tightness, and lung infiltrates, (b) a progression of cough, shortness of breath, fatigue, and chronic lung fibrosis, or (c) an intermediate pattern of acute and chronic lung disease. Industrial hygienists are familiar with many types of HP in workers exposed to organic dusts. For example, farmers, pigeon breeders, cheese makers, wood processors, and mushroom growers are all at risk for HP (Rose, 1994; Rylander and Jacobs, 1994; Yang and Johanning, 1997). However, HP may also occur in various office, residential, and recreational environments.

Physicians diagnose HP on the basis of patient history and the following clinical signs or symptoms (a) abnormal chest radiograph or lung computerized tomography (CT) scan, (b) abnormal respiratory sound (rales) heard through a stethoscope, (c) reduced pulmonary function including reduced diffusing capacity, (d) pattern of exercise-tolerance-test findings compatible with interstitial lung disease, (e) lymphocytic alveolitis on bronchoalveolar lavage, (f) granulomas and interstitial fibrosis on lung biopsy, (g) reproduction of symptoms and signs with bronchial challenge with sterile extracts of implicated environmental organisms, or (h) reduction of symptoms and signs with corticosteroid medications or removal from bioaerosol exposure [see also 8.2.1.3 and 25.2.2.3]. A clinician should consider the public health implications of building-related HP and the possibility that a person is an index case when deciding whether to proceed to bronchoscopy to make a definitive diagnosis for patients with building-related chest symptoms (in whom asthma has been excluded) even in the absence of radiographic disease. Persons with building-related HP may require permanent restriction from entering implicated environments because, once sensitized, individuals react to extremely low, often unmeasurable, concentrations of antigenic material.

Precipitins (i.e., precipitating antibodies, usually IgG) to common saprophytic organisms or to an extract of material collected from an implicated environment may be present in the blood of exposed building occupants, even if they are not symptomatic [see 25.2.3.1]. Presence of these precipitins can provide a justification for collecting environmental samples to measure antigen exposures (Reed et al., 1983; Reed and Swanson, 1986). Often the specific biological antigen (e.g., the specific microorganism involved) in an HP outbreak or its precise source (Reed et al., 1983; Kreiss and Hodgson, 1984; Hodgson et al., 1985, 1987) is not known, although there are exceptions (Woodard et al., 1988). In the latter case, only one fungus was isolated from environmental samples and corresponded with specific precipitins in symptomatic individuals.

Investigation of building-related HP outbreaks has been reviewed elsewhere (Kreiss and Hodgson, 1984; Kreiss, 1989). Other publications have attributed HP to bioaerosols from water incursion and water-damaged furnishings (Hodgson et al., 1985; Hoffman et al., 1993), contaminated air-handling units (Bernstein et al., 1983; Hodgson et al., 1987; Woodard et al., 1988), and water sprays in indoor pools (Rose et al., 1998).

3.4 Inhalation Fevers

3.4.1 Humidifier Fever

Humidifier fever is a flu-like illness, characterized by fever, chills, muscle aches, malaise, and chest symptoms [see also 8.2.2.1]. These symptoms usually arise within 4 to 8 hours of exposure and subside within 24 hours, usually without long-term effects. People with humidifier fever rarely consult physicians for this short-lived, flu-like illness and, if they do, seldom undergo extensive diagnostic testing. Consequently, the pathophysiology of this condition is unclear, as it may overlap with radiologically normal HP. Recent work suggests that intermittent, high exposure to endotoxin may cause humidifier fever, whereas daily endotoxin exposure may result in asthma (Milton et al., 1995). In addition, repeated, severe episodes of inhalation fever due to endotoxin exposure may lead to emphysema. Thus, some cases of humidifier fever may be toxic diseases rather than diseases resulting from immunological responses. Organic dust toxic syndrome (ODTS) is a poorly characterized condition similar to humidifier fever. ODTS is diagnosed when short-lived, febrile reactions follow exposures to high dust levels in composting and various agricultural operations.

3.4.2 Pontiac Fever

Legionella spp. are associated with two BRIs: Legionnaires' disease and Pontiac fever. The former is a bacterial pneumonia [see 3.5.1]. The latter is a self-limited, flu-like illness characterized by fever, chills, headache, and myalgia. Pontiac fever was first described in 1968 in a 144-case Michigan epidemic in a county health department building where the attack rate was nearly 100% and the average incubation period was 36 hours (Kaufmann et al., 1981). Contaminated air-conditioning systems, whirlpool spas, steam turbine condensers, and industrial coolants have been associated with outbreaks of Pontiac fever (Friedman et al., 1987). Pontiac fever is diagnosed by an elevated serum antibody titer but is not considered to be an infection because live bacteria have not been recovered from clinical cases and the disease has apparently been caused by non-culturable bacteria (APHA, 1995) [see also 8.2.2.2]. Why these bacteria cause two distinct clinical syndromes has been attributed to (a) the inability of some legionellae to multiply in human tissue (for a variety of reasons, including virulence, host range, and viability of the bacteria), (b) unusual characteristics of their lipopolysaccharides and possible toxic effects of dead bacteria (Kaufmann et al., 1981; Horwitz, 1993), and (c) differences in host susceptibility (Fields, 1997).

3.4.3 Other Diseases from Irritant or Toxic Exposures

Conjunctivitis, rhinitis, and asthma may occur by irritant as well as immunological hypersensitivity mechanisms (Kipen, et al., 1994). For example, asthma has been observed in the industrial setting in high prevalence among workers with elevated daily endotoxin exposures from contaminated water sprays (Milton et al., 1995). In this industrial setting, it is likely that much of the asthma was due to irritant rather than allergic mechanisms. Building-related exacerbations of pre-existing allergic rhinitis or asthma may be triggered by exposure to irritant or toxic compounds, and these exposures may have a significant impact on disease severity (Michel et al., 1996).

Cancer occurs in clusters among office workers as it does in other occupations and among residential neighbors. Environmental cancer generally has a long latency period (i.e., the disease does not appear until years after the responsible exposure has occurred). Clusters of cancers of the same cell and organ type are rare and have not been clearly linked with indoor air pollution. Although it does not present in work-related clusters, radon-associated lung cancer is also an indoor air risk (Pershagen et al., 1994). Other carcinogens in indoor air to which workers may be exposed include asbestos, certain mycotoxins (e.g., aflatoxin, fumonisin, ochratoxin, patulin, and sterigmatocystin) (see 24.2), and environmental tobacco smoke.

The potential for toxigenic fungi to cause pulmonary hemorrhage and death in susceptible individuals (infants) has received much recent attention (CDC, 1995a; Jarvis et al., 1996; Sorenson et al., 1996; Montaña et al., 1997). However, the evidence to support a plausible exposure route and inhalation of a significant dose of mycotoxin remains weak. Exposures to agents other than mycotoxins were also significantly associated with the single reported outbreak (CDC, 1995a; Burge, 1996; Montaña et al., 1997). Cancer has been associated with ingestion of mycotoxins, but the only association of cancer with inhalation exposure to mycotoxins has been in heavily contaminated agricultural environments [see 24.2.4.1].

To date, no studies have been conducted to investigate the role of mycotoxins in the development of cancer for persons exposed in non-agricultural indoor environments. The toxic endpoints (i.e., various types of cancer) and the potency of carcinogenic mycotoxins have been characterized in animal and a few human studies. Currently available methods for measuring indoor exposures to mycotoxins do not yet allow a conclusion as to whether sufficient exposure occurs in non-agricultural settings to initiate or promote cancer [see 24.2]. On the other hand, some bioaerosols are known to reduce the risk of cancer. Epidemiologic and experimental studies have demonstrated that high levels of endotoxin exposure may be protective against lung cancer. However, no studies have been conducted of cancer prevention and endotoxin exposure in residential or office environments [see 23.2].

3.5 Infectious Diseases

The term *infection* is used to describe the entry and multiplication of a biological agent in a host's body. An *infectious disease* is a clinically manifest reaction to an infection [see 9.3.2]. The term *infectious* may also mean able to transmit an infectious agent, for example, an infectious person is one who not only is infected but is also able to transmit the responsible agent to others. Infectious diseases resulting from agents transmitted primarily from person-to-person are called contagious diseases.

3.5.1 Legionnaires' disease

Legionnaires' disease is a bacterial pneumonia most often caused by *Legionella pneumophila* and first recognized in a 1976 outbreak that resulted in 182 cases (29 fatal) among Legionnaires attending a convention at the Bellevue Stratford Hotel in Philadelphia. Since then, epidemic and sporadic cases have been associated with buildings and traced to aerosols from cooling towers and evaporative condensers as well as sprays from whirlpools, shower heads, supermarket vegetable misters, and potable water supplies. There is evidence that not all cases are due to airborne transmission of the bacteria and that drinking contaminated water followed by aspiration is the route of exposure in certain cases [see 9.4.2]. *Legionella* spp. are ubiquitous in nature. The incubation period for this pneumonia is 2 to 10 days, and only a small percentage (~5%) of exposed persons contract the disease. In addition to pneumonia, the infection can produce gastrointestinal tract, kidney, and central nervous system symptoms. Legionnaires' disease is diagnosed as outlined in Section 8.2.3.1.

3.5.2 Other Infectious Diseases

The risk of contagious disease transmission in buildings with low outdoor air ventilation rates or high rates of air recirculation has attracted much public interest. Outbreaks of tuberculosis, varicella (chicken pox), measles, and smallpox have confirmed the importance of airborne disease transmission (LaForce, 1986). Measles epidemics have spread by aerosol transmission through ventilation systems (Riley, 1980), and epidemic pneumococcal disease has been documented in an inadequately ventilated jail (Hoge et al., 1994). Swiss investigators found that absenteeism due to respiratory illness was greater in a fully air-conditioned building than in a naturally ventilated one housing a similar population (Guberan, 1985). Brundage et al. (1988) reported increased rates of febrile respiratory infection in army trainees housed in modern, tightly sealed barracks compared with trainees in older, draftier barracks. Military personnel in Operation Desert Shield who slept in tight, air-conditioned buildings (which were preferred and therefore crowded) complained more often of sore throat and cough (Richards et al., 1993). Troops who slept in naturally ventilated tents complained more often of rhinorrhea, possibly due to higher exposures to outdoor air pollutants and allergens.

Unusual infections sometimes appear as building-associated epidemics in environments such as hospitals and research laboratories. For example, the Q-fever agent (*Coxiella burnetii*—a rickettsia) has been disseminated through ventilation systems in buildings housing infected sheep, goats, or cattle and in buildings in which the organism was being grown in a laboratory. Q-fever symptoms include high fever, chills, headache, and myalgia and sometimes pneumonia, hepatitis, and endocarditis. Systemic fungal infections (e.g., histoplasmosis and cryptococcosis) have occurred when contaminated bird droppings or construction/demolition dusts were disseminated in indoor environments (Rose, 1994; CDC, 1995b). The viral agent of hantavirus pulmonary syndrome (the Sin Nombre or Four Corners virus) may be spread by inhalation from infected rodent droppings and urine (Jackson, 1994). Finally, many opportunistic bacterial and fungal pathogens can cause infectious disease in immunocompromised persons occupying buildings where these common organisms are present [see 8.4].

3.6 Other Building-Related Symptoms

Bioaerosols are often suspected as causative agents for BRSs when other causes are not readily identified even if, in fact, another cause is responsible. For example, detergent residues left in carpets after cleaning can cause cough and dry throat symptoms (Kreiss et al., 1982; CDC, 1983) and, though rare in office environments, carbon monoxide poisoning can cause headache, fatigue, and nausea.

Persons with BRSs whose complaints are not addressed and whose exposure persists sometimes develop a more general awareness of or sensitivity to other environments. These reactions may continue despite improvements in the initial environment (Rosenstock and Cullen, 1994). Such persons may also find that their symptoms become more severe with time and do not resolve after leaving the problem building or environment as they did initially. Over the last 20 years, an increasing number of such patients have been labelled as having environmental illness or multiple-chemical sensitivity (MCS), a syndrome with a variety of definitions (Fiedler and Kipen, 1997). Although several etiologic theories have been postulated for MCS, no agreement has been reached on diagnostic criteria for or causes of such reactions. In fact, some affected persons may have developed asthma or irritant or allergic rhinitis with chronic sinusitis that has gone unrecognized and untreated. Whatever the proximate cause, it seems clear from clinical experience that an im-

portant contributing factor in the development of these chronic conditions is a delayed response to complaints of BRSs and a sense on the part of the affected workers that management does not believe or value them.

Early investigations of BRS complaints occasionally resulted in a diagnosis of mass psychogenic illness (mass hysteria) when specific causes were not readily found to explain symptoms. Certainly, the occupants of buildings with high complaint levels are often angry and fearful after encountering managerial resistance to investigations, inconclusive results from studies, or ineffectual remedies. However, the diagnosis of psychogenic illness has specific criteria, as do most other medical diagnoses (Guidotti et al., 1987), and this diagnosis is not appropriate merely because other causes are not apparent. Mass psychogenic illness is characterized by subjective symptoms with features difficult to explain on an organic basis. Hyperventilation is common, and epidemics are transmitted from person to person via a visual or verbal chain (Baker, 1989; Boxer, 1990). Incidence rates of mass psychogenic illness may shift throughout an occupied space and rarely correlate with measurable air quality parameters. Chronic health complaints related to IEQ are unlikely to be due to mass psychogenic illness, even though a psychologic overlay is common. Poor management, boring work, incorrect lighting, temperature variations, uncomfortable work stations, and noise may lower peoples' complaint thresholds. However, air quality and thermal comfort complaints are likely to have a preventable basis, which can often be determined from epidemiological evidence and environmental inspection.

3.7 Epidemiological Investigations

Epidemiology is the study of the occurrence and distribution of diseases and other health-related conditions in populations (Kelsey et al., 1996). Many epidemiological studies of building-associated problems are *descriptive stud-*

TABLE 3.3. Components of an Epidemiological Study

• Define what constitutes a case of illness (i.e., establish a case definition).

• Consider possible confounding factors (i.e., variables that are related to both risk and outcome, for example, the presence of multiple antigens in a contaminated environment).

• Select appropriate control groups (i.e., unexposed or unaffected workers).

• Carefully design a questionnaire that can be administered with or without an interviewer.

• Select and administer appropriate diagnostic tests to screen for or confirm disease or condition.

• Determine how data will be analyzed before collecting data.

ies, intended to provide information on patterns of symptom occurrence in populations according to various characteristics such as work activities and location, worker age and gender, and onset and timing of symptom occurrence. *Analytical studies* test hypotheses and are carried out when leads about the causes of a condition are already available. Examples of analytical investigations are case–control studies and cohort studies, which may be prospective or retrospective. Of these study designs, common epidemiological approaches include investigations of the prevalence of health and environmental conditions in problem environments as well as comparisons of cases and controls within problem environments or in apparent problem and non-problem areas.

An epidemiological investigation can often clarify whether there is a building-related problem and, if so, its nature as well as possible means for resolution. Mapping the distribution of cases in time and space within a building may lead to hypotheses regarding probable sources and pathways and may suggest appropriate sampling protocols. Table 3.3 outlines the components of a well-designed epidemiological study.

Generic questionnaires are available for investigating IEQ complaints [see 8.5.2]. Using a standardized questionnaire facilitates comparison of findings with other studies. For example, the USEPA and NIOSH have an IEQ questionnaire and have compiled a database derived from investigations that used the questionnaire (http://www.cdc.gov/niosh/ieqwww.html). However, investigators may decide to tailor a questionnaire to gather information that will enable them to test a specific hypothesis not studied previously. A good questionnaire avoids leading questions and eliminates ambiguities. Information on risk factors (e.g., contact lens use, pre-existing medical conditions, and smoking habits) may be necessary, as well as demographic information, identification of health complaints (with their time of onset and occurrence), and location of respondents within a building. Information on possible exposures outside the work environment may be pertinent for some types of complaints. Precoded forms facilitate data analysis in investigations involving large numbers of people.

Medical epidemiologists, often in collaboration with other specialists, can arrange appropriate diagnostic testing to substantiate or further define medical problems suggested by questionnaires. However, objective tests are not available for BRS outbreaks, except in a research context. For hypersensitivity diseases, pre- and post-shift spirometry, peak flow measurements, diffusing capacity measurements, chest radiographs, and serum precipitin testing can be used, with clinical referral of suspected cases for more definitive diagnostic tests [see 8.6.1].

A major concern for epidemiologists is measurement, for example, measurement of exposures, diseases, confounding variables, and effect modifiers (Kelsey et al., 1996).

Epidemiologists need to carefully choose the most reliable sources of data, the correct study design, representative study populations, and the appropriate methods of measuring the parameters of possible importance. Unfortunately, epidemiological investigations are difficult in many workplaces and other indoor settings because the numbers of cases and populations available for study are often small and the methods available to measure exposures to biological agents may be limited. Additionally, it may be difficult to categorize workers into exposure categories if their jobs require them to spend time in more than one part of a building or to engage in multiple activities. Comparisons between populations from separate work areas or buildings have been conducted, but this approach is not always possible and, when it is, investigators may encounter problems if they cannot eliminate possible confounders or measure exposures adequately.

3.8 Summary

Medical assessment of occupants with IEQ complaints is an ideal way to begin an investigation. A study to confirm that a biological agent is the cause of a BRI (e.g., a hypersensitivity disease, inhalation fever, or infection) may differ substantially from a search for the cause of BRS complaints. Admittedly, physicians, epidemiologists, and toxicologists are rarely involved in investigations from the beginning. However, other disciplines can approach building-related complaints using the conceptual framework outlined here. Once the causes of building-related complaints are understood and corrected, epidemiological surveillance tools can be used to demonstrate that the problem has been resolved or to document its reappearance. Even in the absence of complaints from building occupants, investigators can use information about known associations between biological agents and health outcomes to judge the safety of conditions and practices they observe in buildings.

3.9 References

APHA: Legionnaires' Disease, Legionnaires' Pneumonia, Pontiac Fever. In: *Control of Communicable Diseases Manual*, pp. 256-258. A.S. Benenson, Ed. American Public Health Association, Washington, DC (1995).

Baker, D.B.: Social and Organizational Factors in Office Building-Associated Illness. Problem Buildings: Building-Associated Illness and the Sick Building Syndrome. Occup Med State of Art Rev 4:607-624 (1989).

Bernstein, R.S.; Sorenson, W.; Garabrant, D.; et al.: Exposures to Respirable, Airborne *Penicillium* from a Contaminated Ventilation System: Clinical, Environmental and Epidemiologic Aspects. Am. Ind. Hyg. Assoc. J. 44(3):161–169 (1983).

Boxer, P.A.: Indoor Air Quality: A Psychosocial Perspective. J. Occ. Med. 32:425-428 (1990).

Brundage, J.F.; Scott, R.McN.; Lednar, W.M.; et al.: Building-Associated Risk of Febrile Acute Respiratory Diseases in Army Trainees. J. Am. Med. Assoc. 259(14):2108–2112 (1988).

Burge, P.S.; Finnegan, M.; Horsfield, N.; et al.: Occupational Asthma in a Factory with a Contaminated Humidifier. Thorax 40:248–254 (1985).

Burge, S.; Hedge, A.; Wilson, S.; et al.: Sick Building Syndrome: A Study of 4373 Office Workers. Ann. Occup. Hyg. 31(4A): 493–504 (1987).

Burge, H.: Health Effects of Biological Contaminants. In: Indoor Air and Human Health, pp. 171-178. R.B. Gammage and B.A. Berven, Eds. CRC Press, Boca Raton, FL (1996).

CDC: Respiratory Illness Associated with Carpet Cleaning at a Hospital Clinic. MMWR (CDC) 32:378–384. Centers for Disease Control, Virginia (1983).

CDC: Acute Pulmonary Hemorrhage/Hemosiderosis Among Infants Cleveland, January 1993-November 1994. MMWR (CDC) 44:4. Centers for Disease Control, Cleveland, OH (1995a).

CDC: Histoplasmosis. MMWR (CDC) 44:701-703. Centers for Disease Control, Kentucky (1995b).

Cullen, M.R.: The Worker with Multiple Chemical Sensitivities: An Overview. State Art Rev. Occup. Med. 2:655-661 (1987).

Fanger, P.O.; Lauridsen, J.; Bluyssen, P.; Clausen, G.: Air Pollution Sources in Offices and Assembly Halls, Quantified by the Olf Unit. Energy Build. 12:7–19 (1988).

Fiedler, N; Kipen, H: Chemical Sensitivity: The Scientific Literature. Environ. Health Perspect. 105 (Suppl 2):409-415 (1997).

Fields, B.S.: Legionellae and Legionnaires' Disease. pp. 666- 675. In: Manual of Environmental Microbiology, pp. 629-640. C.J. Hurst, G.R. Knudsen, M.J. McInerney, et al., Eds. American Society for Microbiology Press, Washington, DC (1997).

Fink, J.: Hypersensitivity Pneumonitis. In: Principles and Practice of Allergy. C.V. Mosby, St. Louis, MO (1983).

Finnegan, M.J.; Pickering, C.A.C.: Building-Related Illness. Clinical Allergy 16:389–405 (1986).

Finnegan, M.J.; Pickering, C.A.C.; Burge, P.S.: The Sick Building Syndrome: Prevalence Studies. Br. Med. J. 289:1573–1575 (1984).

Friedman, S.; Spitalny, K.; Barbaree, J.; et al.: Pontiac Fever Outbreak Associated with a Cooling Tower. Am. J. Public Health 77:568–572 (1987).

Gravesen, S.; Larsen, L.; Gyntelberg, F.; Skov, P.: Demonstration of Microorganisms and Dust in Schools and Offices. Allergy 41:520–525 (1986).

Guberan, E.: Letter to the Editor. Brit. Med. J. 290:321 (1985).

Guidotti, T.L.; Alexander, R.W.; Fedoruk, M.J.: Epidemiologic Features that May Distinguish Between Building-Associated Illness Outbreaks Due to Chemical Exposure or Psychogenic Origin. J. Occup. Med. 29:148-150 (1987).

Gyntelberg, F.; Suadicani, P.; Nielsen, J.W.; et al.: Dust and the Sick Building Syndrome. Indoor Air. 4:223-238 (1994).

Harrison, J.; Pickering, A.C.; Finnegan, M.J.; et al.: The Sick Building Syndrome Further Prevalence Studies and Investigation of Possible Causes. In: Indoor Air '87: Proceedings of the 4th International Conference on Indoor Air Quality and Climate, August 17-21, 1987, Berlin, West Germany. 2:487-491 (1987).

Hedge, A.: In Defence of "The Sick Building Syndrome." Indoor Environ. 4:251-253 (1995).

Hodgson, M.J.; Morey, P.R.; Attfield, M.; et al.: Pulmonary Disease Associated with Cafeteria Flooding. Arch. Environ. Health 40:96–101 (1985).

Hodgson, M.J.; Morey, P.R.; Simon, J.S.; et al.: An Outbreak of Recurrent Acute and Chronic Hypersensitivity Pneumonitis in Office Workers. Am. J. Epidemiol. 125(4):631–638 (1987).

Hodgson, M.: The Sick-Building Syndrome. Occup. Med. State Art Rev. 10:167-175 (1995).

Hoffman, R.E.; Wood, R.C.; Kreiss, K.: Building-Related Asthma in Denver Office Workers. Am. J. Public Health 83:89-93 (1993).

Hoge, C.W.; Reichler, M.R.; Cominguez, E.A.; et al.: An Epidemic of Pneumococcal Disease in an Overcrowded, Inadequately Ventilated Jail. New Eng. J. Med. 331:643-648 (1994).

Horwitz, M.A.: Toward an Understanding of Host and Bacterial Molecules Mediating *Legionella pneumophila* Pathogenesis. In: *Legionella: Current Status and Emerging Perspectives*, pp. 55-62. J.M. Barbaree, R.F. Breiman, A.P. Dufour, Eds. American Society for Microbiology, Washington, DC (1993).

Jaakkola, J.J.K.; Miettinen, P.: Type of Ventilation System in Office Buildings and Sick Building Syndrome. Am. J. Epidemiol. 141:755-765 (1995).

Jackson, G.W.: Viruses. In: Physical and Biological Hazards of the Workplace, pp. 271-317. P.H. Wald and G.M. Stave, Eds. Wiley, New York, NY (1994).

Janko, M.; Gould, D.C.; Vance, L.; et al.: Dust Mite Allergens in the Office Environment. Am. Ind. Hyg. Assoc. J. 56:1133-1140 (1995).

Jarvis B.B.; Zhou Y.; Jiang, J.; et al.: Toxigenic Molds in Water-Damaged Buildings: Dechlorogriseofulvins from *Memnoniella echinata*. J. Natural Prod. 59(6):553-4 (1996).

Kaufmann, A.F.; McDade, J.E.; Patton, C.M.; et al.: Pontiac Fever: Isolation of the Etiologic Agent (*Legionella pneumophila*) and Demonstration of its Mode of Transmission. Am. J. Epidemiol. 114:337–347 (1981).

Kelsey, J.L.; Whittemore, A.S.; Evans, A.S.; et al.: Methods in Observational Epidemiology, 2nd Ed. Oxford University Press, New York, NY (1996).

Kipen, H.M.; Blume, R.; Hutt, D.: Asthma Experience in an Occupational and Environmental Medicine Clinic: Low Dose Reactive Airway Dysfunction Syndrome. J. Occup. Env. Med. 36(10):1133-7 (1994).

Kreiss, K.: The Epidemiology of Building-Related Complaints and Illness. In: Problem Buildings: Building-Associated Illness and the Sick Building Syndrome. Occup. Med. State Art Rev. J.E. Cone and M.J. Hodgson, Eds. 4:579–592 (1989).

Kreiss, K.; Gonzalez, M.G.; Conright, K.L.; et al.: Respiratory Irritation Due to Carpet Shampoo: Two Outbreaks. Environ. Intl. 8(1-6):337–342 (1982).

Kreiss, K.; Hodgson, M.J.: Building-Associated Epidemics. In: Indoor Air Quality, pp. 87–108. P.J. Walsh, C.S. Dudney, and E.D. Copenhaver, Eds. CRC Press, Inc., Boca Raton, FL (1984).

LaForce, F.M.: Airborne Infections and Modern Building Technology. Environ. Intl. 12:137–146 (1986).

Lundblad, F.P.: House Dust Mite Allergy in an Office Building. Appl. Occup. Environ. Hyg. 6:94-96 (1991).

Mendell, M.J.: Non-Specific Symptoms in Office Workers: A Review and Summary of the Epidemiologic Literature. Indoor Air 3:227-236 (1993).

Mendell, M.J.; Smith, A.H.: Consistent Pattern of Elevated Symptoms in Air-Conditioned Office Buildings: A Reanalysis of Epidemiologic Studies. Am. J. Public Health 80:1193–1199 (1990).

Menzies, R.; Tamblyn, R.; Farant, J.P.;et al.: The Effect of Varying Levels of Outdoor-Air Supply on the Symptoms of Sick Building Syndrome. New Eng. J. Med. 328:821-7 (1993).

Menzies, D.; Bourbeau, J.: Building-related Illnesses. N. Engl. J. Med. 337:1524-1531 (1997).

Menzies, D.; et al: The "Sick Building" — A Misleading Term that Should be Abandoned. IAQ '94 — Engineering Indoor Environments. Atlanta, GA, pp. 37-48 (1994).

Michel, O., Kips, J.; Duchateau, J.; et al.: Severity of Asthma is Related to Endotoxin in House Dust. Am. J. Respir. Crit. Care Med. 154: 1641-1646 (1996).

Milton, D.K.; Amsel, J.; Reed, C.E.; et al.: Cross-Sectional Follow-up of a Flu-like Respiratory Illness among Fiberglass Manufacturing Employees: Endotoxin Exposure Associated with Two Distinct Sequelae. Am. J. Indust. Med. 28:469-488 (1995).

Montaña E.; Etzel, R.A.; Allan, T.; et al.: Environmental Risk Factors Associated with Pediatric Idiopathic Pulmonary Hemorrhage and Hemosiderosis in a Cleveland Community. Pediatrics. 99(1 Suppl S):E51-E58 (1997).

Morey, P.R.: Microorganisms in Buildings and HVAC Systems: A Summary of 21 Environmental Studies. In: Engineering Solutions to Indoor Air Problems, Proceedings of IAQ '88, Atlanta, GA. American Society for Heating, Refrigeration, and Air-Conditioning Engineers, Inc. Atlanta, GA (1988).

Morey, P.R.; Hodgson, M.J.; Sorenson, W.G.; et al.: Environmental Studies in Moldy Office Buildings: Biological Agents, Sources and Preventive Measures. Ann. Am. Conf. Govt. Ind. Hyg. 10:21–35 (1984).

Nelson, N.A.; Kaufman, J.D.; Burt, J.; et al.: Health Symptoms and the Work Environment in Four Non-problem United States Office Buildings. Scand. J. Work Environ. Health 21:51-59 (1995).

Pershagen, G.; Akerblom, G.; Axelson, O.; et al.: Residential Radon Exposure and Lung Cancer in Sweden. N. Engl. J. Med. 330:159-164 (1994).

Redlich, C.A.; Sparer, J.; Cullen, M.R.: Occupational Medicine: Sick-Building Syndrome. Lancet. 349:1013-1016 (1997).

Reed, C.E.; Swanson, M.C.: Indoor Allergens: Identification and Quantification. Environ. Intl. 12:115–120 (1986).

Reed, C.E.; Swanson, M.C.; Lopez, M.; et al.: Measurement of IgG Antibody and Airborne Antigen to Control an Industrial Outbreak of Hypersensitivity Pneumonitis. J. Occup. Med. 25(3):207–210 (1983).

Richards, A.L.; Hyams, K.C.; Watts, D.M.; et al.: Respiratory Disease Among Military Personnel in Saudi Arabia During Operation Desert Shield. Am. J. Pub. Health. 83:1326-1329 (1993).

Riley, E.C.: The Role of Ventilation in the Spread of Measles in an Elementary School. Ann. N.Y. Acad. Sci. 353:25–34 (1980).

Robertson, A.S.; Burge, P.S.: Building Sickness. The Practitioner 229:531–534 (1985).

Robertson, A.S.; Burge, P.S.; Hedge, A.; et al.: Comparison of Health Problems Related to Work and Environmental Measurements in Two Office Buildings with Different Ventilation Systems. Br. Med. J. 291:373–376 (1985).

Rose, C.S.; Martyny, J.W.; Newman, L.S.; et al.: "Lifeguard Lung": Endemic granulomatous pneumonitis in an Indoor Swimming Pool. Am. J. Public Health 88:1795-1800 (1998).

Rose, C.S.: Fungi. In: Physical and Biological Hazards of the Workplace, pp. 392-416. P.H. Wald; G.M. Stave, Eds. Wiley, New York, NY (1994).

Rosenstock, L.; Cullen, M.R.: 1994. Textbook of Clinical, Occupational and Environmental Medicine. WB Saunders, Philadelphia, PA (1994).

Rylander, R.; Persson, K.; Goto, H.; et al.: Airborne Beta-1,3-glucan May be Related to Symptoms in Sick Buildings. Indoor Environ. 1:263-267 (1992).

Rylander, R.; Jacobs, R.R., Eds.: Organic Dusts: Exposure, Effects, and Prevention. Lewis Publishers, Boca Raton, FL (1994).

Sieber, W.K.; Stayner, L.T.; Malkin, R.; et al.:The National Institute for Occupational Safety and Health Indoor Environmental Evaluation Experience, Part Three: Associations between Environmental Factors and Self-Reported Health Conditions. Appl. Occup. Environ. Hyg. 11:1387-1392 (1996).

Skov, P.; Valbjørn, O.: Danish Indoor Climate Study Group: The "Sick" Building Syndrome in the Office Environment: The Danish Town Hall Study. Environ. Intl. 13:339–349 (1987).

Solomon, W.R.: Fungus Aerosols Arising from Cold-Mist Vaporizers. J. Allergy Clin. Immunol. 54:222–228 (1974).

Sorenson B.; Kullman, G.; Hintz, P.: NIOSH Health Hazard Evaluation Report. HETA 95-0160-2571 (NIOSH — CDC), National Center for Environmental Health (1996).

Sundell, J.; Lindvall, T.: Indoor Air Humidity and the Sensation of Dryness as Risk Indicators of SBS. In: Proceedings of Indoor Air '93, 6th International Conference on Indoor Air Quality and Climate, July 4-8, 1993, Helsinki, Finland. 1:405-410, (1993).

Teeuw, K.B.; Vundenbroucke-Grauls, C.M.J.E.; Verhoef, J.: Airborne Gram-negative Bacteria and Endotoxin in Sick Building Syndrome. Arch. Intern. Med. 154:2339-2345 (1994).

Woodard, E.D.; Friedlander, B.; Lesher, R.J.; et al.: Outbreak of Hypersensitivity Pneumonitis in an Industrial Setting. J. Amer. Med. Assoc. 259(13):1965–1969 (1988).

Yang, C.S.; Johanning, E.: Airborne Fungi and Mycotoxins. In: Manual of Environmental Microbiology, pp. 651-660. C.J. Hurst, G.R. Knudsen, M.J. McInerney, et al., Eds. ASM Press, Washington, DC (1997).

Chapter 4

The Building Walkthrough

4.1   General Considerations
    4.1.1   *What Investigators Can Learn from Walkthrough Inspections*
    4.1.2   *Conducting a Walkthrough Inspection*
    4.1.3   *Areas Frequently Associated with Biological Contamination or Bioaerosol Entry*
    4.1.4   *Moisture and Other Factors*
4.2   Outdoor Sources of Bioaerosols
4.3   The HVAC System
    4.3.1   *HVAC System Design*
    4.3.2   *Outdoor Air*
        4.3.2.1   Outdoor Air Supply Rate
        4.3.2.2   Outdoor Air Quality
    4.3.3   *Filtration*
    4.3.4   *Heat Exchangers*
        4.3.4.1   Surface Condensation within HVAC Systems
        4.3.4.2   Drain Pans and Standing Water
        4.3.4.3   Air Washers and Humidifiers
    4.3.5   *Supply Air*
        4.3.5.1   Supply Air Plenums and Ductwork
        4.3.5.2   Supply Air Distribution
4.4   The Occupied Space
    4.4.1   *Supply Air Diffusers*
    4.4.2   *Fancoil and Induction Units*
    4.4.3   *Water Leaks, Spills, and Elevated RH*
    4.4.4   *People, Animals, and Plants as Sources of Biological Agents*
4.5   Return Air
4.6   Protecting Investigators and Building Occupants During Walkthrough Inspections
    4.6.1   *Assessing the Risk of Bioaerosol Exposure*
    4.6.2   *PPE Selection and Use*
    4.6.3   *Protecting Building Occupants During Walkthrough Inspections*
4.7   References
Appendix 4.A   Building Walkthrough Checklist

4.1 General Considerations

4.1.1 What Investigators Can Learn from Walkthrough Inspections

On-site building inspections are conducted for many reasons, among them (a) to assess the suitability of a space for a particular group of occupants and type of activity, and (b) to attempt to identify possible causes of complaints from current building occupants. During preliminary building walkthroughs, investigators collect environmental information of a fairly general nature [see Figure 2.4]. A preliminary inspection may identify the need for an in-depth building evaluation, during which investigators focus on potential problem areas and gather data to address specific questions. Obtaining evidence of worker exposure to biological agents and proving causal associations between specific exposures and symptoms is often difficult. However, a thorough walkthrough inspection by a team of experienced investigators may yield valuable insight about the existence of sources of biological agents and possible pathways from sources to building occupants [see 2.1.2].

4.1.2 Conducting a Walkthrough Inspection

During initial walkthroughs, investigators collect primarily observational data (i.e., information obtained by visual inspection of a building and interviews with the building operator and occupants). Investigators (a) examine the physical structure, maintenance activities, and occupancy patterns of a building, (b) look for potential sources of biological agents, (c) look for evidence of current or past water damage or excess moisture, (d) note sources of other indoor air contaminants, and (e) as needed, formulate plans for an in-depth investigation or for control and remediation of noted problems. Building management, heating, ventilating, and air-conditioning (HVAC) system operators, maintenance personnel, and employee representatives should accompany investigators on building walkthroughs.

Sources of biological agents may be found (a) outside buildings, (b) in attics and below-grade spaces, (c) within wall cavities, (d) in HVAC systems (Figure 4.1), and (e) in the occupied space. Investigators should obtain and study building blueprints, if available, including those for any retrofitting and remodeling that has been done. As-built plans, which incorporate changes made during construction, can be important to identify where deviations from the original building design occurred and where problems may arise. Although seldom available or complete, maintenance and reconstruction records can provide valuable information regarding the history and current status of HVAC system operation.

4.1.3 Areas Frequently Associated with Biological Contamination or Bioaerosol Entry

Table 4.1 summarizes specific sites and equipment that investigators may check and some of the problems they may find while conducting a building walkthrough. Appendix 4.A provides a sample checklist that emphasizes identifying sources of biological agents. Investigators should view the example checklist as a companion to Chapters 10 and 15, which identify areas where biological growth often occurs in buildings and what can be done to avoid and correct problems. The table and appendix are coded by letter to correspond with the HVAC system components shown in Figure 4.1.

4.1.4 Moisture and Other Factors

In addition to the features highlighted in Table 4.1 and Appendix 4.A, investigators should note moisture anywhere in a building that may have supported biological growth in the past or may do so in the future. Temperature must also be considered, because of its role in moisture transfer and condensation [see 10.2]. The use of portable, hand-held moisture meters may enable investigators to pinpoint areas of potential biological growth during a walkthrough inspection. Many of these meters detect the presence of moisture by measuring the electrical conductance between two probes inserted into the medium to be tested. When set to the proper resistance for the material to be tested, a moisture meter can give a reliable indication of the amount of moisture present. Foarde et al. (1996) used a moisture meter in a number of field investigations and found that the meter detected significant moisture in building materials when moisture was not otherwise evident to the investigators. In several of these studies, the investigators found biological growth associated with these damp areas as well as reports of symptoms by the building occupants. Moisture meters can be used to survey the moisture associated with any non-conductive porous material to which the probes can be applied (e.g., ceiling tiles, gypsum board, and carpeting) (Foarde et al., 1996). These devices indicate internal moisture content, a difficult measurement to make without a meter. Moisture meters provide qualitative information but do not indicate the actual amount of water available to microorganisms for growth, as defined by

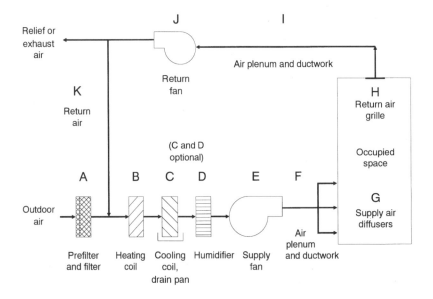

FIGURE 4.1. Typical HVAC system components for a large building.

water activity, a_w, [see 10.3]. Other instruments use radiowaves to detect temperature and moisture differences remotely (e.g., at wall surfaces).

Identification of a source of a biological agent does not necessarily implicate bioaerosols as a substantial factor contributing to occupants' illness or complaints. Even in assessments of suspected biological contamination, investigators should consider other reasons for poor air quality and occupant dissatisfaction with the indoor environment. Various documents can aid investigators in conducting systematic building walkthrough inspections (AIHA, 1990; USEPA/NIOSH, 1991; USEPA, 1994; Morey, 1995; ISIAQ, 1996). Godish (1995) describes several protocols for investigating problem buildings.

4.2 Outdoor Sources of Bioaerosols

Microorganisms are normal inhabitants of the outdoor environment. Outdoor air often contains fungal spores,

TABLE 4.1. Potential Sources of Biological Agents or Bioaerosol Entry Routes into Buildings and Factors Related to Microbial Growth or Bioaerosol Dissemination (letters in parentheses identify components illustrated in Figure 4.1)

Source/Route	Related Factors
Outdoors	
Outdoor air	Crop planting and harvesting—exposed soil, soil turning, or disturbance of plant materials; excavation or construction operations; wastewater treatment or irrigation; textile mills; slaughterhouses or rendering plants; composting operations
Building exterior	Poor grading or water drainage; evidence of water intrusion (discoloration); blocked rain gutters; penetrations in siding or veneer; damage to building envelope; wood rot in structural timbers; animal infestation near building or in crawl space beneath building; automatic lawn sprinklers that wet exterior walls
HVAC System	
Outdoor air intakes (OAIs)	Bioaerosol sources near OAIs (e.g., plant debris, feathers and bird droppings, insect or rodent infestations, sanitary air vents, cooling towers or evaporative condensers, standing water); below-grade OAIs
Filters (A)	Dampness; microbial growth on filters; gaps between filters and housings; low efficiency filters
Heat exchangers (B and C)	Dirty heating or cooling coils; excessive water in condensate pans—inadequate drainage from collection pans; blow-through of water droplets onto surfaces downstream of coils; dampness and microbial growth on acoustical lining; poorly maintained air washers or humidifiers; stagnant water in air washers or humidifiers
Supply air plenums and ductwork (F)	Excessive surface deposits; dampness and surface microbial growth; inaccessible humidifiers
Supply air diffusers (G)	Surface deposits, rust, or microbial growth on louvers; soiling of adjacent ceilings and walls; poor air mixing
Occupied Space	
Water-damage	Evidence or history of plumbing or roof leaks, water intrusion or spills, high indoor humidity (>70%), attempts to clean or disinfect carpets and other materials, musty or moldy odors
Chronic condensation	Inadequate insulation or intrusion of humid outdoor air that results in chronic condensation on windows, perimeter walls, or other cool surfaces
Window air conditioners and evaporative air coolers	Location inconvenient for maintenance; dirty grilles; standing water in condensate pans or sumps; dampness and surface microbial growth in or near units
Fancoil and induction units	Dirty heating or cooling coils or filters; excessive water in condensate pans—inadequate drainage from collection pans; dampness and surface microbial growth near units
Potted plants	Microbial growth on leaves, soil, plant containers, or surfaces in contact with containers; excess moisture from overwatering
Carpet	Poorly maintained or water-damaged carpet that serves as a source for dirt accumulation or microbial growth
Fabric office partitions, wall coverings, drapes; upholstered furniture	Poorly maintained or water-damaged fabric-covered and upholstered items that serve as sources for dirt accumulation or microbial growth
Portable (console) humidifiers	Poorly maintained units with microbial growth in the water reservoirs; spray or mist units
Return air plenums (I)	Excessive surface deposits; dampness and surface microbial growth

pollen, bacteria, and algae as well as fine plant and insect fragments (Muilenberg, 1995). Fungal spores may account for a few percent to approximately a quarter of particulate matter less than 10 μm (PM10). Pollen can also be a significant component of outdoor air on a mass basis. Except during periods of snow cover, fungal spores and bacteria are abundant in outdoor air, and levels may increase dramatically during and immediately following rain events. In addition to knowing about regional factors that contribute to the atmospheric bioaerosol load, it is essential that investigators evaluate the neighborhood immediately surrounding a building. Possible local sources of strong bioaerosol emissions include farming, animal handling, composting, and certain industrial operations. Examples of outdoor bioaerosol sources that may be immediately adjacent to a building are cooling towers, standing water on roof tops, and accumulated organic matter (e.g., leaves, grass clippings, feathers, bird or bat droppings, insect colonies, and rodent nests). Bioaerosols from such sources, once disseminated into the ambient air, may find their way into buildings through OAIs, windows, doors, vents, and inadvertent openings such as joints and cracks in the envelopes of negatively pressurized buildings.

4.3 The HVAC System

4.3.1 HVAC System Design

Ventilation systems in large buildings mix outdoor and recirculated air, heat or cool this mixture, and distribute it to the occupied space (Figure 4.1). Ventilation systems in residences and similar small buildings usually have simpler means to condition indoor air and typically do not provide for mechanical intake of outdoor air. Rather, these systems circulate indoor air through heating or cooling units and distribute the air through ductwork or directly into a space. In some cases, no mechanical circulation is provided and the indoors is heated radiantly. In such buildings, window evaporative coolers or air conditioners may be used for cooling. In some climatic regions, humidifiers of various types are used to add moisture to dry indoor air. Where indoor humidity conditions are too high, dehumidifiers may be used to remove excess water. The finding of a humidifier or dehumidifier in a building should prompt investigators to inquire why it is needed. To diagnose problems in HVAC system performance, building inspectors must be familiar with the climatic conditions of the geographic area in which a building is located as well as what building designs, construction materials, and HVAC equipment are appropriate for the region.

4.3.2 Outdoor Air

4.3.2.1 Outdoor Air Supply Rate An inadequate supply of outdoor air often leads to building-related complaints and symptoms, regardless of the presence or absence of inappropriate bioaerosols (Seitz, 1990). A properly maintained and operated mechanical ventilation system can reduce indoor bioaerosol concentrations by limiting infiltration of outdoor bioaerosols and by diluting those from indoor sources (Parat et al., 1994, 1996). The American Society of Heating, Refrigerating, and Air-Conditioning Engineers (ASHRAE, 1989) has recommended a minimum of 10 L/s (20 ft³/min) of outdoor air per person for office environments. Provided the outdoor air being introduced is of high quality and properly conditioned, the introduction of adequate volumes of outdoor air may improve indoor air quality by reducing the concentration of indoor contaminants through dilution [see 10.5.2]. Residences and similar small buildings rely on natural ventilation for outdoor air entry (i.e., open windows and doors and cracks in the building envelope). Closing and sealing these openings can make heating and cooling more efficient, but may reduce outdoor air infiltration to levels too low to adequately dilute indoor contaminants. Opening windows immediately increases the outdoor air supply, but allows penetration of outdoor aerosols and can waste energy and be uncomfortable.

4.3.2.2 Outdoor Air Quality The location of the OAI can significantly affect the nature of outdoor bioaerosols that enter a building (Bearg, 1993) [see 10.5.4]. A bioaerosol source with a clear pathway to an OAI may also result in contamination of the interior of an HVAC system. Rooftop OAIs are vulnerable to bioaerosol sources such as cooling towers, sanitary vents, building exhausts, and standing water. Streetlevel and below-grade OAIs are susceptible to entrainment of moisture, vehicle emissions, and other ground-level pollutants including odors and dust from decaying vegetation and accumulated animal matter.

Cooling towers and evaporative condensers located close to or directly upwind of OAIs should be evaluated as potential bioaerosol sources, especially in legionellosis investigations [see 10.6]. Films, foam, and algae on wet surfaces may indicate poor maintenance. Water samples and scrapings of biofilms can be collected from these water reservoirs for analysis. Usually such analyses are only necessary where specific BRIs (e.g., Legionnaires' disease, Pontiac fever, or HP) have been identified and specific biological agents are suspected.

Stagnant water, soil, and plant and animal matter near or in an OAI can support the growth of bacteria and fungi which can subsequently enter a building. In addition, birds and bats may use OAIs, window ledges, eaves, and attics as roosting and nesting sites. Accumulated bird or bat droppings may harbor the infectious fungi *Cryptococcus neoformans* and *Histoplasma capsulatum* as well as other fungi and bacteria. Similar problems can occur where large flocks of birds nest or roost in trees near occupied buildings. Rodent and other animal infestations near OAIs are also of concern [see 9.2.1]. Investigators may collect bulk samples of material near OAIs if necessary to confirm the biological nature or content of the material.

4.3.3 Filtration

A primary function of filters within HVAC systems (A in Figure 4.1) is to protect heating and cooling coils (B and C). Filters may also reduce maintenance requirements by limiting the accumulation of dirt throughout a building and its HVAC system and may contribute to worker health by removing potentially harmful particles, including ambient aerosols and those from local sources. However, such particles may enter an HVAC system and possibly penetrate to the occupied space if a filter has poor efficiency for fine particles or if there are gaps between the filter and the filter housing (Pasanen, 1995). HVAC system filters are not intended to protect equipment and occupants from heavily contaminated air such as may enter an OAI located near a strong outdoor bioaerosol source [see 10.4.4.1 and 10.5.3].

Filters may become damp during the air-conditioning season or when an OAI is not adequately protected from rain, snow, or fog. Microorganisms may grow on a damp filter itself or on collected dust (Martikainen et al., 1990; Kemp et al., 1995; Morey, 1996; Parat et al., 1996; Simmons et al., 1997). Microscopic analysis of material accumulated on filters may help distinguish between a normal accumulation of material of biological origin on a filter and actual microbial growth.

4.3.4 Heat Exchangers

In many large buildings, the mixture of outdoor and return air passes a heat exchanger, which generally consists of heating and cooling coils (B and C in Figure 4.1). Heat and moisture are either added to or removed from the supply air to maintain comfortable conditions. During the air-conditioning season, water condenses from warm air as it passes over cooling coils. This condensate collects in a drain pan beneath the coils and should promptly exit the air-handling unit through drain lines (Figure 10.7). If present, humidification devices are located downstream of the heat exchanger (D in Figure 4.1). In some large buildings, cooling may be accomplished in part by passing air through a water spray (air washer) (Morey, 1988). Section 10.4.4 describes problems investigators may find with heat-exchange equipment.

4.3.4.1 Surface Condensation within HVAC Systems

An increase (up to 90%) in the relative humidity (RH) of air downstream of cooling coils is a natural result of the energy transfer between the air and the coils. Moisture may condense on cool surfaces in contact with this damp air or may "wick" off cooling coils. Particles (e.g., soil,

organic matter, and microorganisms) not removed by the filtration system can collect on surfaces within HVAC systems. Such organic matter may support microbial growth under wet or damp conditions [see 10.4.4.2]. In these areas of HVAC systems, materials that retard water evaporation or provide increased surfaces on which dirt can accumulate may present particular problems. Investigators may sample surfaces in HVAC systems if microbial contamination is suspected and needs to be confirmed.

4.3.4.2 Drain Pans and Standing Water

Microorganisms can grow in the stagnant water that results when drain pans are inadequately pitched toward an outlet or a drain is blocked (Morey et al., 1984) (C in Figure 4.1). The pressure differential between the inside and outside of a ventilation system may also affect water drainage [see 10.4.4.3 and Figure 10.7]. The presence of a film or foam in standing water may indicate microbial growth or other contamination. Investigators may collect water samples for testing if they suspect and need to confirm microbial growth. However, interpretation of test results requires experience because such waters will always yield some culturable bacteria and fungi. In addition, investigators must consider if there is a mechanism for dissemination of contaminants into the occupied space. Microorganisms may release odorous compounds, which can readily be distributed throughout a building, but the release of particles from contaminated water requires mechanical disturbance or other action sufficient to produce aerosols.

4.3.4.3 Air Washers and Humidifiers

Sumps for air washers and humidification devices that use recirculated cold water are always microbially contaminated to some degree (D in Figure 4.1) [see also 10.4.4.4]. Because these devices generate aerosols, they should always be considered potential bioaerosol sources. Growth may also occur in a heat-exchange plenum if the surfaces are sufficiently cool to allow condensation. Steam humidification systems have been shown to emit fewer bacteria and fungi than water spray systems (Burkhart et al., 1993), but condensation and related microbial growth may still be a problem. Samples can be collected from air washers, humidifiers, and damp surfaces to confirm contamination and to identify contaminating organisms.

4.3.5 Supply Air

Conditioned air leaving a heat exchanger is transported to the occupied spaces of a building through a system of ducts (F in Figure 4.1). In large buildings, the main supply air plenum or duct is usually constructed of sheet metal, which may be internally or externally insulated for noise control and to minimize heat exchange. Air from the main supply air plenum travels until it enters a branch duct or a terminal (Morey and Shattuck, 1989). Air eventually enters the occupied space through diffusers which, if properly designed, promote mixing of the conditioned supply air with room air [see 4.4.1].

4.3.5.1 Supply Air Plenums and Ductwork Supply air ducts are often internally lined with glass fiber to reduce noise and/or to insulate so as to minimize heat exchange with surrounding materials. Any surfaces within supply air plenums and ductwork can accumulate dirt that, in the presence of adequate moisture, can serve as a substrate for microbial growth. Although ductwork is never sterile or entirely free of debris, it should not be coated with a thick layer of deposited material. For example, glass-fiber ductwork should be undamaged and should show its original color and metal ductwork should still look metallic.

If dirt and debris have accumulated on ductwork, microbial growth can occur when sufficient moisture becomes available [see also 10.4.4.5]. Humidification devices may be located in supply air ductwork (rather than near a heat exchanger) where they cannot easily be reached for cleaning or examination. The reservoirs of these humidifiers are often contaminated, and microorganisms may grow on downstream ductwork if water condenses there [see 4.3.4.3].

Investigators can collect surface samples of ductwork deposits suspected of being microbial growth. Collection of samples from apparently clean ductwork is less important, but may provide laboratory personnel with a reference against which to judge other samples. Identification of the same biological agent in an HVAC system and in air in the occupied zone may implicate the ventilation system as the source or disseminator of the contaminant. However, many of the fungi and bacteria that grow as contaminants within ventilation systems are also common in outdoor air. Therefore, investigators should compare the types and relative concentrations of microorganisms in indoor and outdoor air to identify whether the biological agents arise from outside the building (in which case, better filtration may be needed) or if the source for the air contaminants is within the HVAC system.

4.3.5.2 Supply Air Distribution The distribution of conditioned air to occupied spaces may also disseminate airborne particles. Particles may originate from inadequately filtered outdoor air, contamination within an HVAC system, or contaminants from the occupied space that enter the return air [see 4.5]. Thus, an HVAC system that moves air and particles throughout a building may become a major distribution pathway for both chemical and biological agents. In a "healthy" indoor environment (i.e., one with no unusual sources of biological agents), bioaerosol concentrations vary as a function of outdoor bioaerosol concentration, HVAC system characteristics, the number of building occupants present, and the activities conducted in the building.

4.4 The Occupied Space
4.4.1 Supply Air Diffusers
Conditioned air enters the occupied space through supply air diffusers (G in Figure 4.1), which are either the terminations of a ductwork system or openings from a pressurized ceiling plenum. Contaminant concentrations and relative humidity rapidly come to equilibrium when conditioned air mixes with room air. For example, during the air-conditioning season, comparatively dry air from a diffuser reduces the overall RH in a room. Likewise, the addition of clean, conditioned air reduces the concentration of pollutants generated within an occupied space.

Particles entrained in the supply air stream can also deposit on a diffuser's louvers and on adjacent ceiling and wall surfaces. Particles from the occupied space may also be entrained in eddy currents induced by the supply air stream. These currents travel along the ceiling, walls, and floor and then back into the diffuser air stream from which the dirt may deposit on adjacent surfaces. Often, this soiling is only a cosmetic concern. However, in air-conditioned buildings, cold air leaving a diffuser may cool adjacent surfaces below the dew point, and the resulting condensation may support microbial growth. Investigators may collect samples of such deposits for microscopic examination to differentiate soil and soot deposits from microbial growth.

4.4.2 Fancoil and Induction Units
Fancoil and induction units may handle a significant portion of the heating and cooling load in a large building (Morey and Shattuck, 1989). These units are located in enclosures, usually beneath the windows in perimeter areas or offices or in columns in interior spaces. Fancoil units contain a fan, heating and cooling coils, and a filter with which they recondition and recirculate indoor air. Induction units mix a portion of room air with conditioned air provided from a central air-handling system. Like central ventilation system components, fancoil and induction units are susceptible to dust accumulation and may become sites for microbial contamination. If condensate collection pans in these units do not drain properly, humidity may become sufficiently high that dirt on filters and other surfaces can support microbial growth. In addition, the water may overflow onto floors, carpets, and wall systems.

4.4.3 Water Leaks, Spills, and Elevated RH
Important water sources in the occupied building zone are plumbing and roof leaks, water spills and overflows, and high RH [see Figure 10.2]. Water-damaged materials provide sites for microbial growth (e.g., ceiling tiles, wall coverings, carpet and other floor coverings, wickerware, upholstered furniture, books and paper, as well as leather and wood items) (ISIAQ, 1996). Section 10.3 describes different measures of moisture in air and in building mate-

rials and how researchers use these measurements to identify when microbial growth can occur.

The microclimate at exterior wall surfaces is distinct from the more homogeneous climate of a building's interior walls. The transfer of heat between indoor air and exterior walls or windows will produce condensation if the dew point is reached [see 10.4.3]. This water then becomes available to microorganisms on these surfaces (Lstiburek, 1994; Morey, 1995; ISIAQ, 1996). Often, water damage and resultant microbial growth are obvious. If necessary, suspected growth can be confirmed by collecting surface or bulk samples for examination under a microscope.

4.4.4 People, Animals, and Plants as Sources of Biological Agents

People themselves are sources of viruses (e.g., the agents of influenza and the common cold) and of human-source bacteria. Therefore, the more densely occupied a building, the greater workers' exposure to such microorganisms. Some pathogenic viruses and bacteria are spread by coughing and sneezing [see 9.2.2], and many bacteria are shed with skin sloughed off by abrasion (e.g., by clothing rubbing against skin) [see 14.2.3.2 and 18.5.2].

Animal occupants of buildings (e.g., cats, dogs, rodents, cockroaches, and other arthropods) shed allergens. Indoor plants may also harbor pests, and moisture from overwatering may support microbial growth. Therefore, investigators should note the presence of animals and plants indoors. People may also carry particles into a workplace on their clothing (e.g., animal danders) [see 25.5.4]. In cases where allergies are suspected, but sources cannot be identified, workers should be questioned about contact with animals outside the workplace. Consistently elevated RH (especially within porous materials) can support dust-mite populations. Carpeting installed on uninsulated concrete floors at ground level is of particular concern because of the potential for dampness, condensation, and both microbial and dust mite growth. Dust samples collected with vacuum devices can be tested for dust-mite and other common allergens (e.g., cockroach, bird, cat, and dog antigens) as well as microorganisms [see 22.4 and 25.4].

4.5 Return Air

In most office and institutional buildings, air exits from the occupied space by passing through a common return plenum or an open space above a suspended ceiling (I in Figure 4.1). Air in the common return plenum enters the main return air duct and is transported back to an air-handling unit for reconditioning (K in Figure 4.1). In some buildings, air in the occupied space enters branch return ducts (without a common return plenum) before transport to the main return air duct. Thus, bioaerosols produced in an occupied space may be circulated to other parts of a building, and particles in this air stream may settle on surfaces in the ducts or plenums through which the air travels. Backflow from a return or exhaust air system could re-aerosolize these particles. Water leaks in return air systems may also contribute to microbial contamination in this part of an HVAC system. Thus, backflow of air could also deliver fungal spores to the occupied zone.

4.6 Protecting Investigators and Building Occupants During Walkthrough Inspections

4.6.1 Assessing the Risk of Bioaerosol Exposure

During any building evaluation, investigators must be aware of the potential for exposure to biological agents and the possibility that their activities may increase exposures for building occupants. Actions that may release high concentrations of fungal spores and other biological particles include accessing the interiors of HVAC systems, exposing contamination when removing wall or floor coverings or lifting ceiling tiles, and collecting bulk samples of contaminated building materials. Inspecting and collecting water samples from operating cooling towers in legionellosis investigations may also place investigators at risk of exposure to airborne bacteria. Therefore, systems should be shut down during inspection, if possible. The uncertainties about the risks associated with potential exposure to biological agents during walk-through building inspections are similar to those faced by remediation workers [see 15.2.3.6].

Disturbance of large accumulations of bird, bat, or rodent droppings presents an exposure risk for building inspectors because of the large numbers of fungal spores and hyphal fragments as well as other potentially infectious or allergenic materials and organic dust that may be released. Investigators should use reasonable precautions when entering areas where there may be animal infestations and when examining and removing materials from such areas (Lenhart, 1994; Lynn et al., 1996; Lenhart et al., 1997). Given the difficulties of testing animal droppings for infectious and antigenic biological agents, investigators should assume that exposure to animal droppings and nesting materials in buildings or air-handling systems may be hazardous and should take appropriate precautions [see 15.6].

4.6.2 PPE Selection and Use

While conducting building inspections, investigators may encounter hazardous biological agents as well as chemical and physical hazards. The best protection against exposure is the use of common sense and the anticipation and avoidance of hazardous situations. In addition, investigators should use personal protective equipment (PPE) appropriate for the suspected hazards to which they may be exposed. Such decisions require *a priori* awareness of potentially hazardous agents, significant exposure routes (e.g., inhalation, skin or mucous membrane contact, or ingestion), and probable concentrations of the biological agents in the materials known to be or suspected of being contaminated [see also 9.2.1, 15.2, and 15.6]. For example, investigators closely inspecting and sampling small, localized patches of fungal growth on a wall may decide they do not need PPE or that respirators and disposable gloves are sufficient. Other situations may warrant the additional use of reusable or disposable coveralls and protective eyewear and footwear. Investigators entering an attic with a large accumulation of bird or bat droppings or extensive fungal growth may need full-face-piece, powered air-purifying respirators; disposable protective clothing with hoods; disposable latex gloves under cotton work gloves; and disposable shoe coverings (Lenhart, 1994).

Airborne fungal spores generally range from 1 to 50 μm, and most other bioaerosols are in a similar size range. Therefore, in many circumstances, a disposable N-95 NIOSH-approved respirator (which removes 95% of particles ≥0.3 μm) offers adequate protection (Chen et al., 1994; Jonson et al., 1994; Willeke et al., 1996; Qian et al., 1997, 1998; Wake et al., 1997). However, the facepiece must fit tightly to ensure that contaminants do not enter through leaks between the respirator and a wearer's face. Respirator use has been recommended to prevent exposure to the agents of legionellosis (CDC, 1997b) and hantavirus pulmonary syndrome (CDC, 1993c). Specific environments may require more protective respirators [see Chapter 15].

Investigators must appreciate what is involved when they decide they need respiratory protection. Respirator use must follow a complete respiratory protection program as specified by the Occupational Safety and Health Administration (OSHA) Standard 29, Code of Federal Regulations 1910.134 (NIOSH, 1996). OSHA requires that respiratory protection programs include written standard operating procedures; respirator selection on the basis of hazard; user instruction and training; respirator cleaning, disinfection, storage, and inspection; surveillance of work area conditions; evaluation of the respirator protection program; medical review; and use of certified respirators.

4.6.3 Protecting Building Occupants During Walkthrough Inspections

Building occupants may be affected by the activities of investigators who are conducting a building inspection if biological agents are disturbed and become airborne. When possible, inspections that may release bioaerosols should be conducted when the study site is unoccupied. If this is not possible, all work should be conducted carefully to minimize the dissemination of contaminants.

Investigators should consider their own health and safety and wear PPE whenever they decide it is necessary. However, they also should consider how building occupants may react to the presence of visitors in protective clothing and respirators. Therefore, building occupancy should be kept to a minimum during inspections for which the investigators have decided that PPE is required. This will prevent accidental exposure of building occupants and avoid raising anxiety among them.

4.7 References

AIHA: The Practitioner's Approach to Indoor Air Quality Investigations. D.M. Weeks and R.B. Gammage, Eds. American Industrial Hygiene Association, Fairfax, VA (1990).

ASHRAE: Ventilation for Acceptable Air Quality, ASHRAE Standard 62-1989. American Society of Heating, Refrigerating and Air-Conditioning Engineers, Atlanta, GA (1989).

ATS: American Thoracic Society, Medical Section of the American Lung Association. Respiratory Protection Guidelines. Am. J. Respir. Crit. Care Med. 154:1153-1165 (1996).

Bearg, D.W.: Individual Components of HVAC Systems. In: Indoor Air Quality and HVAC Systems. Lewis Publishers, Boca Raton, FL, pp. 39-60 (1993).

Burkhart, J.E.; Stanevich, R.; Kovak, R.: Microorganism Contamination of HVAC Humidification Systems: Case Study. Appl. Occup. and Environ. Hyg. 8(12):1010-1014 (1993).

CDC: Hantavirus Infection — Southwestern United States: Interim Recommendations for Risk Reduction. CDC — MMWR 42(RR-11), Centers for Disease Control (1993).

CDC: Guidelines for Prevention of Nosocomial Pneumonia. CDC — MMWR 46(RR-1), Centers for Disease Control (1997).

Chen, S.K.; Vesley, D.; Brosseau, L.M.; et al.: Evaluation of Single-Use Masks and Respirators for Protection of Health Care Workers against Mycobacterial Aerosols. Am. J. Infect. Control 22:65-74 (1994).

Foarde, K.; VanOsdell, D.; Leese, K.E.; et al.: Field Moisture Measurement and Indoor Mold Growth. In: Proceedings of the ASHRAE Conference, IAQ 96, pp.19-24. American Society of Heating, Refrigerating, and Air-Conditioning Engineers, Atlanta, GA (1996).

Godish, T.: Diagnosing Problem Buildings. In: Sick Buildings: Definition, Diagnosis and Mitigation, pp. 205-271. Lewis Publishers, Boca Raton, FL (1995).

ISIAQ: Control of Moisture Problems Affecting Biological Indoor Air Quality. International Society of Indoor Air Quality and Climate, Ottawa, Canada (1996).

Jonson, B.; Martin; D.D.; Resnick, I.G.: Efficacy of Selected Respiratory Equipment Challenged with *Bacillus subtils* subsp. *niger*. Appl. Environ. Microbiol. 60:2184-2186 (1994).

Kemp, S.J.; Kuehn, T.H.; Pui; D.Y.H.; et al.: Growth of Microorganisms on HVAC Filters under Controlled Temperature and Humidity Conditions. ASHRAE Trans. 101:305-316 (1995).

Lenhart, S.W.: Recommendations for Protecting Workers from *Histoplasma capsulatum* Exposure During Bat Guano Removal from a Church's Attic. Appl. Occup. Environ. Hyg. 9:230-236 (1994).

Lenhart, S.W.; Schafer, M.P.; Singal, M.; et al.: Histoplasmosis: Protecting Workers at Risk. Publication No. 97-146. Centers for Disease Control and Prevention. National Institutes for Occupational Safety and Health, Cincinnati, OH (1997).

Lstiburek, J.W.: Mold, Moisture, and Indoor Air Quality. Building Science Corporation, Chestnut Hill, MA (1994).

Lynn, M.; Vaughn, D.; Kauchak, S.; et al.: An OSHA Perspective on Hantavirus. Appl. Occup. Environ. Hyg. 11:225-227 (1996).

Martikainen, P.J.; Asikainen, A.; Nevalainen, A.; et al.: Microbial Growth on Ventilation Filter Materials, Vol. 3. In: Proceedings of Indoor Air '90, pp. 203-206. D.S. Walkinshaw, Ed. Canada Mortgage and Housing Corporation, Ottawa, Canada (1990).

Morey, P.R.: Microorganisms in Buildings and HVAC Systems: A Summary of 21 Environmental Studies. In: Proceedings of the ASHRAE Conference, IAQ 88, pp. 10-24. American Society of Heating, Refrigerating, and Air-Conditioning Engineers, Atlanta, GA (1988).

Morey, P.R.: Control of Indoor Air Pollution. In: Occupational and Environmental Respiratory Disease, pp. 981-1003. P. Harber, M. Schenker, and J. Blames, Eds. Mosby, St. Louis, MO (1995).

Morey, P.R.: Suggested Guidance on Prevention of Microbial Contamination for the Next Revision of ASHRAE Standard 62. In: Proceedings of the ASHRAE Conference, IAQ 95. American Society of Heating, Refrigerating, and Air-Conditioning Engineers, Atlanta, GA (1995).

Morey, PR: Biocontaminant Control for Acceptable IAQ. In: Improving Indoor Air Quality Through Design, Operation and Maintenance, pp. 257-268. M. Meckler, Ed. Fairmont Press, Lilburn, GA (1996).

Morey, P.R.; Shattuck. D.E.: Role of Ventilation in the Causation of Building-Associated Illness. Occup. Med. State Art Rev. 4(4):625-642 (1989).

Morey, P.R.; Hodgson, M.J.; Sorenson, W.G.; et al.: Environmental Studies in Moldy Office Buildings: Biological Agents, Sources, and Prevention Measures. In: Evaluating Office Environmental Problems, Ann. Am. Conf. Gov. Ind. Hyg.10:21-35 (1984).

Muilenberg, M.L.: The Outdoor Aerosol. In: Bioaerosols, pp. 163-204. H.A. Burge, Ed. Lewis Publishers, Boca Raton, FL (1995).

NIOSH: NIOSH Guide to the Selection and Use of Particulate Respirators Certified under 42 CFR 84. U.S. Department of Health and Human Services, Public Health Service, Centers for Disease Control and Prevention, National Institute for Occupational Safety and Health. DHHS (NIOSH) Publication No. 96-101, NIOSH, Cincinnati, OH (1996).

Parat, S.; Fricker, H.; Perdrix, A.; *et al.*: Airborne Microorganisms — Comparing an Air-Conditioned Building and a Naturally Ventilated Building. In: Proceedings of the ASHRAE Conference, IAQ 93, Vol. 4, pp. 201-206. American Society of Heating, Refrigerating, and Air-Conditioning Engineers, Atlanta, GA (1994).

Parat, S.; Fricker-Hidalgo, H.; Perdrix, A.; et al.: Airborne Fungal Contamination in Air-Conditioning Systems: Effect of Filtering and Humidifying Devices. Am. Ind. Hyg. Assoc. J. 57:996-1001 (1996).

Pasanen, P.: Impurities in Ventilation Ducts. In: Proceedings of the ASHRAE Conference, IAQ 94. American Society of Heating, Refrigerating, and Air-Conditioning Engineers, Atlanta, GA (1995).

Qian, Y.; Willeke, K.; Grinshpun, S.A.; et al.: Performance of N95 Respirators: Reaerosolization of Bacteria and Solid Particles. Am. Ind. Hyg. Assoc. J. 58:876-880 (1997).

Qian, Y.; Willeke, K.; Grinshpun, S.A.; et al.: Performance of N95 Respirators: Filtration Efficiency for Airborne Microbial and Inert Particles. Am. Ind. Hyg. Assoc. J. 59:128-132 (1998).

Seitz, T.A.: NIOSH Indoor Air Quality Investigations — 1971 through 1988. In: The Practitioners Approach to Indoor Air Quality Investigations, D.M. Weekes and R.B. Gammage, Eds. American Industrial Hygiene Association, Akron, OH (1990).

Simmons, R.B.; Price, D.L.; Noble, J.A.; et al.: Fungal Colonization of Air Filters from Hospitals. Am. Ind. Hyg. Assoc. J. 58:900-904 (1997).

USEPA & NIOSH: Building Air Quality: A Guide for Building Owners and Facility Managers. U.S. Environmental Protection Agency and the National Institute for Occupational Safety and Health, Washington, DC (1991).

USEPA: A Standardized EPA Protocol for Characterizing Indoor Air Quality in Large Office Buildings. U.S. Environmental Protection Agency, Washington, D.C. (1994).

Wake, D.; Bowry, A.C.; Crook, B.; et al.: Performance of Respirator Filters and Surgical Masks Against Bacterial Aerosols. J. Aerosol Sci. 28:1311-1329 (1997).

Willeke, K.; Qian, Y.; Donnelly, J.; et al.: Penetration of Airborne Microorganisms through a Surgical Mask and a Dust/Mist Respirator. Am. Ind. Hyg. Assoc. J. 57:348-355 (1997).

Appendix 4.A. Building Walkthrough Checklist
(letters in parentheses identify components illustrated in Figure 4.1)

1. Surrounding Area

Type of location (urban, suburban, rural): _____

Surrounding land use (business, residential, agricultural): _____

Animal-confinement operations? .... Yes   No

Construction or agricultural activity? .... Yes   No

Water sprays (e.g., fountains, irrigation)? .... Yes   No

Cooling towers present? .... Yes   No

Shallow groundwater areas (e.g., marshes, bogs)? .... Yes   No

Site drainage adequate? .... Yes   No

Vegetation surrounding building? .... Yes   No

2. Heating, Ventilating, and Air-Conditioning System

a. *General Characteristics*

Type of ventilation system: _____

Location of air-handling units: _____

Cooling method used: _____

Heating method used: _____

Locations served by individual air handlers: _____

b. *Outdoor Air Intake*

Location: _____

Bird screen present? .... Yes   No

Feathers or bird droppings near or in OAI? .... Yes   No

Other organic matter near or in OAI (e.g., leaves, plant down, insects) .... Yes   No

OAI protected from rain, snow, fog? .... Yes   No

Standing water or evidence of standing water near or in OAI? .... Yes   No

Cooling tower within 7.5 m (25 ft)? .... Yes   No

Exhaust air outlet within 7.5 m (25 ft)? .... Yes   No

c. *Filters* (A)

Filters present and free of organic debris and microbial growth? .... Yes   No

d. *Mixing Chamber of Air-Handling Unit*

Mixing area clean and free of debris and microbial growth? .... Yes   No

Malodors? .... Yes   No

Evidence of water damage or intrusion? .... Yes   No

e. *Heating and Cooling Coil Area* (B and C)

Coils clean and free of organic material and microbial growth? .... Yes   No

Condensate pan and drain present? .... Yes   No

Condensate pan well drained (i.e., no standing water, biofilm, or residue)? .... Yes   No

Corrosion on pan? .... Yes   No

Malodors? .... Yes   No

Evidence of water transport from coil area to other areas? .... Yes   No

f. *Spray Humidifiers, Evaporative Coolers, or Air Washers* (D)

Type of unit: _____

Chemicals or additives used: _____

Maintenance schedule: _____

Type of medium, if any: _____

Microbiological samples of the water taken routinely? .... Yes   No

If yes, results: _____

Recirculated water used? .... Yes No

Biofilm, dirt, or microbial growth in sump area? .... Yes No

Malodors? .... Yes No

Water leakage from humidifier into duct system? .... Yes No

Water pooled near unit? .... Yes No

Unit enters airspace directly (or ducted to other areas)? .... Yes No

g. *Supply Side of Air-Handling Unit* (E, F, and G)

Where do ducts enter building (e.g., at ceilings, below floors): _____

Type of supply ducts (lined or unlined): _____

Supply area clean and free of debris and microbial growth? .... Yes No

Malodors? .... Yes No

Evidence of water damage or intrusion? .... Yes No

h. *Return Side of Air-Handling Unit* (H, I, J, and K)

Type of return (ducted or plenum): _____

Porous lining on ducts or plenums? .... Yes No

Return area clean and free of debris and microbial growth? .... Yes No

Malodors? .... Yes No

Evidence of water damage or intrusion? .... Yes No

3. Building/Occupied Space

Number of floors: _____

General building uses: _____

Attic present? .... Yes No

   If yes, condition: _____

Basement or crawlspace present? .... Yes No

   If yes, condition: _____

Water features (e.g., fountains, sprays, indoor waterfalls)? .... Yes No

Malodors? .... Yes No

Visible microbial growth? .... Yes No

History of water damage? .... Yes No

Evidence of water damage (stained or discolored ceiling tiles, walls, floors, carpeting)? .... Yes No

Condensation on walls and windows? .... Yes No

Window air conditioners? .... Yes No

Evaporative air coolers? .... Yes No

Sump pump used? .... Yes No

Fancoil and induction units? .... Yes No

Potted plants? .... Yes No

Portable air cleaners? .... Yes No

   If yes, why? _____

Console humidifiers? .... Yes No

   If yes, why? _____

Console dehumidifiers? .... Yes No

   If yes, why? _____

Typical RH levels in the building: _____

Chapter 5

Developing A Sampling Plan

5.1 Introduction
    5.1.1 *The Sampling Plan*
    5.1.2 *Environmental Sampling*
5.2 Sample Collection
    5.2.1 *Where and When to Sample*
        5.2.1.1 Typical and Worst-Case Exposure Assessments
        5.2.1.2 Averaging Times for Bioaerosol Sampling
        5.2.1.3 Practical Considerations
        5.2.1.4 Example of Where and When to Sample
    5.2.2 *Choosing an Air Sampler*
    5.2.3 *Sample Volume and Sample Collection Time*
        5.2.3.1 Direct-Agar Impaction
        5.2.3.2 Spore-Trap Samples
        5.2.3.3 Calculating Optimal Sampling Time
    5.2.4 *Sample Number*
5.3 Record Keeping
5.4 Summary
5.5 References
Appendix 5.A   Random Selection of Sampling Times or Sites

5.1 Introduction

5.1.1 The Sampling Plan

Often the hypotheses investigators develop to explain possible bioaerosol exposures can be tested by collecting environmental samples. The hypotheses explain *why* the investigators have decided to sample a work environment for biological agents and also describe *what* agents they intend to study. A sampling plan describes *where*, *when*, and *how* investigators propose to collect samples to test their hypotheses. The following discussion outlines a rigorous and scientifically sound approach to studying biological agents in indoor environments. Unfortunately, investigators often must reach a compromise between an ideal sampling plan and one that is feasible. Constraints on sampling may involve the time, manpower, instruments, analytical resources, and the access to the building. Each investigation will have its own practical constraints, and investigators must decide case by case how to reach their goals without jeopardizing the credibility, representativeness, and integrity of the data they collect.

Table 5.1 outlines typical plans to sample four biological agents. In this table, sample analysis is intentionally placed before (to the left of) sample collection to remind investigators that these steps are coupled. Once an investigation team has decided what biological agents to study, the investigators should contact a laboratory and

discuss sample analysis before collecting samples. Readers can also follow this table from right to left, studying first the column marked Sampling Plan and then reading the columns Sample Collection and Sample Analysis to see how a plan can be executed.

A sampling plan includes specification of (a) the biological agents under study, (b) typical sources and reservoirs of the suspected agents, (c) the anticipated concentration and variability in space and time for each agent, (d) the analytical methods that will be used to detect, identify, and quantify the agents, (e) the sampling methods to be used, (f) the operating parameters of all sampling instruments, and (g) the locations and times at which samples will be collected.

Table 5.2 outlines the primary steps in the process of collecting environmental samples to test hypotheses.

5.1.2 Environmental Sampling

Investigators may decide to sample for biological agents in indoor and outdoor environments for a variety of reasons. Among these purposes are determining what kinds and relative concentrations of biological agents are present for comparison with another location. Investigations may also be undertaken to monitor changes that may occur over time and to detect existing or emerging problems. Sampling for specific biological agents may be undertaken to locate their sources as well as to estimate exposures or document that exposure may occur. If problems are found, sampling can help investigators develop and implement control strategies and assess their effectiveness. Sampling results may be used to determine what degree of PPE is needed during a building inspection or a remediation operation. Occasionally, environmental sampling is undertaken in response to an accident or emergency to evaluate possible consequences of the unusual event. Some investigations are conducted as research efforts to assess or validate sampling methods or to better understand environmental conditions and occupational exposures.

The word "sample" means different things in different contexts. At times, investigators use the term to designate an individual measurement (e.g., an air or source sample). However, a sample may also designate a set of measurements (e.g., multiple measurements of some parameter that comprise a sample of size n). In either case, the goal of sampling is to learn about entire populations by looking at subsets of the members of the population. For example, investigators may attempt to characterize exposure conditions in a building by measuring concentrations of environmental contaminants at selected places and times chosen to be as similar as possible to where and when workers may be exposed. The key to accomplishing this goal is to select samples so that they (a) accurately and completely represent the entire population, or (b) represent the conditions most likely to be associated with adverse effects.

TABLE 5.1. Example: Bioaerosol Sampling Plans

Agent	Sources	Sample Analysis	Sample Collection	Sampling Plan
Fungal allergens	Outdoor air, Fungal growth in HVAC system	Culture • Identify major taxa Immunoassay • Assay for specific antigens Microscopy • Identify major taxa	Bulk samples • Pieces of material Surface samples • Adhesive tape or swab samples	Collect bulk and surface samples from materials with suspected fungal growth
		Culture • Count colonies and identify major taxa Immunoassay • Assay for specific antigens Microscopy • Count spores and identify major taxa	Air samples • Agar impactor • Filter cassette • Spore trap	Collect air samples outdoors, near suspected contamination, and in the occupied space with the HVAC system off then on
Legionella sp.	Cooling tower, Hot water supply	Culture Direct fluorescent antibody stain • Count colonies or cells	Bulk samples • Water, biofilm (Air samples not recommended)	Collect water from cooling tower sump, hot water tank, and municipal water supply
Endotoxin	Humidifier	Limulus amebocyte lysate assay • Assess concentrations of endotoxin	Bulk samples • Water	Collect water from reservoir before and after cleaning
			Air samples • Filter cassette	Collect air samples with humidifier off then on before and after cleaning water reservoir
Stachybotrys toxins	Wet wallboard, *Stachybotrys chartarum*	GC-MS toxin analysis • Macrocyclic trichothecenes	Bulk samples • Pieces of material	Collect bulk and surface samples from materials with suspected funal growth
		Microscopy • Identify spores • Identify and count *Stachybotrys chartarum* spores	Surface samples • Adhesive tape Air samples • Spore trap	Collect air samples while area is unoccupied then during normal use

As much as possible, the relative proportion or concentration of a biological agent in a sample should reflect that in the original material. To achieve this goal, a sample must be handled so that it does not deteriorate or become contaminated before analysis (i.e., biological agents that are present should not be lost and agents that are not present in the original sample should not be introduced). Viable organisms in samples should change as little as possible after removal from a test environment (i.e., they should not multiply appreciably nor lose culturability). Table 5.3 outlines points to consider in designing a sampling plan.

Environmental sampling may provide clear evidence supporting a hypothesis that investigators have formulated. For example, for the first situation in Table 5.1, the bulk samples may show high concentrations of a particular fungus and the air samples may show a higher concentration of the same fungus when the HVAC system is operating than when it has been off overnight and the building has been unoccupied. These findings may convince the investigators that the fungus is growing within and disseminated by the HVAC system. Conversely, negative results may persuade investigators to abandon this hypothesis and to construct others. For the second situation, failure to identify *Legionella* spp. in the water systems of a building may suggest that exposure for a patient with Legionnaires' disease may not have occurred at the workplace and that the investigator should look elsewhere for the exposure source to prevent other infections.

Investigators are cautioned that sampling results are not always easy to interpret and that incomplete bioaerosol data may confuse and complicate an investigation rather than provide clarity. For example, the types and relative proportions of bioaerosols in indoor and outdoor air are highly variable, occasionally making it difficult to see differences between test and control sampling sites. Also, many microorganisms to which everyone is exposed throughout life are opportunistic pathogens. A laboratory report that indicates that many of the microorganisms isolated from a workplace may cause

disease (even if this is unlikely in healthy persons) can cause unnecessary alarm and require lengthy explanation. Therefore, study design is often the single most important element of an investigation. Even the most careful and sophisticated sample collection and data analysis will not salvage a poor study design. Investigators must define their hypotheses as clearly as possible at the beginning of a study and determine what kinds of data they need and what information they can expect to obtain with a proposed sampling plan.

The following discussion focuses on air sampling for biological agents, the intention of which is usually to identify what agents are present and to estimate concentrations quantitatively or semi-quantitatively. Investigators should not underestimate the potential value of source samples even though this chapter does not discuss bulk and surface sampling as extensively as air sampling. All bioaerosols have a source, the identification of which is often key to eliminating or minimizing the presence of a biological agent in indoor air.

Tracing the origin of industrial pollutants in a workplace is often fairly straight forward because the materials used in a manufacturing process are generally the sources of the compounds to which workers are exposed. Therefore, it may be known that a particular, potentially hazardous material will be present in a workplace atmosphere. In such case, the purpose of sampling would be to determine the air concentration of that material. However, the biological agents of potential interest in BRIs are usually not part of a manufacturing process but environmental contaminants for which there may be more than one origin. Identifying which single or multiple biological or nonbiological agents are responsible for workers' complaints and identifying sources to allow control is often a challenging undertaking.

5.2 Sample Collection

When developing sampling strategies, IHs, EHPs, and IEQ consultants must consider possible sources of error, identify the desired precision and accuracy of the measurements to be made, and specify the degree of confi-

---

TABLE 5.2. Fundamental Steps When Collecting Environmental Samples to Test Hypotheses

- State the objective of the study and outline the goals and expectations of the sampling program.
- State the explicit questions to be addressed and the information needed to answer them.
- Design a sample collection plan to obtain the required information.
- Collect and analyze samples.
- Compile and summarize data.
- Perform statistical tests, if appropriate.
- Determine if the objectives of the sampling program have been met and if the questions posed at the outset of the investigation can be answered.

TABLE 5.3. Points to Consider When Designing a Sampling Plan

- The nature and expected concentrations of biological materials.
- The relevance of average or worst-case concentration measurements.
- The cost and availability of various types of sample analyses.
- Constraints that analytical methods may impose on sample collection.
- The suitability, cost, and availability of sample collection devices and related supplies.
- Constraints the time and method of sample transport to an analytical laboratory may impose.
- The technical expertise required of field and laboratory personnel.

dence they need to interpret the results (Gross and Morse, 1996). These decisions are coupled with the practical questions of *where, when,* and *how* to measure biological agents as well as *how many* samples to collect, as discussed in this section. Investigators base their decisions on their training, experience, and the approaches other investigators have used successfully in similar settings.

5.2.1 Where and When to Sample

5.2.1.1 Typical and Worst-Case Exposure Assessments

Workplace investigations are driven by the need either to (a) establish typical worker exposures and exposure conditions, or (b) measure worst-case exposures. For comparison, background, baseline, or best-case measurements may be collected at outdoor or control locations or during non-work periods. Direct-reading instruments would make identification of ideal sampling sites and times easier. However, although a few such devices exist, they are not widely available to measure occupational bioaerosol exposures. Therefore, in deciding where and when to collect appropriate bioaerosol samples, investigators must learn to combine information they gathered from site visits with information they have acquired from interviews and literature reviews. A site visit provides information on specific worker activities and work processes. Background knowledge, prior experience, and discussions with workers and managers provide information on potential agents, their sources, and factors that may lead to exposure. Many investigations begin only after workers have developed BRSs or a BRI, at which time exposures may no longer be occurring or occurring at the original level. Therefore, it may be necessary to try to recreate environmental conditions that existed previously.

Personal samples are the best way to estimate inhalation exposures. However, few personal samplers are available for collecting biological agents. Therefore, samples collected in a worker's breathing-zone are often used to estimate inhalation exposures. Area (or static) samples may be adequate to estimate average bioaerosol concentrations. For example, air samplers may be placed at the desks of stationary workers or located centrally in an area that several mobile workers share. To establish typical exposures that represent an entire population, the samples should represent the entire range of exposures and conditions. One way to achieve this is to select sampling sites and times randomly (i.e., all persons and work periods should have an equal chance of being monitored) (Appendix 5.A).

To estimate maximum or worst-case exposures, samples are collected near suspected bioaerosol sources, symptomatic workers, or the workers assumed to have the highest exposures. Sampling to measure maximum or worst-case exposures could be considered a form of stratified sampling, where the strata are defined according to the investigators' expectations. A stratified sampling plan can also be used to compare exposures at two or more sites. In such cases, the investigators decide in advance to sample subgroups of the population separately. Within the zones of anticipated high and low exposures, the sites or workers to be monitored can be chosen using a random selection process so that all areas are adequately represented (Appendix 5.A). Stratified sampling is also used when particular subgroups of interest are small and the investigators want to ensure that a study includes at least some members from each subgroup. For example, investigators may suspect that exposures for a few part-time workers are different than for full-time workers. However, given the small number of part-time workers, few or none of them may be included in a study if participants were chosen strictly randomly. Therefore, investigators may choose initially to study the two groups separately, collecting equal numbers of samples to represent exposures for the two populations. If exposures are determined to be the same, the data for the part-time workers could be combined with that for the other workers.

Sampling goals need to be clearly defined to develop a sampling plan, and the sampling plan should be clearly developed before samples are collected. Some measurements of worst-case exposures might correctly be included in a random or representative sample. However, it is difficult, if not impossible, to estimate average exposures from samples intentionally chosen to represent worst-case conditions. Table 5.4 summarizes the decisions involved in determining where to collect bioaerosol samples. The steps outlined in this table apply to bioaerosol sampling in research studies as well as problem-building investigations.

5.2.1.2 Averaging Times for Bioaerosol Sampling

Occupational exposure limits are often established as allowable air concentrations for averaging times appropriate to particular health hazards. An averaging time is the period over which an air concentration (exposure) is measured. A time-weighted average concentration can be calculated when concentrations and their time durations are known. Typical averaging times are (a) a conventional 8-hour work shift and a 40-hour work week, or (b) a 15-min., short-term exposure period. The appropriate averaging time for occupational exposure assessment depends on whether health effects are anticipated to be cumulative or acute. Cumulative health effects are those that result from repeated (chronic) exposures (e.g., lead poisoning), whereas acute health effects arise from single, sufficiently high exposures (e.g., carbon monoxide poisoning). Monitoring for chronic hazards is usually conducted for the duration of one or more work shifts (Jensen and O'Brien, 1993). Establishing a mean air concentration, with appropriate confidence limits, may provide a good basis for estimating exposures associated with long-term or chronic health risks (Rock, 1995). On the other hand, an upper confidence limit or 95th percentile of an exposure distribution may more accurately indicate a worker's acute health risk (i.e., that associated with infrequent but high exposures). Section 5.2.4 discusses the minimum numbers of samples recommended for estimating confidence intervals and sample variance.

Analogous averaging times have not been established for chronic and acute bioaerosol exposures, except for those biological agents listed in the TLV booklet [see 1.2]. Some biological agents can be considered to pose both chronic and acute risks. For example, repeated exposure to an antigen may lead to the development of HP in previously unsensitized workers whereas a single, acute exposure could cause immediate allergic reactions in already sensitized workers. Mycotoxins are another group of biological agents known to manifest different effects for chronic and acute exposures, at least by ingestion. To study long-term or chronic health risks, continuous spore-trap samplers for microscopic examination; 8-hour, filter-cassette samples for allergen assay; or multiple, discrete, agar-impaction samples for culture of viable microorganisms may be used to assess worker exposure. For example, Robins et al. (1997) used sequential sampling with filter-cassettes to characterize the health effects of chronic exposures to bacterial aerosols (including endotoxins). To study acute exposures to biological agents, investigators would want to characterize peak, short-term exposures, perhaps while maximizing activity in the space. For example, worst-case sampling might help identify what allergen exposures provoke acute asthma attacks in sensitized workers.

Mechanical disturbances to potential sources of biological agents and human activity during sample collection may affect the type and amount of material in the air. Therefore, during sample collection, the operation of ventilation system equipment and the activities of occupants should reflect typical patterns of building use. Exceptions would be monitoring of actual or simulated worst-case conditions during which equipment operation may be modified and occupant activity may be maximized. Simulating worst-case circumstances has been termed "aggressive sampling," which should be consistent with what might actually occur in an occupied, operating building [see 14.2.4.1]. Table 5.5 summarizes the types of samples discussed above and gives examples of time patterns for collecting air samples.

5.2.1.3 Practical Considerations For many investigations, resources are a major limiting factor and determine what type of study can be conducted. Investigators must decide how best to use available field personnel, sampling equipment, and laboratory services to meet a client's needs. A clear understanding of the questions at issue and the information a client expects to receive can help an investigator decide how to design a study. The services an investigator will provide and the scope of the responsibilities an investigator will undertake should be defined clearly in a written contract, memoran-dum or letter, or other applicable type of signed agreement.

When deciding where, when, how, and how many samples to collect, investigators must consider the number of sampling devices available for a study. If mul-

---

TABLE 5.4. Deciding Where to Collect Bioaerosol Samples to Compare Anticipated High and Low Exposures

- Tentatively identify bioaerosol sources (e.g., HVAC system components, building materials, or furnishings that are visibly contaminated with microbial growth or that show signs of water damage) and estimate their source strengths (i.e., bioaerosol generation rates). Predict spatial and temporal bioaerosol concentration gradients.

- Identify zones with expected differences in bioaerosol kind or concentration (e.g., indoor and outdoor sites as well as areas near suspected sources and control areas).

- Identify occupants anticipated to receive the highest and lowest exposures and show the strongest and weakest reactions based on their proximity to potential sources, types of activities conducted, and medical condition.

- Identify areas to which investigators will be allowed access. Identify areas that can be monitored without disrupting typical occupant activities.

- Select at least one, preferably three, sampling locations in each of the following areas
    - An anticipated high-exposure area
    - An anticipated low-exposure area
    - Outdoors near air intakes for the building.
    If applicable, also sample at the following locations
    - Outdoors near potential sources of bioaerosols that may enter a building
    - Outdoors high above grade and away from potential bioaerosol sources.

tiple, comparable instruments are available, they remain at different sampling locations throughout a study. Otherwise, investigators need to arrange a rotation system to monitor all selected sampling sites in a way that does not compromise the study design. For example, it may not be correct to compare test and control samples if they were collected at markedly different times of day due to the time required to disassemble, move, and reassemble the sampling equipment. Therefore, equipment and field personnel need to minimize the time between collection of samples at sites that will be compared, or the investigators should collect a sufficient number of samples at each site to characterize the different locations well. Randomization of sample sequence can help minimize bias [see Appendix 5.A]. Investigators should also determine if their presence and activities may change the way work is conducted and, if so, how these changes may affect bioaerosol exposures.

5.2.1.4 Example of Where and When to Sample Air samples are typically collected outdoors and indoors, both before and after potential sources of microbial growth are disturbed. Outdoor samples are usually collected at the highest point of a building (e.g., the rooftop) facing into the wind as well as near building air intakes. For naturally ventilated buildings, reference samples should be collected where they will sample outdoor air representative of the air that may enter buildings through open windows and doors. Outdoor samples should be collected throughout the period of indoor sampling (Table 5.4). Table 5.6 outlines an example sampling plan designed to assess the contribution to the indoor atmosphere of fungi growing in ventilation ducts.

5.2.2 Choosing an Air Sampler

An investigator selects one or more sample collection methods and gathers samples based on (a) the agents of interest, (b) the analytical methods by which the laboratory will identify and perhaps quantify the biological agents or indicators of the active agents, and (c) the sites and times at which samples will be collected. Chapter 2 describes how investigators identify the biological agents they wish to collect by combining information gathered during preliminary interviews and site inspections. The locations and time patterns chosen for sample collection (Tables 5.4 and 5.5) depend on what information the investigators need to obtain to evaluate the hypotheses they have formed about the setting.

Chapter 11 describes bioaerosol samplers, their operation principles, applications, and limitations. Investigators often adopt the same collection device others have used for a particular biological agent in a similar setting. Familiarity with recent published literature describing collection devices and study designs can help investigators in the sampler-selection process (Burge, 1995; Cox and Wathes, 1995; Health Canada, 1995; AIHA, 1996; ISIAQ, 1996; ASM, 1997). Table 11.1 summarizes commercially available samplers and their applications. Table 5.7 outlines a process for selecting an air sampler to collect bioaerosols.

5.2.3 Sample Volume and Sample Collection Time

Sample volume and collection time are related through the minimum air volume needed to detect a target material present at a given concentration. In turn, sample volume and collection time depend on the volumetric flowrate of a bioaerosol sampler. The *lower detection limit* (LDL) for a bioaerosol sampling method can be deter-

---

TABLE 5.5. Time-Related Patterns for Collecting Bioaerosol Samples

CAVEAT: Stated sampling methods and sample collection times are given for illustration only.

Time Pattern	Description and Example (samples collected at least in duplicate)
Continuous	Sample continuously or in series throughout the time interval of interest, for example, (a) a continuous, 8-hour, filter sample, or (b) eight, sequential, 1-hour samples with a moving-slide, slit sampler.
Periodic	Divide the total time period into equal segments and collect samples during the same portion of each segment, for example, (a) 1-min., direct-agar impaction samples during the last 10 min. of every hour for 8 hour, or (b) 10-min., portable spore-trap samples at 10 a.m. and 2 p.m. each day for three, consecutive days.
Random	Use a random selection process (Appendix 5.A) to choose sampling periods from all intervals in a chosen averaging time, for example, 5-min., direct-agar impactor samples during eight, randomly selected, 5-min intervals in an 8-hour work day.
Worst-case	Collect samples before, during, and after activities known or expected to create worst-case exposures, for example, 15-min., liquid impinger samples (a) before disturbance begins, (b) during peak activity, and (c) after the effects of the disturbance have subsided.

mined from (a) the minimum amount of material that an assay can detect, (b) the recovery efficiency for the assay method and any concentration steps that may apply, (c) an air sampler's flow rate, and (d) the maximum sample collection time. The minimum detectable amount of a biological agent may be 1 CFU for a culturable microorganism on an agar plate, 1 fungal spore per microscope field for a spore-trap sample, or a minimal mass amount of allergen per milliliter of a dust extract. Laboratory personnel may use blank samples to determine the background amount of test material that may be present on unexposed collection media or in storage containers. *Upper detection limits* (UDLs) can be determined similarly, taking into account minimum sample collection times and any dilution steps that may apply [see 5.2.3.1 and 5.2.3.2].

Investigators aim for sampling times that are (a) sufficiently long to collect a detectable and representative numbers of particles or other amount of material (Nevalainen et al., 1992, 1993), but (b) short enough to avoid masking (i.e., the overlap of particles on microscope specimens or the contact of colonies on culture media) (Chang et al., 1994, 1995). Laboratories should report the UDL and LDL of a method along with sample results. In particular, LDLs should be specified when a target biological agent is not found because all that can be concluded is that the agent was not present at or above the LDL. Table 11.1 lists the airflow rates for several bioaerosol samplers and Table 11.5 identifies approximate air concentrations for which these samplers are suitable. The American Industrial Hygiene Association (AIHA) *Field Guide* also provides ranges and sensitivities for commonly used bioaerosol samplers (AIHA, 1996).

Short- and long-term variations in bioaerosol concentration along with typically short collection times dictate collection—at a minimum—of sequential or simultaneous replicate samples to obtain confident estimates of bioaerosol concentrations. Where feasible, longer sample collection times may reduce the variability seen with shorter sampling times. For example, Stanevich and Petersen (1990) found that increasing sample collection time from 1 to 5 min. (at 28 L/min) reduced intra-sample variability. Therefore, sampling at a lower airflow rate for a longer time is preferable to sampling at a higher rate for a shorter time, provided the longer sampling time does not adversely affect recovery of the biological agent of interest. Collection of sequential, short-term samples can provide information on concentration fluctuations while also providing an estimate of the long-term average.

5.2.3.1 Direct-Agar Impaction The density of colonies growing on a culture plate affects the reliability of the information that can be obtained. Too few CFUs inaccurately reflect what is present, whereas too many CFUs are difficult to count and examine. Therefore, when using direct-agar impactors, it is especially important to anticipate bioaerosol concentration correctly and to collect samples that will yield an appropriate colony surface density. In general, 25 to 250 bacterial colonies and 10 to 60 fungal colonies are considered optimal for accurate counting and identification of CFUs on standard, 100-mm plates. Target colony counts are adjusted proportionally for smaller and larger plates. Therefore, another rule of thumb microbiologists have used is a maximum surface density of 1 colony/cm². For multiple-hole impactors, investigators should also consider the greater

---

TABLE 5.6. Example: Sampling Plan to Assess the Contribution of Fungal Growth in a Ventilation System to the Presence of Fungal Spores in Indoor Air

(Repeat series at least twice on two or more days; collect at least duplicate samples at all sites)

Outdoors	• Site remote from obvious bioaerosol sources: initial background or baseline (initial) measurement of highest quality general air
	• Site near outdoor air intake: initial background or baseline measurement of air entering the building
Indoors	• Site near a supply air diffuser: air delivered from ventilation system, as follows:
	- When the air distribution system has been turned off at least overnight (preferably for a weekend)
	- Immediately after the air distribution system is restarted, with the sampler oriented so that air from the diffuser directly enters the sampler
	- After the air distribution system has been operating for 30 min.
	- During gentle or vigorous mechanical agitation of the ductwork, as appropriate (Caution: aggressive sampling should only be conducted when a space is unoccupied and should be consistent with what might occur during normal building operation and use)
	• At least three locations in the space served by the supply air diffuser (sampler at occupant breathing height): occupant exposure
Outdoors	• Site remote from obvious bioaerosol sources: final background or baseline (final) measurement of highest quality general air (see above)
	• Site near outdoor air intake: final background or baseline measurement of air entering the building (see above)

variability associated with higher plate counts when using a positive-hole correction [see 11.5.1.1].

Table 5.8 identifies the maximum and minimum air concentrations for three samplers based on an optimal density of fungal colonies and commonly used sample collection times. Obviously, it is possible to count a single colony on a culture plate. However, until colony counts reach ~10, variability between simultaneous samples is very large, and samples with so few CFUs should be avoided. Therefore, 10 CFU/plate is listed as the LDL in Table 5.8. Investigators can perform similar calculations for other air samplers. This table also illustrates how, for a given collection time, investigators can identify a suitable sampler (based on airflow rate) for an anticipated air concentration. For example, compare the concentration ranges suitable for the two multiple-hole impactors for a 1-min sampling time.

Knowing the air flow rate and plate size of an available device, an investigator would determine how long to sample by anticipating in what range the air concentration is likely to fall. For example, consider an anticipated air concentration of approximately 100 CFU/m^3. An investigator could operate the 90-L/min multiple-hole impactor for 1 min, the 28-L/min multiple-hole impactor for 10 min, and the 50-L/min slit-to-agar impactor for 15 min. Similarly, for an air concentration of approximately 1000 CFU/m^3, the 28-L/min and 50-L/min impactors could be run for 1 min. However, the 90-L/min sampler would not be appropriate in this environment because of its high flow rate and limited agar surface (i.e., the resulting colony density would exceed the recommended limit). A sampling time sufficiently short to avoid overloading the agar surface would be too short to be practical. For unknown air concentrations, multiple sampling times are needed. For example, an investigator could collect samples for 0.5, 3, and 10 min. with the first multiple-hole impactor to cover the respective concentration ranges of approximately 35-275, 120-1000, and 700-5500 CFU/m^3.

5.2.3.2 Spore-Trap Samples Considerations similar to those discussed in Section 5.2.3.1 for direct-agar impactors apply to particle counting by microscope. Investigators should consult a knowledgeable laboratory analyst to learn the collection surface area, ideal particle surface density, and fraction of the total collection surface that is typically examined. Knowing the airflow rate and estimating the air concentration, the investigators can determine appropriate sample collection times. A potential problem with microscopic analysis is that non-biological particles can be as or more abundant than bioaerosols, causing masking. If suspected, such interference can be estimated by conducting a pilot study. Investigators can also collect preliminary samples to estimate air concentration before conducting more detailed sample collection.

5.2.3.3 Calculating Optimal Sampling Time Nevalainen et al. (1993) discuss calculation of optimal sampling time for bioaerosol samplers. Equation 5-1 gives a method to calculate optimal sampling time, t, when the desired surface density, δ, deposit area, A, average air concentration, C_a, and airflow rate, Q, are known:

TABLE 5.7. Selecting an Air Sampler for Bioaerosol Collection

---

Determine what specific biological agents may be present in a test environment and which of these are of interest for the current investigation. Determine what indicators will be used to measure the biological agents of interest, and confirm that the laboratory can analyze samples to identify and quantify these agents or indicators.

Determine what information is needed about the targeted biological agents or indicators.
- Concentration (e.g., colony-forming units (CFU)/m^3, number/m^3, or mass/m^3).
- Identification of specific microorganisms (e.g., presence/absence) or identification of all isolates and their relative concentrations.
- Average air concentration, worst-case concentration, or personal-exposure assessment.
- Particle size distribution; separate collection of respirable/non-respirable particle fractions.

Visit each test site and identify possible constraints on bioaerosol sample collection.
- High bioaerosol concentrations - collection surfaces may be overloaded unless the airflow rate is sufficiently low or the collection time is sufficiently short.
- Low bioaerosol concentrations - insufficient sample material may be collected unless airflow rate is sufficiently high or the collection time is sufficiently long.
- Temperature extremes - only short collection times may be possible for direct-agar impaction and liquid impingement in hot, dry areas or where collection media may freeze (e.g., outdoors in summer or winter, respectively).
- Collection from a moving air stream where inlet velocity and orientation will be a concern (e.g., within ventilation ducts, outdoors, at supply air diffusers).
- Access to electrical power.
- Need to collect samples without disturbing occupants or interfering with their activities.

Review literature to learn what devices other investigators have used satisfactorily to collect the target biological agent or indicator in similar work environments.

$$t = \frac{\delta A}{C_a Q} \qquad \text{(5-1)}$$

Sampling time is typically calculated in minutes, surface density in CFU or number of particles per unit of collection surface area (e.g., 1 CFU/cm² on an agar surface or 10⁴ particles/cm² on a surface to be examined by microscope), deposit area in cm², air concentration in CFU/m³ or particles/m³, and airflow rate in L/min or m³/min.

5.2.4 Sample Number

Factors that investigators consider when deciding how many samples to collect include (a) the objective of sample collection, (b) the spatial and temporal variability of the parameter being measured, (c) equipment limitations, and (d) manpower limitations. The number of samples investigators should collect depends on how they intend to use the data. Other things being equal, the larger the sample size the better (i.e., measurements and differences between sets of measurements are more precise for greater sample sizes). Here sample "size" means the number of measurements or samples collected, not the mass or volume of a sample—although this may also be important in determining the representativeness of a source sample and providing sufficient material for laboratory testing. Statisticians can help investigators calculate the number of samples needed to discern a difference of a given magnitude with a chosen probability of detection at a specified level of statistical significance. Smaller differences, higher probabilities of detection, and higher levels of statistical significance require larger sample sizes. For a given sample size, a significant difference may only be detectable if the difference is large, a low probability of detection is acceptable, or a low degree of significance is convincing.

For example, consider a biological agent whose mean concentration is 100 with a standard deviation of 25 in some arbitrary units. Investigators are interested in differences they estimate may be on the order of 30 between this sampling site and another (i.e., the second site could be ≤70 or ≥130). The investigators choose a high significance level (α = 0.05, that is, only once in 20 times would they declare a difference significant when it is not—a false-positive result). To see a difference under these conditions, the investigators would need to collect 5 to 6 samples. Two to three samples would suffice if the investigators were willing to accept a higher false-positive rate (e.g., 0.20 rather than 0.05, that is, they would falsely declare that a difference is significant four times as often as before or approximately once in every five tests). However, to see a smaller difference (±10 instead of ±30) under the original conditions, approximately 50 samples would be needed.

Besides incorrectly declaring that a difference is significant when it is not, investigators can also fail to observe a difference when there is one (a false-negative result). For a given sample size, significance level, sample mean, and sample variance, a statistician can calculate the power of a test. Alternatively, an investigator can decide what probability of detecting a significant difference they wish to have (typically ≥80%; with a 20% chance of missing a real difference) and determine how many samples they would need to collect to achieve this. Standard statistics texts can explain how to calculate sample size and power. Investigators generally assume that important differences will be large and that they will be able to detect them with few samples. However, investigators should keep in mind that with a small sample there often is a fairly high likelihood of missing a real difference.

TABLE 5.8. Calculated Air Concentration Ranges Suitable for Direct-Agar Impactors and Fungal Culture for Selected Sample Collection Times

Sampler	Plate Area (cm²)	Airflow Rate (L/min)	Collection Time (min)	Volume Collected (m³)	Air Concentration (CFU/m³) LDL (10 CFU/plate)	UDL (1 CFU/cm²)
Multiple-hole impactor	78	28	0.5	0.014	710	5570
			1.0	0.028	360	2790
			5.0	0.142	70	549
			10.0	0.283	35	276
Multiple-hole impactor	17	90	0.33	0.030	330	567
			0.66	0.059	170	288
			1.0	0.090	110	189
Slit-to-agar impactor	177	50	1.0	0.050	200	3540
			15.0	0.750	10	236
			60.0	3.0	3	59

Unfortunately, investigators seldom have prior data to use to calculate sample size, although a few preliminary samples may suffice to provide an estimate. Leidel et al. (1977) recommended that exposures to substances that have ceiling limits be monitored by sampling during 3, nonrandom, worst-case exposure periods. Rock (1995) has written that, for general occupational exposure assessments, a minimum of 6 samples is required to obtain a valid assessment of the confidence interval around a mean, and a minimum of 11 samples is needed to estimate the variance of a data set. Another recommended approach to characterize worker exposures is to collect samples 3 times a day for 3 consecutive, representative days (Table 5.9). To characterize exposures completely, investigators may need to repeat such sampling seasonally as well as during different modes of building operation and different patterns of worker activity.

In this discussion, it is assumed that all samples will be within a useful concentration range. If there is doubt what that range may be, investigators should collect the recommended number of samples for each air volume they test to cover the anticipated range of air concentration. To whatever number of samples investigators determine they need, they must add a sufficient number of blank or control samples to establish the quality of the data they collect. For example, investigators should collect at least one field blank for every type of sample collected, every different area studied, and every new batch of collection medium, such as filters or culture media [see 13.4]. The early inclusion of an epidemiologist or statistician on a project can help investigators avoid errors in study design and data interpretation. While consultation with an epi-demiologist or statistician is not always possible, the references in Chapter 13 and standard texts on epidemiological and statistical methods can guide investigators.

Lest there be any doubt, when speaking of "sample number," what is meant is the number of samples for each type of measurement, not the number of different types of samples to be collected. Therefore, direct-agar and spore-trap samples (respectively for fungal culture and microscopic spore examination) are two separate measurements, each of which requires an appropriate number of samples for each environment to be tested. Investigators must collect multiple samples to characterize exposures at individual test sites. Investigators must collect multiple samples from multiple sites to characterize exposures within a building.

Sampling preceding remediation efforts requires a sufficient number of samples to determine the extent of the area requiring cleaning. Sampling following clean-up requires a sufficient number of samples to demonstrate that the remediation process accomplished its goals. Sampling to demonstrate or rule out exposure to a biological agent is a difficult undertaking and may require the greatest number of samples and a rigorous sampling plan. While a single sample may establish the presence of a biological agent, sampling to demonstrate absence usually requires extensive testing. Therefore, a well-designed sampling plan is particularly important in attempts to demonstrate the absence of a biological agent. Similarly, sampling to convincingly rule out unusual or potentially harmful conditions may be more difficult than demonstrating their presence when they exist.

5.3 Record Keeping

Written procedures for sample collection and analysis ensure consistency from study to study and among field and laboratory personnel over time. Clear and complete records of how an investigation was conducted, by whom, and when; how field personnel collected environmental samples; how laboratory staff analyzed samples; and clear summaries and interpretations of test results are necessary to compile accurate reports, prepare publications, and respond to possible litigation.

Investigators should assign unique identifiers to samples and use these identification numbers throughout sample processing and data reporting. Samples submitted for analysis should be accompanied by a form detailing the numbers, sources, and types of samples collected along with the analyses requested (Table 5.10). Sample forms should provide space for chain-of-custody recordkeeping to track samples from the time of collec-

TABLE 5.9. Suggested Sample Numbers

Purpose	Suggestion (for each sampling site and each sample type)
To estimate worst-case inhalation exposures	Monitor ≥3, non-random, worst-case exposure periods; collect at least duplicate samples for all analyses
To estimate average inhalation exposures	Monitor ≥3 times a day for ≥3 consecutive, representative days; collect at least duplicate samples for all analyses
To estimate the confidence interval around a mean exposure	Collect ≥6 samples
To estimate the variance of a data set	Collect ≥11 samples

tion, through transport, to receipt at a laboratory (Madsen, 1991). The components of a chain of custody may include (a) attaching a sample label and seal, (b) recording data in a field log book, (c) completing a chain-of-custody record form, (d) filling in a sample analysis request sheet, (e) delivering a sample to a laboratory, (f) receipt and logging of the sample at the laboratory, and (g) assignment of the sample to a laboratory technician for analysis. At the laboratory, the chain of custody continues through the analysis and reporting processes and includes a note of the final storage location of any remaining sample material or the date and manner of its disposal. Reasons for maintaining chain-of-custody records for samples are (a) to ensure that field and laboratory personnel do not lose track of or exchange samples, (b) to prevent tampering with samples, (c) to track mishandling that might compromise a sample's integrity, and (d) to qualify laboratory results as evidence in legal cases.

5.4 Summary

Table 5.11 expands the steps in Table 5.2 for collecting air samples for biological agents and outlines the components of a sampling report. Often, assessments of possible bioaerosol-related hazards aim to describe (a) what biological or other agents may be present, (b) the environmental conditions contributing to their presence, (c) how the agents may affect building occupants, (d) the means by which the bioaerosols reach these human receptors, and (e) what can be done to prevent exposure. Chapter 13 describes how data collected during an investigation can be summarized and analyzed. Chapter 7 discusses how investigators interpret environmental data

TABLE 5.11. Steps in Bioaerosol Sample Collection

Preliminary Data Gathering
- Visit test site.
- Identify the purpose of testing, determine data to be collected, contact laboratory personnel, and select sample collection and analytical methods (see Table 5.7).
- Select sampling locations and time patterns (see Tables 5.4 and 5.5). Determine the number of samples to be collected (see Table 5.9).

Sample Collection
- Schedule sampling activities and assemble equipment. Make sample labels and label sample containers. Prepare sample data sheets.
- Calibrate sampler airflow rates, check for air leaks, prepare blank samples, and conduct other quality assurance tests.
- Run preliminary tests, analyze preliminary data, and modify procedures, as needed.
- Collect samples. Check samples and sample containers. Repeat questionable samples.

Sample Analysis
- Check sample integrity and completeness of sample data sheets. Analyze samples.

Data Reporting
- Assemble results. Perform necessary calculations. Prepare tabular and graphical summaries of sample data.
- Write report. A report may include a transmittal letter, title page, table of contents, list of figures, list of tables, brief summary, and the following sections

  - Introduction
  - Background, purpose, and objectives
  - Materials and methods
  - Test results
  - Discussion of results
  - Conclusions
  - Recommendations
  - References
  - Appendices — copies of all field data, sample and analytical results, field observations, chain-of-custody affirmations, calibrations, and names of field and laboratory personnel.
- Present results to clients relative to the hypotheses that were being tested, and decide a course of action.

TABLE 5.10. Sample Information (adapted from Harding et al., 1996)

- Collector's name
- ✓ Date and time of sample collection
- Study site (e.g., street address or building name)
- ✓ Sample identification number
- ✓ Sample type (e.g., air, surface, liquid, or bulk material)
- Sample collection site: mark location on a map or drawing of the test area, or photograph site with sampling equipment in place
- Number of persons and types of activities at sampling site
- ✓ Sample transportation method and conditions, sample storage conditions
- ✓ Type of analysis requested
- ✓ Date and time samples received at laboratory

Air Samples
- ✓ Sampler type (model number)
- Air sampler and sampling pump identification numbers
- ✓ Sampling air flow rate, sampling start and finish times, volume of air collected

As needed:
- Indoor and outdoor air temperature, RH, CO_2 concentration, and so forth
- Weather conditions, barometric pressure, wind direction and velocity

Material and Water Samples
- Sample temperature, pH, turbidity
- Sample amount or volume
- Total amount, volume, or area of contaminated material at sampling site

✓ Information that should be transmitted to the analytical laboratory along with samples

on specific biological agents to answer the questions posed at the beginning of a study.

5.5 References

AIHA: Viable Fungi and Bacteria in Air, Bulk, and Surface Samples. In: Field Guide for the Determination of Biological Contaminants in Environmental Samples, pp. 37-74. H.K. Dillon, P.A. Heinsohn, and J.D. Miller, Eds. American Industrial Hygiene Association, Fairfax, VA (1996).

ASM: Manual of Environmental Microbiology. C.J. Hurst, G.R. Knudsen, and M.J. McInerney, et al., Eds. American Society for Microbiology, Washington, DC (1997).

Burge, H.A., Ed: *Bioaerosols*. Lewis Publishers, Boca Raton, FL (1995).

Chang, C.W.; Hwang, Y.H.; Grinshpun, S.A.; et al.: Evaluation of Counting Error Due to Colony Masking in Bioaerosol Sampling. Appl Environ Microbiol. 60:3732-3738 (1994).

Chang, C.W.; Grinshpun, S.A.; Willeke, K.; et al.: Factors Affecting Microbiological Colony Count Accuracy for Bioaerosol Sampling and Analysis. Am Ind Hyg Assoc J. 56:979-986 (1995).

Cox, C.S.; Wathes, C.M., Eds: Bioaerosols Handbook. Lewis Publishers, Boca Raton, FL (1995).

Gross, E.R.; Morse, E.P.: Evaluation. In: Fundamentals of Industrial Hygiene, 4th Ed., pp. 453-483. B.A. Plog, J. Niland, P.J. Quinlan, Eds. National Safety Council. Itasca, IL. (1996).

Harding, L.; Fleming, D.O.; Macher, J.M.: Biological Hazards (Chapter 14). In: *Fundamentals of Industrial Hygiene*, 4th Ed., pp. 403-449. B.A. Plog, J. Niland, P.J. Quinlan, Eds. National Safety Council, Itasca, IL (1996).

Health Canada: Fungal Contamination in Public Buildings: A Guide to Recognition and Management. Federal-Provincial Committee on Environmental and Occupational Health. Environmental Health Directorate, Ontario, Canada (1995).

ISIAQ: Control of Moisture Problems Affecting Biological Indoor Air Quality. TFI-1996. International Society of Indoor Air Quality and Climate, Ottawa, Canada (1996).

Jensen, P.A.; O'Brien, D.: Industrial Hygiene. In: Aerosol Measurement: Principles, Techniques and Applications, pp. 537-559. K. Willeke and P.A. Baron, Eds. Wiley, New York, NY (1993).

Leidel, NA; Busch, KA; Lynch, JR: Occupational Exposure Sampling Strategy Manual. National Institute for Occupational Safety and Health, Cincinnati, OH (1977).

Madsen, F.A.: Quality Control. In: Patty's Industrial Hygiene and Toxicology, pp. 423-460. Wiley, New York, NY (1991).

Nevalainen, A.; Pastuszka, J.; Liebhaber, F.; et al.: Performance of Bioaerosol Samplers: Collection Characteristics and Sampler Design Considerations. Atmos. Environ. 26A:531-540 (1992).

Nevalainen, A.; Willeke, K.; Liebhaber, F.; et al.: Bioaerosol Sampling. In: Aerosol Measurement: Principles, Techniques and Applications, pp. 471-492. K. Willeke and P.A. Baron, Eds. Wiley, New York, NY (1993).

Robins, T.; Seexas, N.; Franzblau, A.; et al.: Acute Respiratory Effects on Workers Exposed to Metalworking Fluid Aerosols in an Automotive Plant. J. Ind. Med. 31:510-524 (1997).

Rock, J.C.: (1995) Occupational Air Sampling Strategies. In: Air Sampling Instruments for Evaluation of Atmospheric Contaminants, 8th Ed., pp. 19-44. B.S. Cohen and S.V. Hering, Eds. American Conference of Governmental Industrial Hygienists, Cincinnati, OH (1995).

Stanevich, R.; Petersen, M.: Effect of Sampling Time on Airborne Fungal Collection. Proc. Ind. Air. Qual. 90:91-95 (1990).

Appendix 5.A. Use of Randomization to Select Sampling Times or Sites

A simple random sample is one in which all possible items have the same probability of being selected for testing. Random selection of sample collection sites and times as well as the sequence in which samples are collected is critical for representative testing. Likewise, laboratory personnel may need random selection methods to examine or analyze samples in an unbiased and representative manner. Occasionally, investigators have little choice in where, when, or in what order they collect samples, but whenever possible, selection should be based on less arbitrary criteria and not solely on judgement or convenience. When in doubt, completely randomized selection may be a good way to minimize bias.

A standard method for random selection of units for testing begins with assigning a number to all units in the population of test units. For example, an 8-hour day could be divided into 20-min segments, numbered from 1 through 24. Likewise, 24 possible sampling sites could be listed in any order and assigned the numbers 1 through 24 (see Column A, Index Number, in Table 5.A.1). Any available method can be used to generate 24 random numbers (Column B). For example, many computer programs and hand calculators can generate random numbers. The random numbers are then sorted from smallest to largest (Column C) while retaining their original index numbers (Column D). This relisting indicates the order in which the 24 units are chosen for testing. For example, if five times or sites are to be tested, sampling would be conducted at the first five in Column D (i.e., those numbered 5, 17, 24, 12, and 4). If all units are to be tested, they would be sampled in the order in which they appear in Column D. Investigators should not re-use a random number set, but should generate new sets for every sampling series.

Random sampling may not always be possible or practical. For example, if the sites available for sampling cannot be identified in advance or the duration of a work period is not known, it may not be feasible to list all possible sampling locations or times and assign them random numbers before beginning to collect samples. In such cases, a systematic or periodic sampling design may be better (e.g., selection of every fifth office down a hallway from a random starting point or collection of samples for the first 15 min. of each hour). Systematic sampling may also ensure that samples are spread over an entire time period, whereas random sampling can lead to clumps and gaps. Investigators are cautioned that such systematic sampling may be biased if there is a regularity or periodicity to exposure or contaminant presence that the systematic pattern misses.

TABLE 5.A.1. Example: The Randomization Process

Index Number (A)	Random Number (B)	Sorted Random Number (C)	Sample Order (D)
1	7012	0475	5
2	9103	0727	17
3	7622	2378	24
4	2625	2470	12
5	0475	2625	4
6	7361	2727	20
7	3282	3282	7
8	6326	3653	11
9	7564	4364	21
10	9910	4777	23
11	3653	6316	18
12	2470	6326	8
13	9826	6515	16
14	7227	7012	1
15	7534	7227	14
16	6515	7361	6
17	0727	7534	15
18	6316	7564	9
19	8847	7622	3
20	2727	7665	22
21	4364	8847	19
22	7665	9103	2
23	4777	9826	13
24	2378	9910	10

Chapter 6

Sample Analysis

6.1   Introduction
    6.1.1   *Choosing an Analytical Method*
        6.1.1.1   Broad Methods
        6.1.1.2   Indicator Methods
        6.1.1.3   Focused Methods
    6.1.2   *Biases in Analytical Methods*
        6.1.2.1   Biases in Culture-Based Methods
        6.1.2.2   Biases in Microscopic Examination of Environmental Samples
        6.1.2.3   Biases in Analytical Methods Other than Culture and Microscopy
6.2   Microorganism Classification and Identification
    6.2.1   *Prokaryotic Microorganisms*
    6.2.2   *Eukaryotic Microorganisms*
    6.2.3   *Viruses*
6.3   Fungal and Bacterial Culture
    6.3.1   *Preparation of Environmental Samples for Culture Analysis*
        6.3.1.1   Agar-Impaction Samples
        6.3.1.2   Liquid Samples
        6.3.1.3   Bulk Material and Surface Samples
    6.3.2   *Culture Media*
    6.3.3   *Incubation Conditions*
        6.3.3.1   Incubation Temperature
        6.3.3.2   Incubation Atmosphere
6.4   Microscopy
6.5   Bioassays
    6.5.1   *Infectivity Assays*
    6.5.2   *Immunoassays*
    6.5.3   *Bioassays for Toxicity*
6.6   Polymerase Chain Reaction (PCR)
    6.6.1   *The PCR Process*
    6.6.2   *Use of PCR for Environmental Sampling*
6.7   Chemical Assays
6.8   Sampling and Analytical Methods
6.9   Qualifications of Laboratories and Laboratory Personnel
    6.9.1   *Choosing an Analytical Laboratory*
    6.9.2   *Laboratory Proficiency*
    6.9.3   *Quality Control in the Laboratory*
6.10   Summary
6.11   References

6.1 Introduction

This overview chapter on sample analysis outlines what investigators can realistically learn from air, bulk, and surface samples assayed for various biological agents. Such samples are used qualitatively and quantitatively to (a) identify biological agents and understand the environmental conditions that lead to their presence indoors, (b) demonstrate possible pathways from environ-

mental sources of biological agents to workers, and (c) measure worker exposure to bioaerosols and learn about exposure–response relationships. This chapter briefly describes how microorganisms are classified taxonomically and the primary methods used for their identification. Also discussed are some of the useful features, as well as limitations, of traditional culture methods. Newer analytical techniques are introduced that may circumvent some of the shortcomings of traditional procedures. These approaches are particularly useful for identifying certain environmental fungi and bacteria and for studying microbial communities. stigators need this kind of information to understand how organisms survive in indoor environments and how their multiplication can be controlled. Table 6.1 in Section 6.8 outlines the assay methods available for various biological agents and the information these methods provide.

6.1.1 Choosing an Analytical Method

Among the first questions investigators ask about indoor environments are: "What types of biological agents are present?" and "How much of each agent is present?" Analytical methods can generally be categorized as one of three approaches, which will be referred to as broad, indicator, and focused methods. Often, microorganisms (e.g., environmental fungi or bacteria) are the biological agents of primary interest but an investigation has not yet targeted any specific microorganisms. For these investigations, *broad analytical methods* that can identify a range of microorganisms are needed. In other cases, particular biological agents are of interest but, for various reasons, an analytical method for an indicator of the agent is chosen rather than a method to detect the agent itself. These are referred to as *indicator methods*. Occasionally, information is needed on one or a few specific biological agents, which narrows an investigation's focus. Therefore, analytical methods for specific biological agents are called *focused methods*.

6.1.1.1 Broad Methods
Many investigations of indoor environments aim to evaluate the microbial status of a space. While this goal is never fully achieved, such investigations seek information on the types and concentrations of microorganisms that are present to determine if they appear to be typical or unusual. In these cases, investigators choose methods for sample collection and analysis that provide the broadest possible information. At present, culture-based methods are used most commonly in broad studies and will be described in some detail in this and other chapters. Direct microscopic examination of sample material is also widely applicable for the identification of many fungi. Methods that assess microbial diversity in an environment are increasingly used in research settings. These approaches rely on molecular detection methods and chemical analyses to describe microbial populations and to monitor changes in microbial communities.

6.1.1.2 Indicator Methods
Besides microorganism culture and direct microscopic examination of fungi, another means of evaluating the microbial status of an environment is to measure quantitative indicators of the presence of large groups of microorganisms. Examples of this approach are the specific identification of *Escherichia coli* (as an indicator of contamination with raw sewage), the analysis of the glucan or ergosterol content of samples (as a measure of fungal biomass), or the detection of guanine as an indicator of dust-mite presence. Elsewhere in this book, the term "indicator" is used in a similar context when referring to microorganisms or chemical markers whose detection may reflect the simultaneous occurrence or presence of the actual biological agents responsible for adverse health effects (e.g., allergens, toxins, or other cell products) [see 6.7]. Some fungi have also been suggested as indicators of excessive moisture or health hazards [see 19.5.1 and 19.5.3.2]. Indicator methods are relatively straightforward and provide general information on the types of biological agents in an environment. However, these assays may not provide information that is sufficiently specific to make correlations with some types of health effects because these methods measure an indicator or surrogate for a biological agent rather than the agent responsible for a health effect.

6.1.1.3 Focused Methods
A third type of investigation seeks to document the presence of specific biological agents associated with particular health effects. For example, when studying an outbreak of Legionnaire's disease, an investigator would use analytical methods specific for detecting legionellae in environmental and clinical samples. Typical analyses would include culture to isolate the bacteria from environmental water samples and specimens from potentially infected persons. Nonculture-based methods are also used to detect the bacteria in environmental and clinical samples (e.g., direct fluorescent antibody stains and polymerase chain reaction). Blood samples could also be drawn to determine if persons with a history of an illness compatible with legionellosis had elevated antibody titers to *Legionella* spp. Similarly, if the specific health effects of aflatoxin were of concern, an investigator would use an assay method suitable for detecting and quantifying that toxin. Methods that focus on single organisms, agents, or health outcomes include nucleic acid probes, chemical assays, and bioassays. The decision to use a focused analysis requires that investigators hypothesize that specific biological agents may be present. Such hypotheses may be based on observed health effects (associated with specific agents) or environmental conditions (that lead investigators to suspect the potential presence of specific biological agents).

6.1.2 Biases in Analytical Methods
6.1.2.1 Biases in Culture-Based Methods
All of the broad methods that are currently available for analysis of biological agents are biased in some way. Culture is the most commonly used method for assessing broad groups of microorganisms in indoor environments, but

this method misses fungi and bacteria that are not culturable. Some of the culture methods in popular use to detect and count microorganisms in environmental samples identify such limited subgroups of microorganisms that one might question if these methods may not be more misleading than helpful. Culture-based methods allow detection of only those organisms that are alive, in a condition to grow in culture, and able to successfully compete with the other organisms in an environmental mixture. It is possible (and perhaps likely) that the majority of microorganisms in a particular sample are not identifiable with culture-based methods. The species most readily cultured from a given environment may not be the most prevalent or the most important species present.

The method by which microorganisms were collected may also affect microorganism culturability [see 11.2.3]. Several researchers have studied the effects of air sample collection on the ability of microorganisms to be isolated in culture (Juozaitis et al., 1994; Stewart et al., 1995; Terzieva et al., 1996). Collection methods that expose bacterial cells to drying (e.g., filters or dry cyclones) are generally not suitable for analyses that require that bacteria be viable (Jensen, et al., 1992). However, these sample collection methods may be appropriate for analytical approaches other than culture [see Table 11.1].

6.1.2.2 Biases in Microscopic Examination of Environmental Samples Microscopic examination of spore-trap samples can provide an accurate estimate of total fungal spore concentrations, a broad picture of the kinds of spores in an air sample and, when combined with culture-based analysis, quite accurate estimations of fungal populations. However, used alone, microscopic examination of air samples allows specific identification of only a few fungal species, and most spores must be placed in general categories [see 19.4.2]. Direct microscopic examination of source samples may allow identification of individual fungi, but this method is not usually quantitative and does not allow identification of unstained bacteria. Alone, neither culture nor microscopy provides an accurate assessment of the spectrum of biological agents in environmental samples. Together, these methods can provide relatively complete information on fungal populations. For use with bacteria, investigators must decide which genera and species are of interest and select appropriate stains and viewing techniques [see 6.4 and 18.4].

6.1.2.3 Biases in Analytical Methods Other than Culture and Microscopy In various ways, analytical methods other than culture or microscopy also restrict the information that can be obtained about biological agents in environmental samples. Chemical, immunological, and biological assays can be useful indicators or focused tools. To be used as broad methods, these assays require the development of extensive data libraries for agent identification.

Development of such databases requires that the potential constituents of the biological populations of interest be known, which is not always the case.

6.2 Microorganism Classification and Identification

When biological agents are examined in environmental samples, it is often important to identify the organisms (more or less specifically) to allow comparisons of the agents present in different locations or at different times. All living organisms (bacteria, fungi, plants, and animals) are identified in the context of some previously agreed upon classification scheme. Such systems define the characteristics of organisms that are grouped together at different levels of similarity, beginning with kingdoms and proceeding to species. There is no universally accepted definition of a "species," which is simply the taxonomic group that subdivides a genus (Stahl, 1997) [see Chapter 17].

In large part, classification depends on morphological and physiological similarities and differences among related organisms, such as those described in the sections on *Characteristics* in the chapters on bacteria, fungi, amebae, viruses, and house dust mites. Members belonging to the same classification or taxon share certain key characteristics (i.e., combinations of features or properties that are unique to a group). Key characteristics may be morphological, physiological, biochemical, immunological, or molecular. Whole dust mites, amebae, and some fungal spores and fungi in culture can be readily identified by microscopic examination. However, identifying bacteria to genera and species generally requires subculturing and biochemical testing of the pure cultures. Environmental isolates that do not exactly match the key characteristics listed in reference texts may present identification problems. Therefore, making a reliable identification of the entire range of microorganisms that may be found in environmental samples can be a time-consuming undertaking. Before sampling, investigators should carefully consider what level of organism identification they will need to test a study's hypotheses and develop recommendations for control and remediation.

6.2.1 Prokaryotic Microorganisms

Bacteria and archaea are prokaryotes (Gr. before cells) and are usually classified based on evolutionary (phylogenetic) relationships. Observed similarities in characteristics of morphology (physical size and shape) and physiology (processes and functions) are also used for categorization [see Chapter 18]. Many classification schemes for bacteria have been proposed over the years, and it is important to note that there is no "official," universally agreed upon system. One of the best known and most widely accepted classification systems is that described in *Bergey's Manual of Systematic Bacteriology* (1989). In this scheme, classification is based, in part, on the probability of matching defined morphological, physiological, and

TABLE 6.1. Collection and Analysis of Environmental Samples for Biological Agents (for acronyms and abbreviations see list)

Biological Agent	Primary Sample Collection Methods	Sample Analysis	Data Obtained	Chapter
AMEBAE	Bulk (water) Air: impingers, impactors	Culture Microscopy: cell morphology Bioassay: immunoassay — labeled antibody stains	Concentration: number per m³ or mL Isolate identification Confirmation of the presence of a specific ameba	20
ALLERGENS/ANTIGENS (cockroach, mammalian, avian; see also bacteria, fungi, and dust mites)	Bulk (settled dust)	Bioassay: immunoassay (ELISA)	Concentration (allergen/antigen): μg/g or units of allergen/g or antigen/g	25
BACTERIA	Air: impactors, impingers, wetted cyclones, filters Bulk/Surface samples	Direct microscopy/total counts Bioassay: immunoassay — labeled antibody stain Molecular assay: nucleic acid probe (molecular hybridization), nucleic acid amplification (PCR) Culture: colony morphology Microscopy following culture: cell morphology, Gram-stain characteristics Biochemical assay following culture: substrate utilization assay	Concentration: cells per m³, g, or cm² Confirmation of the presence of a specific bacterium Confirmation of the presence of a specific bacterium Concentration: CFU per m³, g, or cm² Isolate identification (general) Isolate identification (specific)	18
Bacterial antigens	Bulk (settled dust)	Bioassay: immunoassay (ELISA)	Concentration (bacterial antigen): μg/g	25
Bacterial cell-wall components: Endotoxin (Gram-negative bacteria) Other (total bacteria)	Air: filters, impingers Bulk (settled dust) Air: filters Bulk (settled dust)	Bioassay: LAL Chemical assay: GC-MS or HPLC Chemical assay: GC-MS or HPLC	Biological Activity (endotoxin): endotoxin units per m³ or g Concentration (LPS): ng per m³ or g Concentration (muramic acid, diaminopimelic acid - PG): μg per m³ or g	23
Bacterial whole-cell lipids; phospholipids	Air: filters, impingers Bulk: surface samples	Chemical assay: GC, GC-MS	Community profile	18
FUNGI Bulk/Surface samples Microscopy following culture: spore and hyphal morphology	Air: impactors, filters, impingers, wetted cyclones Culture: colony morphology	Direct microscopy/total counts Concentration: CFU per m³, g, or cm² Isolate identification	Concentration: spores per m³, g, or cm²; spore identification	19
Yeasts (whole-cell lipids, fatty acids)	Air: impactors, impingers, wetted cyclones Bulk/Surface samples	Chemical assay: GC Biochemical assay: substrate utilization assay	Isolate identification	19

TABLE 6.1. Collection and Analysis of Environmental Samples for Biological Agents (for acronyms and abbreviations see list) (cont.)

Biological Agent	Primary Sample Collection Methods	Sample Analysis	Data Obtained	Chapter
Fungal allergens	Bulk (settled dust)	Bioassay: immunoassay (ELISA)	Concentration (fungal allergen): ng or units per g	25
Fungal cell-wall components	Air: filters Bulk samples	Bioassay: LAL, immunoassay Chemical assay: CG-MS, HPLC	Biological activity (glucan): units or μg per m³ Concentration (glucan or ergosterol): units or μg per g	24
Fungal toxins	Air: impactors, filters, impingers, wetted cyclones Bulk samples	For toxigenic fungi: (see Direct microscopy, Culture, and Microscopy following culture, above) For toxins: Chemical assay (TLC, HPLC, GC-MS); Bioassay: immunoassay, cytotoxicity assay	Confirmation of toxin presence; Concentration (toxin): ng per m³ or g Confirmation of toxin presence Detects toxic activity without toxin identification	19 24
DUST MITES	Bulk (settled dust)	Microscopy: mite morphology	Concentration (number mites): per g of dust or m² of sampled surface Mite identification	22
Dust-mite allergen	Bulk (settled dust)	Bioassay: immunoassay (ELISA)	Concentration (mite allergen): ng per g of dust or m² of sampled surface	
Guanine (marker)	Bulk (settled dust)	Chemical assay: LC, colorimetric	Concentration (guanine): per g of dust or m² of sampled surface	
VIRUSES	Air: impingers, cyclones, impactors, filters	Bioassay: cell culture, Immunoassay — labeled antibody stains Electron microscopy Molecular assay: nucleic acid probes (molecular hybridization), nucleic acid amplification (PCR)	Concentration (cytopathic units): number/m³; Isolate identification Confirmation of the presence of a specific virus Isolate identification Confirmation of the presence of a specific virus	21
MVOCs	Air: sorbent or whole air sampling	Chemical assay: GC-MS	Concentration (compound): mg/m³ Compound identification	26

phenotypic characteristics, the latter being visible properties of an organism that are produced by interactions between a genotype and the environment [see 18.1.2]. Considerable laboratory effort may be needed to identify a bacterium isolated from the environment. Occasionally, it is not possible to assign a given organism to any described genus; other times, an isolate may be assigned to a genus, but it may not be possible to make a definite species identification.

6.2.2 Eukaryotic Microorganisms

The eukaryotic organisms (Gr. true cells) of interest in investigations of indoor environments include many fungi, a few amebae, and the plants and animals that produce allergens responsible for building-related hypersensitivity diseases [see Chapters 19, 20, 22, and 25]. Fungi may reproduce both sexually and asexually but are generally classified by the morphology of their sexual structures [see 19.1.2 and 19.1.3]. The sexual and asexual forms of a fungus may differ morphologically and even have different names. These multiple names are a potential source of confusion in laboratory reports and the literature. Fungal isolates that do not sporulate in laboratory culture may be very difficult or impossible to identify. In addition, fungal characteristics used for identification may vary depending on growth conditions. In the past, for routine bioaerosol investigations, fungi often were identified only to the genus level. Increasingly, investigators are recognizing the importance of identifying fungi to the species level [see 19.4.1]. Such identification is becoming more widely available from commercial laboratories, but some identifications require the experience and expertise of a mycologist who has specialized in the study of a particular fungal group.

6.2.3 Viruses

Viruses are classified according to the structure and arrangement of their proteins and by the kind of nucleic acid (DNA or RNA—but never both) in their cores [see 21.1]. Viruses are not included in routine bioaerosol characterizations, in part, because the detection and identification of viruses require very specific methods and may be difficult. However, modern molecular methods have reduced the necessity for viral cultivation. Specifically designed molecular probes now permit the screening of environmental samples for some viral agents, primarily in water and soil samples but increasingly in air samples [see 6.6].

6.3 Fungal and Bacterial Culture

Culture-based assays for fungi and bacteria involve providing a growth environment that encourages the microorganisms in a sample to multiply. For air samples, this is often done by collecting the bioaerosols directly onto agar-based media in culture plates or by inoculating plates with small portions of air samples collected in a liquid. For source samples, culturing is done by directly inoculating culture plates with measured portions of dust or liquid samples or by making suspensions of bulk or surface samples and plating portions of these.

Knowing how to optimize the information that culture-based methods can provide requires familiarity with the range of fungi and bacteria that may be found in indoor environments. The *Sample Analysis* sections of the chapters on bacteria, fungi, amebae, and viruses describe how microbiologists identify microorganisms that have been grown in laboratory culture. Analysis of culture data is discussed in Chapter 13.

6.3.1 Preparation of Environmental Samples for Culture Analysis

Air samples are possibly the most common type of environmental sample that investigators collect to study bioaerosols. Investigators should know what type of sample analysis they plan to use when collecting source samples so that the collection procedure is compatible with the requirements of the analytical procedure. Typical source samples are settled dust, pieces of potentially contaminated material, water samples, and surface washes or contact samples. Preparation of source samples for assays other than culture may be similar to the methods of sample preparation described here for culture-based analysis.

6.3.1.1 Agar-Impaction Samples No processing is needed prior to incubation for air samples collected by direct impaction onto agar-based culture media. However, laboratory personnel should examine plates upon receipt and note any growth that occurred prior to incubation. Laboratory personnel should also look for debris on the agar surface or evidence of excess moisture or agar dehydration that could invalidate samples.

6.3.1.2 Liquid Samples Liquid samples (e.g., bulk water samples, impinger and cyclone fluids, and surface and bulk material washings) may be plated directly on the culture media of choice. If bulk samples are viscous or semisolid (i.e., sludge like) or if liquids were collected from heavily contaminated areas, it may be necessary to make serial dilutions in sterile buffered saline or diluted broth prior to plating. Such dilutions are usually made in multiples of ten (i.e., 10^{-1}, 10^{-2}, and so forth).

6.3.1.3 Bulk Material and Surface Samples Dust samples may be sieved and weighed portions (e.g., 30 mg) inoculated directly onto the surfaces of culture media where the samples are spread with a sterile glass rod or other implement. Dust samples and bulk samples of solid items (e.g., contaminated fabrics, carpet, wallboard, ceiling tile, or insulation material) may also be weighed and washed or extracted in sterile buffered saline or diluted broth. Addition of a wetting agent, such as 0.01% Tween 80 (J.T. Baker Inc., Phillipsburg, NJ), helps to separate microbial cells from each other and from material to which the cells may adhere. Washings or suspensions are processed as liquid samples and may be diluted prior to plating. More types of culturable fungi have been isolated when dust samples were plated directly onto agar than when the same dust was suspended in a liquid before plating. However, overgrowth and interference is more common with the direct dust inoculation method than with suspension and dilution methods. Suspending bulk samples in liquid and making serial dilutions of the suspensions has been found to yield higher estimates of the concentration of fungi in bulk samples, probably by reducing problems with overgrowth and enhancing spore or cell separation.

6.3.2 Culture Media

There are many varieties of culture media and many specialized uses for them. Some media are fairly general, that is, able to support a large variety of microorganisms, sometimes both fungi and bacteria. Other media are selective, differential, or both. A *selective medium* is formulated to give some microorganisms a physiological or nutritional advantage over others. Some nonselective media can be made selective by adding antibiotics to inhibit the growth of certain microorganisms. A *differential medium* may support the growth of more than one microorganism but contains ingredients that produce differences in the appearance of various fungi or bacteria. An example of a medium that is both selective and differential is the eosin-methylene blue agar used to isolate and differentiate enteric bacteria (i.e., those found in the intestines). Gram-positive bacteria (GPB, e.g., *Staphylococcus aureus*) will not grow on this medium. Among the Gram-negative bacteria (GNB) that will grow, *Escherichia coli* produces blue-purple colonies with a metallic sheen whereas *Shigella sonnei* colonies are colorless. Selective and differential media form the basis of the classic methods used to identify bacteria [see 18.4.1.3].

In a general survey of environmental microorganisms, the culture medium of choice would likely be a nonselective, broad-spectrum medium able to support the growth of a wide variety of microorganisms. Soybean-casein digest agar (also known as tryptic or trypticase soy agar), nutrient agar, and R2A are examples of nonselective media used to culture bacteria. Various formulations of malt extract agar are broad-spectrum media often used for fungi. Rose bengal agar is a selective medium that favors the multiplication of some slow-growing fungi from environments such as soil, in which many rapid-growing fungi are also present. This medium contains a dye that inhibits bacteria and some rapid-growing fungi. Fungi adapted to

dry conditions (xerotolerant or xerophilic fungi) may require a medium such as DG18 (with dichloran and glycerol) or other medium with high solute concentration that limits the amount of available water. Tables 18.4 and 19.5 provide the formulations for these media. Obviously, an investigator's choice of a culture medium is a critical part of a sampling and analysis plan. There is no agreement on the best all-purpose medium for isolating either total culturable fungi or bacteria. Investigators and microbiology consultants can refer to the literature on environmental sampling and check other references for direction on choosing culture media for specific purposes (Atlas, 1993; AIHA, 1996a; ASM, 1997).

6.3.3 Incubation Conditions

6.3.3.1 Incubation Temperature Microorganisms growing in indoor environments are generally adapted for growth at ambient temperatures. Most fungi grow optimally over a temperature range from 18° to 25°C, whether in the environment or in laboratory culture. The majority of fungi grow poorly, if at all, at temperatures above 35°C. A few fungi (e.g., *Aspergillus fumigatus*) grow well at higher temperatures and can be selectively isolated by incubation at 40°C.

Environmental bacteria also grow well in the 18° to 25°C temperature range, but many bacteria can tolerate temperatures above 30°C. Cultures for environmental bacteria are generally held at ~28°C [see 18.4.1.2]. This temperature is lower than that typically used in clinical laboratories for isolating pathogenic microorganisms, which grow at human body temperature. Incubation of environmental samples at ~37°C is only appropriate to isolate bacteria that cause infectious diseases. Cultures for thermophilic actinomycetes are incubated above 50°C, which encourages the growth of these bacteria and excludes nonthermotolerant fungi and bacteria. During sample transport, the multiplication of microorganisms that can grow at room temperature must be prevented by keeping inoculated plates cool enough to inhibit microbial multiplication. Samples should be delivered to the laboratory within 24 hours of collection. Temperature extremes during sample transport and storage can also cause dehydration of water-based media and condensation in culture plates, both of which should be avoided.

6.3.3.2 Incubation Atmosphere Atmosphere is another incubation condition that an analytical laboratory must consider. Most analysis of air samples for culturable microorganisms has been limited to obligately or facultatively aerobic species (i.e., those that require molecular oxygen or those that can grow with or without it). Therefore, no special care is taken to control the atmosphere in which samples are incubated. Essentially all fungi and many bacteria are aerobic. However, anaerobic bacterial species may persist in the environment as spores or as vegetative forms growing in sequestered areas without oxygen. Selective analysis for anaerobic bacteria requires special sample transport and incubation to exclude oxygen (e.g., anaerobic liquid media in sealed containers or anaerobic chambers or jars for culture plates). Growth of some bacteria is enhanced by incubation in 2.5% to 10% CO_2 (e.g., *Legionella* spp.), and special incubators may be required to maintain these conditions.

6.4 Microscopy

Microbiologists and other biologists use microscopes extensively as described in Sections 18.4.2, 19.4.2, 20.4.3, 22.4.2, and 24.4.1.2. Total counts of fungal and bacterial cells can be made by microscopic examination of environmental air samples. Fungal cells and spores can be counted with or without staining. Many fungi have morphologically distinctive spores and can be identified by direct microscopic examination to the genus level (e.g., *Cladosporium* and *Alternaria* spp.) or even the species level (e.g., *Stachybotrys chartarum* and *Epicoccum nigrum*).

For bacteria in culture, staining is one of the first steps performed to identify cell morphology and staining characteristics. For identification of bacteria by direct microscopic examination (without culture), the information obtained is the concentration of total cells or total cell counts. Fluorescing stains [e.g., acridine orange (AO), 4',6-diamidino-2-phenyl-indole (DAPI), or fluorescein isothiocyanate (FITC)] are often used with epifluorescence microscopy (Atlas and Bartha, 1993; APHA, 1995; Chapin, 1995). Until recently, a disadvantage of some direct bacterial stains was the lack of differentiation among cells of similar size and shape. For example, many rod-shaped bacteria, live or dead, Gram-negative or Gram-positive, appear alike with some stains. However, fluorescent Gram-stain kits are now commercially available as are reagents that differentially stain live and dead organisms (Hensel and Petzoldt, 1995; Madelin and Madelin, 1995; Morris, 1995; Terzieva et al., 1996; Heidelberg et al., 1997; Lawrence et al., 1997). Image processing may facilitate counting of microorganisms in some types of samples (Kildesø and Nielsen, 1997). Labelled antibody stains are useful to identify selected bacteria (e.g., a direct fluorescent antibody stain may be used to detect *Legionella pneumophila* in filtered water samples) (AIHA, 1996b).

6.5 Bioassays

Bioassays are quantitative methods in which the end point is some observable effect on a biological system or organism. In this book, the term bioassay is used in a broad sense to include *in-vivo* and *in-vitro* tests based on biological systems or organisms. Animal and cell culture assays for viruses and bacteria are bioassays as are assays that expose cells or whole organisms to samples that may contain tox-

ins. Immunoassays (which depend on the response of biological molecules, that is, antibodies) and the *Limulus* amoebocyte lysate assay for endotoxin (which depends on the response of a lysate of cells from the blood of the horseshoe crab) can also be considered bioassays.

6.5.1 Infectivity Assays

Viruses cannot function outside of living cells, therefore, laboratory assay for viruses involves inoculating tissue cultures or whole animals with sample material, the former being more widely used [see 21.4]. In a tissue culture, each viral particle results in a focus of infection in the confluent cell layer as evidenced by formation of a plaque or lesion. Tissue culture assays provide information on the number of viral particles that were able to attack the type of cell provided. Applying tissue culture analysis to environmental samples is especially difficult because of the likelihood that other microorganisms present in the samples will also grow and obscure the results. Thus viral culture is limited as a tool for assay of environmental samples.

6.5.2 Immunoassays

Immunoassays are biological assays based on the specificity of the antigen-antibody reaction. This specific affinity has been exploited for a great number of applications, among them the detection of biological agents in environmental samples (e.g., dust-mite, animal, and cockroach allergens; microbial antigens; fungal glucans; aflatoxin; and latex) [see 22.4.3, 24.7.3, and 25.4].

An enzyme-linked immunosorbent assay (ELISA) is a widely-used method to quantify allergen concentration in dust samples. Briefly, the wells of a polystyrene microtiter plate are coated with capture antibodies, which are specific for particular antigens (Ab_1; Step 1, Figure 6.1). Dust samples are sieved to obtain the fine dust fraction, soluble proteins are extracted from the dust in an appropriate buffer, and the extracts are serially diluted. The same is done for an allergen standard, and aliquots of diluted standards and dust sample extracts are added to the antibody-coated wells (Step 2). The plates are incubated and washed to remove unbound materials. Bound allergen is detected with a second enzyme-conjugated antibody (Ab_2) that recognizes a cross-reacting site (Step 3). An enzyme substrate/chromogen solution is added, and the optical absorbance (color change) is read with a microtiter plate reader (Step 4). The concentrations of target antigens are obtained by comparing the optical density readings for dust samples with standard curves of optical densities generated from allergen standards of known concentration. Assays for more than one allergen (e.g., fungus, cockroach, dust mite, bird, cat, dog, or rodent allergens) can be performed on the same dust suspensions by substituting appropriate antibodies (IOM, 1993) [see 25.4].

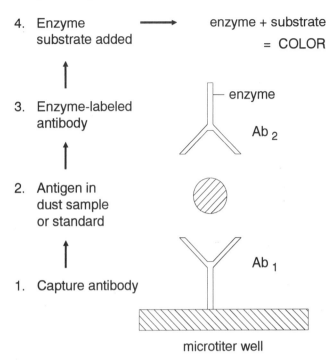

FIGURE 6.1. Enzyme-linked immunosorbent assay (ELISA).

6.5.3 Bioassays for Toxicity

The *Limulus* amoebocyte lysate assay is a bioassay in which the observable effect is a clot, increase in turbidity, or a chromogenic reaction. This method is used primarily to measure endotoxin concentrations but has also been used for fungal glucans [23.4.2 and 24.7.3]. Cytotoxicity assays are used to detect materials toxic to cells [see 24.4.3]. For example, the brine shrimp assay is used to assess the relative toxicity of purified toxins or environmental samples that may contain toxins. The endpoint measured in this test is the death of these microscopic animals. Various other tests are available to identify compounds that are mutagenic in bacterial or mammalian cell assays. These assays can be used on specific compounds or mixtures of compounds.

6.6 Polymerase Chain Reaction (PCR)
6.6.1 The PCR Process

Sometimes, rather than a survey of all microbial species in an environment, what is needed is a method to screen samples for a specific genus or species [see 6.1.1.3]. Polymerase chain reaction (PCR) is an example of such a method. PCR can be very sensitive and specific and has seen application for detecting the presence of some airborne infectious agents. Advantages of PCR over culture are its speed (results can be obtained in a matter of hours rather than days or weeks) and its ability to detect difficult-to-grow, slow-growing, and even nonculturable microorganisms (Gao and Moore, 1996; Buttner et al., 1997).

PCR is based on *in-vitro* replication of selected nucleic acid sequences and, theoretically, can copy and recopy the

DNA from a single organism millions of times. DNA consists of two complementary strands with specific nucleotide bases always pairing from opposite pieces of a genome (A in Figure 6.2). The strands separate (B) and are copied, with each strand acting as a template to replicate its complement (C).

The PCR process uses temperature to separate two DNA strands (B in Figure 6.2). Enzymes (polymerases) are added along with the appropriate nucleotide bases to be used as building blocks for the copying. If the appropriate physical and chemical conditions are repeated in a fixed cycle, the strands are repeatedly copied, separated, and recopied, resulting in an exponential increase in the number of identical DNA pieces. In a few hours, there is enough material for analysis. Nucleic acid probes are short nucleotide sequences able to bind (hybridize) with species-specific regions of targeted DNA (Figure 6.3). Specialized probes are tagged with radioactive or fluorescent labels, and these labelled probes are permitted to react with the amplified DNA. A probe will bind in detectable quantities if the organism of interest was present in the initial sample and a targeted segment of its DNA was successfully amplified.

A. Double-stranded DNA

```
        -------G - A - A - T - T - C-------        strand 1
                || || ||   || || ||
        -------C - T - T - A - A - G-------        strand 2
```

B. Strands separated

```
        -------G - A - A - T - T - C -------       strand 1


        -------C - T - T - A - A - G-------        strand 2
```

C. Strands copied   -------G - A - A - T - T - C-------   strand 1

```
                || || ||   || || ||
        -------C - T - T - A - A - G-------        strand 2-c


        -------G - A - A - T - T - C-------        strand 1-c
                || || ||   || || ||
        -------C - T - T - A - A - G-------        strand 2
```

FIGURE 6.2. Short section of a DNA sequence showing paired nucleotide bases and DNA replication.

6.6.2 Use of PCR for Environmental Sampling

PCR has been used for clinical diagnosis of certain infectious diseases without the necessity of culturing the responsible pathogens. Some clinical PCR tests have been adapted to also detect infectious agents in environmental samples. Likewise, PCR systems designed for environmental water samples have been used to detect specific

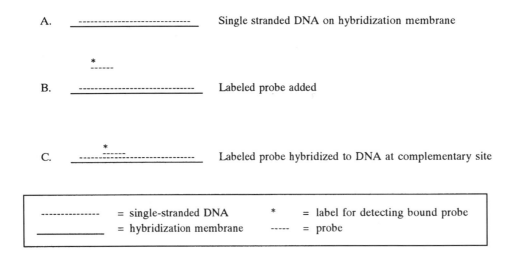

A.   ----------------------------   Single stranded DNA on hybridization membrane

B.   ----------------------------   Labeled probe added

C.   ----------------------------   Labeled probe hybridized to DNA at complementary site

---------------   = single-stranded DNA   *   = label for detecting bound probe
_____   = hybridization membrane   -----   = probe

FIGURE 6.3. Illustration of hybridization of a DNA probe to a complementary section of a nucleic acid sequence.

organisms in air samples collected with impingers, cyclones, filters, or other apparatus (Alvarez et al., 1994; Mukoda et al., 1994; Palmer et al., 1995; AIHA, 1996b; Mastorides et al., 1997).

The PCR method is especially useful for the rapid detection of organisms that are difficult or impossible to culture in the laboratory. It is also an alternative method for the detection of organisms for which culture and manipulation might pose special hazards for laboratory personnel. The PCR method amplifies DNA from live or dead organisms. A variation of the method permits the analysis of gene expression, which provides information about the biological activity of microorganisms as well as their presence.

At the time of this publication, some of the limitations of the method are that it is a focused method and only predetermined agents can be detected, if present. PCR is occasionally subject to interference from compounds found in some environmental samples. The method is not as widely available as other analyses, being more a research than investigative tool, and is primarily qualitative (a measurement of presence or absence) rather than quantitative (a measurement of the exact amount of the target organism). Despite these limitations, PCR and other forms of nucleic acid amplification are finding increasing use in environmental microbiology, and new methods appear in the literature as investigators adapt procedures for wider applications (Amann et al., 1995; Podzorski and Pershing, 1995; Stahl, 1997). PCR can be used to analyze bulk material, water, soil, and bioaerosol samples for the presence of specific microorganisms or for environmental surveillance of indicator organisms.

6.7 Chemical Assays

Dissatisfaction with analytical methods that depend on the growth of microorganisms in laboratory culture has led to the application of the tools and methods of analytical chemistry to organism identification. High performance liquid chromatography (HPLC), gas chromatography (GC), and mass spectrometry (MS) are used—singly or in combination—to identify certain "chemical markers." Chemical markers are biological molecules that may be either components or products unique to certain groups of microorganisms (Fox et al., 1990). These markers include (a) structural molecules (e.g., carbohydrates such as peptidoglycans from bacterial cell walls and beta-glucans from fungi), fatty acids (components of bacterial cell walls), lipids (ergosterol from fungal membranes), and lipopolysaccharides (endotoxin from GNB), which contain both lipid and carbohydrate moities, (b) metabolic products (e.g., mycotoxins from fungi, alcohols, aldehydes, and other volatile organic compounds, from bacteria and fungi) (c) secreted macromolecules (e.g., specific enzymes and other proteins), and (d) excreted macromolecules (e.g., guanine in dust-mite fecal pellets). Depending on the specific method used, the information derived may be general (e.g., measurement of glucan or ergosterol to estimate total fungal biomass), agent specific (e.g., chemical assays for specific mycotoxins), or microorganism specific (e.g., determination of fatty acid profiles to identify specific bacteria). Methods for analysis of some of these compounds and the kinds of data these assays can provide are discussed in Sections 19.4.3, 22.4.4, 23.4.3, 24.7.3, and 26.5.

Much effort is currently being invested in the development of analytical microbiology. However, at this time, the work is being performed primarily in research laboratories, and the commercial availability of such testing is limited. The development of computer-directed automation of some of the methods may result in simpler, reproducible protocols for instrumental analyses of chemical markers. Fatty acid analysis allows measurement of total biomass in a sample (a quantitative description of community structure) and does not require that organisms be culturable or even viable (Macnaughton et al., 1997; White et al., 1997). GC has been used to analyze the methyl ester derivatives of fatty acids from whole-cell extracts. This method was evaluated for identification of bacteria from the family *Micrococcaceae*, which often comprise a major fraction of culturable airborne bacteria in buildings (Pendergrass and Jensen, 1997). Some investigators have used carbon-source analysis to characterize microbial communities based on metabolic profiles and made inferences about community diversity (Garland and Mills, 1991).

6.8 Sampling and Analytical Methods

As described above, a wide assortment of analytical methods are available to investigators to study biological agents in indoor environments. Occasionally, multiple methods are available for a single biological agent (e.g., a bioassay and a chemical assay). Table 6.1 summarizes the information in the *Sample Analysis* sections of the chapters on specific biological agents in Part III of this book. The table lists biological agents, methods for sample collection, methods for sample analysis with the corresponding information investigators can gain from the analyses, and the chapters that discuss these methods. Note that more than one sample collection method may be listed for each biological agent. The listed collection methods apply to all of the analytical methods that follow for a particular agent. For example for amebae, both bulk water samples and air samples can be analyzed by culture, microscopy, and bioassay. Note, however, that the column on data obtained for each analysis is listed for a particular analysis. Respectively, the three analyses listed for amebae in water or air samples provide information on organism con-centration (culture), isolate identification (microscopy), or confirmation of the presence of a specific ameba (bioassay).

6.9 Qualifications of Laboratories and Laboratory Personnel

Few IHs, EHPs, or IEQ consultants have laboratories capable of analyzing bioaerosol samples. Therefore, commercial laboratories are employed to analyze most environmental samples for biological agents. Investigators should contact a laboratory before collecting samples to ensure that the laboratory can handle the samples in a timely manner when received. Until investigators gain experience and confidence in bioaerosol sampling, laboratory personnel should participate in the study design phase of investigations and in the development of sample collection plans. Such participation will help ensure that investigations are conducted reasonably and in accordance with current recommendations from appropriate professional associations.

6.9.1 Choosing an Analytical Laboratory

Many factors contribute to an investigator's choice of what type of analytical laboratory to use and selection of a particular laboratory within a chosen type. Among the factors that may influence this choice are (a) the services the laboratory can provide, (b) what the laboratory charges for analyses, (c) proximity of the laboratory to the sampling location, (d) availability of rapid sample transport, and (e) turn-around time for delivery of laboratory reports. However, equally important are the laboratory's qualifications to conduct the requested analyses, their record of past performance, and the willingness of the laboratory personnel to work with an investigator in designing a sampling plan and interpreting test results.

At a minimum, a laboratory must have the necessary equipment and staff to carry out the tests requested. For example, if asked to identify indoor and outdoor bacteria, a laboratory should have the basic equipment needed to incubate and identify bacteria and should provide the services of a qualified environmental bacteriologist. Likewise, if fungal samples are to be analyzed, it is necessary that an experienced environmental mycologist be available to examine them. Laboratories should have written analytical methods and procedures, and laboratory personnel should keep careful and complete records of what tests they perform and the results. In addition, quality assurance programs should be routine in all laboratories, and written policies and records of compliance should be available [see 5.3 and 13.4].

6.9.2 Laboratory Proficiency

As requests for indoor air analyses have increased, more laboratories have offered bioaerosol testing. Until recently, most laboratories conducting environmental microbiology analyses performed only the most basic tests (e.g., the fecal coliform test to assess water quality). There are certification programs to evaluate such laboratories, but the criteria on which water testing is judged may not be appropriate for evaluating bioaerosol analysis. The American Industrial Hygiene Association (AIHA, Fairfax, VA; Laboratory Accreditation Department: 703-849-8888) accredits laboratories for testing metals, asbestos, silica, and organic solvents to help investigators locate laboratories that have demonstrated their ability to analyze workplace samples accurately. In 1996, AIHA began the Environmental Microbiology Proficiency Analytical Testing (EMPAT) program with participants from the U.S. and Canada. The program is designed for laboratories specializing in detection of microorganisms commonly found in air, fluids, and bulk samples. Plans are that the EMPAT program will lead to an accreditation program for environmental microbiology laboratories; however, at the time of this publication, no laboratories have been accredited.

Until laboratory certification procedures are in place for laboratories analyzing environmental samples for biological agents, testing laboratories should participate in appropriate programs for performance evaluation or proficiency testing. If no such programs are available, laboratories should consider conducting inter-laboratory evaluations in collaboration with other researchers and commercial testing groups.

6.9.3 Quality Control in the Laboratory

Quality control during sample analysis in the laboratory is as important as quality control during sample collection in the field. Laboratory personnel should clearly understand an investigators record-keeping requirements and should be prepared to assume responsibility for samples placed in their custody [see 5.3]. The ability with which investigators can draw accurate conclusions from observations and measurements depends on the quality and defensibility of their data. Data defensibility involves the use of proper procedures (written protocols that include descriptions of the laboratory's quality control practices); protection of samples from inappropriate alteration; use of proper record collection, handling, and security; and accurate documentation of all sample-related information. Investigators often submit field blanks and duplicate samples to check the quality of sample handling in the field and sample analysis in the laboratory [see 5.3, 6.9, 7.2.2, and 13.4].

A laboratory is responsible for samples after they are received from the field, but not before that point. A laboratory may include a disclaimer in a test report to limit the laboratory's liability for inaccurate results or inappropriate use of the data they supply because of their lack of control over the method of sample collection, the age or condition of samples when received, or the limitations of the analytical methods used. The disclaimer may state that the field investigator is ultimately responsible for interpretation of test results because the laboratory personnel did not see the sampling site first hand and were not present when the samples were collected [see 2.1.4].

6.10 Summary

Many analytical methods are available to investigators to study biological agents in indoor environments. However, there clearly is still much to be accomplished to improve the assessment of indoor biological contaminants, demonstrate the pathways by which bioaerosols travel from sources to workers, and measure worker exposures. Many newer assay methods are useful for analyzing and characterizing microorganisms in bulk samples. However, some of these methods do not have the sensitivity required to analyze air samples. Some methods may offer sensitivity, but lack specificity, whereas others may be both sensitive and specific, but not quantitative. At the time of this publication, it is still very much the task of a field investigator or expert consultant to assess each unique situation and determine, first, if sampling is needed and, if so, which sample collection and analysis methods would provide the best answers to the questions the circumstances pose. Therefore, it is important that field investigators clearly communicate their specific needs and goals to the laboratory personnel with whom they work. Precise objectives will help ensure design of the best possible study, development of suitable protocols for sample collection and analysis, optimal use of resources, and accurate interpretation of the findings.

6.11 References

AIHA: Viable Fungi and Bacteria in Air, Bulk, and Surface Samples. In: Field Guide for the Determination of Biological Contaminants in Environmental Samples, pp. 37-74. H.K. Dillon, P.A. Heinsohn, and J.D. Miller, Eds. American Industrial Hygiene Association, Fairfax, VA (1996a).

AIHA: Legionella Bacteria in Air and Water Samples. In: Field Guide for the Determination of Biological Contaminants in Environmental Samples, pp. 97-117. H.K. Dillon, P.A. Heinsohn, and J.D. Miller, Eds. American Industrial Hygiene Association, Fairfax, VA (1996b).

Alvarez, A.J.; Buttner, M.P.; Toranzos, G.A.: Use of Solid-Phase PCR for Enhanced Detection of Airborne Microorganisms. Appl. Environ. Microbiol. 60(1):374-376 (1994).

Amann, R.I.; Ludwig, W.; Schleifer, K.: Phylogenetic Identification and in situ Detection of Individual Microbial Cells without Cultivation. Microbiological Reviews. 59(1):143-169 (1995).

APHA: Methods for the Examination of Water and Wastewater, 19th Ed. American Public Health Association, American Water Works Association and Water Pollution Control Federation, Washington, DC (1995).

Atlas, R.M.; Bartha, R.: Microbial Communities and Ecosystems. In: Microbial Ecology: Fundamentals and Applications, pp. 130-162. Benjamin/Cummings Publishing Co., Inc., Boca Raton, FL (1993).

Atlas, R.M.: Bartha, R.: Measurement of Microbial Numbers, Biomass, and Activities. In: Microbial Ecology: Fundamentals and Applications, pp. 162-211. Benjamin/Cummings Publishing Co., Inc., Boca Raton, FL (1993).

Atlas, R.M.; Bartha, R.: Statistics in Microbial Ecology. In: Microbial Ecology: Fundamentals and Applications, pp. 485-505. Benjamin/Cummings Publishing Co., Inc., Boca Raton, FL (1993).

Atlas, R.M.: Handbook of Microbiological Media. CRC Press, Boca Raton, FL (1993).

ASM: Manual of Environmental Microbiology. C.J. Hurst, G.R. Knudsen, M.J. McInerney, et al., Eds. American Society for Microbiology, Washington, DC (1997).

Krieg, N.R.; Hold, J.G., Eds.: Bergey's Manual of Systematic Bacteriology. Williams and Wilkins, Baltimore, MD (1984).

Buttner, M.P.; Willeke, K.; Grinshpun, S.A.: Sampling and Analysis of Airborne Microorganisms. In: Manual of Environmental Microbiology, pp. 629-640. C.J. Hurst, G.R. Knudsen, M.J. McInerney, et al., Eds. American Society for Microbiology, Washington, DC (1997).

Chapin, K.: Clinical Microscopy. In: Manual of Clinical Microbiology, 6th Ed., pp. 33-51. P.R. Murray, Ed. American Society for Microbiology, Washington, DC (1995).

Fox, A.; Gilbart, J.; Morgan, S.L.: Analytical Microbiology Methods, p. 1. A. Fox, S.L. Morgan, L. Larsson, et al., Eds. Plenum Press, New York, NY (1990).

Gao, S.-J.; Moore, P.S.: Molecular Approaches to the Identification of Unculturable Infectious Agents. Emerg. Infect. Dis. 2:159-167 (1996).

Garland, J.L.; Mills, A.L.: Classification and Characterization of Heterotrophic Microbial Communities on the Basis of Patterns of Community-Level Sole-Carbon-Source Utilization. Appl. Environ. Microbiol. 57:2351-2359 (1991).

Heidelberg, J.F.; Shahamat, M.; Levin, M.; et al.: Effect of Aerosolization on Culturability and Viability of Gram-Negative Bacteria. Appl. Environ. Microbiol. 63:3585-3588 (1997).

Hensel, A.; Petzoldt, K.: Biological and Biochemical Analysis of Bacteria and Viruses. In: Bioaerosols Handbook, pp. 335-360. C.S. Cox and C.M. Wathes, Eds. Lewis Publishers, Boca Raton, FL (1995).

IOM: Assessing Exposure and Risk. In: Indoor Allergens: Assessing and Controlling Adverse Health Effects, pp. 185-205. A.M. Pope,

R. Patterson, and H. Burge, Eds. Institute of Medicine, National Academy Press, Washington, DC (1993).

Jensen, P.A.; Todd, W.F.; Davis, G.N.; et al.: Evaluation of Eight Bioaerosol Samplers Challenged with Aerosols of Free Bacteria. Am. Ind. Hyg. Assoc. J. 53:660-667 (1992).

Juozaitis, A.; Willeke, K.; Grinshpun, S.A.; et al.: Impaction onto a Glass Slide or Agar Versus Impingement into a Liquid for the Collection and Recovery of Airborne Microorganisms. Appl. Environ. Microbiol. 60:861-870 (1994).

Kildesø, J.; Nielsen, B.H.: Exposure Assessment of Airborne Microorganisms by Fluorescence Microscopy and Image Processing. Ann. Occup. Hyg. 41:201-216 (1997).

Lawrence, J.R.; Korber, D.R.; Wolfaardt, G.M.; et al.: Analytical Imaging and Microscopy Techniques. In: Manual of Environmental Microbiology, pp. 29-51. C.J. Hurst, G.R. Knudsen, M.J. McInerney, et al., Eds. American Society for Microbiology, Washington, DC (1997).

Macnaughton, S.J.; Jenkins, T.L.; Alugupalli, S.; et al.: Quantitative Sampling of Indoor Air Biomass by Signature Lipid Biomarker Analysis: Feasibility Studies in a Model System. Am. Ind. Hyg. Assoc. J. 58:270-277 (1997).

Madelin, T.M.; Madelin, M.F.: Biological Analysis of Fungi and Associated Molds. In: Bioaerosols Handbook, pp. 361-386. C.S. Cox and C.M. Wathes, Eds. Lewis Publishers, Boca Raton, FL (1995).

Mastorides, S.M.; Oehler, R.L.; Greene, J.N.; et al.: Detection of *Mycobacterium tuberculosis* by Air Filtration and Polymerase Chain Reaction. Clin. Infect. Dis. 25:756-757 (1997).

Morris, K.J.: Modern Microscopic Methods of Bioaerosol Analysis. In: Bioaerosols Handbook, pp. 285-316. C.S. Cox and C.M. Wathes, Eds. Lewis Publishers, Boca Raton, FL (1995).

Mukoda, T.J.; Todd, L.A.; Sobsey, M.D.: PCR and Gene Probes for Detecting Bioaerosols. J. Aerosol Sci. 25(8):1523-1532 (1994).

Palmer, C.J.; Bonilla, F.G.; Roll, B.; et al.: Detection of Legionella Species in Reclaimed Water and Air with the EnviroAmp Legionella PCR Kit and Direct Fluorescent Antibody Staining. Appl. Environ. Microbiology 61(2):407-412 (1995).

Pendergrass, S.M.; Jensen, P.A.: Application of the Gas Chromatography-Fatty Acid Methyl Ester System for the Identification of Environmental and Clinical Isolates of the Family *Micrococcaceae*. Appl. Occup. Environ. Hyg. 12:543-546 (1997).

Podzorski, R.P.; Pershing, D.H.: Molecular Detection and Identification of Microorganisms. In: Manual of Clinical Microbiology, pp. 130-157. P.R. Murray, E.J. Baron, M.A. Pfaller, et al., Eds. American Society for Microbiology, Washington, DC (1995).

Stahl, D.A.: Molecular Approaches for the Measurement of Density, Diversity, and Phylogeny. In: Manual of Environmental Microbiology, pp. 102-114. C.J. Hurst, G.R. Knudsen, M.J. McInerney, et al., Eds. American Society for Microbiology Washington, DC (1997).

Stewart, S.L.; Grinshpun, S.A.; Willeke, K.; et al.: Effect of Impact Stress on Microbial Recovery on an Agar Surface. Appl. Environ. Microbiol. 61:1232-1239 (1995).

Terzieva, S.; Donnelly, J.; Ulevicius, V.; et al.: Comparison of Methods for Detection and Enumeration of Airborne Microorganisms Collected by Liquid Impingement. Appl Environ Microbiol. 62:2264-2272 (1996).

White, D.C.; Pinkart, H.C.; Ringelberg, D.B.: Biomass Measurements: Biochemical Approaches. In: Manual of Environmental Microbiology, pp. 91-101.. C.J. Hurst, G.R. Knudsen, M.J. McInerney, et al., Eds. American Society for Microbiology, Washington, DC (1997).

Chapter 7

Data Interpretation

7.1 Introduction
7.2 Rationale and Assumptions
    7.2.1   *Data Collection*
    7.2.2   *Data Quality*
    7.2.3   *Data Analysis*
    7.2.4   *Data Interpretation*
7.3 Approaches to Data Interpretation
    7.3.1   *Dealing with Visible Microbial Growth*
    7.3.2   *Comparison with Existing Standards or Guidelines*
7.4 Interpretation of Data on Biological Agents
    7.4.1   *Bacteria*
            7.4.1.1   Bacteria from Outdoor Air
            7.4.1.2   Human-Source Bacteria
            7.4.1.3   Gram-Negative Bacteria (GNB)
            7.4.1.4   Significant or Substantial Health Risk
    7.4.2   *Fungi*
            7.4.2.1   Indoor/Outdoor Comparisons
            7.4.2.2   Indicator Species
            7.4.2.3   Potentially Pathogenic (Infectious) Fungi
    7.4.3   *Amebae*
    7.4.4   *Viruses*
    7.4.5   *Environmental/Allergens*
            7.4.5.1   House Dust Mite Allergens
            7.4.5.2   Fungal Allergens
            7.4.5.3   Bacterial Antigens
            7.4.5.4   Cockroach Allergens
            7.4.5.5   Cat and Dog Allergens
    7.4.6   *Endotoxin*
    7.4.7   *Peptidoglycan (PG)*
    7.4.8   *Fungal Toxins*
    7.4.9   *β-(1→3)-D-Glucans*
    7.4.10   *Microbial Volatile Organic Compounds (MVOCs)*
7.5 Summary
7.6 References

7.1 Introduction

Data interpretation should be considered when developing an investigation strategy and designing a sampling plan. In turn, investigation strategies and sampling plans evolve after investigators have compiled enough information to formulate hypotheses to explain available facts and have determined that environmental sampling or an epidemiological study are required. Ideally, investigators establish the criteria by which they will judge samples for biological agents before collecting them. Criteria for interpreting environmental sampling results can be drawn from reference documents and research publications. Identifying the criteria by which data will be judged can help investigators determine the usefulness of the data they propose collecting. Anticipating potential test outcomes can help investigators determine if they have sufficiently clear and defensible criteria for judging test results and if the data they propose collecting will provide a firm basis for further decisions that may need to be made.

7.2 Rationale and Assumptions

7.2.1 Data Collection

"Data" may consist of the simple observation of fungal growth on a wall, analytical measurements from hundreds of environmental samples, or the results of a survey of building occupants with and without particular building-related conditions. During building evaluations, investigators may collect observational data, data on building performance, and bioaerosol data. Investigators gather observational data when they see visible sources of biological agents or observe environmental conditions that may allow the growth of microorganisms or arthropods. To collect information about a building's performance, investigators often measure such environmental parameters as temperature, humidity, air movement, and pressure. Investigators may collect air or source samples to gather bioaerosol data.

7.2.2 Data Quality

The ability of qualified persons to draw accurate conclusions from observations and measurements depends in large part on the quality and reliability of the original data. Some of the criteria used to judge data quality are (a) representativeness, (b) completeness, (c) reproducibility, (d) accuracy, (e) precision, and (f) integrity. Data are not useful if their quality cannot be assessed because investigators cannot be sure what weight to give the information. For example, the interpretation of a single air sample is difficult without information on the variability of the concentrations of biological agents in the sampled environment. One of the greatest difficulties with design and interpretation of bioaerosol studies is the fact that variability in the measurement of airborne contaminants is almost always large. Most approaches to bioaerosol sampling involve collecting short-term, grab samples which, alone, provide no information on the variability of the concentration of the agent being measured. Some bioaerosol samplers allow time-discriminated or time-weighted-average sampling. However, such sampling is uncommon in indoor environments because of the longer sampling times involved (e.g., hours to days rather than minutes). Among the most difficult decisions investigators face is identifying what environmental data they need and then determining if they have the time and resources to obtain data of sufficient quality to meet their needs. If adequate means are not available, investigators must decide if they will collect data of lesser quality and trust that it will be interpretable in some meaningful way. Alternatively, investigators may decide not to collect environmental samples using instead other information (e.g., observational data) to support their conclusions and recommendations.

7.2.3 Data Analysis

Data analysis is the step where an investigator (a) checks and validates data (i.e., reviews it for consistency and to identify particularly striking observations), (b) reduces or summarizes data (i.e., compiles the results in tables or figures and calculates summary statistics), and (c) estimates the data's reliability and looks for statistically significant associations and differences. The ultimate goal in data analysis is to test the validity of the hypotheses the investigators have formulated about the study site. For example, in an HP outbreak, an investigation team may have hypothesized that disease is related to exposure to thermophilic actinomycetes. The investigation team gathers environmental data in areas with and without HP cases and analyzes the data to decide if there is a demonstrable and significant difference between the case and control environments. This information may help the investigators decide how to improve the work environment and reduce workers' risk for HP.

Investigators should carefully consider the criteria by which they will judge environmental sampling data and information from epidemiological surveys. It may be difficult to demonstrate statistically significant differences in relative concentrations of biological agents or prevalence rates for specific symptoms. Therefore, investigators should not allow failure to demonstrate statistical significance to persuade them to ignore suspected hazards. Conversely, statistical significance does not automatically imply relevance or importance in the real world in terms of worker health and safety. There may be real differences between the types and relative concentrations of biological agents in environments or the rates of symptom reporting in groups of workers. However, these differences may not indicate that one environment is less safe than the other or that one group of workers is being affected by a biological agent or other hazard in their environment. The challenge for an investigator is to apply knowledge, expert advice, training, and experience when analyzing data and to recognize the strengths and limitations of various types of data and methods of data analysis.

7.2.4 Data Interpretation

Data interpretation is the step where investigators make decisions on (a) the relevance to human exposure of environmental observations and measurements, (b) the strength of associations between exposure and health status, and (c) the probability of current or future risks. These interpretation steps are followed by decisions on what measures can be taken to interrupt exposure and prevent future problems. Table 7.1 summarizes these steps for three examples. Case A illustrates the interpretation of data that is primarily observational. Case B presents air sampling data collected for an evaluation of a single worker with HP, who may be a sentinel case indicating a need to evaluate co-workers. Case C discusses the interpretation of environmental sampling data when epidemiological and medical data are also available.

For the examples in Table 7.1, several questions, possible interpretations, and alternative outcomes are presented. The possible interpretations illustrate that a study may reach more than one conclusion and investigators should be prepared to handle whatever interpretation the data best support. Among the possible outcomes are "yes, this agent from this source is responsible for adverse health effects," or "no, there is no evidence that this environment is a source for disease-causing agents." Note that the statement "these data indicate the tested environment is safe" cannot be made. Data may support the assertion that an environment is "safe" (i.e., presents little or no risk of exposure to biological agents), but investigators cannot claim to have proven this.

7.3 Approaches to Data Interpretation

Figure 7.1 illustrates the factors necessary for an exposure to a biological agent to result in a response, for example, a symptom or a disease. Beneath each primary

TABLE 7.1. Example: The Interpretation Process for Several Kinds of Data

Data	Questions Addressed	Possible Interpretations	Possible Outcomes
A. Surface contamination (possibly fungal growth) observed in a building in which the occupants complain of allergic symptoms	What is the chance the observation is not true (i.e., that the material is not microbial growth or is not releasing bioaerosols)? How representative is the observation of the general environment?	The material is microbial contamination and is releasing bioaerosols. The growth is likely to be producing biological agents that are known to cause BRI. The growth is relevant to human exposure and existing disease.	Yes, the evidence appears to support these relationships. No, the evidence does not support these relationships. The evidence is insufficient to reach a conclusion.
B. Results of 100 air samples for culturable thermophilic actinomycetes in a workplace in which a case of HP has been confirmed	What is the nature of the variability within the data set? What are the qualitative characteristics of the data set (i.e., what kinds of bacteria are present; are any thermophilic actinomycetes present at unusual or unexpected concentrations)?	Thermophilic actinomycetes are present in sufficient quantity to cause disease. There are statistically significant patterns that could explain disease distributions. The data represent an unusual and undesirable exposure. The data satisfactorily explain the existence of disease.	Yes, the data represent an unusual exposure. Yes, the data explain the existence of disease. Yes, the data represent an unusual exposure, but do not explain disease. No, the data do not support these relationships. The data are insufficient to reach a conclusion.
C. Five documented cases of Legionnaires' disease; *Legionella pneumophila* recovered from a water source that may produce aerosols	What are the chances that five cases of Legionnaires' disease would occur by chance among the occupants of a building of this type? Is there commonality among the cases with respect to (i) *L. pneumophila* type or (ii) possible exposure? Do any of the *L. pneumophila* isolates from patients match those from environmental sources?	Water in an environmental source is supporting the *L. pneumophila* serogroup that was isolated from one or more patients. There is a logical pathway linking a suspected source and infected workers. The data support bacterial transmission from a specific environmental source. The epidemiological and environmental evidence is sufficient to link the outbreak to the building.	Yes, the infections are building related and the probable source has been identified. Yes, the infections are building related, but the source has not been identified. Yes, a source has been identified, but is not clearly associated with disease. No, the data do not support a building-related source for the apparent outbreak. The data are insufficient to reach a conclusion.

factor are related parameters that determine if the progression can continue and if the outcome will be a change in health status.

Ideally, documentation that an agent causes a particular disease includes identification of (a) an agent, (b) its immediate, local source and environmental reservoir, and (c) exposure sufficient to cause the response observed. Data may consist of the identification of a source (observational data), measurement of the concentration of a biological agent at the source (bulk sampling data), or measurement of the concentration of the agent in air at a particular time (air sampling data). To establish a clear connection between an exposure and an outcome, an investigator must work through all the steps in Figure 7.1. Unfortunately, data are seldom available on release of biological agents from sources, bioaerosol decay, airway deposition of bioaerosols, agent release to the body from inhaled and deposited particles, or the human factors determining host response. Further, so little is known about these factors for most biological agents that extrapolation from available general information to the specifics of a given case may be uncertain. However, good quality data from environmental investigations can be interpreted as being representative of environmental conditions, and extrapolation of these data to indicate exposure is also often possible, as discussed in the next sections.

7.3.1 Dealing with Visible Microbial Growth

Various authors have recommended that microbial growth in occupied interiors, in HVAC systems, and on building materials and furnishings, especially if extensive, should be avoided and that any contamination that exists should be removed and further contamination should be prevented (Samson et al., 1994; Health Canada, 1995; Maroni et al., 1995; ISIAQ, 1996). "Extensive" visible fungal growth has been defined as surface areas greater than 3 m^2 (32 ft^2) (NYCDH, 1993; Health Canada, 1995; ISIAQ, 1996).

It is well established that bioaerosols cause infectious and hypersensitivity diseases and that bioaerosols in the indoor environment may cause toxic effects, although data on inhalation exposure is limited. Therefore, it is reasonable to use indicators of environmental contamination as a basis for evaluating the need for improved maintenance or remediation in a preventive context. It is also reasonable to use this approach as a means of making decisions in response to outbreaks of BRIs and BRSs. However, caution should be taken in ascribing specific causal links, as described in Chapter 14. At the time of this publication, there is no scientific basis for applying specific exposure limits for concentrations of total or specific culturable or countable bioaerosols [see 1.2].

7.3.2 Comparison with Existing Standards or Guidelines

In the U.S., no federal agency has clear authority to regulate exposure to biological agents associated with BRIs. The OSHA General Duty Clause and Hazard Communication Standard have been used to resolve IEQ problems, for example, to protect remediation workers and building occupants during clean-up operations and to inform building occupants of probable exposure to significant amounts of potentially harmful biological agents (Morey, 1992). The situation may differ in other countries and investigators should be familiar with federal and local regulations relating to bioaerosol exposures.

Workplace exposure limits are based on epidemiological and measurement data (Vincent, 1995). Epidemiological studies examine dose–response relationships and lead to health-based exposure criteria. The dose needed to produce a given response must be related to many factors, for example, the air concentration of an agent, the time period workers are exposed to the agent, the deposition and retention of particles within the respiratory tract, the concentration of the active agent at the target tissue, and the potency of the agent. Measurement-based studies used to establish exposure criteria provide comparative or relative data for problem and control environments. From comparisons of the resulting data, exposures are derived that may be considered acceptable, tolerable, or unlikely to cause harm. If available, data from studies of controlled human or experimental animal exposures may also be considered in the establishment of limit values.

By far, comparison of an environmental measurement with an existing standard is the simplest method to interpret data. Providing data are collected in the manner that

ENVIRONMENT		HOST	
SOURCE → OF AGENT	EXPOSURE → TO AGENT	DOSE → OF AGENT	RESPONSE TO AGENT
Dissemination of agent	Air concentration of agent Time spent in environment	Breathing rate Particle deposition Release of biologically active agent	Host sensitivity/ susceptibility Adequacy of host defenses

FIGURE 7.1. Steps connecting a biological agent and a host response.

was used to establish a standard, a measurement and a standard can be compared directly and an immediate determination can be made as to whether the sample exceeded the limit. However, there are no standards that specify acceptable concentrations for airborne materials of biological origin other than those that have been studied in certain manufacturing environments [see 1.2]. Guidelines and recommendations are available for bioaerosol control in the commercial biotechnology industry and microbiology laboratories (USDHHS, 1993; WHO, 1993; Sayre et al., 1994), but these recommendations have little bearing on other indoor environments.

Although numerical guidelines and recommendations have been published (e.g., for total concentrations of fungi and bacteria), these criteria vary over orders of magnitude. No consensus, health-based guidelines exist nor are any likely to be developed until more data are available on dose–response relationships for specific agents and health outcomes and more baseline data have been collected from randomly selected environments. The reader is referred to Rao et al. (1996) for a review of guidelines on bioaerosol exposure, to Health Canada (1995) for a review of guidelines for mycological indoor air quality, and to Maroni et al. (1995) for a discussion of IEQ guidelines and standards. Other references provide recommendations for controlling moisture problems (ISIAQ, 1996), assessing and remediating indoor fungal growth (NYCDH, 1993), and interpreting environmental sampling data (AIHA, 1996). Some baseline data on typical, acceptable, or tolerable bioaerosol concentrations may result from the analysis of data such as that from the USEPA Building Assessment Survey and Evaluation (BASE) studies of non-problem office buildings (USEPA, 1994).

Until guidelines on acceptable concentrations of biological agents are developed for particular environments and human populations, it is imperative that investigators not invoke previously published numbers, most of which the original authors no longer support. Although, ACGIH has given numerical guidelines for data interpretation in earlier documents (ACGIH, 1986; 1987; 1989), at the time of this publication, ACGIH does not support any existing numerical criteria for interpreting data on biological agents from source or air samples in non-manufacturing work environments. Instead, ACGIH recommends gathering the best data possible and using knowledge, experience, expert opinion, logic, and common sense to interpret information and design control and remediation strategies.

> Although, ACGIH has previously published numerical guidelines, at the time of this publication, ACGIH does not support any existing numerical criteria for interpreting data on biological agents from source or air samples in non-manufacturing environments.

7.4 Interpretation of Data on Biological Agents

The following sections briefly describe interpretation of data on air or source samples for the primary biological agents to which workers in non-manufacturing occupations may be exposed. These sections are brief summaries of the data interpretation sections of the individual chapters in Part III, which provide more complete discussions for specific agents and references for the information included here.

7.4.1 Bacteria

Few guidelines or recommendations are available to interpret data on bacterial aerosols. The information obtained from environmental sampling for bacteria may help investigators determine if environmental bacteria are multiplying in a building and, if so, may identify their source. Sampling may also help investigators determine if human-source bacteria are present at concentrations that should cause concern and if the kinds and relative concentrations of bacteria identified indicate a risk to health.

7.4.1.1 Bacteria from Outdoor Air Some bacteria that are common in outdoor air may penetrate to building interiors and may also grow indoors. Decisions on whether identification of bacteria and bacterial agents in indoor air represents simple entry with outdoor air or actual growth in the indoor environment depend on the relative indoor and outdoor concentrations. The dominance of a single species may reflect indoor growth, whereas the presence of mixed bacterial species may reflect intrusion of outdoor air or shedding from building occupants. Actinomycetes (especially thermophilic ones) are relatively rare in indoor and outdoor air. Therefore, the presence of actinomycetes in indoor air may be considered an indication of an indoor source.

7.4.1.2 Human-Source Bacteria Human-source bacteria typically dominate indoor air in occupied buildings and are abundant in dust and on surfaces. Air concentrations of these bacteria depend on the number of persons present, their activity levels, the types of clothing they wear, and the ventilation rate to the space. Extremely high concentrations of human-source bacteria in low-activity spaces may indicate overcrowding or inadequate ventilation.

7.4.1.3 Gram-Negative Bacteria (GNB) A predominance of GNB in an indoor environment may suggest the presence of sources that would be considered unusual in clean, properly ventilated buildings. For example, bacteria typical of fecal contamination may be found where plumbing has leaked or toilet exhausts are improperly vented. However, finding a few of these bacteria mixed with normal human-source organisms is not a cause for concern. Contamination of standing water with GNB is common and may be a concern if the bacteria produce odors or can become aerosolized. Dominance of GNB in air may suggest the presence and aerosolization of contaminated water.

7.4.1.4 Significant or Substantial Health Risk Some information is available on concentrations of infectious bacteria that have been associated with disease outbreaks. For example, elevated concentrations of *Legionella* spp. in cooling towers have been linked epidemiologically with disease outbreaks. Likewise, tuberculosis transmission has been related to source strength and ventilation rate. However, little is known about the potential effects for healthy adults and children of exposure to the majority of bacteria recovered during routine air and source sampling. Often, it is only possible to establish that an unusual exposure situation does or does not exist with respect to a control environment.

7.4.2 Fungi

Many fungi produce allergens and some fungi produce toxins. Fungal growth in buildings is undesirable and may cause health problems for building occupants. Although it may be difficult to establish that exposure to fungal aerosols occurs or that exposure presents a hazard, indoor fungal growth is inappropriate and should be removed. Further, steps should be taken to correct conditions that led to fungal growth so that it does not recur. Visible contamination that is confirmed by source sampling to be fungal growth is evidence of indoor contamination. Air sampling (culture or spore-trap sampling) may also indicate indoor fungal growth but should be followed by inspection and source sampling to identify the location of fungal contamination.

In the presence of the inevitable background concentration, the challenge for environmental sampling is to detect indoor fungal growth or entry of fungal aerosols from sources near OAIs and to document the contribution of such sources to occupant exposure. Interpretation of possible indoor fungal exposure has been addressed using (a) indoor/outdoor total concentration ratios, (b) comparisons of the species compositions indoors and out, and (c) the presence of indicator species in the indoor environment.

7.4.2.1 Indoor/Outdoor Comparisons The concentration of fungi in indoor air typically is similar to or lower than the concentration seen outdoors. Exceptions are enclosed agricultural and other specialized environments (where indoor fungal concentrations may be much higher). Outdoor concentrations may exceed those measured indoors even where indoor fungal growth is obvious. If outdoor fungal concentrations are very high, indoor/outdoor concentration ratios for total fungi may be low, even in the presence of significant indoor growth. On the other hand, outdoor fungal concentrations may be reduced during times of snow cover or other conditions that suppress the release of fungal spores from outdoor sources, at which times, indoor measurements may be higher than those outdoors even in the absence of sig-

nificant indoor sources. Finally, if the variability of the data is high (which is common), extensive sampling may be required to establish that two locations differ. The species of fungi found in indoor and outdoor air typically are similar if outdoor air is the primary source for the fungi in indoor air. Comparisons of the species compositions of indoor and outdoor populations requires accurate identification of fungal species not simply identification to the genus level.

7.4.2.2 Indicator Species Fungi whose presence may indicate excessive moisture or a specific health hazard have been termed indicator organisms. Interpreting the presence or absence of an indicator species requires the ability to identify fungi to the species level and a knowledge of the prevalence of the indicator species in both indoor and outdoor environments. The mere presence of a few CFUs or spores of any fungus should be interpreted with caution. Identification of a particular fungus in an indoor environment does not allow an investigator to conclude that building occupants are being exposed to allergenic or toxic agents. Investigators should also recognize that fungi that have been named indicator species are not the only fungi of significance. Many fungi other than those specifically listed by various groups may cause problems for building occupants exposed through inhalation of aerosols or by other contact.

7.4.2.3 Potentially Pathogenic (Infectious) Fungi Some fungal pathogens should be assumed to be present when materials known to support their growth are found (e.g., *Histoplasma capsulatum* and *Cryptococcus neoformans* in bird and bat droppings). Removal of such materials should be conducted as if they contained pathogenic fungi. Disturbance of soil or other material that may contain fungal pathogens (e.g., compost containing *Aspergillus fumigatus* or material enriched with bird or bat droppings) should be conducted with consideration that occupants of neighboring buildings may be exposed if airborne fungal spores enter the buildings.

7.4.3 Amebae

Investigators should be aware that amebae can cause inhalation fever (e.g., humidifier fever), severe eye and wound infections, and fatal encephalitis, although these conditions are rare. The size of amebic trophozoites and cysts causes them to fall rapidly from air and greatly diminishes the risk of infection when these amebic forms are discharged as aerosols. Physicians should consider the presence of pathogenic amebae when a patient's infection does not respond to traditional antibiotic treatment. Potential environmental sources of amebae should be tested for pathogenic types if infections are identified. These tests should be conducted by laboratories with experience in the assay of pathogenic amebae. Matching of amebae from clinical specimens and environmental samples may implicate a water supply as the source of an infection.

7.4.4 Viruses

The number of virus particles required to cause infection in a susceptible individual is not known for most viruses. However, evidence suggests that for some viruses one particle may be sufficient to initiate infection. The data retrieved from environmental sampling is usually the number of viruses in a sample that are able to infect cultured cells or laboratory animals. This number is not necessarily the same as that needed to infect a human host. Therefore, data from such samples may indicate only (a) the presence and concentration of the virus, or (b) the probability of the absence of the virus. Investigators must establish their own criteria for interpreting such data with respect to risks for humans.

7.4.5 Environmental /Allergens

A lack of well-characterized, standardized reagents poses problems in the testing of environmental samples for the presence of many antigens (which stimulate the production of antibodies) or allergens (which are antigens associated with hypersensitivity disease). The precision of immunoassays for indoor allergens is rather poor and sample variation can be large. In addition, people have varying sensitivities to allergens. Therefore, investigators should not interpret too strictly comparisons of recommended allergen thresholds and actual environmental measurements.

7.4.5.1 House Dust Mite Allergens Limits for dust-mite allergen concentrations in settled dust have been proposed for residences. Exposure to dust containing 2 µg/g of group-1 mite allergen (roughly equivalent to 100 mites/g or 0.6 mg/g of guanine) is considered to increase the risk of dust-mite sensitization as well as the development of asthma and bronchial hyperreactivity in affected persons. Exposure to dust containing 10 µg/g of group-1 mite allergen or >2.5 mg/g of guanine represents an increased chance of acute asthma attacks. Epidemiological studies have found better correlations between symptoms and ELISA measurements of mite allergen than mite counts. However, measurement of dust allergen content on a per-gram basis may be misleading if diverse substrates are compared because the density of dust from various sources may differ markedly (cf. dust from bedding or upholstered furnishing, which contains lint and fine feathers, and carpet dust, which contains soil, sand, and coarse fibers).

7.4.5.2 Fungal Allergens Over 60 species of fungi are known to produce allergens that cause allergic rhinitis (hay fever) and asthma, and many more are probably also in-

volved. Purified and characterized allergens have been prepared for *Alternaria alternata, Aspergillus fumigatus,* and *Cladosporium herbarum* and a few other fungi. However, none of these allergens has been adequately tested for the analysis of environmental samples. To date, the limited analysis of fungal allergens in indoor environments precludes estimation of a risk level for symptom exacerbation or even determination of what is a high level. Researchers have also attempted to determine threshold concentrations of fungal spores associated with immunological sensitization or symptom development. While there may be threshold concentrations of fungal allergens below which no symptoms are experienced, this is probably not an absolute value but a gradient based on individual sensitivities and the kind of fungus.

7.4.5.3 Bacterial Antigens Exposure to bacterial allergens has been associated with work-related asthma, HP, humidifier fever, and a disease resembling ABPM. Like fungi, bacteria secrete enzymes that can act as allergens. *Bacillus* spp. are used to produce proteases that are added to laundry detergents for stain removal. ACGIH has adopted a ceiling limit for subtilisins (100% pure crystalline enzyme) of 0.00006 mg/m^3 (ACGIH, 1997). However, data on bacterial allergens is limited, many bacterial allergens are poorly characterized, and exposure to bacterial allergens in non-manufacturing indoor environments is interpreted on a case-by-case basis.

7.4.5.4 Cockroach Allergens Cockroaches are the only insects repeatedly recognized as a common source of indoor allergens. However, significant levels or thresholds of cockroach allergen exposure have not been determined. Any cockroach allergen detected in an environment (i.e., any amount above the LDL of the assay used) identifies an indoor area that places cockroach-allergic persons at some risk for symptoms and places allergy-prone (susceptible) persons at some risk for sensitization and symptom development.

7.4.5.5 Cat and Dog Allergens Cat and dog allergens have been detected in schools and offices where these animals were not kept, suggesting that persons with animal contact may carry allergens on their clothing. Moderate to high concentrations of cat allergen (≥1 µg *Fel d* 1/g dust) have been found, most often in the offices of persons who have cats at home. *Fel d* 1 concentrations above 8 µg/g have been proposed as a threshold for sensitization. Fewer persons appear to be sensitive to dog than to cat allergens, perhaps because more dogs are housed outdoors and owners may bathe dogs fairly frequently whereas cats generally groom themselves. Concentrations of dog allergen (*Can f* 1) sufficient to result in the development of symptoms have not been established. In one study, *Can f* 1 concentrations ranged from <0.3 to

23 μg/g in dust from houses without dogs and from 10 to 10^4 μg/g in those with dogs.

7.4.6 Endotoxin

Environmental endotoxin measurements from different laboratories are not necessarily comparable because the *Limulus* amebocyte lysate (LAL) assay is a comparative rather than analytical method and because of the lot-to-lot variation in LAL reagent sensitivity. At present, it appears that only data from one laboratory assaying samples from one environment while using one LAL lot can be considered entirely comparable. Experimental studies of human exposure to cotton dust and the majority of recent field studies suggest an endotoxin threshold for acute airflow obstruction in the range of 45 to 330 EU (endotoxin unit)/m³. Thus, most current data support a threshold for acute airflow obstruction within a 10-fold range. These values are generally 10 to 100 times the background endotoxin levels reported in the studies.

For the time being, the imprecision of the LAL assay over time and among laboratories makes it impossible to establish a TLV at a given endotoxin level. Therefore, ACGIH proposes a practical alternative approach for limiting endotoxin exposure using relative limit values. This approach is based on comparison of endotoxin activity levels in the environment in question with simultaneously determined background levels. Investigators must ensure that background endotoxin levels are determined under appropriate and representative conditions.

Relative limit values (RLVs) are proposed because of the observation that endotoxin levels between 10- and 100-fold higher than background are frequently associated with adverse health effects. Therefore, it should be assumed that endotoxin plays a role and action should be taken to reduce exposure when (a) there are health effects consistent with endotoxin exposure (e.g., fatigue, malaise, cough, chest-tightness, and acute airflow obstruction), and (b) endotoxin exposures exceed 10 times simultaneously determined, appropriate background levels. Thus, 10 times background is proposed as an RLV action level in the presence of respiratory symptoms. In environments with a potential for endotoxin exposure but no current complaints, endotoxin levels should not exceed 30 times the appropriate background. Thus, 30 times background is a maximum RLV in the absence of symptoms. When exposures exceed the RLV action level or maximum RLV, appropriate remedial actions should be taken. Other than as described here, it is premature to make recommendations regarding low-level endotoxin exposures such as may occur in homes and offices.

7.4.7 Peptidoglycan (PG)

Sampling and analysis for PG is an interesting area worthy of further research. However, there currently is no significant body of data relating airborne PG exposure to health effects. Therefore, guidelines for interpreting sampling results and recommendations regarding the significance of observed PG concentrations cannot be made.

7.4.8 Fungal Toxins

Investigators may encounter six indicators of potential exposure to mycotoxins:

1. Observation of fungal growth in an indoor environment.
2. In source samples, identification of fungi known to produce mycotoxins.
3. In air samples, identification of fungi known to produce mycotoxins.
4. Identification of mycotoxins in source samples.
5. Identification of mycotoxins in air samples.
6. Detection of biomarkers for mycotoxins or fungi in biological specimens (e.g., urine, breast milk, or blood).

The first of these provides the weakest evidence of possible mycotoxin exposure; the next five provide progressively stronger evidence, with the possible exception of biomarkers. Specific associations between any of these indicators and any illness cannot be made. However, all may indicate inappropriate conditions that should be corrected. See Section 7.4.2 for a discussion of the interpretation of data in the first three of these categories.

7.4.9 β-(1→3)-D-Glucans

Exposure to some concentration of β-(1→3)-D-glucans can be assumed whenever fungi are isolated from an environment. β-(1→3)-D-glucans have been proposed as agents possibly responsible for some cases of BRSs. However, it is difficult to imagine how BRSs could routinely be associated with such ubiquitous agents unless other factors also play a role. Even in problem environments, fungal concentrations (and presumably, although not studied, β-(1→3)-D-glucan concentrations) are likely lower than concentrations frequently encountered outdoors. Workers exposed to high concentrations of β-(1→3)-D-glucan (during drybagging of the compound produced for use as a food additive in processed cheese) showed no irritant symptoms. In studies that have noted associations between BRSs and glucans, it is possible that glucan measurements were acting as surrogates for exposure to other fungal products. It is also possible that exposure to a mixture of fungal products (including glucans) such as may occur in some indoor environments plays an important role in the development of BRSs.

7.4.10 Microbial Volatile Organic Compounds (MVOCs)

Studies indicate that many microorganisms produce VOCs some of which may also be released from non-microbial sources. A few MVOCs (e.g., 1-octen-3-ol, 3-methyl-1-butanol, 2-hexanone, and 2-heptanone) may originate specifically from fungi but not solely from any single fungus. Other MVOCs (e.g., ethanol, which microorganisms may produce in large amounts) have many additional non-microbial sources. The fraction of the total VOC burden in buildings that may arise from microorganisms is not known, but it can be expected to vary with the nature and extent of indoor microbial growth and the presence of other sources. The few available laboratory measurements indicate that microorganisms may contribute significant amounts of VOCs. However, at the time of this publication, no simple, rapid, inexpensive, and reliable method is available to identify the nature and extent of indoor microbial growth from measurements of indoor VOC concentrations. Also, no library is available that gives MVOC profiles for combinations of microbial species and substrates. Some degree of fungal and bacterial growth can be found in most indoor environments and humans also contribute VOCs. Researchers need to distinguish typical MVOC emissions from such background sources.

7.5 Summary

Chapter 2 describes how investigators gather information with which to formulate hypotheses, collect data to test these hypotheses, and draw conclusions from the information they have compiled. This chapter summarized key points in the interpretation of data on biological agents. The separate chapters in Part III elaborate on these points for the individual agents. Chapter 13 describes how data are handled to facilitate interpretation and Chapter 14 describes how investigators use experimental, epidemiological, and environmental data to establish cause–effect relationships for exposures and responses.

Of the types of data outlined in Figure 2.4 (environmental, bioaerosol, medical, and epidemiological data), this chapter primarily described interpretation of bioaerosol data. Investigators who need to interpret environmental data should review Chapters 4 and 10. Investigators may need to consult an engineer, architect, or HVAC specialist for advice and direction on conducting and interpreting tests of building performance. Investigators who need to interpret medical or epidemiological data should review Chapters 3 and 8. Investigators may also need to consult a physician, toxicologist, or epidemiologist for advice and direction on evaluating medical or epidemiological data.

7.6 References

ACGIH: Bioaerosols — Airborne Viable Microorganisms in Office Environments: Sampling Protocol and Analytical Procedures. Appl. Indust. Hyg. 1(4):R19-R23 (1986).

ACGIH: Guidelines for Assessment and Sampling of Saprophytic Bioaerosols in the Indoor Environment. Appl. Indust. Hyg. 2(5):R10-R16 (1987).

ACGIH: Guidelines for the Assessment of Bioaerosols in the Indoor Environment. American Conference of Governmental Industrial Hygienists. Cincinnati, OH (1989).

ACGIH: 1998 TLVs and BEIs. Threshold Limit Values for Chemical Substances and Physical Agents. Biological Exposure Indices. American Conference of Governmental Industrial Hygienists Cincinnati, OH (1998).

AIHA: Field Guide for the Determination of Biological Contaminants in Environmental Samples. H.K. Dillon, P.A. Heinsohn, and J.D. Miller, Eds. American Industrial Hygiene Association, Fairfax, VA (1996).

Health Canada: Fungal Contamination in Public Buildings: A Guide to Recognition and Management. Federal-Provincial Committee on Environmental and Occupational Health. Environmental Health Directorate, Ottawa, Ontario, Canada (1995).

ISIAQ: Control of Moisture Problems Affecting Biological Indoor Air Quality. TFI-1996. International Society of Indoor Air Quality and Climate, Ottawa, Ontario, Canada (1996).

Maroni, M.; Axelrad, R.; Bacaloni, A.: NATO's Efforts to Set Indoor Air Quality Guidelines and Standards. Am. Ind. Hyg. Assoc. J. 56:499-508 (1995).

Morey, P.: Microbiological Contamination in Buildings; Precautions During Remediation Activities. In: Proceedings ASHRAE IAQ '92, Environments for People. Atlanta, 1992, pp. 171-177. American Society of Heating, Refrigerating, and Air Conditioning Engineers. (1992).

New York City Department of Health, New York City Human Resources Administration, and Mount Sinai-Irving J. Selikoff Occupational Health Clinical Center: Guidelines on Assessment and Remediation of *Stachybotrys atra* in Indoor Environments. New York, NY (1993).

Rao, C.Y.; Burge, H.A.; Chang, J.C.S.: Review of Quantitative Standards and Guidelines for Fungi in Indoor Air. J. Air Waste Manage. Assoc. 46:899-908 (1996).

Samson, R.A.; Flannigan, B.; Flannigan, M.E.; et al., Eds.: Health Implications of Fungi in Indoor Environments. Elsevier, New York, NY (1994).

Sayre, P.; Burckle, J.; Macek, G.; et al.: Regulatory Issues for Bioaerosols. In: *Atmospheric Microbial Aerosols: Theory and Applications*, pp. 331-364. B. Lighthart and A.J. Mohr, Eds. Chapman & Hall, New York, NY (1994).

USDHHS: Biosafety in Microbiological and Biomedical Laboratories, 3rd Ed. HHS (CDC) Publication No. 93-8395. Public Health Service, Centers for Disease Control and Prevention and National Institutes of Health, Washington, DC (1993).

USEPA: A Standardized EPA Protocol for Characterizing Indoor Air Quality in Large Office Buildings. U.S. Environmental Protection Agency, Washington, DC (1994).

WHO: Laboratory Safety Manual, 2nd Ed. World Health Organization, Geneva, Switzerland (1993).

Vincent, J.H.: Standards for Health-Related Aerosol Measurement and Control. In: Aerosol Science for Industrial Hygienists, pp. 204-237. Elsevier, New York, NY (1995).

Chapter 8

Medical Roles And Recommendations

Cecile S. Rose with Kathleen Kreiss, Donald K. Milton, and Edward A. Nardell

8.1 Background
8.2 Approach to Patients with Building-Related Illness (BRI)
    8.2.1 *Hypersensitivity Diseases*
        8.2.1.1 Rhinitis and Sinusitis
        8.2.1.2 Asthma
        8.2.1.3 Hypersensitivity Pneumonitis (HP)
        8.2.1.4 Medical Management of Hypersensitivity Diseases
    8.2.2 *Inhalation Fevers*
        8.2.2.1 Humidifier Fever
        8.2.2.2 Pontiac Fever
    8.2.3 *Infections*
        8.2.3.1 Legionnaires' Disease
        8.2.3.2 Tuberculosis
        8.2.3.3 Viral Infections
        8.2.3.4 Immunization
8.3 Approach to Patients with Building-Related Symptoms (BRS)
    8.3.1 *Physiological Tests*
    8.3.2 *Management of Building-Related Symptoms*
8.4 Approach to Immunocompromised Building Occupants and Remediation Workers
8.5 Sentinel Health Event Followback of a BRI
    8.5.1 *Environmental Monitoring*
    8.5.2 *Active Case Finding*
    8.5.3 *Reportable Diseases*
8.6 Investigation of Possible BRI in a Problem Building
    8.6.1 *Case Definitions and Referral Criteria*
    8.6.2 *Physical Examinations*
    8.6.3 *Prospective Surveillance*
8.7 References

8.1 Background

Medical personnel can play important roles in identifying bioaerosol-related illnesses and responding to them. Medical expertise should be sought for (a) diagnosis and management of an individual worker with a possible building-related illness (BRI), typically building-related hypersensitivity disease, inhalation fever, or infection, and (b) investigations of populations of workers with building-related sympotns (BRSs). Medical personnel usually participate as members of a team that may include an IH, EHP, IEQ consultant, engineer, epidemiologist, or toxicologist. In the first circumstance, diagnostic confirmation of a BRI is a sentinel health event that raises concern for others with shared building exposures. Such events that involve serious morbidity, mortality, or risk to the community may require notification of public health authorities. In the second circumstance, investigation of symptomatic building occupants may require medical surveillance and/or diagnostic evaluation.

This chapter outlines the medical evaluation of possible bioaerosol-related conditions and provides guidelines for physicians and other health-care professionals on (a) identifying and managing BRIs and BRSs due to bioaerosol exposure, and (b) the appropriate roles of health-care professionals in investigating such complaints. The health effects of exposure to specific biological agents are also discussed in the individual chapters of Part III.

8.2 Approach to Patients with Building-Related Illness (BRI)

8.2.1 Hypersensitivity Diseases

Hypersensitivity (allergic) respiratory diseases are among the most common illnesses associated with bioaerosol exposures in the indoor environment. Untreated, hypersensitivity diseases can cause significant decrements in workers' productivity and quality of life and can even be life threatening. A complete medical history—including information on possible workplace, residential, and recreational exposures (Table 8.1)—is the first step in accurate diagnosis and management of hypersensitivity diseases (Menzies and Bourbeau, 1997). In addition, physicians use a number of diagnostic tools to assess the presence of hypersensitivity diseases caused by bioaerosols [see also 25.2.3].

A key factor in hypersensitivity diseases is that disease is not apparent with the first exposure to an antigen (allergen). A latency period is required, during which the immune system becomes sensitized to the antigen [see 25.2.1]. The sensitization process requires three factors (a) a genetic capability to respond to the antigen, (b) sufficient exposure to the antigen or to a combination of the antigen and an immunostimulating substance, and (c) time for the specific lymphocytes and production of effector molecules, such as antibodies [see 25.2.2].

TABLE 8.1. Components of an Occupational and Environmental History for Patients with BRI Possibly Due to Bioaerosol Exposures

Occupational History
- Chronology of current and previous occupations
- Description of job processes and specific work practices
- List of specific chemical, dust, and other aerosol exposures (e.g., grain dust; animal handling; food and plant processing; cooling towers, fountains, and other water sprays; and metalworking fluids)
- Review of Material Safety Data Sheets (MSDSs) to identify known chemical sensitizers and irritants
- Records or reports of industrial hygiene evaluations or environmental testing at the workplace
- Symptom improvement away from work or symptom increase with specific workplace exposures
- Presence of persistent respiratory or constitutional symptoms in exposed co-workers

Environmental and Residential History
- Pets and other domestic animals (especially birds and cats)
- Hobbies and recreational activities (especially those involving chemicals, feathers or fur, plant materials, and organic dusts)
- Presence of humidifiers, dehumidifiers, swamp coolers, clothes dryers vented indoors, and other humidity sources
- Use of hot tubs or saunas
- Leaking or flooding in a basement or attic
- Water damage to carpets or furnishings
- Visible fungal growth
- Feather pillows, comforters, or clothing

8.2.1.1 Rhinitis and Sinusitis In allergic rhinitis, inflammation of the nasal lining causes rhinorrhea (runny nose), congestion, itching, and sneezing. Allergic sinusitis—as a complication of allergic rhinitis—is associated with thickening of the mucosal lining, leading to purulent (puslike) nasal and pharyngeal drainage and cough. Headache, facial pain, earache, sore throat, halitosis, and fever are also common. Rhinitis and sinusitis may be evaluated using rhinomanometry, radiography or CT of the nasal sinuses, and assessment of inflammatory cells in the nasal mucosa. Testing a person before and after exposure to an implicated environment may be useful. Sections 3.3.1 and 25.2.2.1, on allergic asthma, rhinitis, and sinusitis, also address these conditions.

Besides biological agents, chemical irritants can also cause inflammation of the nasal mucosa—leading to rhinitis—as well as irritation of the eyes and upper and lower respiratory tract. Chemical irritants may be inhaled as gases or mists or via particles to which they have adsorbed. Such irritation is generally reversible after short-term exposure. However, temporary tissue damage may make a worker susceptible to irritants that the person otherwise would tolerate. To identify and control exposures, it is necessary to consider all possible causes of nasal and sinus irritation, that is, biological (antigenic), chemical, and physical agents (Kipen et al., 1994).

8.2.1.2 Asthma Diagnosis of asthma is usually made on the basis of symptoms (wheezing, chest tightness, cough, and shortness of breath) and the presence of airway hyperreactivity [see also 3.3.2 and 25.2.2.1]. The latter response is measured by either tests of bronchial hyperresponsiveness or a positive reaction to a bronchodilator on pulmonary function testing. Exposure challenge in the suspect environment, with careful physiological assessment, is useful in determining whether asthma symptoms can be causally linked to a work-related exposure. Workers themselves can collect peak-flow measurements using small, portable instruments. Serial measurements may be useful to document airflow changes. A worker can take peak flow measurements throughout the day, both in and out of the implicated environment (i.e., at work during the day and at home in the evening and on weekends). Serial tests of lung function (such as spirometry measurements made by a trained technician) at the beginning and end of the work week can also be helpful [see 25.2.3.2]. Four patterns of lung function response to occupational exposures are recognized (i.e., equivalent daily decline, weekly decline, first day only decline, and persistent obstruction) (Burge, 1993). A pulmonologist or occupational medicine physician experienced in diagnosis of occupational asthma is often required to make such a diagnosis. In some cases it may be necessary to remove an individual from a work environment before improvement in lung function occurs.

Less informative are laboratory challenge and immunological tests. Laboratory challenge with a suspect antigen is poorly standardized, difficult to interpret, and usually unnecessary to confirm the diagnosis of building-related asthma. Skin prick testing for specific IgE antibodies to common indoor allergens confirms the allergic status of an individual but is seldom helpful in diagnosing building-related asthma. Section 25.2.3.1 describes immunological tests.

As mentioned for rhinitis in Section 8.2.1.1, asthma may also have chemical and physical causes in addition to biological (antigenic) ones (Bernstein, 1997). Therefore, to identify and control asthma-related exposures, it is necessary to consider all possible work-related agents. Skin testing may be useful for diagnosis of allergen-induced asthma but not for identification of irritant-induced asthma.

8.2.1.3 Hypersensitivity Pneumonitis (HP) Clinicians often base the diagnosis of HP on chest radiographs or CT scans, tests of lung function, the finding of specific serum precipitating antibodies, and a history of exposure to materials known to be associated with HP. However, these tests may be normal in early disease, and antibody testing may be negative when the responsible antigen is not among the standard panel of tested substances [see also 3.3.3 and 25.2.2.3]. Referral to a pulmonary disease subspecialist for exercise testing or fiberoptic bronchoscopy with bronchoalveolar lavage and transbronchial biopsy is usually necessary to confirm the diagnosis of HP and should be obtained when symptoms or exposures are compelling (Rose, 1996). Testing a person before and after exposure to an implicated environment may also be useful in the acute form of HP.

8.2.1.4 Medical Management of Hypersensitivity Diseases The pharmacological treatment of individuals with hypersensitivity BRIs does not differ from the management of patients with such illnesses due to causes unrelated to building exposures. Antihistamines, decongestants, intranasal corticosteroids, and nasal saline irrigation are the mainstays of treatment for allergic rhinitis and sinusitis. Chronic asthma is treated with inhaled steroids, bronchodilators, other inhaled medications with antiinflammatory effects, and oral corticosteroids for more severe or acute attacks. HP may resolve completely following removal from exposure without pharmacological treatment, although a course of oral corticosteroids is often used in circumstances of severe or persistent symptoms.

Environmental control strategies to eliminate antigen exposures are key to the medical management of building-related hypersensitivity diseases as these conditions may worsen with continued exposure despite aggressive pharmacological treatment. Chapter 25 on antigens discusses common indoor aeroallergens and measures to control them. Management of hypersensitivity diseases often includes temporarily removing individuals from the implicated environment while abatement efforts are underway or to assess symptom improvement [see 25.2.3.3]. Medical follow-up to assess response to treatment is important prior to building reoccupancy. The risks of building reoccupancy and disease reactivation may preclude an individual's return to an implicated building if assurance of adequate antigen abatement is difficult. This is especially true in HP cases where low-level exposure may cause disease progression in the absence of acute symptoms (i.e., damage may continue even though the patient does not notice it).

8.2.2 Inhalation Fevers

Inhalation exposure to microbially contaminated humidifiers, moldy grain dust during silo unloading, and aerosols containing legionella bacteria have been associated with similar, febrile, flu-like illnesses known collectively as inhalation fevers. Inhalation fevers are self-limited, and respiratory impairment is unusual. The syndromes are notable for very high attack rates among those heavily exposed, even if the persons previously were naive (unexposed) to the inhalant. The incubation period is typically short (e.g., a few hours to less than three days following exposure). Treatment is supportive, and antibiotics have no role because, even if microorganisms are involved, they are not acting as infectious agents. Exposure to the organic dusts seen in agricultural and related settings is uncommon in nonmanufacturing work environments. However, humidifier and Pontiac fevers may be seen in office, residential, and recreational settings.

8.2.2.1 Humidifier Fever Humidifier fever is an influenza-like illness with constitutional symptoms of fever, chills, headache, malaise, and, less prominently, respiratory symptoms [see also 3.4.1]. The syndrome occurs 4 to 12 hours after exposure to aerosols generated from air-conditioning or humidification systems. Humidifier fever may be associated with return to work after time away from a contaminated environment (so-called "Monday miseries"), and symptoms may improve over the work week. The specific pathogenic factors leading to humidifier fever remain to be identified, but the excessive growth of microorganisms within humidification systems seems to be common to all outbreaks. Implicated organisms include amebae (e.g., *Naegleria gruberi*), Gram-positive bacteria (GPB) (e.g., actinomycetes), Gram-negative bacteria (GNB) (e.g., *Pseudomonas* spp.), and bacterial endotoxins.

8.2.2.2 Pontiac Fever Airborne transmission of *Legionella* spp. from environmental sources can cause two forms of legionellosis, that is, Legionnaires' disease [see 8.2.3.1] and the nonpneumonic, flu-like illness known as Pontiac fever. An inhalation fever should be suspected when workers describe a flu-like illness without respiratory symptoms following exposure to aerosolized water. Several outbreaks of Pontiac fever have been identified retrospectively, including the one that gave the disease its name [see also 3.4.2]. The first recognized outbreak occurred in a county health department building in Pontiac, Michigan, in July, 1968. This outbreak later was traced to the airborne spread of *Legionella pneumophila*, serogroup 1, from a contaminated air-conditioning system (Glick et al., 1978). Seroconversion to this legionella strain was shown in symptomatic persons but not in controls. Diagnostic confirmation of Pontiac fever rests on the finding of typical symptoms, symptom onset after a short incubation period, short illness duration, absence of pneumonia, and demonstration of seroconversion.

8.2.3 Infections

Chapter 9 describes the transmission and control of airborne infectious diseases. Here issues related to infectious disease surveillance and treatment are discussed. Fungal infections are not discussed because these are rare relative to bacterial and viral infections and are unlikely to be seen in normal hosts as a consequence of occupying a building in which fungi are growing. Section 19.2.1 and other references (Rose, 1994) describe work-related fungal infections. Likewise, amebic infections are unusual and are covered in Section 20.2.

8.2.3.1 Legionnaires' Disease

Outbreaks of Legionnaires' disease (also called Legionnaires' pneumonia) are relatively uncommon, and most cases are believed to occur sporadically (WHO, 1990; Stout and Yu, 1997). However, Legionnaires' disease may be the most common airborne bacterial infection associated with environmental contamination of office or commercial buildings and one associated with mortality as well as morbidity. The incubation period for legionella pneumonia averages 5 to 6 days from exposure. One to five percent of exposed persons develop illness, and the case-fatality rate may exceed 15%. A person's risk of acquiring Legionnaires' disease following exposure to contaminated water depends on the type and intensity of exposure and the person's health status.

Diagnosis of Legionnaires' disease is based on clinical illness, pneumonia confirmed by chest radiograph, and laboratory evidence of recent legionella infection. A diagnosis of Legionnaires' disease may be confirmed by a positive finding on one or more of the following tests: (a) isolation of the organism from a clinical specimen; (b) demonstration of a four-fold or greater rise in serum antibody titer; (c) demonstration of the organism in a clinical specimen by direct fluorescent antibody stain; or (d) detection of antigen in urine (available only for *L. pneumophila*, serogroup 1) (CDC, 1997a). Legionnaires' disease responds well to specific antibiotic treatment (e.g., erythromycin).

Legionella bacteria are relatively common in natural and man-made water systems. Whirlpools, leisure pools, evaporative condensers, grocery store misters, and cooling towers have been implicated as sources in legionellosis outbreaks (Spitalny et al., 1984; Friedman et al., 1987; Fenstersheib et al., 1990). Besides inhalation of aerosols containing the bacteria, ingestion or aspiration of water containing high concentrations of legionellae has been linked with nosocomial transmission of Legionnaires' disease. Isolation of a *Legionella* sp. from an infected person (a clinical isolate) may facilitate identification of the probable source of work-related infection. Legionellae may be found in more than one water system in a facility, but evidence of causality requires that an environmental isolate match one from a patient (ASTM, 1996).

8.2.3.2 Tuberculosis

Tuberculosis (TB) is a contagious disease most often transmitted from person to person by droplet nuclei containing *Mycobacterium tuberculosis*. TB outbreaks have been documented in many indoor environments including health-care settings, prisons, shelters for the homeless, and following naval and air travel (Houk, 1980; CDC, 1989, 1995; Nardell et al., 1991). Transmission is most likely to occur from persons with unrecognized lung or laryngeal disease who are not on adequate therapy and have not been placed in TB isolation. Because individuals with multi-drug resistant TB (MDR-TB) may remain infectious for long periods, outbreaks of MDR-TB have heightened concern about the risk of nosocomial or occupational transmission. TB incidence increased in the U.S. in the 1980s and 1990s, in part related to the high risk for TB among immunosuppressed persons, particularly those infected with the human immunodeficiency virus (HIV).

Active TB should be treated with a combination of medications following the most recent treatment recommendations from the Centers for Disease Control and Prevention (CDC) (CDC, 1994a; www.cdc.gov). Prophylactic therapy should be considered in individuals with demonstrated tuberculin skin test (TST) conversion—a positive test after a history of negative tests—who do not have evidence of active disease. Employees with active pulmonary TB should be restricted from work until they are shown to be noninfectious. Co-workers and other close contacts of employees with active TB should be screened for TST conversion if there was a likelihood of exposure to aerosolized *M. tuberculosis*.

Typically, surveillance for TB is conducted by periodically skin testing high-risk employee groups (e.g., workers in direct contact with patients, clients, animals, or tissues known to have a high risk of infectivity). However, TSTs are neither sensitive nor specific when applied to low-risk populations (e.g., all persons in a building in which a person with active TB worked rather than just those likely to have had close or prolonged contact with the person). Unfortunately, defining contacts too loosely is a frequent response to administrative pressure to do something following the diagnosis of an active TB case in a work population. Such overtesting may result in unnecessary anxiety and needless preventive treatment. False-positive TST reactions can occur and positive reactions may be due to previous infection or vaccination rather than recent exposure to the sentinel case under investigation. Both prior infection and vaccination are common among workers from countries with a high TB prevalence (e.g., Mexico, southeast Asia, and Africa). OSHA and the CDC can provide information on current recommendations for TB prevention in high-risk and other settings (www.osha.gov; www.cdc.gov).

8.2.3.3 Viral Infections Airborne viral illnesses (e.g., influenza, common colds, measles, rubella, and varicella) are important causes of morbidity in the general population. The transmission of these infections may be increased where large numbers of persons live and work in close quarters in buildings with low outdoor air ventilation rates (e.g., military housing and jails) (Brundage et al., 1988; Richards et al., 1993; Hoge et al., 1994; Tappero et al., 1996). Often, acute viral infections may be considered community-acquired, rather than work-related, even when apparently contracted in office buildings or institutional settings, because the prevalence of these infections is so widespread. Antibiotic treatment is seldom indicated for these viral infections, although ill persons should be excused from work to limit transmission to others. Immunization is the primary means of protecting people from many common viral infections. In the military, vaccination against influenza and common adenoviruses has helped to prevent outbreaks. Increased ventilation may reduce the risk of airborne infections, but there are few studies supporting this assumption [see 10.5.2.1 and 9.8.6.3].

8.2.3.4 Immunization At a minimum, all adults should have current immunizations for the most common communicable diseases (CDC, 1991, 1994b; www.cdc.gov). The airborne or droplet-borne infections for which safe and effective vaccines are available are diphtheria, measles, mumps, pneumococcal diseases, rubella, and varicella. Routine influenza immunization has long been recommended for (a) workers at increased risk for exposure, (b) persons likely to experience complications from infection (e.g., adults with chronic pulmonary or cardiovascular disease and workers ≥65 years of age), and (c) persons who have contact with the latter (CDC, 1996a). However, immunization against influenza is not routine for all occupants of densely populated buildings with recirculated air.

8.3 Approach to Patients With Building-Related Symptoms (BRS)

Mucous membrane irritation of the eyes, nose, and throat,

fatigue, and headache associated with building occupancy but no identified specific cause are considered BRSs [see 3.2]. Clinical evaluation can be helpful in IEQ investigations to rule out allergic, toxic, irritant, or infectious diseases because many BRSs and symptoms of BRIs overlap, and the two syndromes may occur in the same building. Hodgson (1989) recommended an approach similar to the following to evaluate patients with BRSs: (a) identify the organ system with the most severe complaints (e.g., the eyes, nervous system, upper respiratory tract, or skin), (b) examine workers for physiological abnormalities, (c) if physiological abnormality is present, collect occupational and environmental histories to identify potential causes and conduct diagnostic tests, as required, (d) if no physiological abnormality is detected, recommend an engineering review of the building by a group knowledgeable in potential causes of BRSs, and (e) correct problems identified during the building evaluation.

8.3.1 Physiological Tests

A few investigators have identified physiological tests that correlate specific BRSs with measurable abnormalities (Franck, 1986; Morgan and Camp, 1986; Bascom, 1991; Kjaergaard et al., 1992; Koren et al., 1992). Research has provided insight into the potential mechanisms of some BRS complaints, but the tests used in these studies are not routinely available in clinical settings. Franck (1986) associated eye complaints (e.g., dryness, irritation, and inability to wear contact lenses) with decreased stability of the precorneal tear film, epithelial damage, and absence of foam in the eye canthus. Others have reported similar findings of corneal film breakup and inflammatory cells in tear fluid. Measurement of the sensory irritation threshold of the eyes to carbon dioxide may be a sensitive method for evaluating eye irritation complaints related to airborne pollutants (Kjaergaard et al., 1992).

Changes in nasal airflow impedance (i.e., obstruction to air flow) have been found in subjects exposed to vapors from carbonless copy paper, with increases in impedance at exposures below concentrations necessary to cause subjective irritation or congestion (Morgan and Camp, 1986). Nasal challenge, with analysis of cells and mediators in nasal lavage, may be useful to study responses to irritants in indoor air (Bascom, 1991). In one study, 14 subjects were exposed to clean air and low concentrations of a VOC mixture, followed by nasal lavage (Koren et al., 1992). After VOC exposure, but not clean-air challenge, there was a statistically significant increase in neutrophils, both immediately and 18 hours later. This finding suggested a physiological basis for the symptoms of nasal congestion and irritation that some occupants of problem buildings report. Section 26.3 discusses other health effects associated with VOC exposures and possible effects of MVOCs.

8.3.2 Management of Building-Related Symptoms

Long-term management of BRSs is frequently difficult. Medical treatment with pharmacological agents has not proven useful, other than on a symptomatic basis. Where warranted, reassurance by the treating physician that the annoyance complaints are unlikely to persist or lead to more serious illness is usually helpful to a patient. An employer may be willing to reassign a symptomatic worker to another location or job even in the absence of objective adverse effects, with the expectation that a happy employee and a supportive environment will result in greater productivity and less risk of a crisis situation. Unfortunately, many symptomatic workers without measurable physiological dysfunction are left on their own to develop management or avoidance strategies that enable them to continue working in a problem environment. A preferable approach is environmental assessment [see Chapter 4] and, if necessary, epidemiological evaluation of the potential problem building to identify strategies for remediation. Medical follow-up of affected individuals should take place once control strategies have been implemented.

8.4 Approach to Immunocompromised Building Occupants and Remediation Workers

Immunocompromised persons are at greater than normal risk for illness from exposure to opportunistic pathogens and may require special consideration. Individuals with hematological and other malignancies, HIV infection, diabetes mellitus, renal dysfunction, splenic disorders, alcoholism, cirrhosis, transplanted organs, and those receiving immunosuppressive medications (e.g., high-dose corticosteroids or chemotherapy) have variable degrees of immune system dysfunction (Macher and Rosenberg, 1999). Opportunistic fungal pathogens (e.g., *Aspergillus* spp. and other fungi able to grow at body temperatures) can cause invasive disease in immunocompromised hosts. These agents are common on plant and soil substrates and in outdoor air and are also present in most indoor environments (Arnow et al., 1991; Rose, 1994; Dixon et al., 1996; CDC, 1997b). Therefore, it is difficult to completely prevent exposure to opportunistic fungal pathogens.

Immunosuppressed workers at increased risk of infection and those for whom infection may have especially severe outcomes (e.g., pregnant workers) should be advised by their physicians about what types of activities to avoid and educated in techniques to limit their exposure to infectious agents in the workplace. For example, immunocompromised workers should avoid activities that may expose them to sprays of contaminated water (e.g., work near water-cooled heat-transfer equipment that generates water aerosols) or to airborne fungal or bacterial spores (e.g., decontamination and removal of indoor microbial growth or animal droppings) (CDC, 1997c).

These workers should know how to properly use PPE where indicated to limit bioaerosol exposure (Frazier et al., 1994). If potential exposure cannot be eliminated (e.g., TB exposure for a prison or shelter worker), consideration should be given to risk avoidance by temporary or permanent alternative job placement. Immunization may also be advised, although certain vaccines (especially live virus vaccines) may be contraindicated for immunocompromised persons (e.g., measles, mumps, rubella, varicella, and live polio and rabies vaccines) (Frazier et al., 1994; CDC, 1996b).

8.5 Sentinel Health Event Followback of a BRI

The clinical diagnosis of a bioaerosol-related hypersensitivity illness or certain infectious diseases in an individual worker constitutes a sentinel health event and should trigger consideration of an environmental investigation (Rutstein et al., 1983). The recognition of disease in one person may suggest that others who share the same work, residential, or recreational environment may also be at risk (Reingold, 1998). In this circumstance, appropriate intervention extends beyond the individual patient to encompass the at-risk environment and its occupants, with the goals of identifying other cases of disease, providing them appropriate treatment, and eliminating shared hazards. Obtaining a careful occupational and environmental exposure history (Table 8.1) from an affected individual can help direct a subsequent environmental investigation and is part of the hypothesis testing process discussed in Chapter 2.

It is essential to approach an investigation with a team of experienced experts (e.g., physicians, epidemiologists, toxicologists, engineers, IHs, EHPs, and IEQ consultants). Building managers, maintenance personnel, and building occupants themselves should be involved in investigations to ensure that risk factors are carefully sought and accurately identified. An initial inspection of an implicated building is often an early step, during which the investigators look for potential bioaerosol sources and assess the need to collect air or source samples to identify the presence of unusual concentrations or kinds of biological agents [see Chapter 4].

8.5.1 Environmental Monitoring

The decision to proceed with quantitative bioaerosol sampling is guided by the results of the initial worksite survey and the type of illness identified. Bioaerosol sampling to identify relevant exposures in a case of hypersensitivity lung disease may be helpful if positive, that is, if antigens to which a person is known to be sensitive are identified at elevated concentrations. However, while positive results may indicate exposure to potentially antigenic bioaerosols, evidence of exposure does not necessarily identify which of the agents that are present is the one responsible for a worker's hypersensitivity disease. There is little information on bioaerosol dose–response relationships, and few health-based standards exist for defining acceptable workplace concentrations of biological agents. An unusual exposure may occur when indoor bioaerosol concentrations are several orders of magnitude higher than those outdoors or in control environments or when the detected microbial genera and species differ markedly between test and control sites [see Chapter 7]. However, negative results are usually inconclusive and do not disprove exposure. Investigators should keep in mind that it is not possible to culture many infectious agents from the air or collect them in sufficient numbers for ready detection. Further, quantitative bioaerosol samples often must be obtained at least in duplicate and from multiple test and control sites [see 5.2.4]. It is also important to locate a laboratory experienced in recovering and identifying biological agents and one that uses recognized analytical techniques [see Chapter 6].

8.5.2 Active Case Finding
Investigation of a building-related sentinel health event may include active case finding via an epidemiological survey or clinical testing of building occupants [see also 3.7]. Standardized respiratory questions (Ferris, 1978) are sensitive screening tools and may be supplemented with additional questions about exposure, other symptoms, and the timing of symptom onset. Other standardized survey tools available for BRI investigations are the Asthma Questionnaire of the International Union Against Tuberculosis and Lung Disease (IUATLD, Paris, France) (Burney et al., 1989), the questionnaire used in the USEPA's Large Building study (USEPA, 1994; http://www.cdc.gov/niosh/ieqwww.html), and a questionnaire developed by Arnow et al. (1978). For investigating possible legionellosis outbreaks, a surveillance questionnaire and one for interviewing the treating physicians of suspected cases are also available (ASTM, 1996).

Plans should be made at the outset of an investigation for appropriate referral and diagnostic evaluation of potential cases identified through surveillance [see 8.6.1]. The Association of Occupational and Environmental Clinics (Washington, DC, 202-347-4976) and the American College of Occupational and Environmental Medicine (Arlington Heights, IL, 847-228-6850) may be able to identify local physicians available to assist investigators in evaluating potentially affected workers. The American Academy of Allergy, Asthma, and Immunology (Milwaukee, WI, 800-822-2762) and the American College of Allergy and Immunology (Arlington Heights, IL, 800-842-7777) can help investigators identify local allergists.

8.5.3 Reportable Diseases
Health professionals are required to report certain infectious diseases, including TB, Legionnaires' disease, and several of the airborne infectious diseases mentioned in

Chapter 9. The CDC publishes case definitions for 52 nationally notifiable infectious diseases (CDC, 1997a,d), that is, those for which regular, frequent, and timely information on individual cases is considered necessary for prevention and control (CDC, 1996c). Individual states can provide physicians with lists of the infections they must report to their local health department or state or territorial epidemiologist. Local, state, and federal public health departments can often assist in investigations of infectious disease outbreaks by providing laboratory and epidemiological support. Reporting of occupational hypersensitivity diseases is also required by law in some states (Matte et al., 1990).

The NIOSH Sentinel Event Notification System for Occupational Risks (SENSOR) is a cooperative state–federal effort designed to develop local capability for preventing selected occupational disorders (Baker, 1989). The SENSOR system consists of a network of sentinel providers (e.g., individual practitioners, laboratories, and clinics) in each participating state. The members of the provider network recognize and report cases of selected occupational disorders, including asthma and TB, suspected of being work related. The surveillance centers receive reports from and interact with the health-care providers, analyze the data, and direct intervention activities toward the individual cases, co-workers of these individuals, and the worksites from which the cases were reported. Modification of such a system may be appropriate in other settings for surveillance purposes. NIOSH (Surveillance Branch, Cincinnati, OH) can provide further information on this program and identify participating state programs.

8.6 Investigation of Possible BRI in a Problem Building
When studying a cluster of persons who have developed symptoms possibly related to bioaerosol exposure, the first investigative step is information gathering, as discussed in Chapter 2. These inquiries, typically involve (a) interviews with occupants as well as building managers, (b) a building inspection including examination of the HVAC system, and (c) a questionnaire survey of the at-risk population. This approach may directly suggest a cause of the problem and point to corrective actions. However, a more comprehensive evaluation may be needed, requiring medical evaluation of symptomatic workers and collaboration among medical professionals, IHs, EHPs, IEQ consultants, engineers, epidemiologists, and toxicologists.

Analysis of the type, frequency, and nature of reported symptoms often provides clues to what conditions may be present (e.g., mucous membrane irritation, hypersensitivity lung disease, inhalation fever, or infection). A medical investigation may be guided by an epidemiological survey and should be designed to (a) identify patterns in symptoms and diagnosed illnesses, (b) use clinical tools to evaluate the presence of BRIs, and (c) identify the individuals most affected, including those requiring immediate removal from exposure or referral for further diagnosis and treatment (Quinlan et al., 1989).

8.6.1 Case Definitions and Referral Criteria
Determination of epidemiological case definitions and criteria for referral of suspect cases for further medical evaluation should be part of investigation planning. For example, investigators would need a case definition for Legionnaires' disease to identify past cases during surveillance in a building where one or more workers were found to have been infected. The case definition for Legionnaires' disease may be "physician-diagnosed pneumonia with laboratory evidence of legionella infection" [see 8.2.3.1]. To determine if infections continue to occur, the investigators need criteria for referring building occupants for medical evaluation if they show signs of possible Legionnaires' disease. Prospectively, investigators could use "development of fever and nonproductive cough" as criteria for referring workers to a physician for diagnosis of legionella infection and appropriate treatment.

Commitment to confidentiality of worker responses and timely participant notification of their individual test results is essential when performing an epidemiological survey and during subsequent medical evaluation of workers with suspected BRIs. Once aggregate results are obtained, communication with building occupants who participated in an investigation is essential to ensure that their concerns have been identified and addressed.

8.6.2 Physical Examinations
Medical assessment of BRIs depends on the suspected problem, but physical examination is not a sensitive tool in most BRI cases. Therefore, personal physicians and their patients often find the process of diagnosing BRIs and identifying causes frustrating. If medical surveillance is part of an investigation, the physical examination should be guided by the results of clinical tests and the questionnaire survey. For example, reports of itching or rash related to building occupancy should prompt careful dermatological examination. A preponderance of chest symptoms, such as cough, wheeze, shortness of breath, and chest tightness, would suggest a focus on the cardiopulmonary system. Routine, screening laboratory tests (e.g., serum chemistries, complete blood counts, and sedimentation rates) are low-yield, nonspecific evaluations and seldom indicated. Specialized laboratory tests (e.g., measurement of precipitating antibodies to possible antigens that cause HP) are usually more specific but have limitations in terms of reliability and cost [see also 25.2.3.1].

8.6.3 Prospective Surveillance
It is often necessary to follow groups of workers prospectively if BRI has been identified in their workplace. Fol-

low-up questionnaires may be indicated to ensure adequacy of abatement efforts and to detect new cases. Newly identified potential cases should be referred for evaluation and treatment. Arrangements should be made to have on-site medical practitioners or local occupational health clinics provide these services. As with the original investigation, the ongoing surveillance should be integrated with any continuing environmental monitoring. Whenever possible, reaching a decision that clean-up or abatement efforts have succeeded should include an assessment of the acceptability of the environment by the workers who will occupy the space [see also 15.5].

8.7 References
Arnow P.M,; Fink, J.N.; Schlueter, D.P.; et al.: Early Detection of Hypersensitivity Pneumonitis in Office Workers. Am. J. Med. 64:236-242 (1978).

Arnow, P.M.; Sadigh, M.; Costas, C.; et al.: Endemic and Epidemic Aspergillosis Associated with In-hospital Replication of *Aspergillus* Organisms. J. Inf. Dis. 164:998-1002 (1991).

ASTM: Standard Guide for Inspecting Water Systems for Legionellae and Investigating Possible Outbreaks of Legionellosis (Legionnaires' Disease or Pontiac Fever). D 5952-96. American Society for Testing and Materials, West Conshohocken, PA (1996).

Baker, E.: Sentinel Event Notification System for Occupational Risks (SENSOR): The Concept. Am. J. Public Health. 79:S18-20 (1989).

Bascom R.: The Upper Respiratory Tract: Mucous Membrane Irritation. In: Environ Health Perspect 95, pp. 39-44 (1991).

Bernstein, D.I.: Allergic Reactions to Workplace Allergens. JAMA. 278:1907-1913 (1997).

Brundage, J.; Scott, R.; Ladnar, W.; et al.: Building-Associated Risk of Febrile Acute Respiratory Diseases in Army Trainees. JAMA. 259:2108-2112 (1988).

Burge, P.S.: Use of Serial Measurements of Peak Flow in the Diagnosis of Occupational Asthma. Occup. Med. 8:279-294 (1993).

Burney, P.G.J.; Chinn, S.; Britton, J.R.; et al.: What Symptoms Predict the Bronchial Response to Histamine? Evaluation in a Community Survey of the Bronchial Symptoms Questionnaire (1984) of the International Union against Tuberculosis and Lung Disease. Int. J. Epidemiol. 18:165-173 (1989).

CDC: *Mycobacterium tuberculosis* Transmission in a Health Clinic Florida. CDC — MMWR 256-264. Centers for Disease Control (1989).

CDC: Update on Adult Immunization: Recommendations of the Immunization Practices Advisory Committee (ACIP). CDC — MMWR 40(RR-12). Centers for Disease Control (1991).

CDC: Guidelines for Preventing the Transmission of *Mycobacterium tuberculosis* in Health-Care Facilities. CDC — MMWR 43(RR-131-132). Centers for Disease Control (1994a).

CDC: Recommendations of the Advisory Committee on Immunization Practices (ACIP): General Recommendations on Immunization. CDC — MMWR. 43(RR-1). Centers for Disease Control (1994b).

CDC: Exposure of Passengers and Flight Crew to *Mycobacterium tuberculosis* on Commercial Aircraft, 1992-1995. CDC — MMWR 44(137-140). Centers for Disease Control (1995).

CDC: Prevention and Control of Influenza. CDC — MMWR 45(RR-5). Centers for Disease Control (1996a).

CDC: Update: Vaccine Side Effects, Adverse Reactions, Contraindications, and Precautions: Recommendations of the Advisory Committee on Immunization Practices (ACIP). CDC — MMWR 45(RR-12). Centers for Disease Control (1996b).

CDC: Notifiable Disease Surveillance and Notifiable Disease Statistics — United States, June 1946 and June 1996. CDC — MMWR. 45(530-536). Centers for Disease Control (1996c).

CDC: Case Definitions for Infectious Conditions under Public Health

Surveillance.CDC — MMWR. 46(RR-10). Centers for Disease Control (1997a).

CDC: Guidelines for Prevention of Nosocomial Pneumonia. CDC — MMWR. 46(RR-1):34-37; 58-62. Centers for Disease Control (1997b).

CDC: 1997 USPHS/IDSA Prevention of Opportunistic Infections Working Group. Guidelines for the Prevention of Opportunistic Infections in Persons Infected with Human Immunodeficiency Virus. CDC — MMWR 46(RR-12). Centers for Disease Control (1997c).

CDC: Summary of Notifiable Diseases, United States, 1996. CDC — MMWR 45(53):1-87. Centers for Disease Control (1997d).

Dixon, D.M.; McNeil, M.M.; Cohen, M.L.; et al.: Fungal Infections, a Growing Threat. Public Health Reports 111:226-235 (1996).

Fensdersheib M.D.; Miller, M.; Diggins, C.; et al.: Outbreak of Pontiac Fever due to *Legionella anisa*. Lancet. 336:35-37 (1990).

Ferris, B.G.: Epidemiology Standardization Project. American Thoracic Society and Division of Lung Diseases of the National Heart, Lung and Blood Institute, Bethesda, MD (1978).

Franck, C.: Eye Symptoms and Signs in Buildings with Indoor Climate Problems (Office Eye Syndrome). Acta. Ophthalmol. 64:306-311 (1986).

Frazier, L.M.; Stave, G.M.; Tulis, J.J.: Prevention of Illness from Biological Hazards. In: Physical and Biological Hazards of the Workplace, pp. 257-270. P.H. Wald, G.M. Stave, Eds. Wiley, New York, NY (1994).

Friedman, S.; Spitalny, K.; Barbaree, J.; et al.: Pontiac Fever Outbreak Associated with a Cooling Tower. Am. J. Public Health 77:568-572 (1987).

Glick, T.H.; Gregg, M.B.; Berman, B.; et al.: Pontiac Fever. An Epidemic of Unknown Etiology in a Health Department: Clinical and Epidemiologic Aspects. Am. J. Epidemiol. 107:149-160 (1978).

Hodgson, M.J.: Clinical Diagnosis and Management of Building-Related Illness and the Sick Building Syndrome. Occup. Med. 4:593-606 (1989).

Hoge, C.W.; Reichler, M.R.; Dominguez, E.A.; et al.: An Epidemic of Pneumococcal Disease in an Overcrowded, Inadequately Ventilated Jail. N. Engl. J. Med. 331:643-648 (1994).

Houk, V.: Spread of Tuberculosis via Recirculated Air in a Naval Vessel: The Byrd Study. Ann. N.Y. Acad. Sci. 353:10-24 (1980).

Kipen, H.M.; Blume, R.; Hutt, D.: Asthma Experience in an Occupational and Environmental Medicine Clinic: Low Dose Reactive Airway Dysfunction Syndrome. J. Occup. Environ. Med. 36(10):1133-7 (1994).

Kjaergaard, S.; Pedersen, O.F.; Mølhave, L.: Sensitivity of the Eyes to Airborne Irritant Stimuli: Influence of Individual Characteristics. Arch. Environ. Health 47:45-50 (1992).

Koren H.S.; Graham, D.E.; Devlin, R.B.: Exposure of Humans to a Volatile Organic Mixture: Inflammatory Response. Arch. Environ. Health 47:39-44 (1992).

Macher, J.M.; Rosenberg, J.: Evaluation and Management of Exposure to Infectious Agents. In: Handbook of Occupational Safety and Health, pp. 287-371. L. DiBerardinis, Ed. Wiley, New York, NY (1999).

Matte, T.; Hoffman, R.E.; Rosenman, K.D.: Surveillance of Occupational Asthma under the SENSOR Model. Chest 98:173S-178S (1990).

Menzies, D.; Bourbeau, J.: Building-related Illnesses. New Engl. J. Med. 337:1524-1531 (1997).

Morgan, M.S.; Camp, J.E.: Upper Respiratory Irritation from Controlled Exposure to Vapor from Carbonless Copy Forms. J. Occup. Med. 28:415-419 (1986).

Nardell, E.; Keegan, J.; Cheney, S.A.; et al.: Theoretical Limits of Protection Achievable by Building Ventilation. Am. Rev. Respir. Dis. 144:302-306 (1991).

Quinlan, P.; Macher, J.M.; Alevantis, L.E.; et al.: Protocol for the Comprehensive Evaluation of Building-Associated Illness. In: Occup. Med. State Art Rev. 4:771-797 (1989).

Reingold, A.L.; Outbreak Investigations: A Perspective. Emerg. Infect. Dis. 4:21-27 (1998).

Richards, A.L.; Hyams, D.K.; Watts, D.M.; et al.: Respiratory Disease Among Military Personnel in Saudi Arabia during Operation Desert Shield. Am. J. Public Health 83:1326-1329 (1993).

Rose, C.S.: Fungi. In: Physical and Biological Hazards of the Workplace, pp. 392-416. P.H. Wald and G.M. Stave, Eds. Wiley, New York, NY (1994).

Rose, C.S.: Hypersensitivity Pneumonitis. In: Occupational and Environmental Respiratory Disease, pp. 201-215. P. Harber, M.B. Schenker, and J.R. Balmes, Eds. Mosby, St. Louis, MO (1996).

Rutstein, D.D.; Mullan, R.J.; Frazier, T.M.; et al.: Sentinel Health Events (Occupational): A Basis for Physician Recognition and Public Health Surveillance. Am. J. Public Health 73:1054-1062 (1983).

Spitalny, K.C.; Vogt, R.K.; Orciari, L.A.; et al.: Pontiac Fever Associated with a Whirlpool Spa. Am. J. Epidemiol. 120:809-817 (1984).

Stout, J.E.; Yu, V.L.: Legionellosis. New Engl. J. Med. 337:682-687 (1997).

Tappero, J.W.; Reporter, R.; Wenger, J.D.; et al.: Meningococcal Disease in Los Angeles County, California, Among Men in the County Jail. New Engl. J. Med. 335:833-840 (1996).

USEPA: A Standardized EPA Protocol for Characterizing Indoor Air Quality in Large Office Buildings. U.S. Environmental Protection Agency, Washington, DC (1994).

WHO: Epidemiology, Prevention and Control of Legionellosis: Memorandum from a WHO Meeting. WHO Bull. OMS. 68:155-164,

Chapter 9

Respiratory Infections — Transmission And Environmental Control

Edward A. Nardell and Janet M. Macher

9.1   Introduction
9.2   Reservoirs of Infectious Agents
    9.2.1   *Animal and Environmental Sources of Infectious Agents*
    9.2.2   *People as Sources of Infectious Agents*
9.3   Pathogenesis of Respiratory Infections
    9.3.1   *Infectious Dose*
    9.3.2   *Infectivity and Susceptibility*
    9.3.3   *Relative Hazards of Infectious Agents*
    9.3.4   *Inhalation and Fate of Airborne Infectious Agents*
    9.3.5   *Initiation of Infection*
    9.3.6   *Signs and Symptoms of Respiratory Infections*
9.4   Modes of Transmission for Respiratory Infections
    9.4.1   *Transmission Via Droplets and Droplet Nuclei*
        9.4.1.1   Droplet or Contact Transmission
        9.4.1.2   Airborne Infectious Diseases
    9.4.2   *Transmission Mode and Control of Respiratory Diseases*
9.5   Determinants of Airborne Infections
9.6   Concentrations of Infectious Droplet Nuclei
    9.6.1   *Tuberculosis Ward Study*
    9.6.2   *Variability of Infectious Agent Generation Rates*
9.7   Airborne Transmission Within Buildings
    9.7.1   *Near-Field and Far-Field Exposures*
    9.7.2   *Dissemination of Infectious Aerosols Within Buildings*
9.8   Control of Airborne Infection
    9.8.1   *Number of Infectious Sources*
    9.8.2   *Generation Rates of Infectious Aerosols*
    9.8.3   *Exposure Times*
    9.8.4   *Number of Exposed Persons*
        9.8.4.1   Altering Susceptibility Status
        9.8.4.2   Preventing Exposure
    9.8.5   *Rate at Which Workers "Sample" Air by Breathing*
    9.8.6   *Particle Removal Rate*
        9.8.6.1   Die-Off of Airborne Infectious Agents
        9.8.6.2   Deposition of Infectious Aerosols
        9.8.6.3   Removal of Infectious Aerosols Via Ventilation
        9.8.6.4   Air Cleaners and Air Disinfection for Preventing Infectious Diseases
9.9   Summary
9.10 References

9.1 Introduction

The discussion that follows is relevant to many types of indoor environments, but focuses on infectious disease transmission in workplaces such as office buildings. Specialized facilities associated with increased infection risks include microbiology laboratories, biotechnology industries, health-care settings, animal husbandry and veterinary-care facilities, agriculture and forestry operations, social service agencies, detention and correctional facilities, and shelters for the homeless (Gantz, 1988; Garibaldi and Janis, 1992; Ellis and Symington, 1994; Gerberding and Holmes, 1994; Wald and Stave, 1994; Collins et al., 1997; Macher and Rosenberg, 1999). We do not explicitly address these specialized facilities, but the principles of infectious agent transmission and control are similar for all settings. Related discussions appear in Sections 3.5 and 8.2.3.

Work-related infections almost certainly are under-reported. Reasons for under-reporting are that many infections are so familiar they do not attract attention (e.g., the common cold is the most frequent cause of work absence in the U.S.) and other infections may not be recognized as work related (Gerberding, 1996). Employees may hesitate to report such incidents for fear of losing their jobs, and employers may ignore them for reasons of liability. Investigation of possibly building-related infections is addressed in Sections 3.7, 8.5, and 8.6.

9.2 Reservoirs of Infectious Agents

The work-related infections discussed here are those spread by inhalation of airborne microorganisms rather than those acquired through ingestion, the mucosa, or skin penetration. Infectious agents may reside in animals, the environment, or people and be transported from these sources to building occupants.

9.2.1 Animal and Environmental Sources of Infectious Agents

People and animals share some zoonotic infectious diseases, which are important in certain occupations (e.g., veterinary care and animal husbandry) but are seldom of concern for workers in office buildings and similar settings. Examples of infectious agents transmitted from the environment to humans are Legionnaires' disease (caused by bacteria growing in water) and mycoses (caused by saprobic fungi growing on organic matter in the environment). Fatal outbreaks of Legionnaires' disease have occurred in office buildings, but appear to be relatively uncommon. However, some hospitals have reported Legionnaires' disease as a common cause of nosocomial infections, which may be fatal in patients with other underlying illnesses. Infections contracted from exposure to fungal growth in buildings are also uncommon for health adults.

Medically important fungi are those that are characterized as primary or opportunistic pathogens. Primary pathogens are those capable of routinely causing disease in otherwise healthy hosts, whereas opportunistic pathogens are those that generally require overt immunosuppression to cause disease (Dixon et al., 1996). People have been exposed to airborne fungal pathogens, such as *Cryptococcus neoformans*—the agent of cryptococcosis, when birds or bats have inhabited a building, fungi have grown in their droppings, and this material has been disturbed (Ruiz and Bulmer, 1981; Levitz, 1991) [see 15.6 and 19.2.1]. Bird and bat droppings may also harbor *Histoplasma capsulatum*, another fungal pathogen, which appears to prefer a mixture of decaying guano and soil as opposed to nest material and fresh excreta (Rippon, 1988). Similar problems can occur where flocks of birds nest or roost in trees near occupied buildings. Exposure to the soil fungi that cause endemic and epidemic coccidioidomycosis and histoplasmosis typically occurs outdoors, but exposure to *H. capsulatum* has been associated with chicken coops and demolition work (CDC, 1995a). Workers involved in building demolition or cleaning may also be exposed to saprobic fungi growing as indoor contaminants (Rautiala et al., 1996). These fungi are a concern as agents of infectious disease for immunocompromised workers and as agents of hypersensitivity disease for all demolition workers and building occupants.

Work on contaminated HVAC systems may expose service personnel to fungal and bacterial pathogens. Rodent and other animal infestations are also of concern. For example, occupational and peridomestic exposure to rodent excreta has been associated with Hantavirus pulmonary syndrome (Zeitz et al., 1995; Khan et al., 1996). Utility company workers, plumbing/heating contractors, and the construction trades are among occupations at risk of exposure to hantavirus-infected rodents (Zeitz et al., 1997). Building inspectors and remediation workers who enter and clean previously vacant rodent-infested structures may also be at risk for such exposures (Armstrong et al., 1995).

9.2.2 People as Sources of Infectious Agents

By far, people are the most important sources of airborne agents that infect humans. The most common human airborne infections (e.g., seasonal colds and influenza) are believed to be transmitted indoors where infected and susceptible persons congregate and where aerosol dispersion and dilution are more limited than outdoors. Airborne infections spread from person to person may be transmitted in any indoor environment that infectious and susceptible persons share. In residential settings, relatively few persons are exposed to an infectious individual. However, the duration of exposure is often long, ensuring transmission of common airborne viral infections within families. At the other extreme, large public gatherings in theaters, auditoriums, subways, and aircraft present the potential for extensive transmission of airborne infectious diseases, but the duration of exposure is often relatively brief (Driver et al., 1994). Although dramatic episodes have been reported of airborne communicable disease transmission in public places (Moser et al., 1979), the overall contribution to respiratory infections of such exposures remains unknown. Between these extremes are countless other indoor transmission sites, ranging from schools and offices to health-care settings, nursing homes, correctional institutions, and shelters for the homeless. In these settings, both the potential number of persons exposed to an infectious individual and the duration of exposure may be great.

9.3 Pathogenesis of Respiratory Infections

Infection occurs when a microorganism (e.g., a virus, bacterium, fungus, or protozoan) invades a host and the host's body reacts to the presence of the microorganism or its toxin (Harding et al., 1996). Infectious diseases usually have characteristic incubation periods, that is, the time from exposure to the onset of signs or symptoms of disease.

9.3.1 Infectious Dose

The infectious dose of a pathogenic agent is the number of organisms or amount of toxin required to produce infection or disease. For epidemiological or research purposes, infectious dose is often expressed in terms of an ID_{50} (i.e., the dose that produces infection in half of exposed susceptible hosts). The infectious dose is not known for most human infections and likely varies greatly depending on agent virulence, host resistance, and other factors. Chance plays an important role in determining who becomes infected because infectious particles may be relatively dilute and unevenly distributed in air. For a given exposure, by chance, some persons will inhale more than one infectious dose while others inhale none. Based on Poisson's law of small chances, Wells coined the term "quantum" to represent the required dose for a given airborne infection (i.e., the average number of microorganisms needed to infect ~63% of susceptible persons) (Wells, 1955).

9.3.2 Infectivity and Susceptibility

Although infected persons are a primary source of infection for others, not all infected individuals are infectious, that is, able to transmit the disease to another person [see 3.5]. An infected person harbors an infectious agent, but the person may respond with either manifest or inapparent disease (e.g., a subclinical infection). Tuberculosis (TB) is an example of a disease in which many infected persons never develop active disease or become infectious. Carriers also harbor infectious agents, and may transmit them to others, without themselves exhibiting signs of disease. For example, some infectious agents may colonize or infect a carrier without causing disease in that person [e.g., carrier prevalence for *Neisseria meningitidis* (a cause of meningitis) may be 25% or greater]

(APHA, 1995). A carrier may also transmit an infectious agent immediately before or after experiencing symptoms (e.g., the period of communicability for varicella, measles, and rubella begin before the onset of rash). Immune persons have specific protective antibodies or cellular immunity to an infectious agent as a result of previous infection or immunization. Susceptible persons lack sufficient resistance to a particular infectious agent to prevent infection or disease if or when exposed to the agent.

9.3.3 Relative Hazards of Infectious Agents

Infectious agents handled in microbiology laboratories are classified into hazard categories based on their risks and the activities conducted in the laboratory (USDHHS, 1993, 1994, 1996). These hazard categories do not apply necessarily to infectious agents when encountered in nature or in workplaces other than laboratories. Nevertheless, the classification reflects the relative hazards of infectious agents taking into consideration an agent's virulence, the dose needed to produce infection, the severity of an infection, the potential for transmission by inhalation, and the availability of immunization and effective treatment against the agent. Agents that may be handled at biosafety levels 1 to 4, respectively, are (a) defined and characterized strains of viable microorganisms not known to cause disease in healthy adult humans, (b) indigenous moderate-risk agents present in the community and associated with human disease of varying severity, (c) indigenous or exotic agents with a potential for respiratory transmission and which may cause serious and potentially lethal infection, and (d) dangerous and exotic agents that pose a high risk of life-threatening disease, that may be transmitted via the aerosol route, and for which there is no available vaccine or therapy.

9.3.4 Inhalation and Fate of Airborne Infectious Agents

Microorganisms are almost always present in air and are inhaled along with other airborne particulate matter and gases. An adult breathing $12.5 \text{ m}^3/\text{day}$ would inhale more than 1250 microorganisms even if the average air concentration was only 100 microorganisms/m^3. The vast majority of inhaled microorganisms cause no harm. The respiratory tract has mechanisms that inactivate and remove deposited particles, including potentially infectious agents (Lippmann, 1995) [see also 25.2.2]. For example, nasal hairs intercept large particles, and the irregularities of the nasal passages prevent particles larger than ~10 μm from reaching the lower respiratory tract. The nasal cavity has a mucociliary lining similar to the mucociliary blanket of the thoracic region. This lining carries inhaled particles to the pharynx where they are swallowed. The alveoli of the lungs have no cilia or mucus, but contain macrophages. These cells scavenge the alveolar surfaces, engulf particles, attempt to kill engulfed microorganisms, and migrate with phagocytized cells to the pulmonary

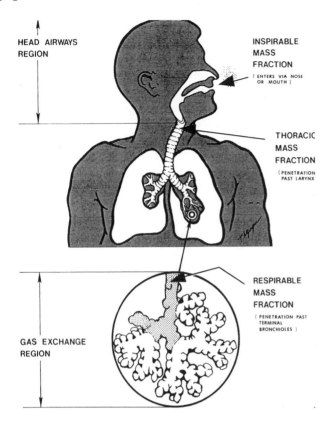

HEAD AIRWAYS REGION

INSPIRABLE MASS FRACTION
(ENTERS VIA NOSE OR MOUTH)

THORACIC MASS FRACTION
(PENETRATION PAST LARYNX)

GAS EXCHANGE REGION

RESPIRABLE MASS FRACTION
(PENETRATION PAST TERMINAL BRONCHIOLES)

FIGURE 9.1. Schematic representative of the major regions of the respiratory tract (Lippman, 1995).

lymph nodes or the mucociliary escalator. Figure 9.1 illustrates these regions of the lungs.

Table 9.1 describes the major regions of the respiratory tract and the anatomy and functions of the primary components. Table 9.2 shows criteria for inhalable, thoracic, and respirable particle deposition. Infectious agents and other bioaerosols are seldom, if ever, measured on a mass basis, but these distributions illustrate where particles of various sizes are most likely to deposit in the respiratory system. From the perspective of airborne respiratory pathogens, inhaled particles >3 μm diameter likely impact in the upper respiratory tract. Of those particles that reach the lower respiratory tract, deposition in the lung is greatest for particles 2 to 5 μm diameter. The diameter of infectious particles is determined not only by the physical size of a single cell or cluster of infectious agents, but also by the presence of salts, proteins, and cellular debris in varying states of hydration that accompany the microorganisms (Morrow, 1980).

9.3.5 Initiation of Infection
Infectious microorganisms that successfully avoid the respiratory tract's anatomical barriers and clearance mechanisms must still encounter cells they can invade. The first step in invasion may be attachment of an infectious agent to a host cell at a specific receptor that binds with proteins on the microorganism's surface. For example,

Mycobacterium tuberculosis may impact on the tracheal mucosa, but bacterial growth normally requires invasion of macrophages or monocytes located predominantly in the alveoli. Therefore, the upper respiratory tract is relatively resistant to TB infection. Damage to the body's defense mechanisms (e.g., due to concurrent infection, exposure to chemicals or particles, or suppression of the immune system) can allow microorganisms that typically would not be pathogenic to become invasive.

9.3.6 Signs and Symptoms of Respiratory Infections
Upper respiratory infections, ranging from viral "colds" to bacterial pharyngitis and otitis, are such universal human afflictions that they require little description. Many of the symptoms of respiratory infections (e.g., fever, muscle aches, stiffness, chills, and headaches) result from circulating immune system activators and foreign proteins and are both systemic and nonspecific. Acute bronchitis and pneumonia are far less common infections of the lower respiratory tract due to viruses or bacteria. These infections are typically accompanied by a productive cough, chest pain, shortness of breath, and fever. Bronchitis is associated with swelling of the airways and excessive mucus secretion, both of which obstruct airflow and lead to labored breathing. Pneumonia represents an infection of the gas-exchange region of the lungs characterized by the filling of the alveoli with secretions. This congestion leads to shortness of breath and low blood-oxygen levels. Fluid-filled or "consolidated" lung tissue is more dense than air-filled lung, usually causing physical

TABLE 9.1. Regions and Functions of the Respiratory Tract (adapted from ICRP, 1994)

Region		Anatomy	Function
Extrathoracic ↓	Extrapulmonary ↓	Anterior nasal passages Nose Mouth Pharynx Larynx	Air conditioning; temperature and humidity control and cleaning; fast particle clearance; air conduction
Thoracic ↓		Trachea Main bronchi	↓
	Pulmonary ↓	Bronchi Bronchioles Terminal bronchioles	
		Respiratory bronchioles	Air conduction; gas exchange; slow particle clearance
		Alveolar ducts Alveolar sacs	Gas exchange; very slow particle clearance
		Lymphatics	↓

signs on examination and a shadow or "infiltrate" on chest radiograph.

9.4 Modes of Transmission for Respiratory Infections

9.4.1 Transmission Via Droplets and Droplet Nuclei

People with contagious respiratory infections may produce aerosols when they cough, sneeze, or engage in other forced respiratory maneuvers that generate high velocity airflow over the thin fluid layer covering the respiratory mucosa. Many respiratory droplets are large enough to see or feel, and such large particles may contain thousands of microorganisms. However, strobe photographs reveal much larger clouds of smaller droplets. Many of these smaller droplets rapidly dry into droplet nuclei, the 1-to-5-µm residue of the original respiratory droplets. Droplet nuclei contain microorganisms together with solutes, proteins, and cellular debris and are the vehicles of airborne respiratory infection. In still air, terminal settling velocities for 1-to-5-µm particles range from 3×10^{-3} to 8×10^{-2} cm/s. Therefore, these particles can be considered truly airborne. Aerosols from other sources (e.g., water sprays) may also be considered droplets and their dried residue are also droplet nuclei. As discussed in Section 9.3.4, aerodynamic particle diameter determines where particles are most likely to deposit in the respiratory tract. The size of particles likely to reach and infect different regions of the respiratory system is determined by the length and diameter of the passages, the resultant air velocity through the airways, and the vulnerability of the tissues to infection.

9.4.1.1 Droplet or Contact Transmission Direct contact is a common mode of transmission for agents infectious for the mucosa of the upper respiratory tract. For example, contaminated fingers as well as relatively large respiratory droplets generated by coughs or sneezes or water sprays or splashes can deliver infectious agents to the nose and mouth. Agent transmission by large (>5 µm) droplets is considered an extension of direct contact because these droplets settle rapidly and travel only short distances from their sources. Many common upper-respiratory viral and bacterial infections (e.g., mumps and pertussis—whooping cough) are considered to be droplet borne. Amebic encephalitis and meningoencephalitis are also transmitted by mucosal contact with water droplets carrying amebae [see Chapter 20]. The agents responsible for these infections are spread via droplets from an infectious individual or an environmental source to susceptible persons in the immediate vicinity of the source. Influenza appears to be spread either by droplets or droplet nuclei, but the upper respiratory tract is relatively resistant. Therefore, the airborne transmission route is thought to predominate with influenza (APHA, 1995).

9.4.1.2 Airborne Infectious Diseases Infectious agents to which the upper respiratory tract is relatively resistant but the alveoli are vulnerable are considered truly airborne because only infectious particles ≤5 µm reach the alveolar region in appreciable numbers (Table 9.2). TB and measles are considered prototype airborne infections. The following discussion of the transmission and control of these respiratory infections illustrates features that also apply to other airborne infections. Additional examples of common airborne infections include influenza, varicella (chicken pox), legionellosis, and rubella (German measles). Evidence suggests that adenovirus and rhinovirus infections may be spread by airborne particles as well as by direct and indirect contact (Knight, 1980; Dick et al., 1987; Hendley and Gwaltney, 1988; Myint, 1994). Under certain circumstances, workers may also be at risk for anthrax, brucellosis, Q-fever, psittacosis, tularemia, and Hantavirus pulmonary syndrome. Inhalation of some

TABLE 9.2. Inhalable, Thoracic, and Respirable Dust Criteria of ACGIH-ISO-CEN (ACGIH, 1997).

Inhalable		Thoracic		Respirable	
Particle Aerodynamic Diameter (µm)	Inhalable Particle Mass (%)	Particle Aerodynamic Diameter (µm)	Thoracic Particle Mass (%)	Particle Aerodynamic Diameter (µm)	Respirable Particle Mass (%)
0	100	0	100	0	100
1	97	2	94	1	97
2	94	4	89	2	91
5	87	6	80.5	3	74
10	77	8	67	4	50
20	65	10	50	5	30
30	58	12	35	6	17
40	54.5	14	23	7	9
50	52.5	16	15	8	5
100	50	18	9.5	10	1
		20	6		
		25	2		

fungal spores can cause lung infections, for example, blastomycosis, coccidioidomycosis, cryptococcosis, and histoplasmosis. Chapters 18, 19, 20, and 21 describe characteristics of these infectious agents and their sampling and identification.

9.4.2 Transmission Mode and Control of Respiratory Diseases

M. tuberculosis and the measles virus infect the lower respiratory tract directly, reflecting the potential vulnerability of this region to even the small number of organisms carried by a single droplet nucleus. In contrast, many pathogenic bacteria (e.g., *Streptococcus pneumoniae*) first colonize the nasopharynx and reach the lower respiratory tract through aspiration (i.e., inspiration of fluid or other material into the airways). Aspiration of minute amounts of oropharyngeal or nasopharyngeal flora is normal and usually does not result in clinical infection. However, bronchitis or pneumonia can result if a larger volume of infectious material is aspirated (e.g., by the elderly or alcoholics) or if the aspirated organisms are particularly virulent. The mode of respiratory infection may be ambiguous or controversial in some cases, as already suggested for rhinovirus and influenza infections.

Legionellosis has been associated with inhalation of bacterial aerosols, in this case from cooling towers, spray humidifiers, vegetable misters, and other water systems. While airborne infection has been clearly demonstrated, an alternative interpretation of the epidemiological data from the original Legionnaires' disease outbreak holds that contaminated drinking water and ice may have been a common source of exposure. Pneumonia was seen in older legionnaires who were more prone to aspirate fluids and more likely to have ingested ice or water than those without the disease (Stout et al., 1992; Yu, 1993). Transmission similar to that for *S. pneumoniae* and other bacteria would help explain the apparent lack of person-to-person transmission of legionellosis. This observation implies that a large inoculum of organisms (such as could be aspirated) may be a more efficient transfer mechanism than the inhalation of the comparatively small numbers of bacteria that droplet nuclei could carry.

The distinction between aspiration and droplet-nuclei inhalation illustrates the importance of determining an infectious disease's transmission route. The environmental controls that would prevent building-associated infections spread by drinking contaminated water and those transmitted by inhaling aerosols containing infectious agents would be fundamentally different, even though contaminated water was a common factor in both scenarios. Therefore, to identify the source of infection in a Legionnaires' disease outbreak investigation, culturing both drinking water and water-cooled equipment reservoirs would be reasonable. The laboratory could compare any legionellae recovered from environmental sources with clinical isolates from infected workers to identify likely sources of exposure.

9.5 Determinants of Airborne Infections

The physical properties of infectious droplet nuclei are similar to those of other airborne particles of comparable size and density. For example, droplet nuclei settle slowly and, under static conditions, distribute randomly within a space. Such particles follow air currents within rooms and are diluted and removed by ventilation. The dynamics of airborne infections that spread from person to person have been analyzed using mass-balance equations similar to those applied to the study of other environmental contaminants (Riley et al., 1978; Catanzaro, 1982; Remington et al., 1985; Nardell et al., 1991; Gammaitoni and Nucci, 1997). Under steady-state conditions, the expected number of cases, C, among a given number of susceptible persons, S, is proportional to the average concentration of infectious droplet nuclei in a room and the probability that the particles will be inhaled. A steady-state concentration is reached if a source generates infectious agents at a constant rate and the particle removal rate also remains constant.

The concentration of droplet nuclei in a room is directly proportional to the number of infectious persons present, I, and to the generation rate of infectious agents, q (infectious doses/hour or quanta/hour). The concentration of infectious agents is inversely proportional to the rate of droplet nuclei dilution and removal by room ventilation, Q (expressed as room air changes per hour, ACH, or m^3/hour), or other means of particle removal or inactivation (e.g., filtration or ultraviolet air disinfection). The chance that exposed persons become infected is directly proportional to the volume of air they inhale [i.e., their rate of air sampling, p (m^3/hour), and their exposure time, t (hours)]. The steady-state equation describing the interaction of these factors is:

$$C = S\,(1 - e^{-x}) \qquad \text{(9-1)}$$

where: C = expected number of new cases
S = number of exposed susceptible persons.

The term x is defined in Equation 9-2:

$$x = \frac{I\,q\,p\,t}{Q}, \qquad \text{(9-2)}$$

where, I = number of sources of infectious aerosols
q = generation rate of infectious agents (infectious doses/hour or quanta/hour)
p = breathing rate for exposed persons (m^3/hour)
t = exposure time (hours)
Q = ventilation rate (ACH or m^3/hour).

$$x = \frac{(1 \ \textit{infected person}) \ (2 \ \frac{doses}{hour}) \ (1 \ \frac{m^3}{hour}) \ (8 \ \textit{hours})}{125 \ \frac{m^3}{hour}} . \quad (9\text{-}3)$$

For example, we would expect 1 to 2 new cases if 10 active persons spent 8 hours in a 250-m³ room with an infectious person. For this example, we assume that the 10 susceptible persons breathe at a rate of 1 m³/hour, that the room is ventilated at a rate of 0.5 ACH, and that the infectious person generates 2 quanta/hour. Equation 9-3 shows the term x for this example:

This model is provided for illustration only and may not produce accurate estimations because contaminant concentration likely does not reach steady-state conditions during brief exposures, with extremely infectious sources, or in very large rooms. However, the conditions of many infectious exposures have been such that the assumption of steady-state conditions appears to be reasonable. Although only a theoretical model, this approach identifies components important to the process of infectious disease transmission and illustrates the relationships among them.

9.6 Concentrations of Infectious Droplet Nuclei

Information on indoor air concentrations of infectious agents is limited. It is generally impossible or impractical to collect airborne infectious agents quantitatively (Burge, 1995), except for some environmental pathogens such as fungal agents (e.g., *Aspergillus* spp.) and for research purposes (Breiman et al., 1990; Alvarez et al., 1994; Mukoda et al., 1994; Mastorides et al., 1997). In addition, air sampling is usually conducted some time after exposure occurs, because a variable number of days must pass between exposure and the first appearance of signs or symptoms that a person has been infected. Sampling conducted weeks to months after exposure may not accurately reflect the conditions when the worker encountered an infectious aerosol.

9.6.1 Tuberculosis Ward Study

In one of the most extensive air monitoring studies conducted, Riley et al. (1962) followed the spread of TB in an experimental hospital ward. A six-bed ward for active TB patients was made airtight, and all exhaust air (containing whatever airborne *M. tuberculosis* the patients generated) passed through an exposure chamber. This chamber housed a colony of guinea pigs, which are highly susceptible to human TB and served as samplers to collect infectious mycobacteria from the ward air they breathed. Over a four-year period, newly diagnosed pulmonary TB patients were admitted to the ward and started on therapy. The guinea pigs were tested weekly for infection. Positive animals were sacrificed, and their lungs were examined for evidence of TB.

Previous experiments exposing animals to artificially generated *M. tuberculosis* aerosols had established that inhalation of 1 CFU corresponded with the finding of a single focus of infection in an animal's lung (Wells et al., 1948). Infected guinea pigs from the TB-ward study uniformly had single foci of infection in their lung periphery, presumably the result of the implantation and growth of a single droplet nucleus containing an estimated one to three *M. tuberculosis* cells. Guinea pigs became infected by patients during the days and weeks before antibiotic treatment became fully effective (i.e., while the patients were still releasing infectious bacteria). Guinea pig infections could often be linked to specific patients by comparing the drug-resistance patterns of the human and animal isolates.

These observations established that TB was transmitted by inhalation because only airborne particles could traverse the ventilation system and reach the guinea pigs. It also showed the great variability of infectiousness among patients and the relatively low production rate of infectious droplet nuclei under the conditions studied. On average, the concentration of *M. tuberculosis* on the ward was only 1 infectious droplet nucleus in ~300 to 400 m³ of air. However, this may underestimate the rate at which active TB cases generate droplet nuclei because the patients in the study were immediately started on therapy, which quickly reduced their infectiousness. Nevertheless, the situation is identical to that of isolated TB patients in today's hospitals, which has been a subject of great concern. Moreover, this low, estimated concentration of infectious airborne bacteria was consistent with that needed to produce the rate of TB infection observed among student nurses working unprotected on hospital wards in the years before effective TB treatment (Riley et al., 1959). Further, a low generation rate of droplet nuclei is consistent with the observation that TB infection typically requires that infected and susceptible persons spend long periods sharing the same air.

9.6.2 Variability of Infectious Agent Generation Rates

Epidemiological studies of infectious disease outbreaks (where the number of persons infected, their duration of exposure, and the ventilation rate were known) have used Equation 9-2 to estimate average air concentrations and generation rates of infectious agents. In these studies, exposed humans accidentally served as air samplers much like the guinea pigs described in Section 9.6.1. Studies of measles transmission in a school and a physician's office estimated that the index cases generated several hundred to several thousand times the number of infectious doses that an average TB patient produces (Riley et al., 1978; Remington et al., 1985). Similar analyses of TB outbreaks have shown that the generation rate of *M. tuberculosis* is also highly variable. An epidemiological study of TB in an office building estimated that the air concentration averaged 1 quantum in 240 m³ of air (Nardell et al., 1991). In contrast, the estimated air concentration in an explosive TB episode, in which 10 of 13 persons were infected in just 150 min, was 1 quantum in 2 m³ (Catanzaro, 1982). These differences have important implications for environmental control strategies, as discussed in Section 9.8.

9.7 Airborne Transmission Within Buildings

The many outbreaks of person-to-person airborne infections that have been studied in detail suggest two transmission patterns: within-room and beyond-room exposure. Within-room transmission occurs when an infectious individual and susceptible persons occupy the same room, the air is relatively still, and droplet nuclei accumulate and disperse within this confined space. Beyond-room transmission occurs when air contaminated with infectious aerosols moves between adjacent spaces, due to pressure differences and other factors. Such particle movement determines the distribution of infectious agents and infection risk. Particle recirculation describes the entrainment of droplet nuclei into the return air of a mechanical ventilation system from which the infectious particles are distributed throughout a building or ventilation circuit. Although closely related, these patterns have different implications for public health investigations and environmental infection control.

9.7.1 Near-Field and Far-Field Exposures

Within a room, contact between a source and a worker can be described as near field or far field, depending on the distance separating the two. Close proximity to an infectious source implies exposure to a higher local concentration of infectious particles. The degree of risk depends on the balance between the rate at which a source generates particles and the rate at which they are dispersed by air currents within the room and are removed by exhaust air and exfiltration. For example, near-field exposure to a person with measles who is generating a large, local cloud of infectious particles might convey substantial risk even for a brief exposure, whereas far-field exposure would carry less risk because the air concentration would be significantly lower further from the source. In contrast, near-field exposure to a TB case of average low infectiousness may not substantially increase a worker's risk relative to that of far-field exposure for the same duration.

9.7.2 Dissemination of Infectious Aerosols Within Buildings

A dramatic example of beyond-room TB spread involved a patient with a hip abscess who occupied a room on a post-operative hospital floor for three days (Hutton et al., 1990). TB was not suspected, and the patient's wound was irrigated with a high-velocity water jet. This unusual mechanism for *M. tuberculosis* aerosolization created a strong concentration gradient in the facility similar to what might be produced by other sources, such as contaminated mist humidifiers or disturbance of contaminated building materials. The abscess patient's room was under positive pressure relative to a corridor, and patients occupying rooms off this corridor were infected in proportion to their proximity to the source patient's room.

An example of recirculation of an infectious agent was seen in a clinic serving HIV-infected patients where 17 staff members developed tuberculin skin-test (TST) conversions (CDC, 1989). A case–control analysis concluded that the greatest risk for infection was working ≥40 hours anywhere in the clinic. TST conversion was high among staff who performed sputum inductions and pentamidine aerosol treatments—procedures that cause patients to cough. However, the risk associated with these activities was no higher than that associated with simply being in the building (i.e., droplet nuclei dispersion via air recirculation appear to have exposed staff members relatively equally to infection).

On aircraft, people may share air for many hours at close quarters, and transmission of TB and influenza has been documented among passengers and crew members (Moser, et al. 1979; Driver et al., 1994; CDC, 1995b). However, the risk of respiratory infection does not appear to be greater on aircraft than in other close spaces. In a case of passenger-to-passenger TB transmission, proximity to the infectious case was a risk factor (i.e., people in the same section of the aircraft were at greater risk) (CDC, 1995b). However, people elsewhere were protected because return air passed through high-efficiency filters that removed infectious airborne bacteria (Driver et al., 1994).

9.8 Control of Airborne Infection

Control of airborne infection can be approached using the mass-balance model discussed in Section 9.5. It should be apparent by now that transmission factors are highly interdependent. Control measures that affect one factor often directly or indirectly affect several others. Again, the example of person-to-person TB transmission is used for illustration. Hopefully, readers can apply the same principles to infectious diseases transmitted from other human, environmental, and animal sources.

9.8.1 Number of Infectious Sources

Reducing the number of infectious sources, I, (i.e., contagious individuals, Equation 9-2) in a community is the role of medical care in general and public health programs in particular. In the case of TB, physicians diagnose and treat individual patients. However, local public health programs ensure that TB patients remain on therapy until they are no longer infectious and, ultimately, are cured. These programs also trace contacts so that infected family members and co-workers can be identified. These efforts have resulted in fewer infectious TB cases in the community and less TB transmission (Frieden et al., 1995). Therefore, in terms of the transmission model, I has been decreasing.

Immunization can effectively reduce I for many common childhood infectious diseases and for some infections for which there is no good therapy [see 8.2.3.4]. For example, annual influenza vaccination campaigns change the susceptibility status of the persons who are successfully immunized. This reduction in the number of susceptible persons, S, ultimately reduces the number of sources, I, because fewer persons are available to become sources. Similarly, measures to control microbial growth in buildings (e.g., moisture control and proper HVAC system maintenance and operation) reduce the number of environmental sources of opportunistic pathogens and opportunities for people to be exposed to these agents.

9.8.2 Generation Rates of Infectious Aerosols
Detection of infected individuals and initiation of effective TB treatment reduce the number of infectious particles people release, q, by reducing the number of viable bacteria in the body. Similarly, minimizing the concentration of *Legionella* spp. and amebae in cooling towers can reduce the generation rate of infectious particles by lowering the likelihood that aerosols that are released contain these agents [see 10.6 and 20.6].

For diseases spread from person to person, such simple measures as covering coughs and sneezes with a hand or tissue can catch large respiratory droplets before they become droplet nuclei. More broadly speaking, control measures that reduce aerosol generation and dissemination from potentially contaminated water sources can lower generation rates (e.g., avoidance of mist or spray humidifiers and use of drift eliminators on cooling towers). Careful handling of contaminated materials to minimize dust release and spore dissemination would be a primary means to limit bioaerosol generation. In a sense, negative-pressure isolation rooms and other containment measures can be considered secondary means of reducing q, because they prevent the escape of infectious droplet nuclei into adjacent spaces. The containment described for remediation work in Section 15.2 is an example of source control with local exhaust ventilation to prevent contaminant dispersion.

9.8.3 Exposure Times
Exposure time, t, refers to the period when a susceptible person, S, is exposed to an infectious source, I, under conditions where transmission may occur. As noted above, sharing the "same breathing space" with an infectious individual does not require that the two be present in the same room at the same time if contaminated air can move between spaces or be distributed by a mechanical ventilation system. The duration of exposure required for

agent transmission varies greatly (e.g., from seconds to months). Among the factors controlling the time required for a susceptible person to receive an infectious dose of an agent are the number of organisms needed to cause infection and the air concentration of the organism. Host susceptibility and resistance determine the dose of an agent required to cause infection in an individual. The air concentration of infectious agents depends on the number of sources, I, that are operating and the rate, q, at which these sources generate infectious aerosols as well as local exhaust and dilution ventilation rates, Q, acting to remove and dilute the concentration of infectious particles in the air.

Normally, exposure and infection risk end when (a) an infectious person leaves a room or another source stops releasing infectious particles, or (b) susceptible persons leave the contaminated environment. Occasionally, susceptible individuals become infected a relatively short time after a source ceases to generate infectious aerosols (i.e., exposure continues as the contaminant concentration declines). Anecdotal evidence of measles transmission in doctors' offices showed that up to an hour after an infected child left the office a sufficient concentration of viruses persisted to infect other children (Remington et al., 1985; Bloch et al., 1985). However, rapid particle dilution through ventilation makes such protracted risk unlikely for all but the most infectious agents.

9.8.4 Number of Exposed Persons
9.8.4.1 Altering Susceptibility Status The most successful way to reduce the number of susceptible persons who are exposed to infectious agents is by reducing sources or changing people's susceptibility status rather than by attempting to prevent contact between potential sources and potentially susceptible persons. Susceptibility status can be altered through immunization for many infectious diseases transmitted from person to person [see 9.8.1]. This approach has been applied to great effect against a number of viral respiratory infections (e.g., measles and influenza). In some occupations (e.g., health care and earth moving), workers for high-risk assignments are selected from those with evidence of protective immunity acquired from previous infection or vaccination.

9.8.4.2 Preventing Exposure Unfortunately, immunization is not available or practical for many airborne infectious diseases. For these infections, preventing exposure is the best means of preventing disease. In particular, minimizing the number of sources or the generation rate of infectious agents from sources is the best way to reduce the number of persons exposed. Screening, isolation, and effective treatment of infectious human sources is the most important way to reduce the number of susceptible persons who are exposed. Similarly, avoiding microbial contamination is the best way to minimize exposures to saprobic microorganisms that may be opportunistic pathogens.

When infectious sources are not readily identified, as is often the case, worker exposure to infectious aerosols is only reducible by following general precautions of good hygiene and sanitation. Standard precautions against airborne infections aim to avoid any exposure to aerosols from other people, the environment, or animals. Such precautions may be administrative, for example, policies that encourage ill workers to remain home until no longer infectious. Likewise, workers at high risk due to temporary or permanent immunodeficiency should be excluded from assignments that may expose them to opportunistic pathogens [see 8.4]. Another example of an administrative measure to control exposure would be the decision to use many small shelters to house homeless persons rather than fewer large facilities where an unrecognized TB case could expose many high-risk persons.

9.8.5 Rate at Which Workers "Sample" Air by Breathing

Respiratory rate, p, varies considerably with level of worker activity but is not normally subject to control. Rather, duration of exposure, t, is kept to a minimum to limit the volume of potentially contaminated air inhaled. Where necessary, the use of effective personal respiratory protection can greatly reduce the probability that workers inhale contaminated air. Unrecognized exposures are responsible for many respiratory infections and most TB transmission, even for health-care workers who have many opportunities for exposure (CDC, 1994; NIOSH, 1996). However, workers potentially exposed to airborne infectious agents cannot wear respirators at all times and their use is limited to known high-risk situations where effective engineering controls are not feasible or while such controls are being implemented. Respirators are the least satisfactory means of exposure control because they only protect workers if properly selected, fit tested, worn, and replaced (Colton, 1996).

High-risk procedures and the need for respiratory protection may be easier to identify for potential exposures to infectious agents from the environment than for situations involving person-to-person transmission. For example, PPE has been recommended for inspectors examining cooling towers associated with Legionnaires' disease outbreaks, inspectors entering rodent-infested buildings associated with Hantavirus pulmonary syndrome cases, and remediation workers handling microbially contaminated materials (CDC, 1993; Lenhart, 1994; ASTM, 1996; Lynn et al., 1996; Rautiala et al., 1996; Lenhart et al., 1997) [see also 4.6, 15.2, and 15.6].

9.8.6 Particle Removal Rate

In our model, the average concentration of airborne infectious agents is proportional to the number of sources and the rate at which they generate infectious agents, $I\,q$. Agent air concentration is inversely proportional to the rates at which the microorganisms die, deposit, or leave an area.

9.8.6.1 Die-Off of Airborne Infectious Agents Not a great deal is known about the natural death rates of airborne infectious agents under ambient indoor conditions. The effects of humidity on microorganism viability and particle size are potentially important factors (Riley, 1972; Marthi, 1994; Cox, 1995a,b). The measles virus has been found to persist in an infectious state for at least an hour while airborne (Bloch et al., 1985; Remington et al., 1985). For TB, experimental studies under controlled conditions using a rotating drum suggested a viable half-life in air of ~6 hours (Loudon et al., 1969). Fungal and bacterial spores and amebic cysts may be assumed to remain viable and infectious for as long as they are airborne.

9.8.6.2 Deposition of Infectious Aerosols Particles that contact a surface are assumed to adhere to it. Therefore, particle deposition by diffusion, electrostatic precipitation, and gravitational settling remove infectious aerosols from suspending air. However, in most situations, such deposition plays a small role in the removal of infectious aerosols relative to other mechanisms (Wickman, 1994). More important mechanisms of particle removal are local exhaust and dilution ventilation.

9.8.6.3 Removal of Infectious Aerosols Via Ventilation One effective (i.e., well-mixed) room air change removes ~63% of airborne droplet nuclei. A second room air change removes ~63% of the remaining aerosol, leaving ~14% of the original concentration. Indoor air-change rates may vary from a low of a fraction of an ACH in tight, unventilated rooms to ≥15 ACH in hospital operating rooms. Using the mass-balance equation presented in Section 9.5, Nardell et al. (1991) estimated the effect on TB infection rate of increasing a building's ventilation rate. Under steady-state conditions, Figure 9.2 illustrates that doubling the ventilation roughly halves the concentration of infectious agents and, consequently, the number of persons infected.

This figure also shows that the level of protection achievable by feasible increases in ventilation depends on where one starts on the exponential curve. The estimated ventilation rate in this episode of TB transmission in an office building was 7 L/s (15 ft³/min) of outdoor air per person. This ventilation rate has been recommended as a minimum to reduce transmission of infectious agents transmitted from person to person (Morey, 1994). However, the model predicts that even doubling the ventilation rate would not have prevented infection for ~20% of the population. By comparison, in the episode that Catanzaro (1982) reported, the baseline ambient ventilation rate of

Number occupants infected
of 67 susceptibles

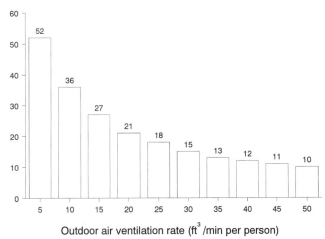

Outdoor air ventilation rate (ft^3/min per person)

FIGURE 9.2. Effect of ventilation rate on TST conversion rate (Nardell et al., 1991).

5.7 L/s (12 ft^3/min) of outdoor air per person was low for a high-risk procedure room [see 9.6.2]. In this setting, an increase to recommended ventilation levels could have greatly reduced the infection risk.

Dilution ventilation in buildings is not intended to prevent transmission of airborne infectious diseases. Rather, recommended outdoor air supply rates are designed for odor control and comfort (ASHRAE, 1989). Nevertheless, meeting recommended ventilation rates may reduce airborne infections (Brundage et al., 1988; Morey, 1994), and higher ventilation rates have been proposed to prevent airborne diseases (ASHRAE, 1996). Morey (1994) has recommended a ventilation rate of (a) at least 7 L/s (15 ft^3/min) of outdoor air per person to reduce the likelihood of transmission of contagious disease agents, especially in indoor environments with dense occupancy, and (b) at least 16.5 L/s (35 ft^3/min) of outdoor air per person for indoor spaces where exposure to *M. tuberculosis* is likely.

9.8.6.4 Air Cleaners and Air Disinfection for Preventing Infectious Diseases Measures to treat recirculated air may also reduce infectious disease transmission and supplement the benefits of ventilation. Germicidal ultraviolet radiation (UVGI) has had some success controlling the spread of infectious viral aerosols. A classic study by Wells et al. (1942) used germicidal lamps in schools to prevent the epidemic spread of measles. Riley and Nardell (1989) describe the merits of UVGI to control other infectious aerosols and discuss considerations for proper lamp placement, installation, and maintenance. Germicidal lamp intensity to achieve good killing of airborne microorganisms must be balanced with the need to protect people from overexposure to ultraviolet radiation. Germicidal lamp fixtures have been placed in HVAC system

ductwork, laboratory areas and airlocks, operating rooms, and crowded waiting rooms and assembly areas (Macher, 1993). Direct irradiation of room air as well as in-place and portable air cleaners that return room air after filtration or UV irradiation have been studied for control of airborne infectious agents (Marier and Nelson, 1993; Rutala et al., 1995; Miller-Leiden et al., 1996). UVGI is used to directly irradiate room air and may be an appropriate means of protecting workers against airborne infectious diseases (Nardell et al., 1996). However, worker eye and skin exposures to UVGI must not exceed recommended exposure limits (NIOSH, 1972; CDC, 1994). Such engineering interventions to control airborne infection have been discussed at length in the recent TB control literature (CDC, 1994; Bates and Nardell, 1995), but the principles are broadly applicable to other airborne infections.

9.9 Summary

Airborne infections can be prevented or reduced by the following.

1. Minimize the number of sources of infectious agents in the workplace.
2. Control the concentration of infectious agents in potential sources.
3. Minimize the rate at which sources generate infectious aerosols.
4. Control the entry of infectious aerosols into buildings.
5. Control the dissemination of infectious aerosols throughout buildings.
6. Maximize removal rates of airborne infectious aerosols through dilution ventilation and use of air cleaners.
7. Limit the number of susceptible persons who are exposed to infectious agents.
8. Change the status of susceptible workers through immunization, where available.
9. Minimize the time susceptible workers are exposed to infectious agents.
10. As needed, inform workers about job-related infectious disease risks, train workers in safe practices, and provide appropriate PPE.

9.10 References

ACGIH: TLVs and BEIs. Threshold Limit Values for Chemical Substances and Physical Agents. Biological Exposure Indices. American Conference of Governmental Industrial Hygienists, Cincinnati, OH (1998).

APHA: Meningitis. In: Control of Communicable Diseases Manual, 16th Ed., pp. 301-310. A.S. Benenson, Ed. American Public Health Association, Washington, DC (1995).

ASHRAE: Standard 62. Ventilation for Acceptable Indoor Air Quality. American Society of Heating, Refrigerating and Air-Conditioning Engineers, Atlanta, GA (1989).

ASHRAE: Standard 62. Ventilation for Acceptable Indoor Air Quality. American Society of Heating, Refrigerating and Air-Conditioning Engineers, Atlanta, GA (1996).

ASTM: Standard Guide for Inspecting Water Systems for Legionellae and Investigating Possible Outbreaks of Legionellosis (Legionnaires' Disease or Pontiac Fever). D 5952-96. American Society for Testing and Materials, West Conshohocken, PA (1996).

Alvarez, A.J.; Buttner, M.P.; Toranzos, G.A.; et al.: Use of Solid-phase PCR for Enhanced Detection of Airborne Microorganisms. Appl. Environ. Microbiol. 60(1):374-376 (1994).

Armstrong, L.R.; Zaki, S.R.; Goldoft, M.J.; et al.: Hantavirus Pulmonary Syndrome Associated with Entering or Cleaning Rarely Used, Rodent-Infested Structures. J. Infect. Dis. 172:1166 (1995).

Bates, J.; Nardell, E., Eds.: Institutional Control Measures for Tuberculosis in the Era of Multiple Drug Resistance: Joint Clinical Consensus Statement. American College of Chest Physicians and The American Thoracic Society. Chest. 108:1690-1710 (1995).

Bloch, A.B.; Orenstein, W.A.; Ewing, W.M.; et al.: Measles Outbreak in a Pediatric Practice: Airborne Transmission in an Office Setting. Pediatric. 75:676-683 (1985).

Breiman, R. F.; Cozen, W.; Fields, B.S.; et al.: Role of Air Sampling in Investigation of an Outbreak of Legionnaires' Disease Associated with Exposure to Aerosols from an Evaporative Condenser. J. Infect. Dis. 161:1257-1261 (1990).

Brundage, J.F.; Scott, R.M.; Lednar, W.M.; et al.: Building-Associated Risk of Febrile Acute Respiratory Diseases in Army Trainees. JAMA 259:2108-2112 (1988).

Burge, HA: Airborne Contagious Disease. In: Bioaerosols, pp. 25-47. H.A. Burge, Ed. Lewis Publishers, Boca Raton, FL (1995).

Cantanzaro, A.: Nosocomial Tuberculosis. Am. Rev. Respir. Dis. 125:559-562 (1982).

CDC: *Mycobacterium tuberculosis* Transmission in a Health Clinic Florida. CDC — MMWR 256-264. Centers for Disease Control (1989).

CDC: Hantavirus Infection — Southwestern United States: Interim Recommendations for Risk Reduction. CDC — MMWR 42(RR-11). Centers for Disease Control (1993).

CDC: Guidelines for Preventing the Transmission of *Mycobacterium tuberculosis* in Health-Care Facilities. CDC — MMWR 43(1-120). Centers for Disease Control (1994).

CDC: Histoplasmosis— Kentucky, 1995. CDC — MMWR 44(701-703). Centers for Disease Control (1995a).

CDC: Exposure of Passengers and Flight Crew to *Mycobacterium tuberculosis* on Commercial Aircraft, 1992-1995. CDC — MMWR 44(137-140). Centers for Disease Control (1995b).

Collins, C.H.; Aw, T.C.; Grange, J.M.: Microbial Diseases of Occupations, Sports and Recreations. Butterworth-Heinemann, Boston, MA (1997).

Colton, C.E.: Respiratory Protection. In: Fundamentals of Industrial Hygiene, 4th Ed., pp. 619-656. B.A. Plog, J. Niland, P.J. Quinlan, Eds. National Safety Council, Itasca, IL (1996).

Cox, C.S.: Stability of Airborne Microbes and Allergens. In: Bioaerosols Handbook, pp. 77-99. C.S. Cox and C.M. Wathes, Eds. Lewis Publishers, Boca Raton, FL (1995a).

Cox, C.S.: Physical Aspects of Bioaerosol Particles. In: Bioaerosols Handbook, pp. 15-25. C.S. Cox and C.M. Wathes, Eds. Lewis Publishers, Boca Raton, FL (1995b).

Dick, E.C.; Jennings, L.C.; Mink, K.A.; et al.: Aerosol Transmission of Rhinovirus Colds. J. Infect. Dis. 156 (3):442-8 (1987).

Dixon, D.M.; McNeil, M.M.; Cohen, M.L.; et al.: Fungal Infections, A Growing Threat. Public Health Reports 111:226-235 (1996).

Driver, C.R.; Valway, S.E.; Meade, M.; et al.: 1994. Transmission of *Mycobacterium tuberculosis* Associated with Air Travel. JAMA 272:1031-1035 (1994).

Ellis, C.J.; Symington, I.S.: Microbial Disease. In: Hunter's Diseases of Occupations, 8th Ed., pp. 545-576. P.A.B. Raffle, P.H. Adams, P.J. Baxter, et al., Eds. Edward Arnold, Boston, MA (1994).

Frieden, T.R.; Fujiwara, P.I.; Washko, R.M.; et al.: Tuberculosis in New York City: Turning the Tide. N. Engl. J. Med. 333:229-233 (1995).

Gammaitoni, L.; Nucci, M.C.: Using a Mathematical Model to Evaluate the Efficacy of TB Control Measures. Emerg. Infect. Dis. 3:335-342 (1997).

Gantz, N.M.: Infectious Agents. In: Occupational Health: Recognizing and Preventing Work-Related Disease, pp. 281-295. B.S. Levy and D.H. Wegman, Eds. Little, Brown and Co., Boston, MA (1988).

Garibaldi, R.; Janis, B.: Occupational Infections. In: Environmental and Occupational Medicine, pp. 607-617. W.N. Rom, Ed. Little, Brown and Co., Boston, MA (1992).

Gerberding, J.L.; Holmes, K.K.: Microbial Agents and Infectious Diseases. In: Textbook of Clinical Occupational and Environmental Medicine, pp. 699-716. W.B. Saunders Co., Philadelphia, PA (1994).

Gerberding, J.L.: Infectious Organisms. In: Occupational and Environmental Respiratory Disease, pp. 561-570. P. Harber, M.B. Schenker, and J.R. Balmes, Eds. Mosby, St Louis, MO (1996).

Harding, L.; Fleming, D.O.; Macher, J.M.: Biological Hazards. In: Fundamentals of Industrial Hygiene, 4th Edition, pp. 403-449. B.A. Plog, J. Niland, P.J. Quinlan, Eds. National Safety Council, Itasca, IL (1996).

Hendley, J.O.; Gwaltney, J.M.: Mechanisms of Transmission of Rhinovirus Infections. Epidemiol. Rev. 10:242-258 (1988).

Hutton, M.D.; Stead, W.W.; Cauthen, G.M.; et al.: Nosocomial Transmission of Tuberculosis Associated with a Draining Abscess. J. Inf. Dis. 161:286-295 (1990).

ICRP: Human Respiratory Tract Model for Radiological Protection. Report of the Committee II of the ICRP. International Commission on Radiological Protection (1994).

Jennison, M.W.: Atomizing of Mouth and Nose Secretions into the Air as Revealed by High-Speed Photography. In: Aerobiology, No. 17, pp. 106-126. American Association for the Advancement of Science, Washington, DC (1947).

Khan, A.S.; Ksiazek, T.G.; Peters, C.J.: Hantavirus Pulmonary Syndrome. Lancet. 347:739-741 (1996).

Knight, V.: Viruses as Agents of Airborne Contagion. In: Airborne Contagion, pp. 147-156. R. B. Kundsin, Ed. New York Academy of Sciences, New York, NY (1980).

Lenhart, S.W.: Recommendations for Protecting Workers from *Histoplasma capsulatum* Exposure During Bat Guano Removal from a Church's Attic. Appl. Occup. Environ. Hyg. 9:230-236 (1994).

Lenhart, S.W.; Schafer, M.P.; Singal, M.; et al.: Histoplasmosis: Protecting Workers at Risk. Centers for Disease Control and Prevention. National Institutes for Occupational Safety and Health, Cincinnati, OH: Publication No. 97-146 (1997).

Levitz, S.M.: The Ecology of *Cryptococcus neoformans* and the Epidemiology of Cryptococcosis. Rev. of Infect. Dis. 13:1163-1169 (1991).

Lippmann, M.: Size-Selective Health Hazard Sampling. In: Air Sampling Instruments for Evaluation of Atmospheric Contaminants, pp. 81-119. B.S. Cohen and S.V. Hering, Eds. American Conference of Governmental Industrial Hygienists, Cincinnati, OH (1995).

Loudon, R.D.; Bumgarner, L.R.; Lacey, J.; et al.: Aerial Transmission of Mycobacteria. Am. Rev. Respir. Dis. 100:165-71 (1969).

Lynn, M.; Vaughn, D.; Kauchak, S.; et al.: An OSHA Perspective on Hantavirus. Appl. Occup. Environ. Hyg. 11:225-227 (1996).

Macher, J.M.; Rosenberg, J: Evaluation and Management of Exposure to Infectious Agents. In: Handbook of Occupational Safety and Health, pp. 287-371. L. DiBerardinis, Ed. Wiley, New York, NY (1999).

Macher, J.M.: The Use of Germicidal Lamps to Control Tuberculosis in Health-Care Facilities. Infect. Cont. Hosp. Epidemiol. 14(12):723-729 (1993).

Marier, R.L.; Nelson, T.: A Ventilation-Filtration Unit for Respiratory Isolation. Infect. Control Hosp. Epidemiol. 14:700-705 (1993).

Marthi, B.: Resuscitation of Microbial Bioaerosols. In: Atmospheric Microbial Aerosols: Theory and Applications, pp. 192-225. B. Lighthart and A.J. Mohr, Eds. Chapman and Hall, New York, NY (1994).

Mastorides, S.M.; Oehler, R.L.; Greene, J.N.; et al.: Detection of *Mycobacterium tuberculosis* by Air Filtration and Polymerase Chain Reaction. Clin. Infect. Dis. 25:756-757 (1997).

Miller-Leiden, S.; Lobascio, C.; Macher, J.M.; et al.: Effectiveness of In-room Air Filtration for Tuberculosis Infection Control in

Healthcare Settings. J. Air Waste Manage. Assoc. 46:869-882 (1996).

Morey, P.R.: Suggested Guidance on Prevention of Microbial Contamination for the Next Revision of ASHRAE Standard 62. In: Proceedings to ASHRAE IAQ '94, Engineering Indoor Environments, pp. 139-148. American Society of Heating, Refrigerating and Air-Conditioning Engineers, Atlanta, GA (1994).

Morrow, P.E.: Physics of Airborne Particles and their Deposition in the Lung. In: Airborne Contagion, pp. 71-80. R. B. Kundsin, Ed. New York Academy of Sciences, New York, NY (1980).

Moser, M.R.; Bender, T.R.; Margolis, H.S.; et al.: An Outbreak of Influenza Aboard a Commercial Airliner. Amer. J. Epidemiol. 110(1):1-6 (1979).

Mukoda, T.J.; Todd, L.A.; Sobsey, M.D.: PCR and Gene Probes for Detecting Bioaerosols. J. Aerosol Sci. 25(8):1523-1532 (1994).

Myint, S.H.: Common Colds, Asthma and Indoor Air Quality. Indoor Environ. 3:274-277 (1994).

Nardell, E.A.; Keegan, J.; Cheney, S.A.; et al.: Airborne Infection: Theoretical Limits of Protection Achievable by Building Ventilation. Am. Rev. Respir. Dis. 144:302-306 (1991).

Nardell, E.A.; Barnhart, S.; Permutt, S.: Control of Tuberculosis in Health Care Facilities: The Rational Application of Patient Isolation, Building Ventilation, Air Filtration, Ultraviolet Air Disinfection, and Personal Respirators. In: Tuberculosis, pp. 873-891. W.N. Rom and S.M. Garay, Ed. Little, Brown and Company, New York, NY (1996).

NIOSH: Criteria for a Recommended Standard: Occupational Exposure to Ultraviolet Radiation. U.S. Department of Health, Education and Welfare. Public Health Service, Washington, DC. Publication No. (HSM)73-110009 (1972).

NIOSH: NIOSH Guide to the Selection and Use of Particulate Respirators Certified under 42 CFR 84. U.S. Department of Health and Human Services, Public Health Service, Centers for Disease Control and Prevention, National Institute for Occupational Safety and Health. DHHS (NIOSH) Publication No. 96-101, NIOSH, Cincinnati, OH (1996).

Rautiala, S.; Reponen, T.; Hyvärinen; et al.: Exposure to Airborne Microbes during the Repair of Moldy Buildings. Am. Ind. Hyg. Assoc. J. 57:279-284 (1996).

Remington, P.L.; Hall, W.N.; Davis, I.H.; et al.: Airborne Transmission of Measles in a Physician's Office. J. Am. Med. Assoc. 253:1574-1577 (1985).

Riley, R.L.; Mills, C.C.; Nyka, W.; et al.: Aerial Dissemination of Pulmonary Tuberculosis: A Two-Year Study of Contagion in a Tuberculosis Ward. Am. J. Hyg. 70:185-196 (1959).

Riley, R.L.; Mills, C.C.; O'Grady, F.; et al.: Infectiousness of Air from a Tuberculosis Ward — Ultraviolet Irradiation of Infected Air: Comparative Infectiousness of Different Patients. Am. Rev. Respir. Dis. 85:511-525 (1962).

Riley, E.C.; Murphy, G.; Riley, R.L.: Airborne Spread of Measles in a Suburban Elementary School. Am. J. Epidemiol. 107:421-432 (1978).

Riley, R.L.; Knight, M.; Middlebrook, G.: Ultraviolet Susceptibility of BCG and Virulent Tubercle Bacilli. Am Rev Resp Dis. 113:413-418 (1972).

Riley, R.L.; Nardell, E.A.: Clearing the Air: The Theory and Application of Ultraviolet Air Disinfection. Am. Rev. Resp. Dis. 139(5):1286-1294 (1989).

Rippon, J.W.: Medical Mycology, the Pathogenic Fungi and the Pathogenic Actinomycetes, 3rd Ed., pp. 381-423. W.B. Saunders Company, Philadelphia, PA (1988).

Ruiz, A.; Bulmer, G.S.: Particle Size of Airborne *Cryptococcus neoformans* in a Tower. Appl. Environ. Microbiol. 41:1225-1229 (1981).

Rutala, W.A.; Jones, S.M.; Worthington, J.M.; et al.: Efficacy of Portable Filtration Units in Reducing Aerosolized Particles in the Size Range of *Mycobacterium tuberculosis.* Infect. Control Hosp. Epidemiol. 16:391-398 (1995).

Stout, J.E.; Yu, V.L.; Muraca, P.; et al.: Potable Water as a Cause of Sporadic Cases of Community-Acquired Legionnaires' Disease. N. Engl. J. Med. 326:151-155 (1992).

USDHHS: Biosafety in Microbiological and Biomedical Laboratories, 3rd Ed. HHS Publication No. (CDC)93-8395, Public Health Service, Centers for Disease Control and Prevention, and National Institutes of Health. CDC, Washington, DC: USGPO (Stock No. 017-040-00523) (1993).

USDHHS, National Institutes of Health. Appendix B. Classification of Etiologic Agents and Oncogenic Viruses on the Basis of Hazard. In: Guidelines for Research Involving Recombinant DNA Molecules (NIH Guidelines). Federal Register July 5, 1994, Separate Part IV, pp. 20-24. (1994).

USDHHS, Centers for Disease Control and Prevention: Additional Requirements for Facilities Transferring or Receiving Select Agents. 42 CFR Part 72, RIN 0905-AE70 (1996).

Wald, P.H.; Stave, G.M., Eds.: Physical and Biological Hazards of the Workplace. Wiley, New York, NY (1994).

Wells, W.F.; Wells, M.W.; Wilder, T.S.: The Environmental Control of Epidemic Contagion. I. An Epidemiologic Study of Radiant Disinfection of Air in Day Schools. Am. J. Hyg. 35:97-121 (1942).

Wells, W.F.; Ratcliffe, H.L.; Crumb, C.: On the Mechanics of Droplet Nuclei Infection. II. Quantitative Experimental Air-Borne Tuberculosis in Rabbits. Am. J. Hyg. 47:11-28 (1948).

Wells, W.F.: Airborne Contagion and Air Hygiene, p. 144. Harvard University Press, Cambridge, MA (1955).

Wickman, H.H.: Deposition, Adhesion, and Release of Bioaerosols. In: Atmospheric Microbial Aerosols: Theory and Applications, pp. 99-165. B. Lighthart and A.J. Mohr, Eds. Chapman and Hall, New York, NY (1994).

Yu, V.: Could Aspiration be the Major Mode of Transmission for Legionella? — Editorial. Am. J. Med. 95:13-15 (1993).

Zeitz, P.S.; Butler, J.C.; Cheek, J.E.; et al.: A Case–Control Study of Hantavirus Pulmonary Syndrome during an Outbreak in the Southwestern United States. J. Inf. Dis. 171:864-870 (1995).

Zeitz, P.S.; Graber, J.M.; Voorhees, R.A.; et al.: Assessment of Occupational Risk for Hantavirus Infection in Arizona and New Mexico. J. Occ. Environ. Med. 39:463-467 (1997).

Chapter 10

Prevention And Control Of Microbial Contamination

Richard J. Shaughnessy and Philip R. Morey with Eugene C. Cole

10.1 Problems of Microbial Contamination in Buildings
10.2 Critical Factors Controlling Indoor Microbial Survival and Growth
    10.2.1 *Nutrient Sources*
    10.2.2 *Temperature*
    10.2.3 *Moisture*
    10.2.4 *Use of Biocides and Antimicrobial Agents*
10.3 Moisture Indices
    10.3.1 *Relative Humidity*
    10.3.2 *Moisture Content*
    10.3.3 *Water Activity and Equilibrium Relative Humidity*
    10.3.4 *Moisture Indices and Control of Microbial Growth*
10.4 Moisture Control
    10.4.1 *Flooding and Water Leaks*
    10.4.2 *Moisture in the Building Envelope*
    10.4.3 *Controlling Relative Humidity, Temperature, and Condensation*
        10.4.3.1 Condensation Control in Cold Climates
        10.4.3.2 Condensation Control in Hot, Humid Climates
    10.4.4 *Moisture Control in HVAC System Components*
        10.4.4.1 Filters
        10.4.4.2 Cooling Coils and Moisture Carryover
        10.4.4.3 Drain Pans
        10.4.4.4 Humidifiers
        10.4.4.5 Airstream Surfaces in Plenums and Ducts
10.5 HVAC System Design and Operation
    10.5.1 *Access*
    10.5.2 *Ventilation*
        10.5.2.1 Ventilation to Control Airborne Infectious Agents
        10.5.2.2 Ventilation to Control Airborne Allergens
    10.5.3 *Filters*
    10.5.4 *Methods Other than Filtration for Controlling Intrusion of Bioaerosols from Outdoor Sources*
    10.5.5 *Intrusion of Moisture from Outdoors*
    10.5.6 *Portable Air Cleaners*
10.6 Control of Legionellae
10.7 Long-Term Prevention and Cleaning Management
    10.7.1 *Long-Term Prevention*
    10.7.2 *Cleaning Management*
10.8 References

10.1 Problems of Microbial Contamination in Buildings

Microorganisms can be found in indoor air and on surfaces, for example, on floors, walls, ceilings, and furnishings as well as within HVAC systems. Many of these microorganisms originate from outdoors. The microorganisms in outdoor air are primarily those that grow in moist soil and on the surfaces of leaves and in plant litter (Morey, 1994; Lacey and Venette, 1995). Outdoor bioaerosol concentrations vary according to geographical location, climatic conditions, season, and surrounding natural and human activities. Fungal spores usually dominate in outdoor air, but pollen, bacteria, algae, and insect fragments also are present (Muilenberg, 1995). These organisms may reach the indoor environment with outdoor air that enters buildings or be transported by people and animals (e.g., on shoes, clothes, skin, and fur). Microorganisms may also already be present on building materials and furnishings when they are brought into a building. However, actual growth of microorganisms indoors is undesirable and should be prevented (Samson et al., 1994; Health Canada, 1995; Maroni et al., 1995; ISIAQ, 1996). Reasons for avoiding indoor microbial growth are (a) to prevent adverse health effects in people exposed to biological agents, and (b) to preserve material integrity.

The growth of bacteria, fungi, protozoa, and arthropods in buildings depends on a number of inter-related variables. Moisture is recognized as the primary factor controlling biological growth. Also important are temperature, atmosphere, and the availability of nutrients or growth substrates. However, these latter factors cannot be considered limiting. Building temperatures and oxygen concentrations that are suitable for humans are also suitable for growth of many microorganisms and arthropods. Suitable nutrients are also abundant in buildings due to the variety of materials organisms can use as energy sources (Flannigan, 1992).

This chapter focuses on fungal and bacterial growth in buildings, mentioning other biological agents where applicable. The general principles for preventing and controlling microbial growth in buildings can be summarized in the following core recommendations.

1. Prevent microbial colonization of materials in the building envelope, HVAC system, and interior space.
2. Limit intrusion of aerosols from outdoor sources.
3. Limit indoor accumulation of aerosols from human and animal sources.

A comprehensive strategy to resolve problems related to microbial growth in buildings entails a series of inclusive steps.

1. **Identification and Control of Sources.** To understand problems and determine how to control microbial contamination, investigators must understand the building dynamics related to moisture accumulation and identify sources of indoor microbial growth.
2. **Removal or Remediation of Existing Contamination.** Contaminated items worth saving should be decontaminated. Contaminated materials that cannot be salvaged must be discarded. Removal procedures depend on the types of materials affected as well as the nature and degree of contamination. All items that may act as potential sources of biological agents should be cleaned. The professional cleaning and restoration industry provides guid-

ance for the proper restoration and cleaning of water-damaged and microbially contaminated indoor environments (IICRC, 1995).

3. **Implementation of a Long-Term Plan to Prevent Problem Recurrence.** Building managers must adopt operation and maintenance practices that will ensure that control strategies are effectively implemented and the potential for problem recurrence is eliminated.
4. **Implementation of a Cleaning Program to Prevent Pollutant Buildup.** Building managers should implement a regularly scheduled program of cleaning to control sources of particles and biological agents. The following guidelines are recommended by the cleaning industry (Berry, 1993; Franke et al., 1997):
   • Clean for health first and appearance second.
   • Maximize the extraction of pollutants.
   • Minimize chemical, particle, and moisture residues.
   • Minimize human exposure to pollutants.
   • Properly dispose of cleaning wastes.

The 1996 draft revision of ASHRAE Standard 62, "Ventilation for Acceptable Indoor Air Quality," (ASHRAE, 1996a) was not accepted and did not replace the 1989 standard (ASHRAE, 1989). However, the draft revision contained useful advice and recommendations. Therefore, sections of this document are referenced in this book in addition to recommendations from the earlier standard.

10.2 Critical Factors Controlling Indoor Microbial Survival and Growth

Buildings are dynamic environments affected by geographic location, season, weather conditions, HVAC system design and operation, moisture intrusion, pest colonization, and human activities. These and other factors continually change the availability of suitable conditions for microbial growth. The interactions of microorganisms with each other and with their environment typically lead to fluctuating microorganism concentrations over time rather than to constant, stable populations.

10.2.1 Nutrient Sources
There are numerous nutrient sources in buildings that can satisfy the needs of environmental microorganisms. Microorganisms have been shown to colonize gypsum board, wood paneling, cellulose ceiling tiles, carpets, upholstered furniture, fiberglass-lined air ducts, and other porous materials where the microorganisms break down the material itself or utilize organic debris that has collected thereon (Foarde et al., 1993; Morey, 1994; Burge, 1995). The nutrients in a material that are available to support microorganisms may change as successions of organisms grow on and metabolize the substrate. The sequence of

microbial colonization and factors that determine which microorganisms are actually growing at a given time depend on how often and for how long the material is wetted and dried or cooled and warmed.

10.2.2 Temperature

For comfort, most buildings are maintained at a temperature of 18° to 24°C (65° to 75°F). This temperature range is also hospitable to many environmental microorganisms, some of which can even survive at temperatures below 10°C and others above 50°C (50° and 122°F) [see 18.1.3 and 19.1.5.3]. However, temperature and water availability are related, and water availability is critical. Temperature can often be controlled in water systems, that is, it may be possible to maintain water at temperatures above or below those that encourage microbial growth [see 10.6].

10.2.3 Moisture

Water is essential to all life, and the chemical reactions that lead to biological growth depend on an adequate water supply. As mentioned in Section 10.2.2, temperature and water availability are related. However, the indoor temperatures desired for human comfort are also suitable for many microorganisms. Thus, control of moisture in buildings becomes the critical variable that will prevent or limit microbial growth and is a primary focus of this chapter [see 10.1, 10.3, 10.4, and 10.5.5].

10.2.4 Use of Biocides and Antimicrobial Agents

Active measures to control microbial growth in buildings may involve the use of biocides and antimicrobial agents in addition to moisture and temperature control. Biocides are used (a) to treat flooding with sewage [see 10.4.1], (b) to control legionellae in water-cooled, heat-transfer equipment [see 10.6] and to treat equipment associated with outbreaks of legionellosis [see 15.7], and (c) to

treat microbial contamination [see 15.2 and 15.3]. Biocides and antimicrobial agents may play a role in treating and preventing microbial growth in other areas, although not as substitutes for proper building construction, maintenance, use, management, and other appropriate control measures [see 10.4.4.3, 15.4, 16.1, and 23.6].

10.3 Moisture Indices

Nutrient availability, indoor temperature, and shelter generally cannot be managed to limit indoor microbial survival and growth. However, moisture is equally necessary to microorganisms, and maintaining available moisture below that necessary for microbial growth generally is possible. Various professions use different methods to measure and describe moisture in materials. For example, microbiologists focus on how much water is available for microbial growth and use the term water activity (a_w; also written as A_w). Engineers use other terms to describe the moisture-holding capacity of building materials. Table 10.1 defines some of these measures.

10.3.1 Relative Humidity

Understanding the significance of RH in an occupied space and the related concept of substrate a_w/ERH is critical to controlling microbial growth indoors. RH is a ratio (expressed as a percentage) of the amount of moisture in air to the maximum amount the air could hold. Warmer air has a greater capacity to hold water in its vapor form than cooler air. Although the RH of room air plays a role in the water content of materials in the room, it is the available moisture in a substrate, not the RH of the room air, that determines if microorganisms can grow and the types of organisms that colonize a material (Pasanen et al., 1991; Flannigan, 1992) [see 10.3.4].

10.3.2 Moisture Content

MC is determined by weighing a material under the conditions of interest and when completely dry. MC is ex-

TABLE 10.1. Terms Used to Describe Water in Air and Materials

Air	
Absolute humidity	The ratio of the mass of water vapor to the mass of dry air; humidity ratio.
Relative humidity (RH)	The ratio of the amount of moisture held in air (vapor pressure) to the maximum amount of moisture that the air can hold at a given temperature and pressure—the saturation vapor pressure; typically determined from a psychrometric chart of dry-bulb, wet-bulb, and dew-point temperatures.
Materials	
Moisture content (MC)	The mass of moisture held in a material; measured as the mass of water as a percentage of the dry mass of a material.
Water activity (a_w)	The ratio of the amount of water in a material at a particular moisture content (vapor pressure) to the maximum amount of water air can hold at the same temperature and pressure (saturation vapor pressure); determined by placing a material in a sealed equilibration chamber, measuring the RH of the air, and expressing RH as a ratio rather than a percentage.
Equilibrium RH (ERH)	The a_w of a material expressed as a percentage.

pressed as a percentage {[(wet mass - dry mass) (100)]/ (dry mass)} or in terms of the mass of water in a given volume of material (e.g., kg/m^3 or lbs/ft^3). Foarde et al. (1993) found that a moisture content >5% permitted growth of *Penicillium glabrum* and *Aspergillus versicolor* on ceiling tiles under controlled laboratory conditions.

10.3.3 Water Activity and Equilibrium Relative Humidity

Available moisture may be expressed in terms of the amount of free water in a material that is available to support microbial growth. The terms a_w and ERH relate the vapor pressure of water in a material with that of free water at the same temperature and pressure (Hunter and Sanders, 1990; Flannigan, 1992). At atmospheric pressure, the a_w of a material varies with the absolute amount of water present and temperature. Both a_w and ERH are measured indirectly, by allowing material to equilibrate in a sealed container and then measuring the RH of the atmosphere in the container (ISIAQ, 1996). A_w is expressed as a fraction ≤ 1; ERH is expressed as a percentage ≥ 1. Thus, an a_w of 0.9 is the equivalent of an ERH of 90%. The term a_w is used primarily in this chapter, but readers can readily convert this measurement to ERH.

Water activity is an important indicator of a material's ability to support microbial growth (Griffin, 1981; Kendrick, 1992) [see 19.1.5.2]. For example, the a_w of foods is important to their quality and safety, and there are some regulations specifying a_w levels in certain foods (Troller and Scott, 1992). For illustration, the a_w in fresh fruit and vegetables and processed cheese is 0.90 to 1.00, that in dried fruit and cheese is 0.60 to 0.70, and that in dried vegetables and milk is approximately 0.20. Theoretical limits for microbial growth lie between a_w values of 0.65 and 1.0 (Grant et al., 1989; Flannigan, 1992; ISIAQ, 1996), with an absolute lower limit of 0.55, the point at which DNA is denatured (Griffin, 1981). Practically speaking, if a_w in materials can be kept below ~0.75, microbial growth will be limited; below an a_w of 0.65, virtually no microbial growth will occur on even the most susceptible materials. However, even under these conditions, fungal and bacterial spores may remain alive for extended periods, and dead cells may continue to present a hazard.

10.3.4 Moisture Indices and Control of Microbial Growth

Limiting the a_w of materials in buildings to ≤ 0.75 (ideally ≤ 0.65) should be a primary goal in the prevention of microbial growth (Morey, 1994). Unfortunately, there are no direct, field measurement techniques to determine a_w. Many investigators attempt to use a material's MC as a direct indicator of its potential to support microbial growth. However, MC can only be related directly to a_w for a specific material. For example, an a_w of 0.80 corresponds to the following MCs in selected construction materials: brick (0.1% to 0.9%), gypsum board (0.7%), cement (1%), wallpaper (11.3%), and soft wood (17%)

(Flannigan, 1992; ISIAQ, 1996). Therefore, reliance on MC alone to assess growth potential is seldom useful due to the disparity (at a given a_w value) between a_w and MC for nonhomo-geneous building materials.

Sorption isotherms can be developed for specific building materials to provide an approximate means of relating equilibrium MC at various RHs (Figure 10.1) (Richards et al., 1992; ISIAQ, 1996). Separate lines show the curves for adsorption of water by dry material and desorption of moisture from the same material when wet. The "humidity" measurement at any point on a sorption isotherm should approximate the a_w of the material (Morey, 1994). For a given building material, a sorption isotherm allows definition of the minimum or threshold MC necessary for fungal growth. For example, if our goal is to maintain a_w below 0.75, the pine wood in Figure 10.1 must be kept below a MC of ~20% and the concrete below 5%.

In the absence of sorption isotherm data, RH in an occupied space is often used to indicate microbial growth potential. Many advisory groups have recommended maintaining RH values in the occupied space below 60% to limit microbial growth (Flannigan, 1992; Morey, 1994; Morey, 1995). By keeping RH values below 60%, it is logical to assume that corresponding a_w values in materials would be limited to ~0.6. However, this assumption may not hold because microorganisms grow on surfaces and in materials, not in air. Therefore, it is the RH in the air adjacent to a surface, not ambient RH, that must be controlled to prevent microbial growth. Maintaining room RH below 60% may keep materials fairly dry, but does not eliminate the possibility of microbial growth because local cold spots and water intrusion may allow the RH of air adjacent to a surface to exceed 70% (Flannigan, 1992). Moisture meters are

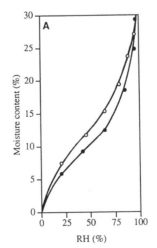 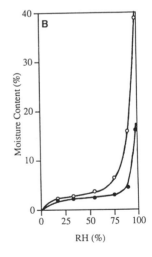

FIGURE 10.1. Sorption isotherms for (A) pine wood and (B) concrete; · = adsorption curve, ° = desorption curve (Richards et al., 1992; ISIAQ, 1996).

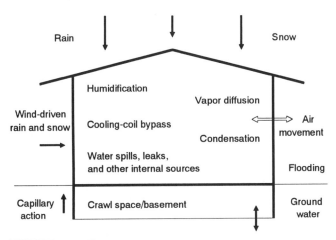

FIGURE 10.2. Means of water entry into a building.

field devices that may help investigators identify damp areas that would otherwise not be evident [see 4.1.4].

10.4 Moisture Control

It is impossible to list all sources of excess moisture that can contribute to indoor biological growth. However, some of the primary means of moisture entry into buildings are water leakage, vapor migration, capillary movement, air infiltration, and bypass of dehumidification cooling coils (Figure 10.2).

10.4.1 Flooding and Water Leaks

Chronic flooding or leaks will almost always result in microbial growth in the indoor environment (Morey et al., 1984). Fungi have been shown to be capable of germination, growth, and sporulation in as little as 24 hours after water damage occurs (Pasanen et al., 1992). Sewage contamination requires prompt and thorough remediation [see 16.2.3 and 16.2.4]. Different procedures are used for restoration of damage by clean (sanitary) and unsanitary water (IICRC, 1995).

To limit microbial growth in materials wet due to leaks or flooding:
1. Within 24 to 48 hours of water damage, dry materials to a moisture level that will not support microbial growth (CMHC, 1993; Foarde et al., 1993; Burge et al., 1994).
2. Discard sewage-contaminated porous materials (IICRC, 1995).
3. Prevent future floods and other moisture incursion (Morey, 1994; Morey, 1995).

10.4.2 Moisture in the Building Envelope

Moisture can accumulate on or inside a building envelope, above or below ground, because of migration from inside or outside of the structure. Among the mechanisms involved are leakage and gravitational flow of water, capillary movement of water through porous

materials, movement of air carrying water vapor, and diffusion of water vapor (ISIAQ, 1996).

Design building envelopes to achieve the following:
1. Reduce the potential for moisture accumulation, including condensation, and provide for egress of water that may accidentally enter the envelope; have an effective drainage plane within wall assemblies to drain rain. (Lstiburek, 1994; ASHRAE, 1996a).
2. Prevent penetration of both surface water (e.g., from rain and snow) and groundwater, including capillary water movement through materials (ASTM, 1994; ISIAQ, 1996).

10.4.3 Controlling Relative Humidity, Temperature, and Condensation

The control of indoor RH levels, surface temperatures, and moisture migration is critical for human comfort as well as being a primary means of minimizing microbial growth (Lstiburek and Carmody, 1991). Air-conditioned buildings should be designed so that the RH of room air does not consistently exceed 60% and the a_w of materials susceptible to biodeterioration does not exceed 0.7 [see 10.3.4] (Lstiburek and Carmody, 1991; ISIAQ, 1996; Morey, 1996). More specific principles useful in controlling condensation are reviewed elsewhere (White, 1990; Burch et al., 1992; ASTM, 1994; ISIAQ, 1996; Morey, 1996).

To control elevated indoor RH:
1. Reduce moisture-generating activities (e.g., by providing local exhaust of humid air to the outdoors).
2. Dilute moisture-laden indoor air with outdoor air at a lower absolute humidity (e.g., during the heating season, in climates where cold, dry outdoor air is available).
3. Dehumidify indoor air by (a) operating the cooling coils in an HVAC system to remove moisture from outdoor or return air, or (b) using console dehumidifiers to maintain RH below 60% (USEPA, 1991).

When the surface temperature of a material equals or approaches the dew-point temperature of the air, condensation conditions are reached and water droplets may form on surfaces. In cold climates, these conditions can occur on poorly insulated windows, walls, and ceilings. In air-conditioned buildings in warm, humid climates, condensation occurs when infiltrating moist air encounters cool surfaces in the air-conditioned zone.

To raise material surface temperature:
1. Increase heat flow to a surface through improved air circulation—thus reducing the RH of air adjacent to surfaces.

2. Reduce heat flow from a surface by insulating exterior walls and controlling wind-washing (i.e., intrusion of air blowing into wall cavities due to cracks in the exterior wall). Wind-washing can be prevented by installing tight sheathing or building paper on a wall's exterior (Lstiburek and Carmody, 1991).

10.4.3.1 Condensation Control in Cold Climates The basic principle behind control of condensation due to vapor migration is to prevent warm moisture-laden air from contacting cool surfaces. In cold climates, condensation can occur within a building envelope when warm, moist indoor air flows outward through cracks and holes. This air cools as it nears the outer boundary of the building shell (White, 1990). RH rises as the air cools, because cooler air has a lower moisture-holding capacity, increasing the risk of condensation in walls.

To control condensation in building envelopes in cold climates:

1. **Insulation.** Install appropriate insulation to prevent large temperature differences between air and surfaces. Make certain that insulation in exterior walls is adequate to prevent thermal bridging or cold spots on interior surfaces.
2. **Air Barriers (Air Retarders).** Control penetration of water into the building envelope via air transport by installing air barriers (if used) on the warm side of the building envelope (Figure 10.3).

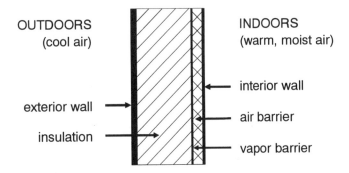

FIGURE 10.3. Location of air and vapor barriers* in a cold climate. *Materials can be selected that act as both air and vapor barriers.

3. **Vapor Diffusion Barriers (Vapor Diffusion Retarders).** Control penetration of water into the building envelope via vapor diffusion by installing vapor barriers (if used) on the warm side of the building envelope (Figure 10.3).
4. **Ventilation.** Use source control (local exhaust ventilation) and dilution ventilation to reduce indoor moisture levels below those that allow condensation to occur (Burch and Thomas, 1991; Burch et al., 1992; Morey, 1994).

10.4.3.2 Condensation Control in Hot, Humid Climates Principles opposite those used in cold climates [see 10.4.3.1] must be employed in air-conditioned buildings in hot, humid areas to prevent the entry of moisture into building envelopes (Burch, 1993; ISIAQ, 1996).

To control condensation in building envelopes in warm, humid regions:

1. **Insulation.** Install appropriate insulation to prevent large temperature differences between air and surfaces.
2. **Air Barriers (Air Retarders).** Install air barriers (if used) in the external portion of building envelopes (Figure 10.4).
3. **Vapor Diffusion Barriers (Vapor Diffusion Retarders).** Install vapor barriers (if used) in the external portion of building envelopes (Figure 10.4).
4. **Wall Coverings.** Avoid impermeable vinyl or other wall coverings and use permeable paints and wall coverings on the interior surfaces of building envelopes (Figure 10.4). •
5. **Pressure Relationships.** HVAC systems in hot, humid climates must be operated to maintain a net positive pressure with respect to the outdoors to avoid entry of unconditioned outdoor air into the building (Figure 10.5). Building pressurization can be achieved by supplying conditioned outdoor air at a rate that exceeds the mechanical exhaust rate and by ensuring that the building envelope is appropriately sealed—to allow pressurization without the need to condition an excess amount of outdoor air.
6. **Excessive Cooling.** Avoid cooling interior spaces below the average, monthly, outdoor, dew-point temperature for the climate in which the building is located (Figure 10.6). This may not be possible in hot, humid climates where the outdoor dewpoint temperature in summer is routinely above that which people find com-

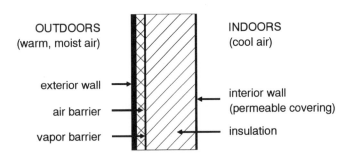

FIGURE 10.4. Location of air and vapor barriers* in an air-conditioned building in a warm, humid climate. *Materials can be selected that act as both air and vapor barriers

fortable for an indoor air temperature. In these cases, proper insulation, correct location of air and vapor barriers, and maintenance of pressure relationships is especially important.

10.4.4 Moisture Control in HVAC System Components
HVAC systems play a key role in (a) providing required ventilation, (b) maintaining desired temperatures, (c) dehumidifying outdoor air (in humid climates), and (d) establishing desired pressure relationships within spaces. However, improper design, operation, or maintenance of an HVAC system can lead to a multitude of problems (NADCA, 1996). Considerations relevant to specific HVAC components follow.

10.4.4.1 Filters
Filters (See A in Figure 4.1) must be maintained to prevent moisture buildup within the filter material. Moisture and the inevitable buildup of dirt on a filter may allow microorganisms to grow on the filter's surface and within the filter medium itself [see 4.3.3]. If kept free of moisture, filter changes can be based on a routine replacement schedule or the pressure drop across the filter as a function of dust buildup.

To limit microbial growth on filters:
1. Protect filters from direct wetting via rain, snow, fog, water leaks, or flooding to avoid microbial growth on the filter medium. Replace filters periodically so they perform to design specifications.
2. Duct humidifiers located upstream of final, higher-efficiency filters should be located at least 4.6 m (15 ft) upstream of the final filters (AIA, 1996).
3. Promptly discard wet filters from HVAC air-handling and fan-coil units.

10.4.4.2 Cooling Coils and Moisture Carryover
HVAC systems should be operated at design airflow rates and

coil temperatures to ensure proper conditioning of air and to prevent moisture entrainment (blow through) from excessive air velocities (ASHRAE, 1996a). Chilled air immediately downstream of the cooling-coil section (See C in Figure 4.1) of an air-handling unit is typically saturated with moisture (RH >90%). Moisture problems occur if an airstream is moved too rapidly through coils to allow sufficient moisture removal. Also, if air moves with too great a velocity over cooling-coil surfaces, water droplets may be released into the ventilation air and may wet surfaces directly downstream of the cooling coils.

To limit microbial growth in or near condenser units:
1. Avoid the use of porous materials on airstream surfaces in persistently wet areas of HVAC systems (Price et al., 1994; Morey, 1995; Morey, 1996).
2. Design and operate dehumidification cooling coils for minimal carryover of water droplets (ASHRAE, 1996a).

10.4.4.3 Drain Pans
The inability of drain pans (See C in Figure 4.1) to properly remove condensation from an air handler is an important control issue because drain pans can be significant sources of microbial growth (ASHRAE, 1989).

To limit microbial growth in or near drain pans:
1. Slope pans to drain completely [e.g., a drop of ~0.2 cm for every 10 cm of pan length (0.25 in/ft)] (ISIAQ, 1996). The slope should direct water toward a drainage point, preferably from the bottom of a pan (Figure 10.7).
2. To allow water to drain properly, isolate the pressure difference between an air-handling unit under negative pressure relative to a mechanical room by installing a water trap in the drain line. The effective height of the water trap should be

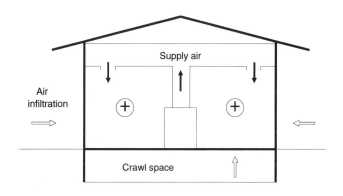

FIGURE 10.5. Recommended pressure relationship between indoors and outdoors in a hot, humid climate (adapted from Lstiburek, 1994).

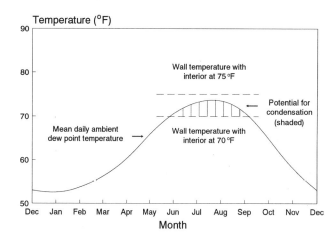

FIGURE 10.6. Potential for condensation when cooling exceeds outdoor dew-point temperature (adapted from Lstiburek, 1994).

40% greater than the expected peak static pressure of the supply air fan (i.e., 1.4 times the peak static pressure in centimeters or inches of water gage) (Figure 10.7) (Bearg, 1993).

3. Keep drain pans clean to avoid extensive microbial growth. Physically remove growth that develops; biocide treatment without removal of microbial growth is inadequate (ASHRAE, 1996a). In humid climates, components should be inspected monthly and cleaned, if necessary.

10.4.4.4 Humidifiers Humidifiers (See D in Figure 4.1) should be used with caution because moisture from humidifiers may support microbial growth on wet surfaces in and around units and in the building envelope where moisture can condense during cold weather (ASHRAE, 1996a; ISIAQ, 1996). Humidifiers that emit droplets from a reservoir of recirculated water are especially prone to microbial contamination and bioaerosol dissemination (Ager and Tickner, 1983; Morey et al., 1984; Tyndall et al., 1989). Potable water supplied to steam humidifiers is less likely to contain viable microorganisms than the recirculated water in air washers. Therefore, steam usually does not contain living organisms. However, steam injection systems do not "sterilize" surfaces within HVAC systems, nor does the high temperature of added steam kill microorganisms already present in the air being humidified (Burkhart et al., 1993).

To limit microbial growth in and dissemination of bioaerosols from humidifiers:

1. If steam is used, supply "clean steam." Raw steam from a central boiler may contain toxic

corrosion inhibitors (e.g., morpholine) (NRC, 1983; Tyndall et al., 1989).

2. Discourage the use of console humidifiers or vaporizers in workplaces because the cleaning and disinfection of these units must be fastidious to avoid microbial colonization (AIA, 1987).

3. Avoid water-spray humidifiers and air washers in non-industrial HVAC systems because such units require frequent maintenance to prevent microbial growth (Burkhart et al., 1993; AIA, 1996). Such units have been associated with outbreaks of humidifier fever and HP (Arnow et al., 1978; Hodgson et al., 1987).

4. Avoid exposed insulation and air cleaners (e.g., filters) in HVAC plenums or ductwork downstream of humidifiers within the absorption distance recommended by manufacturers to allow complete entrainment of humidifier moisture by the ventilation air stream (ASHRAE, 1996a) [see 10.4.4.1].

10.4.4.5 Airstream Surfaces in Plenums and Ducts (F in Figure 4.1). Moisture accumulation, combined with the presence of dirt and organic substrates in HVAC systems, creates a suitable environment for microbial growth (Health Canada, 1995; Morey, 1996). Of potential concern are materials in the vicinity of moisture-producing equipment (e.g., humidifiers or dehumidifying cooling coils) or in other areas subject to moisture incursion (e.g., OAIs).

To limit microbial growth in plenums and ducts:

1. The airstream surfaces of HVAC equipment and ductwork should resist (a) the accumula-

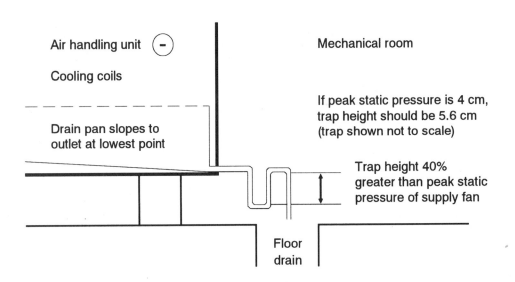

FIGURE 10.7. Detail of water trap at condensate drain line. Drain pan slopes ~0.2 cm for every 10 cm of pan length (0.25 in/ft) (adapted from Bearg, 1993).

tion of dirt—or be readily cleanable or replaceable, (b) moisture absorption or retention, and (c) biodeterioration (ASHRAE, 1996a; ISIAQ, 1996).

2. Maintain all surfaces within the HVAC plenum to prevent the accumulation of moisture or debris.

3. Surfaces in the vicinity of moisture-producing equipment should be smooth and non-absorbent. Keep porous insulation in these areas free of dirt and moisture or protect surfaces with a layer that is impermeable to water (e.g., sheet metal).

10.5 HVAC System Design and Operation

10.5.1 Access

Access is essential for proper maintenance and inspection of HVAC systems. Access doors or panels must be provided for all components, for example: ductwork, plenum walls, drain pans, upstream and downstream areas of filters and fan-coil units, and terminal boxes (AIA, 1996; NADCA, 1997).

10.5.2 Ventilation

Inadequate and improper distribution of ventilation air throughout a space can create a multitude of contaminant-related problems. Proper sizing of HVAC equipment to meet occupancy demands (specifically, air-conditioning and dehumidifying capacity) is critical for moisture removal from ventilation air. Dilution ventilation using filtered outdoor and recirculated indoor air can reduce the concentration of some internally produced bioaerosols [see 4.3.2.1]. However, dilution ventilation is ineffective where the sources of such aerosols are within the HVAC system itself or where sources within a building emit at rates greater than the removal rate provided by the ventilation system. In fact, increased air velocities may raise indoor bioaerosol concentrations if the increased turbulence causes particle release from areas of microbial contamination (Pasanen et al., 1991). In addition, ventilation with excessively moist outdoor air when an HVAC system cannot cope with the latent moisture load may result in serious indoor contamination (Morey, 1993).

10.5.2.1 Ventilation to Control Airborne Infectious Agents Dilution ventilation has historically been used to reduce the incidence of infectious diseases transmitted from person to person through the air (e.g., the common cold and tuberculosis) (Morey, 1994; ASHRAE, 1996a). ASHRAE (1989, 1996a) has recommended a minimum dilution ventilation rate of 7.5 L/s (15 ft³/min) of outdoor air per person to reduce exposure to such human-source microorganisms. Some evidence suggests higher rates of outdoor air supply reduce the incidence of upper respiratory infection (Dick et al., 1987; Brundage et al., 1988). Exposure to *Mycobacterium tuberculosis* can be reduced by the strategies outlined in the most recent CDC and OSHA Guidelines. Guidelines for tuberculosis prevention include engineering controls such as local exhaust and dilution ventilation as well as air cleaning (CDC, 1994). For a discussion of the transmission of airborne infections and ventilation, see Sections 3.5.2, 8.2.3.2, and 8.2.3.3. For a discussion of intervention strategies—including dilution ventilation—for reducing exposure to airborne infectious agents, see Section 9.8.

10.5.2.2 Ventilation to Control Airborne Allergens Airborne allergenic particles (e.g., cat and mite allergens generated in the occupied spaces of buildings) are not effectively removed through ventilation techniques (IOM, 1993). The rate at which cat allergens are generated and fungal spores are released from contaminated surfaces may overwhelm particle removal through dilution ventilation (Luczynaka et al., 1990). Disturbance of settled dust releases mite allergens sporadically and generally close to a person's breathing zone. However, due to their size, particles containing dust-mite and cockroach allergens do not remain airborne sufficiently long to allow practical control through ventilation or the use of portable air cleaners (Antonicelli et al., 1991; Platts-Mills and Carter, 1997) [see 10.5.6 and 19.6.2]. Moisture control, source removal, and other interventions are required to control exposures to allergens [see 25.6].

10.5.3 Filters

Properly installed and maintained filters are essential in HVAC systems to remove particles from outdoor air, for example, mineral dust; soot; pollen, wind-borne seeds, plant hairs, and other plant fragments; flying insects and other animal debris; and large fungal spores (Rose and Hirsh, 1979). Filters have been classified in terms of weight arrestance and dust spot efficiency (ASHRAE, 1992). *Arrestance* describes a filter's ability to capture coarse, synthetic dust and is calculated as a percentage on a weight basis. *Dust spot efficiency* classifies a filter according to its ability to remove finer airborne dusts that can visibly soil interior surfaces. Dust spot efficiency is calculated as a percentage based on light transmission.

ASHRAE has proposed a new standard method to test filter efficiency (Hanley et al., 1995; ASHRAE, 1996b). This method will provide filtration efficiency curves over the 0.3- to 10-μm diameter particle size range for clean and dust-loaded filters. These data should provide a more reliable means of selecting filters for control of respirable-size bioaerosols than the current dust spot and weight arrestance methods. The anticipated release date of the standard is 1998. ASHRAE (1996b) has promulgated a new filtration rating standard based on filter efficiency for particles of various aerodynamic diameters. The draft revision of ASHRAE Standard 62-1989 (1996a) would require a 65% fractional efficiency for particles 3-μm diameter, which would be equivalent to a 30% atmospheric dust spot efficiency.

Filters in HVAC systems should be of the highest grade compatible with the system and the air-handler fan. If higher-efficiency filters are installed, it may be necessary to modify an air-handler's filter housing and to replace the air-handler fan motor to accommodate the increased pressure drop. Most central air handlers can support 30% dust-spot-efficient filters. These filters are significantly more effective at capturing particles in the bioaerosol size range than are standard furnace filters, which are usually <20% efficient. However, a fraction of fungal spores, dirt, and soot smaller than 5 μm can pass through a 30% filter. Filters with a dust-spot efficiency in the 50% to 70% range are ~65% efficient at removing 1- to 3-μm particles and will remove the majority of 1- to 2-μm bioaerosols.

In two-filter systems, the primary (low-efficiency) filter removes larger particles and, subsequently, extends the operating life of the secondary (medium- or high-efficiency) filter. Martiny et al. (1995) studied the occurrence of microorganisms in different filter media from HVAC systems and documented an average reduction ratio of 24 to 1 between primary and secondary filters.

Higher efficiency filters are appropriate in environments that require greater assurance of contaminant removal, such as patient areas in health-care settings. In critical-care areas (e.g., hospitals), it is necessary to provide filtration both upstream and downstream of cooling coils to ensure bioaerosol removal and to protect highly susceptible patients. Guidelines for HVAC systems serving inpatient care, treatment, and diagnosis areas recommend two sets of filters, that is, 30% dust-spot-efficient prefilters upstream of the air-conditioning equipment and ≥90% dust-spot-efficient filters downstream of any fans or blowers (AIA, 1996).

10.5.4 Methods Other than Filtration for Controlling Intrusion of Bioaerosols from Outdoor Sources

The following recommendations should be followed in addition to filtration of outdoor air to prevent and control exposure to outdoor-source bioaerosols.

To limit exposure to bioaerosols from outdoor sources:

1. Locate OAIs at sites with the highest quality outdoor air. This means that inlets should be as far as possible from standing water, soil, plant debris, accumulated bird droppings, and other sources of microorganisms that may become entrained. OAIs should be protected with properly designed bird screens (ASHRAE, 1996a). In large buildings, OAIs should be located as high as practical above grade level.

2. Minimize entrainment into OAIs of cooling-

tower mist that may contain legionellae, amebae, and other microorganisms by locating OAIs and other building openings at least 7.5 m (25 ft) (preferably 15 m, 50 ft) upwind of and horizontally separated from cooling towers and evaporative condensers (AIA, 1987; Broadbent et al., 1992; Morey, 1994). Even OAI placement at these distances does not ensure that aerosols in the exhaust air (drift) will not be entrained (Dondero et al., 1980; Band et al., 1981; Garbe et al., 1985; Friedman et al., 1987). Airflow patterns and prevailing wind direction are important factors affecting drift entrainment (Dennis, 1990). Regardless of the separation distance between a cooling tower and an OAI, fastidious maintenance and treatment programs should be implemented to control growth of *Legionella* spp. and other microorganisms in cooling towers (Barbaree, 1991; HSE, 1991; Broadbent et al., 1992; ASM, 1993; AHSRAE, 1996a; ASTM, 1996; Freija, 1996; ISIAQ, 1996) [see 10.6].

10.5.5 Intrusion of Moisture from Outdoors

OAIs for HVAC systems should be designed and operated to prevent entrainment of rain, snow, and fog (Morey, 1994). The introduction of humid, unconditioned air into a building should be prevented (e.g., during economizer operation or off-hours when cooling coils are deactivated). In humid climates, enthalpy controllers should be used to activate or deactivate the economizers.

10.5.6 Portable Air Cleaners

The use of portable air cleaners for supplemental control of particles (including bioaerosols) has increased in recent years. However, air cleaners are subject to some of the same limitations as dilution ventilation [see 10.5.2.2] and low airflow rates limit the performance of many portable air cleaners (Godish, 1989). Therefore, such units should be appropriately sized for optimum particle removal, that is, the rate of air circulation through a unit (the supply of cleaned air) must be greater than the source emission rate. This may be difficult to achieve for strong sources and in large spaces. Portable air cleaners using high-efficiency particulate air (HEPA) filters or electrostatic precipitators have demonstrated the highest efficiency with respect to particle removal (Offerman et al., 1985; Nelson et al., 1988; Consumer Reports, 1989, 1992; Shaughnessy et al., 1994; First, 1998). Filter air cleaners—even 90%-efficient filters—have shown promise as supplemental control measures to prevent airborne infectious diseases (Rutala et al., 1995; Miller-Leiden et al., 1996). Ozone generators have not been shown to effectively remove bioaerosols (Shaughnessy et al., 1994) and ozone is not an effective gas-phase biocide [see 16.2.5].

10.6 Control of Legionellae

Legionnaires' disease is possibly the most common airborne bacterial infection associated with environmental contamination of office buildings and one associated with mortality as well as morbidity [see 8.2.3.1].

To limit growth of *Legionella* spp. in potable water systems:

1. Properly size hot water tanks to achieve desired water temperature at anticipated usage.
2. Maintain water temperatures at or below 20°C and at or above 60°C (≤68° and ≥140°F) in cold and hot water systems respectively to minimize the multiplication of *Legionella* spp.
3. Prevent water stagnation in plumbing systems and disconnect unused equipment and plumbing sections (dead legs).
4. Where recommended, continuously chlorinate water to maintain a concentration of free residual chlorine of 1 to 2 mg/L at the tap (ASTM, 1996; Freije, 1996; CDC, 1997; Fields, 1997).

To limit growth of *Legionella* spp. in water-cooled, heat-transfer systems and dissemination of the bacteria from such systems:

1. Construct cooling towers and evaporative condensors of materials that can be readily disinfected and will not support microbial growth.
2. Regularly maintain and properly operate equipment (including the use of biocides, where advised) to limit biological contamination and to identify and address problems promptly.
3. Install efficient drift eliminators on cooling towers and evaporative condensors to intercept water droplets where air is discharged.

10.7 Long-Term Prevention and Cleaning Management

10.7.1 Long-Term Prevention

All control programs should include (a) anticipation of conditions that may lead to microbial contamination, and (b) implementation of plans to prevent such conditions. Among these measures are long-term plans to maintain buildings. Three basic strategies should be followed to maintain building performance and prevent microbial contamination (a) routine surveillance inspections and prompt response to problems, (b) adequate preventive maintenance of the building structure as well as HVAC and plumbing systems, and (c) adequate housekeeping including an emphasis on proper and routine cleaning. These activities are primary and secondary preventive measures (i.e., activities undertaken to prevent problems or promptly address situations before they become problematic). These approaches are generally more cost effective than remediation of problems that went unrecognized or were ignored.

10.7.2 Cleaning Management

Cleaning is the process of identifying, containing, remov-

ing, and properly disposing of contaminants from a surface or environment in order to protect human health and valuable materials. Cleaning is the final defense in managing indoor environmental quality. After all other areas of source management and building operations have been addressed, cleaning is still necessary. The importance of cleaning was demonstrated in a year-long study of cleaning effectiveness in a multi-use building without evident problems (Franke et al., 1997). The routine use of high-efficiency vacuum cleaners, damp dusting, and improved cleaning products—particularly in high traffic areas—along with attention to events such as leaks and spills resulted in meaningful decreases in particulate and microbial contamination. After seven months of improved cleaning practices and environmental monitoring, the data showed decreases of 50% in airborne dust mass, 61% in airborne fungi, 40% in airborne bacteria, 40% in fungi in carpet dust, 84% in bacteria in carpet dust, and 72% in endotoxin in carpet dust.

10.8 References

AIA: Guidelines for Design and Construction of Hospitals and Health Care Facilities. American Institute of Architects, Washington, DC (1987).

AIA: Ch. 7. General Hospital. In: Guidelines for Design and Construction of Hospital and Health Care Facilities, pp. 11-61. American Institute of Architects Press, Washington, DC (1996).

Ager, B.; Tickner, J.: The Control of Microbiological Hazards Associated with Air-Conditioning and Ventilation Systems. Ann. Occup. Hyg. 27:341-358 (1983).

Antonicelli, L.; Biol, M.; Pucci, S.; et al.: Efficacy of an Air-Cleaning Device Equipped with a HEPA Filter in House Dust Mite Respiratory Allergy. Allergy 46:596-600 (1991).

Arnow, P.M.; Fink, J.N.; Schlueter, D.P.; et al.: Early Detection of Hypersensitivity Pneumonitis in Office Workers. Am. J. Med. 64:236-242 (1978).

ASHRAE: Ventilation for Acceptable Air Quality, ASHRAE Standard 62-1989. American Society of Heating, Refrigerating and Air-Conditioning Engineers, Atlanta, GA (1989).

ASHRAE: Gravimetric and Dust Spot Procedures for Testing Air Cleaning Devices used in General Ventilation for Removing Particulate Matter, Standard 52.1-1992. American Society of Heating, Refrigerating and Air-Conditioning Engineers, Atlanta, GA (1992).

ASHRAE: Ventilation for Acceptable Indoor Air Quality. ASHRAE Standard 62R. American Society of Heating, Refrigerating and Air-Conditioning Engineers, Atlanta, GA (1996a).

ASHRAE: Method of Testing General Ventilation Air-Cleaning Devices for Removal Efficiency by Particle Size, Standard 52.2. November, 1996, ASHRAE Transactions 3842 (RP-671). American Society of Heating, Refrigerating and Air-Conditioning Engineers, Atlanta, GA (1996b).

ASM: *Legionella.* Current Status and Emerging Perspectives. J.M. Barbaree, R.F. Breiman, A.P. Dufour, Eds. American Society for Microbiology, Washington, DC (1993).

ASTM: Moisture Control in Buildings. H. Trechsel, Ed. American Society for Testing and Materials, Philadelphia, PA (1994).

ASTM: Standard Guide for Inspecting Water Systems for Legionellae and Investigating Possible Outbreaks of Legionellosis (Legionnaires' Disease or Pontiac Fever). D 5952-96. American Society for Testing and Materials, West Conshohocken, PA (1996).

Band, J.; LaVenture, M.; Davis, J.; et al.: Epidemic Legionnaires' Disease; Airborne Transmission Down a Chimney. JAMA 245:2404-

2407 (1981).

Barbaree, J.: Controlling Legionella in Cooling Towers. ASHRAE Journal. June:38-42 (1991).

Bearg, D.W.: Individual Components of HVAC Systems. In: Indoor Air Quality and HVAC Systems, pp. 39-60. Lewis Publishers, Boca Raton, FL (1993).

Berry, M.A.: Protecting the Built Environment: Cleaning for Health. Tricomm 21st Press, Chapel Hill, NC (1993).

Broadbent, C.; Marwood,L.; Bentham, R.: Legionella Ecology in Cooling Towers. Australian Refrigeration, Air-Conditioning and Heating. 46(10):20-34 (1992).

Burch, D.; Thomas, W.; Fanney, A.: Water Vapor Permeability Measurements of Common Building Materials. In: ASHRAE Technical Data Bulletin 8(3):69-77. American Society of Heating, Refrigerating and Air-Conditioning Engineers, Atlanta, GA (1992).

Burch, D.; Thomas, W.: An Analysis of Moisture Acculmulation in a Wood Frame Wall Subjected to Winter Climate. NISTIR 4674. National Institute of Standards and Technology, Gaithersburg, MD (1991).

Burch, P.: An Analysis of Moisture Accumulation in Walls Subjected to Hot and Humid Climates. ASHRAE Transactions. 99(2) (1993).

Burge, H. A.; Sui, H.J.; Spengler, J.D.: Moisture, Organisms, and Health Effects. In: Moisture Control in Buildings, pp. 84-90, MNL 18. H. Trechsel, Ed. American Society for Testing and Materials, Philadelphia, PA (1994).

Burge, H.A.: Bioaerosols in the Residential Environment. In: Bioaerosols Handbook, pp. 579-597. C.S. Cox, C. M. Wathes, Eds. Lewis Publishers, CRC Press, Boca Raton, FL (1995).

Burkhart, J.E.; Stanevich, R.; Kovak, B.: Microorganism Contamination of HVAC Humidification Systems: Case Study. Appl. Occup. Environ. Hyg. 8:1010-1014 (1993).

Brundage, J.F.; Scott, R.McN.; Lednar, W.M.; et al.: Building-Associated Risk of Febrile Acute Respiratory Diseases in Army Trainees. JAMA 259:2108-2112 (1988).

Canada Mortgage and Housing Corporation: Cleanup Procedures for Mold in Houses. Public Affairs Center, Ottawa, Ontario, Canada. NH15-91 (1993).

CDC: Appendix B. Maintenance Procedures Used to Decrease Survival and Multiplication of Legeionella sp. in Potable-Water Distribution Systems, p. 74. Centers for Disease Control, Atlanta, GA. CDC — MMWR. 46(RR-1) (1997).

CDC: Appendix B. Maintenance Procedures Used to Decrease Survival and Multiplication of Legeionella sp. in Potable-Water Distribution Systems, Appendix D. Centers for Disease Control, Atlanta, GA. CDC — MMWR. 46(RR-1) (1997).

CDC: Procedure for Cleaning Cooling Towers and Related Equipment. In: Guidelines for Prevention of Nosocomial Pneumonia, pp. 77-79. Centers for Disease Control, Atlanta, GA. CDC — MMWR. 46(RR-1) (1997).

CDC: Guidelines for Preventing Transmission of Mycobacterium tuberculosis in Health-Care Facilities. Centers for Disease Control, Atlanta, GA. CDC — MMWR. 43(RR-13) (1994).

Consumer Reports: Air Purifiers. Feb., pp. 88-93 (1989).

Consumer Reports: Household Air Cleaners. Oct., pp. 657-662 (1992).

Dennis, P.: An Unnecessary Risk: Legionnaires' Disease. In: Biological Contaminants in Indoor Environments, pp. 84-98. ASTM STP 1071. American Society of Testing and Materials, Philadelphia, PA (1990).

Dick, E.; Jennings, L.; Mink, K.; et al.: Aerosol Transmission of Rhinovirus Colds. J. Infect. Dis. 156:442-448 (1987).

Dondero, T.; Rendtorff, R.; Mallison, G.; et al.: An Outbreak of Legionnaires' Disease Associated with a Contaminated Air-Conditioning Cooling Tower. New Eng. J. Med. 302:365-370 (1980).

Fields, B.S.: Legionellae and Legionnaires' disease. In: Manual of Environmental Microbiology, pp. 666-675. C.J. Hurst, G.R. Knudsen, M.J. McInerney, et al., Eds. American Society for Microbiology, Washington, DC (1997).

First, M.W.: HEPA Filters. J. Am. Biological Safety Assoc. 3:33-42 (1998).

Flannigan, B.: Approaches to Assessment of Microbial Flora of Buildings. In: IAQ '92, Environments for People, pp. 139-145. American Society of Heating, Refrigerating and Air-Conditioning Engineers, Atlanta, GA (1992).

Foarde, K.; Dulaney, P.; Cole, E.; et al.: Assessment of Fungal Growth on Ceiling Tiles under Environmentally Characterized Conditions. In: Proceedings of Indoor Air '93, pp. 357-362. Helsinki Finland (1993).

Franke, D.L.; Cole, E.C.; Leese, K.E.; et al.: Cleaning for Improved Indoor Air Quality: An Initial Assessment of Effectiveness. Indoor Air. 7:41-54 (1997).

Freije, M.R.: Legionellae Control in Health Care Facilities: A Guide for Minimizing Risk. HC Information Resources, Inc., Indianapolis, IN (1996).

Friedman, S.; Spitalny, K.; Barbaree, J.; et al.: Pontiac Fever Outbreak Associated With a Cooling Tower. Am. J. Pub. Health. 77:568-572 (1987).

Garbe, P.; Davis, B.; Weisfeld, J.; et al.: Nosocomial Legionnaires' Disease; Epidemiologic Demonstration of Cooling Towers as a Source. JAMA. 254:521-524 (1985).

Godish, T.: Indoor Air Pollution Control. Lewis Publishers, Chelsea, MI (1989).

Grant, C.; Hunter, C.; Flannigan, B.; et al.: The Moisture Requirements of Moulds Isolated from Domestic Dwellings. Interm. Biodeterioration 25:259-289 (1989).

Griffin, D.H.: Fungal Physiology. Wiley, New York, NY (1981).

Hanley, J.P.; James, T.; Smith, D.D.; et al.: A Fractional Aerosol Filtration Efficiency Test Method for Ventilation Air Cleaners. ASHRAE Transactions. American Society of Heating, Refrigerating and Air-Conditioning Engineers, Atlanta, GA (1995).

Health Canada: Fungal Contamination in Public Buildings: A Guide to Recognition and Management. Federal-Provincial Committee on Environmental and Occupational Health. Environmental Health Directorate, Ontario, Canada (1995).

Hodgson, M.J.; Morey, P.R.; Simon, J.S.; et al.: An Outbreak of Recurrent Acute and Chronic Hypersensitivity Pneumonitis in Office Workers. Am. J. Epidemiol. 125(4):631-638 (1987).

Hunter, C.; Sanders, C.: Mould: Chapter 2 In: Guidelines and Practice, Energy Conservation in Buildings and Community Systems, Vol. 2, Annex XIV. Leuven University, International Energy Agency, Belgium (1990).

HSE: The Control of Legionellosis Including Legionnaires' Disease. Health and Safety Booklet H.S. (G)70. London: Health and Safety Executive, London, England (1991).

IOM: Indoor Allergens: Assessing and Controlling Adverse Health Effects. Institute of Medicine, Washington, DC (1993).

IICRC: Standard and Reference Guide for Professional Water Damage Restoration S500-94. Institute of Inspection, Cleaning and Restoration Certification, Vancouver, WA (1995).

ISIAQ: Control of Moisture Problems Affecting Biological Indoor Air Quality. TFI-1996. International Society of Indoor Air Quality and Climate, Ottawa, Ontario, Canada (1996).

Kendrick, B.: The Fifth Kingdom. Mycologue Publications, Newburyport, MA (1992).

Lacey, J.; Venette, J.: Outdoor Air Sampling Techniques. In: Bioaerosols Handbook, C.S. Cox and C.M. Wathes, Eds. CRC Press, Inc., Boca Raton, FL (1995).

Lstiburek, J.; Carrmody, J.: Moisture Control Handbook: Principles and Practices for Residential and Small Commercial Buildings, pp. 1-14. Wiley, New York, NY (1991).

Lstiburek, J.W.: Mold, Moisture, and Indoor Air Quality. Building Science Corporation, Chestnut Hill, MA (1994).

Luczynaka, C.; Li, Y.; Chapman, M.; et al.: Airborne Concentrations and Particle Size Distribution of Allergen Derived from Domestic Cats. Am. Rev. Resp. Dis. 141:361-367 (1990).

Maroni, M.; Axelrad, R.; Bacaloni, A.: NATO's Efforts to Set Indoor Air Quality Guidelines and Standards. Am. Ind. Hyg. Assoc. J. 56:499-508 (1995).

Martiny, H.; Möritz, M.; Rüden, H.: Occurrence of Microorganisms in Different Filter Media of Heating, Ventilation, and Air-Conditioning (HVAC) Systems. In: Proceedings of the ASHRAE Conference, IAQ 94. American Society of Heating, Refrigerating and Air-Conditioning Engineers, Atlanta, GA (1995).

lanta, GA (1992).

Miller-Leiden, S.; Lobascio, C.; Nazaroff, W.W.; et al.: Effectiveness of In-Room Air Filtration and Dilution Ventilation for Tuberculosis Infection Control. J. Air Waste Manage. Assoc. 46:869-882 (1996).

Morey, P.: Use of Hazard-Communication Standard and General Duty Clause During Remediation of Fungal Contamination. In: Indoor Air '93, pp. 391-395, July 4-8, 1993. Helsinki, Finland (1993).

Morey, P.R.: Suggested Guidance on Prevention of Microbial Contamination for the Next Revision of ASHRAE Standard 62. In: Proceedings to ASHRAE IAQ '94, Engineering Indoor Environments, pp. 139-148. American Society of Heating, Refrigerating and Air-Conditioning Engineers, Atlanta, GA (1994).

Morey, P.: Biocontamination Control Guidelines for the Revision of ASHRAE Standard 62. In: Proceedings of Healthy Building '95. Milan, Italy (1995).

Morey, P.: Mold Growth in Buildings: Removal and Prevention. In: Proceedings of Seventh International Conference on Indoor Air Quality and Climate, Nagoya, Japan (1996).

Morey, P.; Hodgson, M.; Sorenson, W.; et al.: Environmental Studies in Moldy Office Buildings: Biological Agents, Sources and Preventive Measures. Ann. Am. Conf. Gov. Ind. Hyg. 10:21-35 (1984).

Muilenberg, M.L.: The Outdoor Aerosol. In Bioaerosols, pp. 163-204. H.A. Burge, Ed. Lewis Publishers, Boca Raton, FL (1995).

NADCA: Understanding Microbial Contamination in HVAC Systems. National Air Duct Cleaners Association, Washington, DC (1996).

NADCA: Requirements for the Installation of Service Openings in HVAC Systems. NADCA 05-1997. National Air Duct Cleaners Association, Washington, DC (1997).

NRC: An Assessment of the Health Risks of Morpholine and Dimethylaminoethanol. National Research Council, Washington, DC (1983).

Nelson, H.; Hirsch, S.R.; Ohman, J.; et al.: Recommendations for the Use of Residential Air-Cleaning Devices in the Treatment of Allergic Respiratory Diseases. J. Allergy. 10/88: 661-669 (1988).

Offermann, F.J.; Sextro, R.G.; Fisk, W.J.; et al.: Control of Respirable Particles in Indoor Air with Portable Air Cleaners. Atmos. Environ. 19:1761-1771 (1985).

Pasanen, A. L.; Kalliokoski, P.; Pasanen, P.; et al.: Laboratory Studies on the Relationship between Fungal Growth and Atmospheric Temperature and Humidity. Environ. Intern. 17: 225-228 (1991).

Pasanen, A. L.; Heindnen-Tanski, H.; Kolliokoski, P.; et al.: Fungal Microcolonies on Indoor Surfaces —An Explanation for the Base-Level Fungal Spore Counts in Indoor Air. Atmos. Enviro. 26B:117-120 (1992).

Platts-Mills, T.A.; Carter, M.C.: Asthma and Indoor Exposure to Allergens. New Eng. J. Med. 336:1382-1384 (1997).

Price, D.; Simmons, R.; Ezeonu, I.; et al.: Colonization of Fiberglass Insulation Used in HVAC Systems. J. Industr. Micro. 12:154-158 (1994).

Richards, R.F.; Burch, D.M.; Thomas, W.C.: Water Vapor Sorption Measurements of Common Building Materials. ASHRAE Technical Data Bulletin 8:58-68 (1992).

Rose, H.; Hirsh, S.: Filtering Hospital Air Decreases *Aspergillus* Spore Counts. Am. Rev. Resp. Dis. 119:511-513 (1979).

Rutala, W.A.; Jones, S.M.; Worthington, J.M.; et al.: Efficacy of Portable Filtration Units in Reducing Aerosolized Particles in the Size Range of *Mycobacterium tuberculosis*. Infect. Control Hosp. Epidemiol. 16:391-398 (1995).

Samson, R.A.; Flannigan, B.; Flannigan, M.E.; et al,. Eds.: Recommendations. In: Health Implications of Fungi in Indoor Environments, pp. 531-538. Elsevier, New York, NY (1994).

Shaughnessy, R.J.; Levetin, E.; Blocker, J.; et al.: Effectiveness of Portable Indoor Air Cleaners: Sensory Testing Results. Indoor Air 4:179-188 (1994).

Troller, J.A.; Scott, V.N.: Measurement of Water Activity (a_w) and Acidity. In: Compendium of Methods for the Microbiological Examination of Foods, 3rd Ed., pp. 135-151. C. Vanderzant and D.F. Splittstoesser, Eds. Am. Pub. Health Assoc., Washington, DC (1992).

Tyndall, R.; Dudney, C.; Katz, D.; et al.: Characterization of Microbial Content and Dispersion from Home Humidifiers. Report to Consumer Product Safety Commission, Project No. 86-1283 (1989).

USEPA: Building Air Quality Document. U.S. Environmental Protection Agency. Washington, DC (1991).

White, J.: Solving Moisture and Mould Problems. In: Indoor Air Quality and Climate, Proceedings of the 4th Intern. Conf. Indoor Air Quality and Climate, Aug. 17-21, 1987, Berlin, West Germany. 4:589-594 (1990).

Chapter 11

Air Sampling

Klaus Willeke and Janet M. Macher

11.1   General Considerations
11.2   Sampling Efficiency
    11.2.1   *Inlet Efficiency*
    11.2.2   *Particle Removal Efficiency*
    11.2.3   *Biological Recovery Efficiency*
11.3   Principles of Particle Collection
    11.3.1   *Overview*
    11.3.2   *Inertial Impaction*
        11.3.2.1   The Impaction Zone
        11.3.2.2   Particle Stopping Distance
        11.3.2.3   Impactor Cut-Off Diameter
        11.3.2.4   Particle Aerodynamic Diameter
        11.3.2.5   Particle Bounce
        11.3.2.6   Particle Blow Off
        11.3.2.7   Satellite Colonies and Particles
    11.3.3   *Liquid Impingers and Wetted Cyclones*
        11.3.3.1   Impinger Operation
        11.3.3.2   Wetted Cyclone Operation
        11.3.3.3   Collection Fluid
        11.3.3.4   Processing Liquid Samples
    11.3.4   *Filtration*
        11.3.4.1   Capillary Pore Filters
        11.3.4.2   Porous Membrane Filters
    11.3.5   *In Situ or Remote Bioaerosol Detection*
11.4   Equipment Calibration
    11.4.1   *Importance of Airflow Calibration*
    11.4.2   *Calibration Methods*
    11.4.3   *Challenge Aerosols*
11.5   Equipment Selection
    11.5.1   *Widely Used Bioaerosol Samplers*
        11.5.1.1   Multiple-Hole Impactors
        11.5.1.2   Slit Impactors
        11.5.1.3   Centrifugal Samplers
        11.5.1.4   Liquid Impingers and Wetted Cyclones
        11.5.1.5   Filters
    11.5.2   *Particle Size*
    11.5.3   *Performance Criteria for Bioaerosol Samplers*
11.6   Summary
11.7   References

11.1 General Considerations

Investigators collect air samples for biological agents to test hypotheses about indoor environments. Some reasons for collecting air samples are to detect and quantify bioaerosol presence, to identify bioaerosol release from sources, to assess human exposure to biological agents, and to monitor the effectiveness of control measures.

Airborne biological material may consist of (a) individual microorganisms, spores, or pollen grains, (b) aggregates of microorganisms, spores, pollen, or other biological material, (c) products or fragments of microorganisms, plants, arthropods, birds, or mammals, or (d) any of the above carried on other particles. MVOCs are the only biologically derived airborne contaminants considered in this book that are not particulate in nature. Section 26.4 discusses air sampling for MVOCs. The chapters in Part III describe air sampling for individual groups of biological agents. This chapter discusses the mechanisms involved in particle collection and introduces widely used, commercially available air sampling instruments.

11.2 Sampling Efficiency

The physics of removing particles from air and the general principles of good sample collection apply to all airborne materials, whether of biological or other origin. Therefore, many of the basic principles investigators use to identify and quantify other airborne particulate matter can be adapted to bioaerosol sampling. Common to all aerosol samplers is consideration of collection efficiency. For collection of some biological agents, other factors also play a role, and the overall efficiency of a bioaerosol sampler can be divided into three components.

1. *Inlet sampling efficiency* — a measure of the ability of a sampling inlet to entrain particles from the ambient environment without bias as to particle size, shape, or aerodynamic behavior.

2. *Particle removal efficiency* — a measure of a device's ability to separate particles from the sampled air stream and to deposit them on or in a collection medium.

3. *Biological recovery efficiency* — a measure of a sampler's ability to deliver the collected particles to an assay system without altering the viability, biological activity, physical integrity, or other essential characteristics of the biological agents.

11.2.1 Inlet Efficiency

Drawing a representative aerosol sample into an inlet is not a trivial process (Brockmann, 1993). Particle size (diameter), density, and shape; a sampler's face velocity, inlet design, and orientation; and ambient wind velocity determine how efficiently bioaerosols enter a sampling device. Indoor air is often considered to be still, calm, or quiescent (e.g., air velocity below ~0.5 m/s). Therefore, a sampler's inlet may be oriented in any convenient position relative to gravity and ambient airflow direction without compromising the representativeness of the particles collected. However, investigators occasionally need to collect samples from moving air streams (e.g., within ventilation ducts or at outdoor air intakes or supply air diffusers). When sampling from a moving air stream, a sampler's inlet should be oriented *isoaxially* so that the direction of the sample flow aligns with that of the moving airstream's flow. Sampling is said to be *isokinetic* when it is isoaxial and the mean face velocity through the inlet equals the airstream velocity (Brockmann, 1993). Advice for measuring air velocities and collecting samples within ducts is available elsewhere (ACGIH, 1995; Billings and Lundgren, 1995).

Experimentally, Martinez et al. (1991) found that bioaerosol samplers may underestimate the concentration of large particles (>10 μm) (e.g., the single-stage Andersen impactor showed impaction losses at the sampling inlet). Grinshpun et al. (1994) theoretically studied the inlet characteristics of several commercially available bioaerosol samplers, looking at the effects on inlet efficiency of particle size, air velocity, sampling rate, and sampler orientation. These researchers found that, under certain sampling conditions, bioaerosol concentration could be significantly overestimated or underestimated through sample enrichment (i.e., capture of particles not in the true sample airstream) or sample loss (i.e., failure to capture all particles in the true sample airstream).

11.2.2 Particle Removal Efficiency

Section 11.3 discusses how samplers are designed to have good particle-capture characteristics. Critical to representative bioaerosol collection is minimal particle loss (wall loss) during sample transport to a collection device (e.g., an impaction surface or a filter). This is usually achieved by minimizing the distance and transit time between the inlet and the collection device (Brockmann, 1993).

11.2.3 Biological Recovery Efficiency

The ability of microorganisms to survive while airborne depends on many factors, not all of which are well characterized. Pollen grains and microbial spores are better protected than vegetative cells from the detrimental effects of light, cold, heat, dryness, and toxic gases while airborne as well as from the shear forces and mechanical stresses encountered during particle collection (Cox, 1987; Marthi, 1994; Cox, 1995; Tong and Lighthart, 1997). Survival of aerosolized test bacteria has been found to be related to particle content and size as well as the temperature and RH of the suspending air (Lighthart and Shaffer, 1997). Stewart et al. (1995) have suggested that the different impaction velocities used in bioaerosol samplers may result in different metabolic and structural injuries to collected microorganisms. Drying of collected particles by continuous exposure to a sample air stream may damage vegetative cells, as can rapid rehydration when collected into a liquid. Therefore, microorganisms need to recover from these harmful effects to be able to reproduce in laboratory culture and be detected by culture methods.

Besides culture procedures, bioassays (e.g., the *Limulus* assay for endotoxin and immunoassays for antigens) also require that samples retain their biological activity during collection, transport, and storage. For some bioaerosol analyses, the physical properties of collected particles are key to identification. Therefore, for these assays, air samplers should be chosen that preserve the physical integrity of particles and particle clusters. Collection methods for particles that will be examined under a microscope must preserve the key morphological details of the material for accurate identification.

11.3 Principles of Particle Collection

11.3.1 Overview

A particle that has entered a sampler will be collected if its trajectory can be separated from the airstream trajectory. This separation may be achieved through the different physical forces illustrated in Figures 11.1 to 11.4, some of which are explained more fully later. Electrostatic precipitation has also been applied to the collection of bioaerosols, but is not currently as widely used as the methods described below.

1. **Inertial Impaction.** Figure 11.1 illustrates how the inertia of a particle can force it to impact onto a surface, usually an agar-based culture medium or an adhesive-coated microscope slide or tape strip. Applications of the principle of inertial impaction are seen in slit samplers (Sampler 1 in Table 11.1), single-stage and multiple-stage (i.e., cascade) impactors (Samplers 2 to 8), centrifugal samplers (Samplers 10 to 13), and liquid impingers (Samplers 14 and 15). Section 11.3.2 explains inertial impaction further.

agar-based medium that is incubated to recover culturable bioaerosols [e.g., slit-to-agar samplers, (Sampler 1) (Figure 11.3)], or (b) onto a glass slide or tape strip that is examined by microscope [e.g., a pollen and spore trap (Samplers 16 to 20) (Figure 11.4)]. The collection substrate may be stationary or may move continuously or in increments beneath the slit.

4. **Centrifugal Samplers.** Figures 11.5a and 11.5b illustrate the principle of particle separation by centrifugal force. Centrifugal collection also uses a particle's inertial behavior but, in this case, in a radial manner. Table 11.1 describes agar-strip impactors (Sampler 10) and cyclone samplers based on this particle collection mechanism (Samplers 11 and 12).

5. **Liquid Impingement.** Liquid impingement (Figure 11.6) is similar to solid-plate impaction but, in this case, inertial forces impact the particles onto a surface submersed in or washed with liquid (Samplers 14 to 15). Particles removed at the

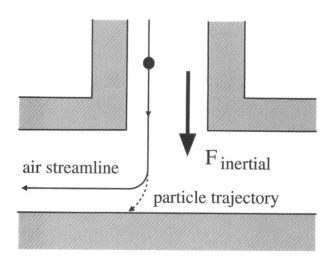

FIGURE 11.1. Mechanisms of particle removal from air — solid-plate impaction (adapted from Nevalainen et al., 1992).

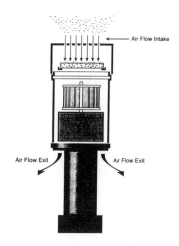

FIGURE 11.2. Single-stage, multiple-hole impactor (Macher et al., 1995)

2. **Multiple-Hole Impactors.** Multiple-hole impactors have several circular nozzles in each stage and have been called "sieve" samplers. However, the nozzles are too large to collect the particles of interest by sieving, and the actual collection mechanism is inertial impaction. Figure 11.2 shows a single-stage, multiple-hole impactor and Table 11.1 describes a variety of commercially available devices (Samplers 2 to 8).

3. **Slit Samplers.** The impaction nozzle in slit samplers is rectangular in shape, usually long and narrow. Again impaction may be (a) onto an

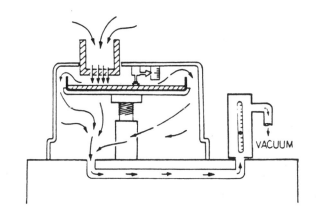

FIGURE 11.3. Slit-to-agar impactor (Macher et al., 1995)

TABLE 11.1. Widely Used, Commercially Available Samplers for Collecting Bioaerosols, adapted from Chapter 23 in *Air Sampling Instruments* (in press)

Sampler[A]	Principle of Operation	Sampling Rate (L/min)	Manufacturer/ Supplier[B]	Commercial Name	Application[C]	d_{50} Cutpoint Calculated	d_{50} Cutpoint Other	Ref
SLIT AGAR IMPACTORS								
1. Rotating slit or slit-to-agar impactors (vp, sc)	Impaction onto agar in 10- or 15-cm plates on rotating surfaces	28	BC	Mattson-Garvin Air Sampler	C	0.53	—	D, E
		175, 350, 525, 700	CAS	Airborne Bacteria Sampler	C	—	0.5	F
		15–55	NBS	Slit-to-Agar Air Samplers: STA-203	C	—	—	—
		15–30		STA-204	C	—	—	—
MULTIPLE-HOLE IMPACTORS								
2. 12-hole impactors (vp, sc)	Impaction onto agar in 10-cm plates	28–71	VAI	Sterilizable Microbiological Atrium (SMA): SMA Micro Sampler	C	—	—	—
		28, 142		SMA MicroPortable Viable Air Sampler	C	—	—	—
3. 100-hole impactor (sc)	Impaction onto agar in 10-cm plates	10	BMC/SBI	Portable Air Sampler for Agar Plates	C	4.18	—	D, E
		20			C	2.56	—	G
		(28)						
4. 219- or 220-hole impactors (sc)	Impaction onto agar in 5.5-cm contact plates	90	PBI/BSI	Surface-Air-Sampler (SAS): Super 90	C	1.97	—	D, E
		90		Super 90 CR	C	1.94	2.24	H
		100		HiVAC Impact	C	—	—	H
		180		High Volume Sampler	C	1.35	1.39	—
							1.45	I
	9-cm plates	100		HiVAC Petri	C	—	—	Mfgr
	5.5-cm contact plates	100	PAR/SBI	MicroBio Air Samplers: MB1, MB2	C	1.8	—	Mfgr
5. 400-hole impactor (vp)	Impaction onto agar in 10-cm plates	28	AND	Andersen Single Stage Viable Particle Sampler; N6 Single Stage Viable Impactor	C	0.57	—	I
						0.58	0.65	D, E
								J
6. Two-stage, 200-hole impactor (vp)	Impaction onto agar in 10-cm plates	28	AND	Andersen Two Stage Viable Sampler/Cascade Impactor	C, stg 0	6.28	8.0	Mfgr
								D, E
					stg 1	0.83	0.95	Mfgr
								D, E
7. Six-stage, 400-hole impactor	Impaction onto agar in 10-cm plates	28	AND	Andersen Six Stage Viable Sampler/ Cascade Impactor	C, stg 1	6.24	7.0	D, E
					stg 2	4.21	4.7	D, E
					stg 3	2.86	3.3	D, E
					stg 4	1.84	2.1	D, E
					stg 5	0.94	1.1	D, E
					stg 6	0.58	0.65	D, E
					C, stg 1	6.61	—	—
					stg 6	0.57	—	—

TABLE 11.1. Widely Used, Commercially Available Samplers for Collecting Bioaerosols, adapted from Chapter 23 in *Air Sampling Instruments* (in press)

Sampler[A]	Principle of Operation	Sampling Rate (L/min)	Manufacturer/ Supplier[B]	Commercial Name	Application[c]	d_{50} Cutpoint Calculated	Other	Ref
8. Eight-stage, personal impactor (vp)	Impaction onto 3.4-cm substrates	2	AND	Personal Cascade Impactor, Series 290 Marple Personal Cascade Impactor	O	stg 1 —	21.3	K
						stg 2 —	14.8	K
						stg 3 —	9.8	K
						stg 4 —	6.0	K
						stg 5 —	3.5	K
						stg 6 —	1.55	K
						stg 7 —	0.93	K
						stg 8 —	0.52	K
FILTERS								
9. Cassette filters (vp)	Filtration, generally 25-, 37-, or 45-mm filters in disposable cassettes or reusable filter holders	1–5	COR, MIL, PGS, SKC	Aerosol Monitor, Air Monitor, Air Monitor Cassette, Particulate Monitor	H,M,O	—	—	—
CENTRIFUGAL SAMPLERS								
10. Centrifugal agar impactors (sc)	Impaction onto agar in plastic strips		BDC	Reuter Centrifugal Samplers (RCS):				
		40		Standard RCS	C	7.5	4	E, F
		50		RCS Plus	C	6	0.82	G, E
11. Wetted cyclone samplers (vp, sc)	Tangential impingement into thin liquid layer	50–55	HG	AEA Technology PLC Aerojet Cyclones	C,M,O	—	—	—
		167			C,M,O	—	0.8	L
		500			C,M,O	—	1.5	L
		167	PAR/SBI	MicroBio MB3 Portable Cyclone	C,M,O	—	—	—
		700–1000	LRI	Aerojet-General Liquid-Scrubber	C,M,O	—	—	—
12. Dry cyclone sampler (sc)	Reverse flow cyclone	≤20	BMC	Cyclone Sampler	H,M,O	—	1.2	Mfgr
13. Three-jet, tangential impactor (vp)	Tangential impaction onto glass or filter surface or impingement into liquid	12.5	SKC	BioSampler	C,M,O	—	—	M
LIQUID IMPINGERS								
14. All-glass impingers (vp)	Impingement into liquid	12.5	AGI, HG/MIL	All-Glass Impingers (AGI):	C,M,O			
				AGI-4, AGI-30		0.30	—	D, E
						0.31	—	I

TABLE 11.1. Widely Used, Commercially Available Samplers for Collecting Bioaerosols, adapted from Chapter 23 in *Air Sampling Instruments* (in press) (cont.)

Sampler[A]	Principle of Operation	Manufacturer/ Supplier[B]	Commercial Name	Sampling Rate (L/min)	Application[C]	d$_{50}$ Cutpoint Calculated	d$_{50}$ Cutpoint Other	Ref
15. Three-stage impingers (vp)	Impingement into liquid	BMC	Multiple-Stage Liquid Impinger	20	C,M,O	stg 1 —	≥10	N
						stg 2 —	4–10	N
						stg 3 —	4	N
		HG	Three-Stage Impinger	10, 55		stg 1 —	≥7	N
						stg 2 —	≥3	N
						stg 3 —	≥1	N
				20	C,M,O	stg 1 —	10	N
						stg 2 —	4	N
						stg 3 —	≥4	N
(see also Sampler 12)								
POLLEN, SPORE, AND PARTICLE IMPACTORS								
16. One- to seven-day tape/slide impactors (sc)	Impaction onto rotating drum with tape strip or glass slide	BMC	1–7-Day Recording Volumetric Spore Trap, 2 x 14 mm slit	10	M	5.2	—	Mfgr
		LAN	Pollen and Particle Sampler	10	M	—	—	—
17. Moving slide impactors (sc)	Impaction onto moving glass slides	ALL	Allergenco Air Sampler (MK-3)	15	M	2.0	—	Mfgr
		BMC	Continuous Recording Air Sampler	10	M	—	—	—
		LAN	Volumetric Pollen/Particle Sampler	10	M	—	—	—
18. Stationary slide impactor (sc)	Impaction onto stationary glass slides	BMC/SBI	Personal Volumetric Air Sampler	10	M	2.52	—	D, I
19. Cassette slide impactor (vp)	Impaction onto stationary glass slides	ZAA/SKC	Air-O-Cell Sampling Cassette	15	M	2.3	—	Mfgr
20. Rotating rod impactors (sc)	Impaction onto rotating rods	STI	Rotorod	48	H,M	—	—	—

A   Letters in parentheses: vp = Requires a vacuum pump and flow control device, which the sampler manufacturer/supplier may provide.
        sc = Self-contained with built-in air mover.

B   See Table 11.A. Manufacturers and /Suppliers of Bioaerosol Samplers.

C   C = Culture of sensitive and hardy microorganisms (e.g., vegetative bacterial and fungal cells and spores).
     H = Culture of hardy microorganisms only (e.g., spore-forming bacteria and fungi).
     M = Microscopic examination of collected particles.
     O = Other assay (e.g., immunoassays, bioassays, chemical assays, or molecular detection methods).

D = Jensen et al., 1994	G = Mehta et al., 1996	J = Andersen, 1958	M = Willeke et al., 1998		
E = Buttner et al., 1997	H = Jensen, 1995	K = Rubow et al., 1987	N = May, 1966		
F = Jensen et al., 1992	I = Nevalainen et al., 1992	L = Griffiths et al., 1997			

wetted impaction surface wash into a collection fluid, and the air stream generally breaks up into bubbles after impaction (Figure 11.6C). Some of the uncollected, small particles may diffuse to bubble surfaces and be transferred to the liquid in this way as well (Grinshpun et al., 1997). Multiple-stage impingers allow size-selective particle collection (Asking and Olsson, 1997) (Sampler 15). Collection fluids from impingers can be divided for analysis by multiple techniques. Section 11.3.3 explains liquid impingement further.

6. **Filtration.** Inertial forces are only one of the mechanisms used to separate particles from an air stream during filtration (Figure 11.7). Interception, gravitational settling, diffusion, and electrostatic attraction enhance deposition on or within filter materials. Filters are usually held in inexpensive cassettes attached to portable pumps, which facilitates sample collection and allows collection of

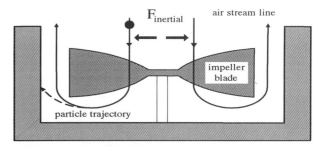

FIGURE 11.5. Mechanisms of particle removal from air — centrifugal impaction (a) agar-strip impactor.

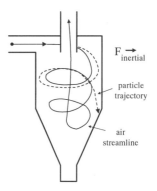

FIGURE 11.5. (b) cyclone sampler

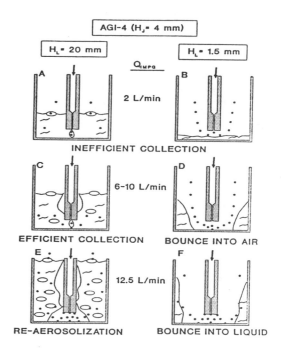

FIGURE 11.6. Schematic representation of the collection process in an impinger

H_J = distance between jet orifice and bottom of the impinger vessel; H = initial liquid level; Q_{IMPG} = impinger flow rate; 11.6A and 11.6B: inefficient particle collection at low flow rates; 11.6C: optimal operating conditions; 11.6D: particle bounce when liquid volume insufficient; 11.6E: reaerosolization of particles from collection liquid; 11.6F: particles bounce into collection liquid pushed to walls of vessel (Grinshpun et al., 1997, with permission).

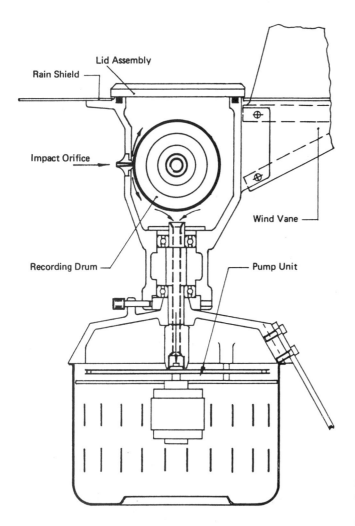

FIGURE 11.4. Pollen and spore trap (Macher et al., 1995)

personal samples (Sampler 9). Larger filters have also been used with high-volume samplers to collect antigenic particles. Section 11.3.4 explains filtration further.

7. **Gravitation or Settling.** The settling of particles by gravitation onto a culture plate or a microscope slide depends highly on particle size and is influenced strongly by air movement. Given the unpredictable and uncontrollable nature of ambient particle movement, investigators cannot directly relate the number of CFUs on a settling plate or the number of spores on a gravity slide to the concentrations of the corresponding particles in the sampled environment. Gravity samples are mentioned in Section 12.2.1 as a means to measure particle deposition on a surface. However, gravity samples are not suitable substitutes for volumetric air samples and should not be used even to determine the relative air concentrations of different microorganisms because of the method's collection bias.

11.3.2 Inertial Impaction

Inertial impaction is the most widely used mechanism for particle collection. The impaction process depends on a particle's inertial properties, such as size, density, and velocity, as well as on an impactor's physical parameters, such as the dimensions of the inlet nozzle and the airflow pathway. Figures 11.8a and b illustrate how these parameters interact in the impaction zone. Particles follow the sampled air as it flows into and through an impactor's inlet nozzle, which typically is round or rectangular. The most important feature of a nozzle is its width, W, which is either

the diameter of a round nozzle or the shorter aspect of a rectangular one. Whether a particle is collected in an impactor depends on the drag force on the particle, its momentum, and its transit time across an impaction surface.

11.3.2.1 The Impaction Zone Particle removal occurs in the impaction zone (Figure 11.8a) where a particle's trajectory may deviate significantly from the air streamlines. This deviation can be predicted based on the shape and width of the impaction nozzle and the jet-to-plate distance, y_2, which is the distance from the bottom surface of a nozzle (where the air jet exits) to the impaction surface (which deflects the air jet). The parameter y_1 is the height of the zone above the impaction surface within which the incoming streamlines are deflected from their original paths. Thus, the air exiting a nozzle is a free jet until it reaches the deflected zone. Impactors are designed with y_2 greater

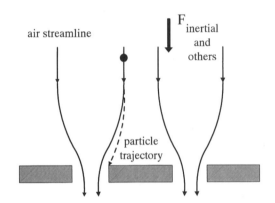

FIGURE 11.7. Mechanisms of particle removal from air - filtration (Adapted from Nevalainen et al., 1992).

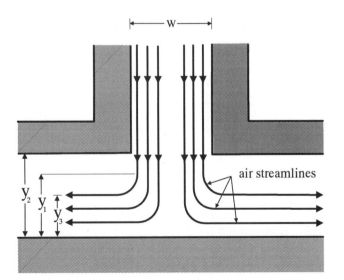

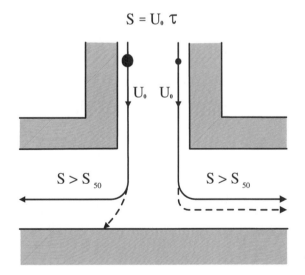

FIGURE 11.8. Cross section of solid-plate impactor (a) critical parameters of the impaction zone, (b) stopping distance and its importance in particle removal (see text and list of abbreviations for explanation of notations) (Adapted from Nevalainen et al., 1992).

than y so that impactor performance will be predictable. The streamline closest to the nozzle wall is termed the "limiting streamline." After deflection from the impaction plate, the limiting streamline defines parameter y_3, the greatest distance from an impaction plate within which the flow of air moves laterally away from the impaction region.

11.3.2.2 Particle Stopping Distance Particle deceleration and collection in an impaction zone is determined by a particle's stopping distance, S. Stopping distance i s the distance a particle travels before stopping when injected into still air and is defined by:

$$S = U_0 \tau . \qquad (11\text{-}1)$$

U_o is the initial particle velocity (usually assumed to equal the incoming airstream velocity), and τ is the relaxation time. Stopping distance is the ratio of particle relaxation time to the transit time of the air flow through an impaction region (Hering, 1995). Thus, stopping distance can be used to predict an impactor's collection efficiency.

Relaxation time, τ, relates to the time required for a particle to adjust or "relax" itself to a new set of forces or conditions. Relaxation time depends on particle diameter, d, particle density, ρ_p, and the viscosity, η, of the fluid medium in which the particle travels, in this case air:

$$\tau = \frac{\rho_p \, d^2 \, C_c}{18 \, \eta} . \qquad (11\text{-}2)$$

C_c is the Cunningham correction factor, which is ~1 for particles larger than 1 μm in diameter and increases with decreasing particle size. C_c corrects the calculation of τ for the fact that very small particles, due to their size, do not move through a continuum of gas molecules, and thus they may slip between air molecules. C_c can be assumed to equal 1 for most bioaerosol calculations.

The non-dimensional Stokes number, Stk, is used to relate the travel of a particle to its confined flow within an impactor:

$$Stk = \frac{S}{W/2} , \qquad (11\text{-}3)$$

where W, again, is either the diameter of a round nozzle or the width of a rectangular one. The Stokes number relates a particle's stopping distance to a relevant spatial dimension (e.g., the radius of a round nozzle) (Hinds, 1982; Baron and Willeke, 1993). The larger the Stokes number, the greater the impaction efficiency. One of the most important uses of the Stokes number is to predict an impactor's cut-off diameter (Hering, 1995).

11.3.2.3 Impactor Cut-Off Diameter Stopping distance is a convenient parameter for determining whether a particle will impact because it relates to a particle's behavior in a non-linear flow field. Figure 11.8b illustrates the spatial relationship of two particles' stopping distances to the likelihood the particles will impact. The larger particle has a stopping distance longer than S_{50} and would be collected. The smaller particle has a shorter stopping distance and would move across some streamlines but would eventually pass through the sampler.

For each streamline through an impaction zone, there is a stopping distance for which all particles above a certain diameter are collected. The flow field composed of all streamlines has a certain composite stopping distance. S_{50} designates the stopping distance for which 50% of particles are collected and 50% pass through an impactor. Similarly, the "cut-off diameter," d_{50}, designates the particle size at which half are removed and half pass through a sampler.

Almost all of the particles of a size somewhat larger than a sampler's cut-off diameter are removed because most impactor stages are designed to have very sharp cut-off characteristics. Therefore, d_{50} is generally assumed to be the diameter above which all particles are collected and is an important characteristic of a particle sampler. Cut-off diameter is related to the 50%-removal Stokes number, Stk_{50}, as shown in Equation 11-4 (Willeke and McFeters, 1975; Marple and Willeke, 1976). The Stk_{50} value is also nozzle specific because the S_{50} value depends on nozzle shape.

$$d_{50} = \sqrt{\frac{9 \eta \, W \, Stk_{50}}{\rho_p \, U_o \, C_c}} \propto \sqrt{Stk_{50}} = \sqrt{\frac{S_{50}}{W/2}} \qquad (11\text{-}4)$$

Table 11.2 and Figure 11.9 compare the design and performance characteristics of commonly used bioaerosol samplers that operate on inertial impaction. The samplers range in volumetric flow rate, Q, from 10 to 180 L/min and have either a single impactor nozzle or several in parallel. The first and last stages of the six-stage Andersen impactor are listed as individual devices. The samplers' 50% stopping distances, S_{50}, and cut-off diameters, d_{50}, were calculated using Equation 11-4 and assuming that (a) air viscosity, η, is 1.81×10^{-5} Pa· s at conditions of normal temperature (20°C) and pressure (1 atm), and (b) bioaerosol density, ρ_p, is 1.0 g/cm^3, that is, unit density—the same as water.

11.3.2.4 Particle Aerodynamic Diameter Aerodynamic diameter, d_a, rather than actual particle diameter, is used to describe the behavior of airborne particles with varying shapes and densities. Aerodynamic diameter is the diameter of a spherical particle with a density of 1 g/cm^3 that has the same settling velocity as the particle in question. Aerodynamic and physical diameters are similar for bioaerosols of approximately unit density and spherical shape. However, for a rod-shaped bacterium or fungal spore, aerodynamic diameter can predict particle behavior better than would a microorganism's actual dimensions.

Figure 11.9 graphically shows how calculated stopping distance varies with aerodynamic particle diameter for selected samplers from Table 11.2. The stopping distance curves terminate on the left at the calculated S_{50} values. For comparison, the experimental sampler cut-off diameters have been included where these have been determined. The calculated d_{50} values, based on the approximate parameters listed in Table 11.2, are somewhat smaller than the experimentally determined cut-off diameters. The exact Stokes number and, therefore, the stopping distance depend on an impactor's geometry and flow characteristics, as described in Section 11.3.2.2.

Figure 11.9 reveals that the cut-off diameters for the various samplers range from <0.5 to >5 µm. An obvious implication of this finding is that none of these samplers is an efficient collector of bioaerosols as small as viruses (unless these units are attached to larger particles, which is typical) and that some samplers will not efficiently collect bioaerosols smaller than 5 µm. Differences in sampler cut-off diameter may explain, at least partly, differences researchers have observed in sampler performance.

Stopping distances and cut-off diameters can also be estimated for other inertial samplers, although for some this is more difficult. For example, the linear impaction parameters described above do not apply readily to centrifugal samplers (Samplers 10 to 13). In these devices, a particle's stopping distance continuously increases as the particle moves toward a sampler's outer wall.

11.3.2.5 Particle Bounce Particle bounce can complicate the collection of bioaerosols and their subsequent analysis. A particle striking a moist agar surface will generally adhere. However, a "bouncy" particle that hits a previously deposited one may rebound and be carried away with the air stream (Juozaitis et al., 1994). Figures 11.6d and 11.6f illustrate how bounce could occur in a liquid impinger.

Significant particle bounce may also occur off an uncoated microscope slide. Coating a slide would improve retention, with bounce then determined by whether the energy stored in particle deformation during impaction overcame the forces of adhesive retention. Commonly used coatings for collecting bioaerosols on glass microscope slides include glycerine jelly, silicone grease, petroleum jelly (sometimes with added liquid paraffin), and polyvinyl alcohol (Madelin and Madelin, 1995). Transparent adhesive tape is used in some spore traps (Sampler 16) and a cassette slide impactor uses an acrylic substrate (Sampler 19). In addition to good particle retention properties, slides, tapes, and coatings must have good optical qualities and be compatible with any stains the analytical laboratory will use on the specimens.

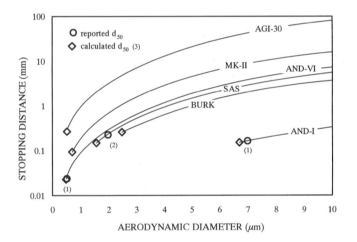

FIGURE 11.9. Stopping distance as a function of aerodynamic particle diameter for selected bioaerosol samplers; cut-off diameters (1) Andersen, 1958, (2) Lach, 1985, (3) calculated, see text and Equation 11-4 (Nevalainen et al., 1992)

TABLE 11.2. Sampling Parameters and Cut-off Diameters for Selected Bioaerosol Impactors
(adapted from Nevalainen et al., 1992)

| | | | | | Nozzle Dimensions | | | | | |
Sampler	Collection Medium	Q (L/min)	U_o (m/s)	Shape	W (mm)	L (mm)	Area (mm²)	Nozzle No.	S_{50} (mm)	Calculate d_{50} (µm)
AGI-30	liquid	12.5	265.2	round	1.0	—	0.79	1	0.125	0.31
AND-VI	agar	28.3	24.02	round	0.25	—	0.05	400	0.031	0.57
MK-II	agar	30	51.02	slit	0.35	28	9.8	1	0.087	0.67
SAS	agar	180	17.36	round	1.0	—	0.79	220	0.125	1.45
BURK	tape	10	11.90	slit	1.0	14	14.0	1	0.25	2.52
AND-I	agar	28.3	1.08	round	1.18	—	1.09	400	0.148	6.61

AGI-30	(Sampler 14)
AND-I	(Sampler 7, stage 1) (Andersen, 1958)
AND-VI	(Sampler 7, stage 6) (Andersen, 1958)

BURK	(Sampler 16, Burkard 1–7-day tape/slide impactor)
MK-II	(Sampler 1, New Brunswick Slit-to-Agar Sampler)
SAS	(Sampler 4, SAS High Volume Sampler) (Lach, 1985)

11.3.2.6 Particle Blow Off A particle captured on a surface may be blown off or re-entrained in an air stream if the drag force acting on the particle is large enough to break the adhesion force that holds it to a surface.

11.3.2.7 Satellite Colonies and Particles A potential source of confusion with impactors are satellite colonies or particles that are occasionally seen at the perimeter of an impaction site. Particles that have bounced or blown off may deposit elsewhere than where they initially impacted. Air motion downstream of an impaction zone may also become turbulent, and previously unimpacted particles in this turbulent air may move sufficiently close to a collection surface to deposit at some distance from the primary impaction site.

11.3.3 Liquid Impingers and Wetted Cyclones

11.3.3.1 Impinger Operation Impingers operate by drawing an air stream into a container with liquid. Airborne particles are transferred to the collection fluid by inertial impaction assisted by particle diffusion within the air bubbles leaving the impaction region (Lin et al., 1997) (Figure 11.6c). Collection in a liquid impinger can help prevent dehydration of biological agents, and the all-glass impinger (AGI) was designed specifically to collect airborne microorganisms (Sampler 14). The suffixes on the AGI-4 and AGI-30 represent the distance between the impinger jet and the base of the liquid container, that is, either 4 or 30 mm. A three-stage impinger is available (Sampler 15), which allows collection of particles in three size fractions. A tangential impactor (Sampler 13) uses three angled jets to generate a swirling air flow similar to that seen in cyclone samplers [see 11.3.3.2] (Willeke et al., 1998). The tangential impactor can be used without a collection fluid as well as with water-based and non-evaporating liquids [see 11.3.3.3].

The nozzle of an impinger can be operated as a critical orifice, thus simplifying the regulation of the airflow rate (e.g., the limiting flow rate for an AGI is ~12.5 L/min but varies with exact nozzle diameter) (Lin et al., 1997). The AGI-4 is a more efficient particle collector than the AGI-30 because of the shorter impaction distance (Lin et al., 1997). However, the greater physical stress of impaction onto an impinger's glass bottom may result in lower biological recovery [see 11.2.3]. Impinger jets are sometimes operated below critical flow, in which case collection efficiency is sacrificed for improved viable recovery. Investigators can identify the optimal flow rate for a particular biological agent and environment (i.e., that which yields the highest overall estimation of bioaerosol concentration) by running a pilot test at the critical impinger flow rate and several flow rates below this.

Other potential problems with impingers are that particles impacting onto the base of a liquid impinger may bounce and be carried away with the effluent air flow

(Figure 11.6d,e) [see 11.3.2.5]. Collected particles may also be re-aerosolized from a bubbling liquid and thus lost (Figure 11.6f) (Juozaitis et al., 1994; Willeke et al., 1995; Grinshpun et al., 1997).

11.3.3.2 Wetted Cyclone Operation A wetted cyclone or cyclone scrubber (Sampler 11) collects particles by tangential impingement into a continuous, thin liquid layer that washes the sampler's inner surface (AIHA, 1996a). The inlet or throat of a glass cyclone is perpendicular to the body of the sampler (Figure 11.5b). The collection fluid enters the cyclone throat via gravity or pump pressure and via aspiration by the air passing the liquid-delivery jet. Fluid carrying collected particles spirals toward the bottom of the sampler where it is collected. The liquid may make a single pass through a cyclone or be recirculated to concentrate a sample. Exhaust air exits through the top of the sampler.

11.3.3.3 Collection Fluid Any liquid compatible with an analytical method and the biological agent under study can be used in an impinger, tangential impactor with collection fluid, or wetted cyclone (e.g., sterile water, buffered salt solutions, or dilute broth). Researchers have used various additives to improve microorganism survival and recovery in collection fluids and culture media (e.g., amino acids, sugars, betaine, pyruvate, catalase, and peptone) (Marthi, 1994; Hensel and Petzoldt, 1995; Lacey and Venette, 1995). Investigators may need to experiment with the targeted bioaerosol to select the best collection fluid and culture medium. A liquid that froths excessively can be a problem if the foam carries over into the vacuum pump, in which case it may be necessary to add an antifoam such as sterile oil or a commercial compound.

The volume of collection fluid in an AGI is typically 15 to 20 mL. The tangential impactor (Sampler 13) is designed for up to 20 mL of collection fluid. Other impingers may use other liquid volumes. Wetted cyclones operate with continuous injection of a fluid to wash collected particles from the walls where they impact. The fluid delivery rate for a cyclone varies with the size of the unit.

Water-based and other volatile collection fluids evaporate during sampler operation, changing the liquid volume in an impinger as well as the concentration of solutes in the sample collection medium. The amount of collection fluid lost to evaporation during sampling depends on the moisture content of the sampled air and sampling duration. Glycerol and other non-evaporating liquids have been tested in the AGI-30 and tangential impactor as possible collection fluids (Willeke et al., 1998).

11.3.3.4 Processing Liquid Samples After sample collection, laboratory personnel measure the fluid volume and assay the collected material in a suitable manner [see 6.3.1.2]. For example, to recover culturable microorganisms, collection fluids can be inoculated onto a growth medium directly and after dilution. This entails additional sample processing not necessary with direct-agar impactors, but it avoids errors due to colony overgrowth or crowding common with agar impactors (Chang et al., 1994, 1995). Clusters of organisms may separate in a collection fluid due to wetting and the turbulent motion of the liquid. Thus, a particle carrying five bacteria that would have formed a single colony if impacted directly onto agar will form five colonies if collected in a liquid. Section 5.2.3 discusses how investigators can adjust sample collection time to achieve the desired surface density of CFUs or particles.

An advantage of a liquid sample is that several different media can be inoculated from one sample (e.g., media for both bacteria and fungi) and multiple plates can be prepared for incubation at different temperatures (e.g., 30°C for mesophilic bacteria and 50°C for thermophilic bacteria). Particles suspended in collection fluids can also be counted in flow cytometers or cell counters or be concentrated by centrifugation or filtration. Concentrated particles can be examined by microscope or inoculated onto culture medium, and whole filters can be placed directly on culture medium [see 11.3.4.2]. Some bioaerosols collected in liquid can also be detected by biochemical or immunological assays [see Table 6.1].

11.3.4 Filtration

Filtration is defined as the separation of particles from the air in which they are suspended by passage through a porous medium (Figure 11.7). The type of filters typically used to collect bioaerosols are membrane filters, which are thin (usually <0.2 mm thick) and contain numerous pores of fairly uniform size (typically 0.2 to 1 μm). Collection of an airborne particle on a filter depends on the aerodynamic diameter of the particle, the filter pore size, and the airflow rate (Lee and Ramamurthi, 1993). Particles larger than a filter's pore diameter are collected by sieving, while particles equal to or smaller than the pore size may be trapped by interception, inertial impaction, or other physical mechanism. Filters with a mean pore size smaller than the diameters of the sampled bioaerosols retain particles on or within a few micrometers of the filter surface.

Bioassays and chemical assays usually require that particles be separated from the filters on which the particles were collected. Section 23.3.1 discusses which filters have proven most compatible with assays for endotoxin. Using filters of the smallest surface area compatible with sampling at the desired airflow rate reduces the volume of wash solution required. This may improve the efficiency with which biological agents are removed from filters and reduce the number of samples that are below an analytical method's lower detection limit due to poor sample recovery from the collection medium.

11.3.4.1 Capillary Pore Filters Capillary pore or straight-through pore filters are made from thin polycarbonate films and have cylindrical holes aligned approximately perpendicular to the filter surface (Hinds, 1982; Lee and Ramamurthi, 1993). These filters have smooth, flat surfaces and are ideal for examining particles with an optical or scanning electron microscope. For other assays, particles may more easily be washed from the surface of a polycarbonate filter than from other membrane filters. Typically, the efficiency of these filters is low for particles smaller than the pore size and greater in the size range where diffusional collection is significant (i.e., particle diameter >0.1 μm) (Lee and Ramamurthi, 1993). Particles greater than a filter's pore size are collected with high efficiency. However, for the same pore size, the flow of air through capillary pore filters may be restricted compared to other filter media, and capillary pores may more readily become blocked, which limits the amount of material they can collect.

11.3.4.2 Porous Membrane Filters Porous membrane filters have complicated and uniform microstructures that provide a tortuous or irregular airflow path. Manufacturers derive pore size for a filter by determining particle filtration from a liquid. This reported pore size often does not match the physical size of the filter's pores or the structural characteristics of a porous membrane filter (Lee and Ramamurthi, 1993). In general, particle collection efficiency is very high, even for particles significantly smaller than a stated filter pore size. Cellulose ester filters have been popular for bioaerosol collection. Bioaerosols can be stained on these filters for microscopic examination or the filters can be cleared with material of the proper refractive index before examination. Immersion oil should not be used to clear filter media when examining particles with high moisture content because this may distort particle morphology and lead to errors in recognizing biological agents. More often, these filters are assayed by placing them face-up directly on an agar surface or an absorbent pad saturated with culture medium. The collected bacteria and fungi that have survived the sampling process form visible colonies on the filter surface.

11.3.5 In Situ or Remote Bioaerosol Detection

The bioaerosol sampling methods discussed in this chapter are all *extractive techniques*, that is, they rely on removing bioaerosols from a suspending air stream for detection and identification. *In situ techniques* are used to study aerosols without separating them from the suspending air (Rader and O'Hern, 1993). Some of these methods have been applied to the study of bioaerosols with varying success (Evans et al., 1994; Pinnick et al., 1995; Hairston et al., 1997; Pinnick et al., 1998). One of the advantages of such methods is rapid detection, often with real-time, direct-reading instruments based on particle light scattering, electrical mobility, or inertia in an accelerated flow (Lacey, 1997). The only commercially available, direct-reading instrument for bioaerosol detection is an ultraviolet, aerodynamic, particle sizer spectrometer (UV-APS; TSI, St. Paul, Minn.). The instrument combines time-of-flight measurement of particle size with light scattering particle detection and single particle UV fluorescence measurement. The latter involves illuminating a particle with a UV laser beam and measuring the emitted fluorescent intensity at certain wavelengths. Such fluorescence is associated with certain molecules that living cells produce. However, this device is expensive, best suited for detecting particles containing clusters of microorganisms rather than single-cell aerosols, and not yet widely used for investigating workplace exposures to biological agents.

11.4 Equipment Calibration

11.4.1 Importance of Airflow Calibration

Regardless of how efficient a sampler is, it is not possible to reliably determine the volume of air collected and bioaerosol concentration if the air flow through the sampler is not measured properly. Therefore, accurate measurement of airflow rate and sample volume must be an integral part of the calibration of air sampling instruments (Lippmann, 1995).

Adherence to a multiple-stage impactor's prescribed airflow rate is important because an incorrect flow rate will alter the cut-off diameter for each stage and lead to inaccurate particle size classification (Heber, 1995) [see 11.3.2.3]. Using the manufacturer's recommended sampling rate is also important with single-stage impactors to ensure comparability of results within and between studies at predictable d_{50} cutpoints. Not only must the airflow rate be exact when sampling with impactors, the distance from the inlet to the surface of the collection medium (y_2 in Figure 11.8a) must also be correct. Adjustment of the nozzle-to-medium (jet-to-plate) distance is fairly easy if an impactor is equipped with a movable stage, as in some slit-to-agar samplers (Figure 11.3). Otherwise, agar should be poured into the culture plates to a predetermined depth, and microscope slides and tape strips must be correctly placed in samplers.

A sampler's airflow rate must be measured or monitored both before and after sample collection. Investigators use the average of the pre- and post-sampling air flows to calculate the sample volume if the flow rate changes (e.g., due to weak batteries). A change in flow rate of more than 10% may invalidate a sample because of changes in sampler collection efficiency and particle cut-off diameter. Table 11.3 provides advice on airflow calibration.

11.4.2 Calibration Methods

A simple and easy-to-use airflow calibration system is a bubble meter consisting of a large, graduated buret inverted on a ring stand (Figure 11.10). The sampler's air pump is started, and a beaker of water (with a small amount of dish detergent) is placed under the inverted buret and lifted momentarily to cover the inlet and form a soap bubble. Volumetric airflow rate can be determined quite accurately by timing the bubble's travel between marks on the buret. More sophisticated, automated calibration systems are available with optical sensors to measure a bubble's travel time.

TABLE 11.3. Advice on Airflow Calibration (adapted from Lippmann, 1995)

- Use standard calibration devices with care and attention to detail.

- Check all calibration instruments and procedures periodically to determine their stability, operating condition, and compliance with a reference standard.

- Understand the operation of an air sampler before attempting to calibrate its airflow rate. Use a calibration procedure that will not change an air sampler's performance or operation characteristics.

- Ensure that all connections in a calibration train are as short and free of constrictions or resistance as possible and that all parts are arranged as they will be assembled during actual sample collection.

- Use extreme care when reading scales, measuring time intervals, and adjusting or leveling calibration devices or air samplers.

- Allow sufficient time for airflow equilibrium to be established, inertia to be overcome, and conditions to stabilize.

- Collect enough data to ensure reliable calibration curves. Collect ≥3 readings for each calibration point.

- Maintain complete records of all procedures, data, and results. Record all data including trial runs and questionable data with appropriate comments. Also record instrument identifications, barometric pressure, temperature, person who conducted calibration, date, and other pertinent information. Attach a tag to secondary standards indicating the date of last calibration and where the data are filed.

- Determine the reasons for changes if a calibration differs from previous records before accepting the new data.

A sampler that cannot be easily attached to a buret can be placed in a sealable, airtight chamber for calibration provided the chamber does not create a pressure drop and the inside pressure remains equal to ambient pressure (Figure 11.11). The manufacturer of the RCS centrifugal sampler (Sampler 10) offers a digital flywheel anemometer to measure airflow rate. Samplers with high flow rates (e.g., >10 L/min) can be calibrated with wet test and dry gas meters, precision rotameters, rotating vane anemometers, critical flow orifices, and commercial airflow calibrators (Lippmann, 1995). However, many such devices create air pressure artifacts in the calibration and sampling train that may affect calibration accuracy. Line or chamber pressure should be monitored near a sampler's inlet to identify pressure-related problems when calibrating samplers. If a pressure drop is noted, air can be supplied to the front of a calibration and sampling train to equalize downstream air pressure (Figure 11.12). For further directions on using such a pressure-regulating system, readers should consult more detailed references on proper calibration techniques (Lippmann, 1973, 1995).

11.4.3 Challenge Aerosols

Prior to sample collection, investigators may need to evaluate bioaerosol sampler performance for the specific biological agents of interest. In the laboratory, researchers have generated test aerosols of various bioaerosols (or safe substitutes with similar particle shape, diameter, and density) using a variety of methods but most often by nebulizing particles from a liquid suspension (Thompson et al., 1994; Reponen et al., 1997; Ulevicius et al., 1997). Efforts have been made to identify appropriate reference materials for assessing bioaerosol sampler performance (Griffiths et al., 1996). The total concentration of aerosolized particles can be measured with a dynamic particle-size spectrometer to provide a reference particle count against which to judge a bioaerosol sampler's results (Brosseau et al., 1994; Qian et al., 1995; Willeke et al., 1996; Gao et al., 1997).

11.5 Equipment Selection

11.5.1 Widely Used Bioaerosol Samplers

This section briefly discusses some of the most popular instruments for collecting bioaerosols to show how an investigator might choose among samplers and to illustrate better how they operate. The column "Application" in Table 11.1 identifies the types of sample analyses with which individual samplers are suitable. Investigators must also decide if they need information on particle size and must anticipate the aerosol concentration. This can be done by collecting preliminary samples to estimate bioaerosol concentration or by reviewing the literature to learn what concentrations other investigators have found in similar environments.

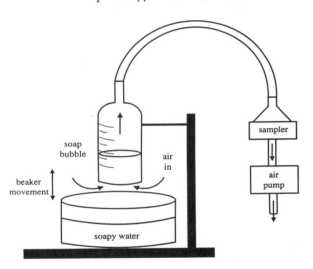

FIGURE 11.10. Soap-bubble-meter (flowmeter) calibration.

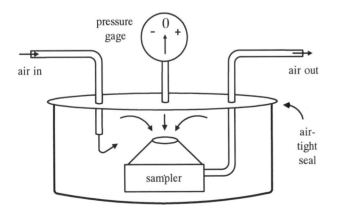

FIGURE 11.11. Calibration chamber.

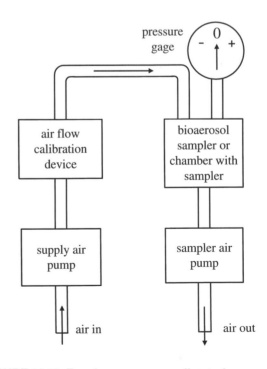

FIGURE 11.12. Equal-pressure sampling train.

Almost all of the listed devices are suitable for sampling intermediate concentrations of culturable particles indoors without sizing them. Filters and impingers are generally less expensive devices for collecting bioaerosols than are impactors, which must be machined precisely. However, impactors generally are more convenient because the specimen is collected directly on the substrate that will be cultured or examined by microscope without further processing [see 6.3.1]. Glass impingers and cyclones usually cannot be manufactured as precisely as machined samplers and are subject to breakage. However, these samplers have the advantage that the collection fluid can be diluted or concentrated before analysis and that each sample can be analyzed in several ways [see 11.3.3.4]. The AIHA field guide (1996b) describes sampling protocols for the most popular bioaerosol samplers. These descriptions include the principle of the method and its range, sensitivity, accuracy, and precision; the method's advantages and limitations; apparatus and reagents needed; sample collection and analysis protocols; and quality control. Section 13.2 describes the calculation of the air concentration of biological agents from various types of samples.

11.5.1.1 Multiple-Hole Impactors

The Andersen six-stage impactor is well characterized and can provide an estimate of the concentration of culturable bioaerosols along with information on the size distribution of viable particles (Sampler 7). There are also one-stage (the N6 version, Sampler 5) and two-stage models of this sampler (Sampler 6). The 100-hole, single-stage, battery-operated sampler for standard culture plates (Sampler 3) compared well with the two-stage impactor (Sampler 6) (Mehta et al., 1996). The Surface Air System (SAS) impactors and the MicroBio air samplers (Sampler 4) are battery or line operated, single-stage impactors. Several of these samplers accommodate a type of culture plate called a contact plate that can also be used to sample surfaces [see 12.2.2.1].

A positive-hole correction has been used with multiple-hole impactors to avoid underestimating bioaerosol concentration (Andersen, 1958; Peto and Powell, 1970; Leopold, 1988; Macher, 1989; Somerville and Rivers, 1994). Respectively, Appendices 11.B(i) to 11.B(iv) list corrections for 100-, 200-, 219-, and 400-hole impactors. These tables include standard deviations on these estimates so investigators can calculate confidence limits on corrected counts (Macher, 1989). Appendix 11.C illustrates a convenient method for calculating corrected counts. In experimental work, where known bacteria were aerosolized and collected, researchers have elected to examine impaction sites and count microcolonies rather than use the positive-hole correction. Some investigators also choose not to use the correction if the data will be evaluated by comparing relative rankings (because the corrected counts would not change the rank order) or if colony counts are below a predetermined number of CFU (because the correction would make little dif-

ference). For example, Appendices 11.B(i) and 11.B(iv) illustrate that the differences between corrected and direct plate counts do not exceed 10% until the number of CFUs reaches ~18 and 75, respectively, for impactors with 100 and 400 holes (Sampler 3 and Samplers 5 and 7, respectively). However, this adjustment becomes increasingly significant as the percentage of filled holes rises. Table 11.C(i) shows that the correction is less than 5% when fewer than 10% of impaction sites show CFUs (the correction factor is 1.054); however, the correction factor is 2.0 when 80% of the impaction sites are filled.

May (1964) showed that the impaction velocity onto the upper two plates in the six-stage sampler was not sufficient to deposit particles in a pattern that directly correlated with jet locations. Therefore, the positive-hole correction is not applied to these stages. For the lower four stages of this impactor, Andersen (1958) advised users to apply the correction but not to count CFUs that do not coincide with the impaction pattern of the sampling holes. May (1964) gave the same advice because these colonies may arise from particles deposited by the turbulence between jets on their way to the next stage [see 11.3.2.7]. Typically, the number of such CFU is small and excluding them should not significantly bias the sampling results.

Investigators should learn what approach the laboratory analyzing multiple-hole impactor samples uses and determine if the system suits the needs of the current investigation. Two relatively easy methods to determine the air concentration of total culturable bacteria or fungi with multiple-hole impactors are as follows.

1. For single-stage and two-stage impactors, count all CFUs that occur at impaction sites and correct these counts for positive-hole coincidence. For the six-stage, multiple-hole impactor, (a) count all CFUs on the upper two stages, (b) count only positive impaction sites for the lower four stages, and (c) correct the latter counts for positive-hole coincidence. Calculate the 90% or 95% confidence interval on all corrected counts, and report air concentration as within this range for each plate or particle size fraction.

2. Establish colony densities for each multiple-hole impactor above which the plate is considered overloaded and colony counting is not sufficiently accurate [see also 5.2.3.1]. For plates below this limit, count all CFUs and calculate air concentration from this colony count. For plates above this limit, calculate air concentration from the UDL and report the concentration as greater than this number.

To determine the types of culturable microorganisms collected for either of these methods of counting total CFUs, examine all colonies, even those not at impaction sites. Individually identify all bacteria or fungi, even those in mixed colonies (i.e., CFUs comprised of more than one morphologically distinguishable type of microorganism). The sum of these colony counts will not necessarily match the total count outlined above.

11.5.1.2 Slit Impactors There are several models of slit impactors. Some of these rotate a culture plate below a long, narrow inlet (Sampler 1), while others impact a sample air stream onto a rotating tape strip (Sampler 16) or a stationary or moving microscope slide (Samplers 17 and 18). A simple impaction unit is also available encased in a cassette holder similar to those used for membrane filter samples (Sampler 19). Impactors with moving collection surfaces are especially useful for observing temporal changes in bioaerosol concentrations. This feature can be used to implicate a specific bioaerosol source when the sampler is run before, during, and after an emission episode. Adjustable rotation speeds on some samplers allow sampling times from 15 min to 1 hour for culture plate samplers and 24 hours to 1 week for spore traps.

11.5.1.3 Centrifugal Samplers The Reuter Centrifugal Sampler (RCS, Sampler 10) is a portable and inconspicuous sampler, which makes it a frequent choice where samples must be taken with minimal disturbance to a room's occupants. An earlier version of the device could not be calibrated easily and collected particles ≤4 µm with low efficiency (Clark et al., 1981; Macher and First, 1983). A second version, the RCS Plus, was found in laboratory and field tests to be equivalent to the SAS Super 90 (Sampler 4) but less efficient than the Andersen two-stage impactor (Sampler 6), the Burkard multiple-hole impactor (Sampler 3), or a slit-to-agar sampler (Benbough et al., 1993; Mehta et al., 1996).

11.5.1.4. Liquid Impingers and Wetted Cyclones Liquid impingers and wetted cyclones collect vegetative cells well but may not be efficient for hydrophobic bacterial and fungal spores [see 19.3.3]. Impaction and rehydration stresses may affect recovery of some microorganisms [see 11.3.3.1].

11.5.1.5 Filters Filtration is an efficient and convenient sampling method for collecting a number of biological agents such as antigens, endotoxin, muramic acid, ergosterol, and other cell components (Sampler 9). In heavily contaminated environments, filters have been used to collect microorganisms for staining and examination by microscope. Vegetative bacterial and fungal cells and even some spores lose culturability due to desiccation. Therefore, if air samples are collected to detect culturable microorganisms, filtration will identify only those cells hardy enough to survive desiccation [see 11.3.4].

11.5.2 Particle Size
Table 11.1 and Figure 11.9 illustrate the range of d_{50} cut-points for widely used samplers. Consideration of particle size and sampler collection efficiency by particle size may limit an investigator's choice of sampler for a particular application. Jensen et al. (1992) demonstrated experimentally that bioaerosol samplers differed in their ability to collect particles with aerodynamic diameters <2 µm. These findings agreed with theoretical calculations of the d_{50} cutpoints for these bioaerosol samplers (Nevalainen et al., 1993) (Table 11.2).

11.5.3 Performance Criteria for Bioaerosol Samplers
Sampler performance has been characterized for very few currently available bioaerosol samplers. Studies comparing samplers with different d_{50} values and operating principles are difficult to interpret, especially when instrument operation periods and sampled air volumes vary widely. ACGIH's Air Sampler Procedures and Air Sampling Instruments committees (Cohen et al., 1993) and others (Bartley and Fischbach, 1993; Bartley et al., 1994; Lidén, 1994; Vincent, 1995; CEN, 1997; Kenny et al., 1997) have discussed performance criteria for air sampling instruments including bioaerosol samplers (Macher and Willeke, 1992; Macher, 1997a). Performance criteria are quantitative, experimentally determined values used to assess the suitability of a collection method for a given purpose.

An important performance feature for an aerosol sampler is collection efficiency relative to particle deposition in the human respiratory tract. Table 9.2 shows deposition by aerodynamic particle diameter for the inhalable, thoracic, and respirable regions of the respiratory tract. For accurate exposure assessment, bioaerosol and other samplers increasingly are designed to approximate this deposition pattern (Upton et al., 1994; Gao et al., 1997; Griffiths et al., 1997). Efforts to develop performance criteria and uniform evaluation methods for bioaerosol samplers will improve the quality of bioaerosol measurements and the reliability of conclusions drawn from environmental samples (Macher, 1997a,b).

11.6 Summary
After reviewing this and other chapters in this book, the reader should understand why no single air sampling method can be used to collect all of the airborne biological agents that may be present in a particular environment. Table 5.7 summarizes the key points in bioaerosol sampler selection. Section 5.2 discusses where and when to sample as well as sample volume, sample collection time, and sample number. The six-stage, multiple-hole impactor and the AGI (Samplers 7 and 14, respectively) have been suggested as "reference" samplers for collecting culturable bioaerosols, but even these samplers cannot collect all airborne microorganisms satisfactorily. Investigators should study the characteristics of their target or suspect bioaerosols and review the literature to identify the sampler that is best-suited and most efficient for the purpose at hand. Other texts and review articles can also guide investigators (Lighthart and Mohr, 1994; Rylander and

Jacobs, 1994; Burge, 1995; Cox and Wathes, 1995; AIHA, 1996b; ASM, 1997; Buttner et al., 1997; Eduard and Heederik, 1998).

11.7 References

ACGIH: Industrial Ventilation: A Manual of Recommended Practice, 22nd Ed. American Conference of Governmental Industrial Hygienists, Cincinnati, OH (1995).

AIHA: Viable Air Sampling Instruments: Description and Operating Procedures. In: Field Guide for the Determination of Biological Contaminants in Environmental Samples, pp. 75-95. H.K. Dillon, P.A. Heinsohn, and J.D. Miller, Eds. American Industrial Hygiene Association, Fairfax, VA (1996a).

AIHA: Field Guide for the Determination of Biological Contaminants in Environmental Samples. H.K. Dillon, P.A. Heinsohn, and J.D. Miller, Eds. American Industrial Hygiene Association, Fairfax, VA (1996b).

Andersen, A.A.: New Sampler for the Collection, Sizing and Enumeration of Viable Airborne Particles. J. Bacteriol. 76:471-484 (1958).

Asking, L.; Olsson, B.: Calibration at Different Flow Rates of a Multistage Liquid Impinger. Aerosol Sci. Technol. 27:39-49 (1997).

ASM: Manual of Environmental Microbiology. C.J. Hurst, G.R. Knudsen, M.J. McInerney, et al., Eds. American Society for Microbiology, Washington, DC (1997).

Baron, P.A.; Willeke, K.: Gas and Particle Motion. In: Aerosol Measurement: Principles, Techniques and Applications, pp. 23-40. K. Willeke and P.A. Baron, Eds. Wiley, New York, NY (1993).

Bartley, D.L.; Fischbach, T.J.: Alternative Approaches for Analyzing Sampling and Analytical Methods. Appl. Occup. Environ. Hyg. 8:381-385 (1993).

Bartley, D.L.; Chen, C.C.; Song, R.; et al.: Respirable Aerosol Sampler Performance Testing. Am. Ind. Hyg. Assoc. J. 55:1036-1046 (1994).

Benbough, J.E.; Bennett, A.M.; Parks, S.R.: Determination of the Collection Efficiency of a Microbial Air Sampler. J. Appl. Bacteriol. 74:170-173 (1993).

Billings, C.E.; Lundgren, D.A.: Sampling from Ducts and Stacks. In: Air Sampling Instruments for Evaluation of Atmospheric Contaminants, 8th Ed., pp. 528-559. B.S. Cohen and S.V. Hering, Eds. American Conference of Governmental Industrial Hygienists, Cincinnati, OH (1995).

Brockmann, J.E.: Sampling and Transport of Aerosols. In: Aerosol Measurement: Principles, Techniques and Applications, pp. 77-111. K. Willeke and P.A. Baron, Eds. Wiley, New York, NY (1993).

Brosseau, L.M.; Chen, S.K.; Vesley, D.; et al.: System Design and Test Method for Measuring Respirator Filter Efficiency using Mycobacterium Aerosols. J. Aerosol Sci. 25:1567-1577 (1994).

Burge, HA, Ed: Bioaerosols. Lewis Publishers, Boca Raton, FL (1995).

Buttner, M.P.; Willeke, K.; Grinshpun, S.A.: Sampling and Analysis of Airborne Microorganisms. In: Manual of Environmental Microbiology, pp. 629-640. C.J. Hurst, G.R. Knudsen, M.J. McInerney, et al., Eds. American Society for Microbiology, Washington, DC (1997).

CEN/TC137: Draft European Standard. Workplace Atmospheres: Assessment of Performance of Instruments for Measurement of Airborne Particle Concentrations. CEN/TC137/WG3/N192 (1997).

Chang, C.W.; Hwang, Y.H.; Grinshpun, S.A.; et al.: Evaluation of Counting Error due to Colony Masking in Bioaerosol Sampling. Appl. Environ. Microbiol. 60:3732-3738 (1994).

Chang, C.W.; Grinshpun, S.A.; Willeke, K.; et al.: Factors Affecting Microbiological Colony Count Accuracy for Bioaerosol Sampling and Analysis. Am. Ind. Hyg. Assoc. J. 56:979-986 (1995).

Clark, S.; Lach, V.; Lidwell, O.M.: The Performance of the Biotest RCS Centrifugal Air Sampler. J. Hosp. Inf. 2:181-186 (1981).

Cohen, B.S.; McCammon, C.S.; Vincent, J.H., Eds.: Proceedings of the International Symposium on Air Sampling Instrument Performance. Appl. Occup. Environ. Hyg. 8:221-411 (1993).

Cox, C.S.: The Aerobiological Pathway of Microorganisms. Wiley, Chichester (1987).

Cox, C.S.; Wathes, C.M., Eds: Bioaerosols Handbook. Lewis Publishers, Boca Raton, FL (1995).

Cox, C.S.: Stability of Airborne Microbes and Allergens. In: Bioaerosols Handbook, pp. 77-99. C.S. Cox and C.M. Wathes, Eds. Lewis Publishers, Boca Raton, FL (1995).

Eduard, W.; Heederik, D.: Methods for Quantitative Assessment of Airborne Levels of Noninfectious Microorganisms in Highly Contaminated Work Environments. Am. Ind. Hyg. Assoc. J. 59:113-127 (1998).

Evans, B.T.N.; Yee, E.; Roy, G.; et al.: Remote Detection and Mapping of Bioaerosols. J. Aerosol Sci. 25:1549-1566 (1994).

Gao, P.; Killon, H.K.; Farthing, W.E.: Development and Evaluation of an Inhalable Bioaerosol Manifold Sampler. Am. Ind. Hyg. Assoc. J. 58:196-206 (1997).

Griffiths, W.D.; Stewart, I.W.; Reading, A.R.; et al.: Effect of Aerosolization, Growth Phase and Residence Time in Spray and Collection Fluids on the Culturability of Cells and Spores. J. Aerosol Sci. 27:803-820 (1996).

Griffiths, W.D.; Stewart, I.W.; Futter, S.J.; et al.: The Development of Sampling Methods for the Assessment of Indoor Bioaerosols. J. Aerosol Sci. 28:437-457 (1997).

Grinshpun, S.A.; Chang, C.W.; Nevalainen, A.; et al.: Inlet Characteristics of Bioaerosol Samplers. J. Aerosol Sci. 25:1503-1522 (1994).

Grinshpun, S.A.; Willeke, K.; Ulevicius, V.; et al.: Effect of Impaction, Bounce, and Reaerosolization on the Collection Efficiency of Impingers. Aerosol Sci. Technol. 26:326-342 (1997).

Hairston, P.P.; Ho, J.; Quant, F.R.: Design of an Instrument for Real-Time Detection of Bioaerosols using Simultaneous Measurement of Particle Aerodynamic Size and Intrinsic Fluorescence. J. Aerosol Sci. 28:471-482 (1997).

Heber A.J.: Bioaerosol Particle Statistics. In: Bioaerosols Handbook, pp. 55-75. C.S. Cox and C.M. Wathes, Eds. Lewis Publishers, Boca Raton, FL (1995).

Hensel, A.; Petzoldt, K.: Biological and Biochemical Analysis of Bacteria and Viruses. In: Bioaerosols Handbook. pp. 335-360. Lewis Publishers, Boca Raton, FL (1995).

Hering, S.V.: Impactors, Cyclones, and Other Inertial and Gravitational Collectors. In: Air Sampling Instruments for Evaluation of Atmospheric Contaminants, 8th Ed., pp. 279-321. B.S. Cohen and S.V. Hering, Eds. American Conference of Governmental Industrial Hygienists, Cincinnati, OH (1995).

Hinds, W.C.: Aerosol Technology. Wiley, New York, NY (1982) (second edition, in press).

Jensen, P.A.; Todd, W.F.; Davis, G.N.; et al.: Evaluation of Eight Bioaerosol Samplers Challenged with Aerosols of Free Bacteria. Am. Ind. Hyg. Assoc. J. 53:660-667 (1992).

Juozaitis, A.; Willeke, K.; Grinshpun, S.; et al.: Impaction onto a Glass Slide or Agar versus Impingement into a Liquid for the Collection and Recovery of Airborne Microorganisms. Appl. Environ. Microbiol. 60:861-870 (1994).

Kenny, L.C.; Aitken, R.; Chalmers, C.; et al.: A Collaborative European Study of Personal Inhalable Aerosol Sampler Performance. Ann. Occup. Hyg. 41:135-153 (1997).

Lacey, J.: Guest Editorial. Proceedings: Sampling and Rapid Assay of Bioaerosols, June 14, 1995, IARC-Rothamsted, England, J. Aerosol Sci. 28:345-538 (1997).

Lacey, J; Venette, J.: Outdoor Air Sampling Techniques. In: Bioaerosols Handbook, pp. 407-471. C.S. Cox and C.M. Wathes, Eds. Lewis Publishers, Boca Raton, FL (1995).

Lach, V.: Performance of the Surface Air System Air Samplers. J. Hosp. Inf. 6:102-107 (1985).

Lee, K.W.; Ramamurthi, M.: Filter Collection. In: Aerosol Measurement: Principles, Techniques and Applications, pp. 179-205.

K. Willeke and P.A. Baron, Eds. Wiley, New York, NY (1993).

Leopold, S.S.: "Positive Hole" Statistical Adjustment for a Two-Stage, 200-Hole-per-Stage Andersen Air Sampler. Am. Ind. Hyg. Assoc. J. 49:A88-A90 (1988).

Lidén, G.: Performance Parameters for Assessing the Acceptability of Aerosol Sampling Equipment. Analyst. 119:27-33 (1994).

Lighthart B.; Mohr, A.J., Eds: Atmospheric Microbial Aerosols. Chapman and Hall, New York, NY (1994).

Lighthart, B.; Shaffer, B.T.: Increased Airborne Bacterial Survival as a Function of Particle Content and Size. J. Aerosol Sci. Technol. 27:439-446 (1997).

Lin, X.; Willeke, K.; Ulevicius, V.; et al.: Effect of Sampling Time on the Collection Efficiency of All-Glass Impingers. Am. Ind. Hyg. Assoc. J. 58:480-488 (1997).

Lippmann, M.: Instruments and Techniques Used in Calibrating Sampling Equipment. In: The Industrial Environment - Its Evaluation and Control, pp. 101-122. NIOSH, U.S. Printing Office, Washington, DC (1973).

Lippmann, M.: Airflow Calibration. In: Air Sampling Instruments for Evaluation of Atmospheric Contaminants, 8th Ed., pp. 139-150. B.S. Cohen and S.V. Hering, Eds. American Conference of Governmental Industrial Hygienists, Cincinnati, OH (1995).

Macher, J.M.: Positive-Hole Correction of Multiple-Jet Impactors for Collecting Viable Microorganisms. Am. Ind. Hyg. Assoc. J. 50:561-568 (1989).

Macher, J.M.; First, M.W.: Reuter Centrifugal Air Sampler: Measurement of Effective Airflow Rate and Collection Efficiency. Appl. Environ. Microbiol. 45:1960-1962 (1983).

Macher, J.M.; Willeke, K.: Performance Criteria for Bioaerosol Samplers. J. Aerosol Sci. 23:S647-S650 (1992).

Macher, J.M.; Chatiny, M.A.; Burge, H.A.: Sampling Airborne Microorganisms and Aeroallergens. In: Air Sampling Instruments for Evaluation of Atmospheric Contaminants, 8th Ed., pp. 589-617. B.S. Cohen and S.V. Hering, Eds. American Conference of Governmental Industrial Hygienists, Cincinnati, OH (1995).

Macher, J.M.: Instrument Performance Criteria: Bioaerosol Samplers. Appl. Occup. Environ. Hyg. 12(11):723-729 (1997).

Macher, J.M.: Evaluation of Bioaerosol Sampler Performance. Appl. Occup. Environ. Hyg. 12(11): 730-736 (1997).

Madelin, T.M.; Madelin, M.F.: Biological Analysis of Fungi and Associated Molds. In Bioaerosols Handbook, pp. 361-386. C.S. Cox and C.M. Wathes, Eds. Lewis Publishers, Boca Raton, FL (1995).

Marple, V.A.; Willeke, K.: Impactor Design. Atmos. Environ. 10:891-896 (1976).

Marthi, B.: Resuscitation of Microbial Bioaerosols. In: Atmospheric Microbial Aerosols. Theory and Applications, pp. 192-225. B. Lighthart and A.J. Mohr, Eds. Chapman and Hall, New York, NY (1994).

Martinez, K.; Todd, W.; Fischbach, T.: Evaluation of Alternative Samplers for Bioaerosols: Physical Sampling Efficiency. Department of Health and Human Services, U.S. Public Health Service, Centers for Disease Control and Prevention, National Institute fro Occupational Safety and Health, Cincinnati, OH. Report No. CT-160-04A, p. 24 (1991).

May, K.R.: Calibration of a Modified Andersen Bacterial Aerosol Sampler. Appl. Microbiol. 12:37-43 (1964).

Mehta, S.K.; Mishra, S.K.; Pierson, D.L.: Evaluation of Three Portable Samplers for Monitoring Airborne Fungi. Appl. Environ. Microbiol. 62:1835-1838 (1996).

Nevalainen, A.; Pastuszka, J.; Liebhaber, F.; et al.: Performance of Bioaerosol Samplers: Collection Characteristics and Sampler Design Considerations. Atmos. Environ. 26A:531-540 (1992).

Nevalainen, A.; Willeke, K.; Liebhaber, F.; et al.: Bioaerosol Sampling. In: Aerosol Measurement: Principles, Techniques and Applications, pp. 471-492. K. Willeke and P.A. Baron, Eds. Wiley, New York, NY (1993).

Peto, S.; Powell, E.O.: The Assessment of Aerosol Concentration by Means of the Andersen Sampler. J. Appl. Bacteriol. 33:582-598 (1970).

Pinnick, R.G.; Hill, S.C.; Nachman, P.; et al.: Fluorescence Particle Counter for Detecting Airborne Bacteria and other Biological Particles. J. Aerosol Sci. 23:653-664 (1995).

Pinnick, R.G.; Hill, SC.; Nachman, P.; et al.: Aerosol Fluorescence Spectrum Analyzer for Rapid Measurement of Single Micrometer-sized Airborne Biological Particles. Aerosol Sci. Technol. 28:95-104 (1998).

Qian, Y.; Willeke, K.; Ulevicius, V.; et al.: Dynamic Size Spectrometry of Airborne Microorganisms: Laboratory Evaluation and Calibration. Atmos. Environ. 29:1123-1129 (1995).

Rader, D.J.; O'Hern, T.J.: Optical Direct-Reading Techniques: *In situ* Sensing. In: Aerosol Measurement: Principles, Techniques and Applications, pp. 345-380. K. Willeke and P.A. Baron, Eds. Wiley, New York, NY (1993).

Reponen, T.; Willeke, K.; Ulevicius, V.; et al.: Techniques for Dispersion of Microorganisms into Air. Aerosol Sci. Technol. 27:405-421 (1997).

Rylander, R.; Jacobs, R.R.; Eds: Organic Dusts. Exposure, Effects, and Prevention. Lewis Publishers, Ann Arbor, MI (1994).

Somerville, M.C.; Rivers, J.C.: An Alternative Approach for the Correction of Bioaerosol Data Collected with Multiple Jet Impactors. Am. Ind. Hyg. Assoc. J. 55:127-131 (1994).

Stewart, S.L.; Grinshpun, S.A.; Willeke, K.; et al.: Effect of Impact Stress on Microbial Recovery when Sampling on an Agar Surface. Appl. Environ. Microbiol. 61:1232-1239 (1995).

Thompson, M.A.; Donnelly, J.; Grinshpun, S.A.; et al.: Method and Test System for Evaluation of Bioaerosol Samplers. J. Aerosol Sci. 25:1579-1593 (1994).

Tong, Y.; Lighthart, B.: Solar Radiation has a Lethal Effect on Natural Populations of Culturable Outdoor Atmospheric Bacteria. Atmos. Environ. 31:897-900 (1997).

Ulevicius, V.; Willeke, K.; Grinshpun, S.A.; et al.: Aerosolization of Particles from a Bubbling Liquid: Characteristics and Generator Development. Aerosol Sci. Technol. 26:175-190 (1997).

Upton, S.L.; Mark, D.; Douglass, E.J.; et al.: A Wind Tunnel Evaluation of the Physical Sampling Efficiencies of Three Bioaerosol Samplers. J. Aerosol Sci. 25:1493-1501 (1994).

Vincent, J.H.: Standards for Health-Related Aerosol Measurement and Control. In: Aerosol Science for Industrial Hygienists, pp. 204-237. Elsevier Science Inc., Tarrytown, NY (1995).

Willeke, K.; McFeters, J.J.: The Influence of Flow Entry and Collecting Surface on the Impaction Efficiency of Inertial Impactors. J. Colloid. Interface Sci. 53:121-127 (1975).

Willeke, K.; Grinshpun, S.A.; Ulevicius, V.; et al.: Microbial Stress, Bounce and Re-Aerosolization in Bioaerosol Samplers. J. Aerosol Sci. 26:S883-S884 (1995).

Willeke, K.; Qian, Y.; Donnelly, J.; et al.: Penetration of Airborne Microorganisms through a Surgical Mask and a Dust/Mist Respirator. Am. Ind. Hyg. Assoc. J. 57:348-355 (1996).

Willeke, K.; Lin, X.; Grinshpun, S.A.: Improved Aerosol Collection by Combined Impaction and Centrifugal Motion. Aerosol Sci. Technol. 28:439-456 (1998).

Appendix 11.A. Manufacturers and Suppliers of Bioaerosol Samplers

AGI   Ace Glass Incorporated
P.O. Box 688
1430 Northwest Boulevard
Vineland, NJ 08360-0688
609-692-3333
800-223-4524
fax: 800-543-6752
web: www.aceglass.com

ALL   Allergen LLC
dba Allergenco/Blewstone Press
P.O. Box 8571
Wainwright Station
San Antonio, TX 78208-0571
210-822-4116
fax: 210-822-4116 *51, 210-805-8518
web: www.txdirect.net/corp/allergen
allergen@txdirect.net

AND   Andersen Instruments
500 Technology Court
Smyrna, GA 30082-5211
770-319-9999
800-241-6898
fax: 770-319-0336
web: www.anderseninstruments.com

BC   Barramundi Corporation
P.O. Drawer 4259
Homosassa Springs, FL 34447-4259
352-628-0200
fax: 352-628-0203
barra@citrus.infi.net
web: www.mattson-garvin.com

BDC   Biotest Diagnostics Corporation
66 Ford Road, Suite 131
Denville, NJ 07834-1300
973-625-1300
800-522-0090
fax: 973-625-9454
web: www.biotest.com

BMC   Burkard Manufacturing Company,
Limited
Woodcock Hill Industrial Estate
Rickmansworth, Hertfordshire
WD3 1PJ
England
44(0)1923-773134
fax: 44(0)1923-774790
web: www.burkard.co.uk

BSI   Bioscience International
11607 Magruder Lane
Rockville, MD 20852-4365
301-230-0072
fax: 301-230-1418
web: www.biosci-intl.com

CAS   Casella Limited
Regent House, Wolseley Road
Kempston, Bedford MK42 7JY
England
44(0)1234-841441
44(0)1234-841468
fax: 44(0)1234-841490
web: www.casella.co.uk

COR   Corning Incorporated
45 Nagog Park
Acton, MA 01720-3413
978-635-2200
800-492-1110
fax: 978-635-2476
web: www.scienceproducts.corning.com

HG   Hampshire Glassware
77-79 Dukes Road, Hampshire
Southampton SO14 0ST
England
44(0)1703-553755
fax: 44(0)1703-553020

LAN   Lanzoni, S.R.L.
Via Michelino 93/B
40126 Bologna
Italy
39(0)51-504810, (0)51-501334
fax: 39(0)51-6331892
web: www.lanzoni.it

LRI   Life's Resources, Incorporated
114 E. Main Street
P.O. Box 260
Addison, MI 49220-0260
517-547-7494
fax: 517-547-5444
web: lifes-resources.com

MIL   Millipore Corporation
80 Ashby Road
Bedford, MA 01730-2271
617-533-2125
800-645-5476
fax: 781-533-8891
web: www.millipore.com

NBS   New Brunswick Scientific
Company, Incorporated
P.O. Box 4005
44 Talmadge Road
Edison, NJ 08818-4005
800-631-5417
732-287-1200
fax: 732-287-4222
bioinfo@nbsc.com
web: www.nbsc.com

PAR   F.W. Parrett Limited
65 Riefield Road
London SE9 2RA
England
44(0)1181-8504226
[UK 0181-8593254]
fax. 44(0)1181-8504226

PGS   Pall Gelman Sciences
600 S. Wagner Road
Ann Arbor, MI 48103-9019
734-665-0651
800-521-1520
fax: 734-913-6114
web: www.pall.com/gelman/

PBI   International PBI
Via Novara, 89
20153 Milan
Italy
39 2-40-090-010
fax: 39 2-40-353695
web: www.wheatonsci.com
pbiexp@interbusiness.it

SBI   Spiral Biotech, Incorporated
7830 Old Georgetown Road
Bethesda, MD 20814-2432
301-657-1620
fax: 301-652-5036
infosbi@spiralbiotech.com
web: www.spiralbiotech.com

SKC   SKC Incorporated
863 Valley View Road
Eighty Four, PA 15330-1301
724-941-9701
800-752-8472
fax: 724-941-1369
800-752-8476
web: www.skcinc.com

STI   Sampling Technologies,
Incorporated
10801 Wayzata Boulevard
Suite 340
Minnetonka, MN 55305-1533
612-544-1588
800-264-1338
fax: 612-544-1977 or
800-880-8040
web: www.rotorod.com
e-mail: rotorod@rotorod.com

VAI   Veltek Associates, Incorporated
Environmental Control
Monitoring Division
1039 West Bridge Street
Phoenixville, PA 19460-4218
610-983-4949
fax:: 610-983-9494
web: www.sterile.com/~vai

ZAA   Zefon Analytical Associates
2860 23rd Avenue North
St. Petersburg, FL 33713-4211
813-327-5449
800-282-0073
fax: 813-323-6965
web: www.zefon.com

APPENDIX 11.B(i). Positive-Hole Correction Table to Adjust Colony Counts from a 100-Hole Impactor for the Possibility of Collecting Multiple Particles Through a Hole

a	b	c		a	b	c
1	1.0	0.1		51	71.3	5.7
2	2.0	0.1		52	73.4	5.9
3	3.0	0.2		53	75.5	6.1
4	4.1	0.3		54	77.7	6.3
5	5.1	0.3		55	79.9	6.5
6	6.2	0.4		56	82.1	6.7
7	7.3	0.5		57	84.4	6.9
8	8.3	0.6		58	86.8	7.1
9	9.4	0.7		59	89.2	7.3
10	10.5	0.7		60	91.6	7.6
11	11.7	0.8		61	94.2	7.8
12	12.8	0.9		62	96.8	8.1
13	13.9	1.0		63	99.4	8.3
14	15.1	1.1		64	102.2	8.6
15	16.3	1.2		65	105.0	8.9
16	17.4	1.2		66	107.9	9.2
17	18.6	1.3		67	110.9	9.5
18	19.8	1.4		68	114.0	9.8
19	21.1	1.5		69	117.1	10.2
20	22.3	1.6		70	120.4	10.5
21	23.6	1.7		71	123.8	10.9
22	24.8	1.8		72	127.3	11.3
23	26.1	1.9		73	130.9	11.7
24	27.4	2.0		74	134.7	12.1
25	28.8	2.1		75	138.6	12.6
26	30.1	2.2		76	142.7	13.1
27	31.5	2.3		77	147.0	13.6
28	32.9	2.4		78	151.4	14.1
29	34.2	2.5		79	156.1	14.7
30	35.7	2.7		80	161.0	15.3
31	37.1	2.8		81	166.1	16.0
32	38.6	2.9		82	171.5	16.7
33	40.0	3.0		83	177.2	17.4
34	41.6	3.1		84	183.3	18.3
35	43.1	3.3		85	189.7	19.2
36	44.6	3.4		86	196.7	20.2
37	46.2	3.5		87	204.1	21.3
38	47.8	3.6		88	212.1	22.5
39	49.4	3.8		89	220.8	23.9
40	51.1	3.9		90	230.3	25.5
41	52.8	4.1		91	240.9	27.3
42	54.5	4.2		92	252.7	29.4
43	56.2	4.3		93	266.1	31.9
44	58.0	4.5		94	281.6	35.0
45	59.8	4.7		95	299.9	39.0
46	61.6	4.8		96	322.4	44.2
47	63.5	5.0		97	351.6	51.7
48	65.4	5.1		98	393.2	64.0
49	67.3	5.3		99	468.2	91.7
50	69.3	5.5		100	>518.7	

a = observed number of CFUs   b = expected number of CFUs corrected for coincidence   c = standard deviation of b

APPENDIX 11.B(ii). Positive-Hole Correction Table to Adjust Colony Counts from a 200-Hole Impactor for the Possibility of Collecting Multiple Particles Through a Hole

a	b	c	a	b	c	a	b	c	a	b	c
1	1.0	0.0	51	58.9	3.1	101	140.6	7.9	151	281.3	18.2
2	2.0	0.1	52	60.2	3.2	102	142.7	8.1	152	285.4	18.6
3	3.0	0.1	53	61.6	3.2	103	144.7	8.2	153	289.6	18.9
4	4.0	0.2	54	62.9	3.3	104	146.8	8.3	154	293.9	19.3
5	5.1	0.2	55	64.3	3.4	105	148.9	8.5	155	298.3	19.7
6	6.1	0.3	56	65.7	3.5	106	151.0	8.6	156	302.8	20.1
7	7.1	0.3	57	67.1	3.5	107	153.1	8.7	157	307.4	20.5
8	8.2	0.4	58	68.5	3.6	108	155.3	8.9	158	312.1	20.9
9	9.2	0.5	59	69.9	3.7	109	157.5	9.0	159	317.0	21.3
10	10.3	0.5	60	71.3	3.8	110	159.7	9.2	160	321.9	21.8
11	11.3	0.6	61	72.8	3.9	111	161.9	9.3	161	327.0	22.2
12	12.4	0.6	62	74.2	3.9	112	164.2	9.5	162	332.2	22.7
13	13.4	0.7	63	75.7	4.0	113	166.5	9.6	163	337.5	23.2
14	14.5	0.7	64	77.1	4.1	114	168.8	9.8	164	343.0	23.7
15	15.6	0.8	65	78.6	4.2	115	171.1	9.9	165	348.6	24.2
16	16.7	0.8	66	80.1	4.3	116	173.5	10.1	166	354.4	24.8
17	17.8	0.9	67	81.6	4.4	117	175.9	10.3	167	360.4	25.4
18	18.9	0.9	68	83.1	4.4	118	178.3	10.4	168	366.5	26.0
19	20.0	1.0	69	84.6	4.5	119	180.8	10.6	169	372.9	26.6
20	21.1	1.1	70	86.2	4.6	120	183.3	10.8	170	379.4	27.3
21	22.2	1.1	71	87.7	4.7	121	185.8	10.9	171	386.2	28.0
22	23.3	1.2	72	89.3	4.8	122	188.3	11.1	172	393.2	28.7
23	24.4	1.2	73	90.8	4.9	123	190.9	11.3	173	400.5	29.5
24	25.6	1.3	74	92.4	5.0	124	193.5	11.5	174	408.1	30.3
25	26.7	1.4	75	94.0	5.1	125	196.2	11.7	175	415.9	31.2
26	27.9	1.4	76	95.6	5.2	126	198.9	11.9	176	424.1	32.1
27	29.0	1.5	77	97.2	5.3	127	201.6	12.0	177	432.6	33.0
28	30.2	1.5	78	98.9	5.4	128	204.3	12.2	178	441.5	34.0
29	31.3	6	79	100.5	5	129	207.1	12.4	179	450.8	35.1
30	32.5	1.7	80	102.2	5.6	130	210.0	12.7	180	460.6	36.3
31	33.7	1.7	81	103.8	5.7	131	212.8	2.9	181	470.8	.6
32	34.9	1.8	82	105.5	5.8	132	215.8	13.1	182	481.6	38.9
33	36.1	1.8	83	107.2	5.9	133	218.7	13.3	183	493.1	40.4
34	37.3	1.9	84	108.9	6.0	134	221.7	13.5	184	505.2	42.0
35	38.5	2.0	85	110.7	6.1	135	224.8	13.7	185	518.1	43.7
36	39.7	2.0	86	112.4	6.2	136	227.9	14.0	186	531.9	45.6
37	40.9	2.1	87	114.2	6.3	137	231.0	14.2	187	546.8	47.7
38	42.1	2.2	88	116.0	6.4	138	234.2	14.5	188	562.8	50.1
39	43.4	2.2	89	117.8	6.5	139	237.5	14.7	189	580.2	52.8
40	44.6	2.3	90	119.6	6.6	140	240.8	15.0	190	599.3	55.8
41	45.9	2.4	91	121.4	6.7	141	244.2	15.2	191	620.4	59.3
42	47.1	2.4	92	123.2	6.8	142	247.6	15.5	192	644.0	63.4
43	48.4	2.5	93	125.1	7.0	143	251.1	15.8	193	670.8	68.3
44	49.7	2.6	94	127.0	7.1	144	254.6	16.0	194	701.8	74.3
45	51.0	2.6	95	128.9	7.2	145	258.2	16.3	195	738.4	82.0
46	52.3	2.7	96	130.8	7.3	146	261.9	16.6	196	783.4	92.2
47	53.6	2.8	97	132.7	7.4	147	265.6	16.9	197	841.8	107.0
48	54.9	2.9	98	134.7	7.6	148	269.4	17.2	198	925.1	131.1
49	56.2	2.9	99	136.6	7.7	149	273.3	17.6	199	1075.1	185.9
50	57.5	3.0	100	138.6	7.8	150	277.3	17.9	200	>1175.6	

APPENDIX 11.B(iii). Positive-Hole Correction Table to Adjust Colony Counts from a 219-Hole Impactor for the Possibility of Collecting Multiple Particles Through a Hole

a	b	c	a	b	c	a	b	c	a	b	c	a	b	c
1	1.0	0.0	51	58.1	2.9	101	135.4	7.2	151	256.1	15.1	201	547.3	43.2
2	2.0	0.1	52	59.4	3.0	102	137.3	7.3	152	259.4	15.4	202	559.8	44.8
3	3.0	0.1	53	60.7	3.0	103	139.2	7.4	153	262.7	15.6	203	573.1	46.5
4	4.0	0.2	54	62.0	3.1	104	141.1	7.5	154	266.0	15.8	204	587.2	48.4
5	5.1	0.2	55	63.3	3.2	105	143.0	7.6	155	269.4	16.1	205	602.3	50.5
6	6.1	0.3	56	64.7	3.2	106	144.9	7.8	156	272.9	16.3	206	618.6	52.8
7	7.1	0.3	57	66.0	3.3	107	146.9	7.9	157	276.4	16.6	207	636.1	55.4
8	8.1	0.4	58	67.4	3.4	108	148.8	8.0	158	279.9	16.9	208	655.2	58.3
9	9.2	0.4	59	68.7	3.4	109	150.8	8.1	159	283.5	17.2	209	676.1	61.6
10	10.2	0.5	60	70.1	3.5	110	152.8	8.2	160	287.2	17.4	210	699.2	65.4
11	11.3	0.5	61	71.5	3.6	111	154.8	8.4	161	291.0	17.7	211	725.1	69.9
12	12.3	0.6	62	72.9	3.7	112	156.9	8.5	162	294.8	18.0	212	754.4	75.3
13	13.4	0.6	63	74.3	3.7	113	158.9	8.6	163	298.7	18.3	213	788.3	81.8
14	14.5	0.7	64	75.7	3.8	114	161.0	8.7	164	302.6	18.6	214	828.5	90.2
15	15.5	0.7	65	77.1	3.9	115	163.1	8.9	165	306.6	19.0	215	877.7	101.4
16	16.6	0.8	66	78.5	4.0	116	165.2	9.0	166	310.7	19.3	216	941.6	117.5
17	17.7	0.8	67	80.0	4.0	117	167.3	9.1	167	314.9	19.6	217	1032.9	143.9
18	18.8	0.9	68	81.4	4.1	118	169.5	9.3	168	319.1	20.0	218	1197.1	203.9
19	19.9	1.0	69	82.9	4.2	119	171.7	9.4	169	323.5	20.3	219	>1307.1	
20	21.0	1.0	70	84.3	4.3	120	173.9	9.5	170	327.9	20.7			
21	22.1	1.1	71	85.8	4.4	121	176.1	9.7	171	332.4	21.1			
22	23.2	1.1	72	87.3	4.4	122	178.3	9.8	172	337.0	21.5			
23	24.3	1.2	73	88.8	4.5	123	180.6	10.0	173	341.7	21.9			
24	25.4	1.2	74	90.3	4.6	124	182.9	10.1	174	346.6	22.3			
25	26.5	1.3	75	91.8	4.7	125	185.2	10.3	175	351.5	22.7			
26	27.7	1.3	76	93.3	4.8	126	187.6	10.4	176	356.5	23.1			
27	28.8	1.4	77	94.9	4.9	127	189.9	10.6	177	361.7	23.6			
28	30.0	1.5	78	96.4	5.0	128	192.3	10.7	178	366.9	24.0			
29	31.1	1.5	79	98.0	5.0	129	194.8	10.9	179	372.4	24.5			
30	32.3	1.6	80	99.6	5.1	130	197.2	11.0	180	377.9	25.0			
31	33.4	1.6	81	101.1	5.2	131	199.7	11.2	181	383.6	25.6			
32	34.6	1.7	82	102.7	5.3	132	202.2	11.4	182	389.4	26.1			
33	35.8	1.7	83	104.3	5.4	133	204.7	11.5	183	395.4	26.7			
34	36.9	1.8	84	106.0	5.5	134	207.3	11.7	184	401.6	27.2			
35	38.1	1.9	85	107.6	5.6	135	209.9	11.9	185	407.9	27.8			
36	39.3	1.9	86	109.2	5.7	136	212.5	12.1	186	414.5	28.5			
37	40.5	2.0	87	110.9	5.8	137	215.1	12.2	187	421.2	29.1			
38	41.7	2.0	88	112.5	5.9	138	217.8	12.4	188	428.2	29.8			
39	42.9	2.1	89	114.2	6.0	139	220.5	12.6	189	435.4	30.5			
40	44.2	2.2	90	115.9	6.0	140	223.3	12.8	190	442.8	31.3			
41	45.4	2.2	91	117.6	6.1	141	226.1	13.0	191	450.5	32.1			
42	46.6	2.3	92	119.3	6.2	142	228.9	13.2	192	458.4	32.9			
43	47.9	2.4	93	121.1	6.3	143	231.8	13.4	193	466.7	33.8			
44	49.1	2.4	94	122.8	6.4	144	234.7	13.6	194	475.3	34.8			
45	50.4	2.5	95	124.6	6.5	145	237.6	13.8	195	484.2	35.7			
46	51.6	2.6	96	126.3	6.7	146	240.6	14.0	196	493.6	36.8			
47	52.9	2.6	97	128.1	6.8	147	243.6	14.2	197	503.3	37.9			
48	54.2	2.7	98	129.9	6.9	148	246.7	14.4	198	513.5	39.1			
49	55.5	2.8	99	131.7	7.0	149	249.8	14.7	199	524.2	40.4			
50	56.8	2.8	100	133.6	7.1	150	252.9	14.9	200	535.4	41.7			

APPENDIX 11.B(iv). Positive-Hole Correction Table to Adjust Colony Counts from a 400-Hole Impactor for the Possibility of Collecting Multiple Particles Through a Hole

a	b	c	a	b	c	a	b	c	a	b	c
1	1.0	0.0	51	54.6	2.0	101	116.4	4.3	151	189.6	7.3
2	2.0	0.1	52	55.7	2.0	102	117.7	4.4	152	191.2	7.3
3	3.0	0.1	53	56.9	2.0	103	119.1	4.4	153	192.8	7.4
4	4.0	0.1	54	58.0	2.1	104	120.4	4.5	154	194.5	7.5
5	5.0	0.2	55	59.2	2.1	105	121.8	4.5	155	196.1	7.5
6	6.0	0.2	56	60.3	2.2	106	123.2	4.6	156	197.7	7.6
7	7.1	0.2	57	61.5	2.2	107	124.5	4.6	157	199.4	7.7
8	8.1	0.3	58	62.7	2.3	108	125.9	4.7	158	201.0	7.7
9	9.1	0.3	59	63.8	2.3	109	127.3	4.7	159	202.7	7.8
10	10.1	0.4	60	65.0	2.4	110	128.6	4.8	160	204.3	7.9
11	11.2	0.4	61	66.2	2.4	111	130.0	4.8	161	206.0	7.9
12	12.2	0.4	62	67.4	2.4	112	131.4	4.9	162	207.7	8.0
13	13.2	0.5	63	68.6	2.5	113	132.8	5.0	163	209.4	8.1
14	14.3	0.5	64	69.7	2.5	114	134.2	5.0	164	211.1	8.2
15	15.3	0.5	65	70.9	2.6	115	135.6	5.1	165	212.8	8.2
16	16.3	0.6	66	72.1	2.6	116	137.0	5.1	166	214.5	8.3
17	17.4	0.6	67	73.3	2.7	117	138.4	5.2	167	216.2	8.4
18	18.4	0.6	68	74.5	2.7	118	139.8	5.2	168	217.9	8.5
19	19.5	0.7	69	75.7	2.8	119	141.2	5.3	169	219.6	8.5
20	20.5	0.7	70	76.9	2.8	120	142.7	5.3	170	221.4	8.6
21	21.6	0.8	71	78.2	2.8	121	144.1	5.4	171	223.1	8.7
22	22.6	0.8	72	79.4	2.9	122	145.5	5.5	172	224.8	8.8
23	23.7	0.8	73	80.6	2.9	123	147.0	5.5	173	226.6	8.8
24	24.8	0.9	74	81.8	3.0	124	148.4	5.6	174	228.4	8.9
25	25.8	0.9	75	83.1	3.0	125	149.9	5.6	175	230.1	9.0
26	26.9	1.0	76	84.3	3.1	126	151.3	5.7	176	231.9	9.1
27	28.0	1.0	77	85.5	3.1	127	152.8	5.8	177	233.7	9.1
28	29.0	1.0	78	86.8	3.2	128	154.3	5.8	178	235.5	9.2
29	30.1	1.1	79	88.0	3.2	129	155.7	5.9	179	237.3	9.3
30	31.2	1.1	80	89.3	3.3	130	157.2	5.9	180	239.1	9.4
31	32.3	1.1	81	90.5	3.3	131	158.7	6.0	181	241.0	9.4
32	33.4	1.2	82	91.8	3.4	132	160.2	6.1	182	242.8	9.5
33	34.4	1.2	83	93.0	3.4	133	161.7	6.1	183	244.6	9.6
34	35.5	1.3	84	94.3	3.5	134	163.2	6.2	184	246.5	9.7
35	36.6	1.3	85	95.6	3.5	135	164.7	6.2	185	248.3	9.8
36	37.7	1.3	86	96.8	3.6	136	166.2	6.3	186	250.2	9.9
37	38.8	1.4	87	98.1	3.6	137	167.7	6.4	187	252.1	9.9
38	39.9	1.4	88	99.4	3.7	138	169.2	6.4	188	254.0	10.0
39	41.0	1.5	89	100.7	3.7	139	170.8	6.5	189	255.8	10.1
40	42.1	1.5	90	102.0	3.8	140	172.3	6.5	190	257.7	10.2
41	43.3	1.5	91	103.2	3.8	141	173.9	6.6	191	259.7	10.3
42	44.4	1.6	92	104.5	3.9	142	175.4	6.7	192	261.6	10.4
43	45.5	1.6	93	105.8	3.9	143	177.0	6.7	193	263.5	10.4
44	46.6	1.7	94	107.2	4.0	144	178.5	6.8	194	265.4	10.5
45	47.7	1.7	95	108.5	4.0	145	180.1	6.9	195	267.4	10.6
46	48.9	1.8	96	109.8	4.1	146	181.7	6.9	196	269.3	10.7
47	50.0	1.8	97	111.1	4.1	147	183.2	7.0	197	271.3	10.8
40	51.1	1.8	98	112.4	4.2	148	184.8	7.1	198	273.3	10.9
49	52.3	1.9	99	113.7	4.2	149	186.4	7.1	199	275.3	11.0
50	53.4	1.9	100	115.1	4.3	150	188.0	7.2	200	277.3	11.1

(cont.)

APPENDIX 11.B(iv). Positive-Hole Correction Table to Adjust Colony Counts from a 400-Hole Impactor for the Possibility of Collecting Multiple Particles Through a Hole (cont.)

a	b	c	a	b	c	a	b	c	a	b	c
201	279.3	11.1	251	395.0	16.7	301	558.5	25.6	351	839.9	44.8
202	281.3	11.2	252	397.7	16.8	302	562.6	25.8	352	848.1	45.5
203	283.3	11.3	253	400.4	16.9	303	566.7	26.1	353	856.5	46.2
204	285.3	11.4	254	403.1	17.1	304	570.8	26.3	354	865.1	46.9
205	287.4	11.5	255	405.9	17.2	305	575.0	26.6	355	873.9	47.6
206	289.4	11.6	256	408.7	17.4	306	579.3	26.8	356	882.9	48.3
207	291.5	11.7	257	411.4	17.5	307	583.5	27.1	357	892.1	49.1
208	293.6	11.8	258	414.3	17.6	308	587.9	27.3	358	901.5	49.9
209	295.7	11.9	259	417.1	17.8	309	592.2	27.6	359	911.2	50.7
210	297.8	12.0	260	419.9	17.9	310	596.7	27.9	360	921.1	51.5
211	299.9	12.1	261	422.8	18.1	311	601.1	28.2	361	931.2	52.4
212	302.0	12.2	262	425.7	18.2	312	605.7	28.4	362	941.6	53.3
213	304.1	12.3	263	428.6	18.4	313	610.2	28.7	363	952.2	54.3
214	306.3	12.4	264	431.5	18.5	314	614.9	29.0	364	963.2	55.3
215	308.4	12.5	265	434.5	18.7	315	619.5	29.3	365	974.5	56.3
216	310.6	12.6	266	437.5	18.8	316	624.3	29.6	366	986.1	57.3
217	312.8	12.7	267	440.4	19.0	317	629.1	29.9	367	998.0	58.5
218	315.0	12.8	268	443.5	19.2	318	633.9	30.2	368	1010.3	59.6
219	317.2	12.9	269	446.5	19.3	319	638.8	30.5	369	1023.0	60.8
220	319.4	13.0	270	449.6	19.5	320	643.8	30.8	370	1036.1	62.1
221	321.6	13.1	271	452.7	19.6	321	648.8	31.2	371	1049.7	63.4
222	323.9	13.2	272	455.8	19.8	322	653.9	31.5	372	1063.7	64.8
223	326.1	13.3	273	458.9	20.0	323	659.1	31.8	373	1078.3	66.3
224	328.4	13.4	274	462.1	20.2	324	664.3	32.2	374	1093.4	67.9
225	330.7	13.5	275	465.3	20.3	325	669.6	32.5	375	1109.1	69.5
226	333.0	13.6	276	468.5	20.5	326	675.0	32.9	376	1125.4	71.3
227	335.3	13.7	277	471.7	20.7	327	680.4	33.2	377	1142.5	73.1
228	337.6	13.9	278	475.0	20.8	328	685.9	33.6	378	1160.2	75.1
229	339.9	14.0	279	478.3	21.0	329	691.5	34.0	379	1178.9	77.2
230	342.3	14.1	280	481.6	21.2	330	697.2	34.4	380	1198.4	79.4
231	344.6	14.2	281	484.9	21.4	331	703.0	34.8	381	1218.9	81.9
232	347.0	14.3	282	488.3	21.6	332	708.8	35.2	382	1240.5	84.5
233	349.4	14.4	283	491.7	21.8	333	714.7	35.6	383	1263.4	87.3
234	351.8	14.5	284	495.2	22.0	334	720.7	36.0	384	1287.7	90.4
235	354.2	14.7	285	498.6	22.1	335	726.8	36.4	385	1313.5	93.8
236	356.6	14.8	286	502.1	22.3	336	733.0	36.9	386	1341.1	97.5
237	359.1	14.9	287	505.6	22.5	337	739.3	37.3	387	1370.8	101.6
238	361.5	15.0	288	509.2	22.7	338	745.7	37.8	388	1402.9	106.2
239	364.0	15.1	289	512.8	22.9	339	752.2	38.2	389	1437.7	111.4
240	366.5	15.2	290	516.4	23.1	340	758.9	38.7	390	1475.9	117.3
241	369.0	15.4	291	520.0	23.4	341	765.6	39.2	391	1518.1	124.2
242	371.5	15.5	292	523.7	23.6	342	772.4	39.7	392	1565.3	132.2
243	374.1	15.6	293	527.5	23.8	343	779.4	40.2	393	1618.9	141.8
244	376.6	15.7	294	531.2	24.0	344	786.5	40.7	394	1680.8	153.7
245	379.2	15.9	295	535.0	24.2	345	793.7	41.3	395	1754.1	168.8
246	381.8	16.0	296	538.8	24.4	346	801.0	41.8	396	1844.1	188.9
247	384.4	16.1	297	542.7	24.7	347	808.5	42.4	397	1960.8	218.0
248	387.0	16.3	298	546.6	24.9	348	816.1	43.0	398	2127.5	265.9
249	389.7	16.4	299	550.5	25.1	349	823.9	43.6	399	2427.5	374.8
250	392.3	16.5	300	554.5	25.3	350	831.8	44.2	400	>2628.0	

APPENDIX 11.C. Calculation of Positive-Hole Correction to Adjust Colony Counts from a Multiple-Hole Impactor for the Possibility of Collecting Multiple Particles Through a Hole

Corrected colony counts, n_c, for a multiple-hole impactor with N_j jets can be calculated from Equation 11-5 (Hinds, in press):

$$n_c = n_f \left(\frac{1.075}{1.052 - f} \right)^{0.483} \quad \text{for} \quad f < 0.95 \qquad \text{(11-5)}$$

where n_f is the number of CFUs or filled impaction sites and $f = n_f/N_j$. Equation 11-5 agrees with the tabulated values in Appendices 11.B(i) to 11.B(iv) within 1% for filled fractions less than 0.95.

TABLE 11.C(i). Correction Factors for Particle Collection with Multiple-Hole Impactors

Filled Fraction[A]	Correction Factor[B]
0.05	1.026
0.10	1.054
0.15	1.084
0.20	1.116
0.25	1.151
0.30	1.189
0.35	1.231
0.40	1.277
0.45	1.329
0.50	1.386
0.55	1.452
0.60	1.527
0.65	1.615
0.70	1.720
0.75	1.848
0.80	2.012
0.85	2.232
0.90	2.559
1.00	>5.878

[A]   Fraction of impaction sites with CFUs.
[B]   Total number of viable particles collected equals the number of filled sites times the correction factor.

Chapter 12

Source Sampling

John W. Martyny and Kenneth F. Martinez with Philip R. Morey

12.1   Bulk Sampling
    12.1.1   *General Considerations*
    12.1.2   *Sample Collection*
    12.1.3   *Sample Analysis*
        12.1.3.1   Culture-Based Analyses
        12.1.3.2   Non-Culture Analyses
        12.1.3.3   Information from Bulk Samples
    12.1.4   *Interpreting Sampling Results*
        12.1.4.1   Differentiating In-Situ Microbial Growth from an Accumulation of Material of Biological Origin
        12.1.4.2   Basing Recommendations on Bulk Sampling Data
12.2   Surface Sampling
    12.2.1   *General Considerations*
    12.2.2   *Sample Collection*
        12.2.2.1   Contact Sampling
        12.2.2.2   Surface-Wash Sampling
    12.2.3   *Sample Analysis*
        12.2.3.1   Culture-Based Analyses
        12.2.3.2   Non-Culture Analyses
    12.2.4   *Interpreting Sampling Results*
12.3   References

12.1 Bulk Sampling

12.1.1 General Considerations

Bulk samples are portions of environmental materials (e.g., settled dust, sections of wallboard, pieces of duct lining, carpet segments, or return-air filters) tested to determine if they may contain or be contaminated with biological agents. The objective of such sampling is to collect a portion of material small enough to be transported conveniently and handled easily in the laboratory while still representing the material being sampled. Testing is done to determine if organisms (e.g., microorganisms or dust mites) have colonized the material and are actively growing as well as to identify surfaces where previously airborne biological particles have deposited and accumulated (Gravesen et al., 1986; Cole et al., 1994; Flannigan et al., 1994). The analysis of surfaces or dust for culturable fungi and bacteria may be considered a necessary adjunct to air sampling because, under some circumstances and especially for quiescent sampling, air samples may fail to detect biological contamination (i.e., yield false-negative findings) (AIHA, 1996). Investigators may also consider bulk sampling an integral part of bioaerosol investigations be-

cause a study's primary purpose is often identifying sources of biological agents so that they can be remediated. Some infectious agents (e.g., *Legionella* spp., amebae, and microorganisms present in low numbers) are difficult to culture from air samples and are best identified from their original sources (Burge and Solomon, 1987). Likewise, some biological agents are identified primarily from bulk dust or water samples (e.g., arthropod, avian, and mammalian antigens as well as microbial toxins, such as endotoxin and mycotoxins).

Bulk samples can provide information about (a) possible sources of microorganisms, antigens, microbial toxins, and other biological agents in buildings, and (b) the general composition and relative concentrations of biological agents in these sources. Investigators have also made extracts of bulk samples or of microorganisms isolated therefrom to test serum samples from workers with hypersensitivity diseases (Nolard et al., 1994). Workers exposed to antigens in biologically contaminated bulk materials may show evidence of immunological sensitization not seen in unexposed or asymptomatic workers [see 3.3, 8.2.1, and 25.2.3]. Bulk sampling may also be used to evaluate the effectiveness of mitigation and remediation efforts (Cole et al., 1994; Morey, 1994) [see 15.5].

Dust, house dust, or settled dust are terms used to describe the material that collects on horizontal surfaces and in textiles such as drapes, upholstered furniture, and carpets. Such dust may contain many types of materials in varying amounts, depending on the location and surroundings of a building and the types of activities that occur in it. Yang et al. (1992) identified by microscopic examination at least 13 components of dust from office buildings, including (in descending order of occurrence) skin flakes (believed to be primarily of human origin), cellulose fibers (from paper and paper products), synthetic fibers (from carpets, fabrics, office furniture, and clothing), human hairs, cat hairs (even though cats were not allowed in the buildings), fibrous glass, pollen, other animal hairs [dog, rabbit (from human clothing), and unidentified], wood fragments, plant materials (partially decayed leaves and trichomes—plant hairs), fungal spores, insect parts, and bird feathers.

It has been hypothesized that settled dust or dust collected on return-air filters would reflect previously airborne biological particles and provide a more representative picture of bioaerosol exposure than short-term air samples. Fox and Rosario (1994) studied dust collected on return-air filters under the assumption that collected particles would reflect the airborne materials to which building occupants had been exposed. Also, filter deposits can serve as a growth substrate if filters become damp. However, bulk samples cannot replace air samples because the former have not been found to accurately reflect past, future, or even current bioaerosol exposures. Researchers who have collected parallel bulk and air samples have seen differences that reflect the contrasting types of biological agents present on surfaces and in air (Chew et al., 1996). An inverse relationship was even observed between the mass of dust in a carpet and the mass of airborne dust (Leese et al., 1997).

While some investigators have found the concentrations and types of microorganisms in dust to be stable over time (Takatori et al., 1994), others have not (Hoekstra et al., 1994; Verhoeff et al., 1994). Microbial growth in contaminated materials is influenced by indoor temperature and moisture availability (Wickman et al., 1992; Becker, 1994) [see 10.2]. Bioaerosol dissemination from sources is influenced by agitation or disturbance of contaminated materials, the presence of aerial spores, indoor air velocity, and ventilation rate (Gravesen et al., 1986; Macher et al., 1991; Flannigan et al., 1993) [see 10.5.2 and 15.2].

12.1.2 Sample Collection

In general, bulk samples are cut or otherwise aseptically removed from a source and placed in clean, new or sterilized containers. Strict aseptic techniques may not be required for some samples if the abundance of biological agents in a sample is anticipated to far outweigh any that the person collecting a sample may add by handling. For laboratory examination, an analyst typically selects a segment of a sample from an area that is least likely to have been handled (e.g., from the center of a piece of material rather than from an outer edge).

Suitable containers for bulk samples are sterile jars for dry items or sterile bottles for water samples. New paper bags may be adequate to transport dry material samples. Sealable plastic bags are useful for samples of ventilation duct lining, ceiling tiles, wallpaper, and similar materials. To preserve the integrity of samples and avoid cross contamination, paper bags may be placed in plastic bags with a packet of desiccant material to keep the sample dry. The amount or volume of sample to collect and the manner in which to remove and transport it depend on the sample type and the analytical methods to be applied (AIHA, 1995; AIHA, 1996; ASM, 1997).

Samples of loose material (e.g., carpet dust for antigen detection) can often be conveniently collected using a suction device (AIHA, 1996). Researchers have designed filters for full-size vacuum cleaners (Hamilton et al., 1992; Verhoeff et al., 1993) or have simply used a new vacuum-cleaner bag for each sample area (Hamilton et al., 1992). Other investigators have used hand-held vacuum cleaners (Flannigan et al., 1993; O'Rourke et al., 1993), sometimes with special filters (Lindstrom et al., 1993). Researchers have also collected material on filters in personal cassettes used open face, as miniature vacuum cleaners, or with an inlet port and a sampling template to define the area sampled (Farfel et al., 1994a; Reynolds et al., 1997). Researchers studying lead and pesticide exposures have standardized methods for collecting household dust samples (Roberts et al., 1991; ASTM, 1994; Farfel et al., 1994a; Wang et al., 1995; Reynolds et al., 1997). Investigators may consider using such equipment and procedures to collect bulk samples to evaluate the presence of biological agents in indoor environments (Gyntelberg et al., 1994; Leese et al., 1997).

Depending on how the results will be used, individual areas may be sampled separately or samples may be composited (during collection or in the laboratory). Compositing samples from a large area reduces the number of samples to be analyzed and may improve the chances of detecting the material of interest (Linter et al., 1994). Some researchers sample a square meter of test surface or vacuum a prescribed area for a specified time and present the results as the amount of biological material per gram of dust for that area. Other researchers have vacuumed an entire room or house and reported the average concentration of biological agents in the collected dust for that unit. Sampling sites can be chosen randomly [see Appendix 5.A] or by using other processes to select sites and sample areas (Schneider et al., 1990). The number of samples collected depends on the purpose of the study [see 5.2.4].

Close consultation with laboratory personnel is vital in planning a bulk sampling program. The type of material to be tested, the biological agents sought, the information needed about the agents, and the expected results determine the appropriate collection method. Laboratory and field staff should discuss how much material is required to conduct particular assays, the number of samples needed to obtain representative results, the number of samples the laboratory can handle per day so that sample processing is not delayed beyond an acceptable holding time, and required sample storage and shipping conditions [see Chapter 6 and Part III]. Samples to be tested for viable microorganism counts generally require overnight delivery and need to be either chilled (i.e., maintained between ~4° and 10°C) or kept at room temperature but protected from extremely high or low temperatures during transport. In some cases, preservatives or an agent to neutralize a biocide (e.g., sodium thiosulfate for chlorine in water samples) are added in the field to stabilize samples and limit changes prior to analysis (Thorne et al., 1994; APHA, 1995). Culture plates can also be inoculated at the collection site prior to sample shipment, and convenient dipslides are available for some types of water testing (Biotest Hycon, Denville, NJ; Difco Laboratories, Detroit, MI).

12.1.3 Sample Analysis

Chapter 6 and the individual chapters in Part III describe methods for analyzing biological agents in bulk samples. These methods include detection of culturable microorganisms and morphologically distinctive particles as well as bioassays and chemical assays (AIHA, 1996). Dust samples are often sieved to collect the finer particles, which are those most likely to have been or to become airborne. Biological agents in weighed portions of collected dust and bulk materials are typically suspended or extracted in a fluid appropriate for the assay to follow. Sections 13.2.2 and 13.2.3 describe the calculation of the concentration of biological agents in bulk samples. Occasionally, bulk samples are collected for archival purposes and are not analyzed immediately. Unused portions of bulk samples may also be stored for later retesting. Storage conditions must minimize changes in the biological agents of interest. In particular, assays for many living organisms are of limited value after prolonged sample storage.

12.1.3.1 Culture-Based Analyses
Culture-based analyses involve growing microorganisms from bulk samples. For example, sieved dust as well as liquid samples, dust suspensions, and washings of other materials may be inoculated onto suitable agar-based culture media or cell cultures (Samson et al., 1994; AIHA, 1996). The type of growth medium or cells used and the incubation conditions (e.g., temperature, humidity, atmosphere, and duration) influence what bacteria, fungi, viruses, or amebae can be isolated. In addition, factors related to the sampled environment, the microbial agents, and the methods of sample collection and handling determine how representative the culture results are of the types and relative proportions of microorganisms present in bulk samples.

12.1.3.2 Non-Culture Analyses
Non-culture analyses for bulk samples may involve identifying biological agents under a microscope or employing biological or chemical assays. The information available from these analyses varies with the biological material under study. Non-culture methods generally do not provide information on the ability of identified microorganisms to propagate or to cause infection. Many fungal spores and pollen grains as well as dust mites and some amebic cysts can be identified by direct microscopic examination because of the organisms' distinctive sizes, shapes, and surface features. However, bacterial cells and viruses are generally much smaller and less distinctive. Few bacteria can be identified by microscope without some sort of staining, and viruses are only visible with an electron microscope. In some cases, bulk samples are analyzed to detect the presence of antigens from fungi, dust mites, cockroaches, birds, or mammals. Bulk samples can also be tested for the presence of microbial products and marker compounds (e.g., endotoxin, muramic acid, ergosterols, VOCs, and specific nucleic acid sequences).

12.1.3.3 Information from Bulk Samples
Some detection methods provide only semiquantitative information and are used only to identify the presence or absence of the biological agent in question. However, even such limited information may help investigators decide if further sampling is indicated and may help them formulate recommendations for remediation.

Bulk samples can also provide material with which to immunologically test exposed workers. Positive reactions only in symptomatic workers or the occupants of a problem area could reflect exposure to environmental antigens even though the specific antigenic material was not identified. Similarly, extracts of microorganisms isolated from environmental samples by culture-based methods can be used to determine workers' immunological sensitivity to specific bacteria or fungi. Tests conducted with extracts of dust samples or microorganisms isolated from an actual workplace may have more value than tests run with a standard extract battery such as an HP panel [see 3.3.3]. The antigen mixtures available in commercial panels may not include the relevant biological agents, or the microorganisms in a work environment may differ from those used to make commercial extracts even if they are identified as the same genus and species.

The concentration of a biological agent in a bulk sample is determined from the number of cells, number of CFU, or amount of target material per area sampled or per mass or volume of material analyzed. For example, bulk sampling results could be expressed as the number of fungal spores per square centimeter of ceiling tile, number of bacterial CFU per gram of insulation material, or nanograms of endotoxin per milliliter of cooling-tower water. Numerical measurements can help investigators compare locations within a workplace and relate current findings with those from other studies.

12.1.4 Interpreting Sampling Results

Investigators need experience to interpret bulk sampling data. In some cases, it may be sufficient to demonstrate that a specific biological agent was found in a study area. However, failure to detect a target material does not necessarily mean it is not present. Laboratories should report such results as less than the test's LDL rather than as zero [see 5.2.3 and Chapter 7]. Because bulk samples are not good estimators of actual exposure, measurements of biological agents in bulk samples do not allow conclusions as to the air concentration of material that would be associated with a given disease outcome. However, measurements of inhalation exposure may be difficult for some biological agents (e.g., some antigens), and positive correlations have been seen between health effects and some antigen concentrations in dust samples [see 22.5 and 25.5].

12.1.4.1 Differentiating In-Situ Microbial Growth from an Accumulation of Material of Biological Origin On microscopic examination or laboratory culture, the finding of predominantly one type of microorganism or the presence of hyphae and spore-bearing fungal structures (rather than a mixture of fragmented hyphae and spores) may indicate that growth is occurring at the sampled site rather than that biological material has simply collected there. However, water in drain pans, cooling tower sumps, or humidifiers is not expected to be sterile (i.e., some microorganisms are likely to be present even if not actively multiplying).

12.1.4.2 Basing Recommendations on Bulk Sampling Data Detecting high concentrations of microorganisms in bulk samples may eliminate the need for air sampling if there is a clear potential for the material to become aerosolized or for building occupants to directly contact the contaminated items. Recommendations for remediation might also be straight forward because such a situation is generally undesirable and should not be allowed to persist [see 14.1.2]. For example, investigators who find biofilm or large areas of visible fungal growth in an air handler should suggest mitigation of the problem rather than further sampling. However, air sampling is needed to establish worker inhalation exposure in terms of amount or type of biological material [see 14.1.3].

Part of the difficulty in determining that material is biologically contaminated and in establishing a connection between bulk sample measurements and health complaints stems from the many factors that can affect sampling results. Large variability has been observed among sampling sites in buildings and even at a single sampling location over time, reflecting seasonal effects and changing environmental conditions. Significant biological agents may be missed, misleading investigators, if the choice of collection and analytical methods were inappropriate or insufficiently sensitive. Therefore, investigators must consider bulk sampling data in conjunction with all available medical and environmental information to decide if the data support the hypotheses under evaluation. Further, investigators must decide what evidence is sufficient to warrant recommendations for remediation of suspected sources of biological agents. Accurate determination of the type and extent of indoor biological contamination is critical in selecting suitable remediation methods and appropriate precautions during the cleanup process [see Chapter 15].

12.2 Surface Sampling
12.2.1 General Considerations
Surface sampling during IEQ investigations is frequently linked to bulk and air sampling. Surface samples can provide information similar to that obtained from bulk samples regarding whether environmental materials may be contaminated beyond background levels and possibly serve as sources of biological agents that may be disseminated as bioaerosols. In addition, building occupants may be exposed to biological agents via skin contact with contaminated indoor surfaces.

Surface sampling may be used to (a) confirm the nature of suspected microbial growth on environmental surfaces, (b) measure the relative degree of biological contamination, and (c) identify the types of microorganisms and other biological agents present. Surface sampling is preferred over bulk sampling when a less destructive method of sample collection is desired. For example, it may be possible to collect samples of fungal growth from the surfaces of valuable furnishings or materials of historical interest without damaging the original items.

The concentration and composition of microorganisms growing on indoor surfaces depend on a number of factors, such as the nature of the material and its moisture history. Surfaces may also become contaminated as a consequence of bioaerosol deposition, for which the relevant parameters are the size, shape, and density of the biological particles; air velocities past surfaces; and air movement in the area. Researchers have collected gravity samples by exposing glass slides or settling plates in sample areas to collect whatever particles fall onto a surface in a given time (Pasanen et al., 1992). However, comparisons of data from gravity or sedimentation samples and active air samples rarely agree in their assessments of either bioaerosol concentration or composition (Hyvärinen et al., 1993; Ren et al., 1993). While gravity samples may reflect surface deposition during a sampling period, they are not suitable substitutes for volumetric air samples [see 11.3].

12.2.2 Sample Collection

Surface samples are collected by removing material (a) with a suction device, (b) by pressing a collection material (e.g., a contact plate or adhesive tape) onto a surface, or (c) by washing a prescribed area with a wetted swab, cheesecloth or gauze swatch, or filter (AIHA, 1995; AIHA, 1996). As compared to bulk dust sampling from floors, upholstered furniture, or other porous or fabric surfaces, the amount of material removed for a surface sample is generally small and the surfaces tested are generally smooth. More exact surface sampling methods, developed for other purposes, could also be used to collect biological materials (Lioy et al., 1993; Freeman et al., 1996; OSHA, 1996; ASTM, 1997; Corrigan and Blehm, 1997; Reynolds et al., 1997). Section 12.1.2 describes suction sampling using vacuum cleaners and other devices. Wipe and vacuum dust collection methods have been found to be highly correlated for removal of lead and other contaminants from floors and window sills as well as linoleum and wood surfaces (Farfel et al., 1994b; Reynolds et al., 1997). Wipe samples generally exceed measurements made with vacuum methods for smooth surfaces (Farfel et al., 1994b), but vacuum methods have been found to be better for removing settled dust from carpets (Reynolds et al., 1997). Tape sampling was found to be a more efficient and reproducible method for collecting glass fibers from smooth, painted metal than a microvacuum method (Corrigan and Blehm, 1997). Sampling sites can be chosen randomly [see Appendix 5.A] or by using other processes to select representative sites and sample areas (Schneider et al., 1990). The number of samples collected depends on the purpose of the study [see 5.2.4].

12.2.2.1 Contact Sampling Loose particles may be collected by pressing a contact plate to a surface or applying an adhesive material to lift off sample material.

Agar-Contact Sampling. Contact plates are special culture dishes or flexible containers with a meniscus of agar extending beyond the container's rim (Biotest Hycon, Denville, N.J.; Difco Laboratories, Detroit, Mich.; Falcon Nunc, Inc., Naperville, Ill.; Becton Dickinson and Company, Lincoln Park, N.J.). Advantages of the contact-plate method are that it is fairly easy to conduct (e.g., no filters or pumps are needed), and the exposed plates are simply packaged and shipped overnight to a laboratory for incubation and examination. Disadvantages of this method are the limitations inherent in all culture-based analyses as well as the possibility that growth on a contact plate may be so heavy that counting and identification of the isolated microorganisms is impossible [see 6.1.2].

Adhesive-Tape Sampling. If information about viable microorganisms on environmental surfaces is not needed, the adhesive-tape (or tape-lift) method can provide information on the types and relative concentrations of biological particles that are present (Morey, 1994). Such samples can be collected using clear adhesive tape or packing tape (e.g., 3M No. 480 polyethylene tape with an acrylic pressure-sensitive adhesive) or commercially available sampling strips (RJ Lee Group, Inc., Monroeville, PA) (ASTM, 1992). For microscopic examination of collected particles, adhesive tapes must be of good optical quality and compatible with any stains the analytical laboratory may use on the specimens.

Easily removed material is collected by gently touching the tape to a test surface and removing the tape with a slow steady force. The investigator should hold the tape by the edges only. A sufficient sample may appear as a light deposit on the sticky side of the tape. Samples with too heavy a deposit or large pieces of particulate matter may be difficult to examine by microscope and should be repeated. Investigators should also submit blank samples of the tape to the testing laboratory.

Following sample collection, exposed tape strips are attached to glass slides (avoiding creases and folds) or placed in protective containers (e.g., scintillation vials) for transport. Alternatively, investigators may prepare slides in advance by attaching one end of a tape strip to the bottom of a cleaned microscope slide, stretching the tape across the top of the slide, and folding the other end of the strip onto itself to form a tab for lifting the tape prior to sample collection. At the laboratory, the tape strips are examined under a light microscope with or without staining (e.g., cotton phenol blue for fungi and carbolfuchsin for pollen). Examination of samples by scanning electron microscope is useful in some cases. The adhesive-tape method works best for fungal spores and hyphae, pollen grains, and other large, morphologically distinctive particles of biological origin. Examination of samples under a microscope may also help investigators distinguish between microbial growth and macroscopically similar accumulations of cloth fibers, paper dust, soot, and other materials.

Adhesive-tape samples for examination by microscope are simple to collect. Many samples can be collected in a short amount of time; the results do not depend on the culturability of collected microorganisms; and samples that show hyphal fragments and reproductive structures can provide evidence of microbial growth, not just the presence of settled spores. However, the value of the information obtained depends on the field investigators' decisions on where to sample. Usually several fungi contribute to visible growth, and multiple samples from such areas may be necessary to accurately assess the kinds of fungi present. Tape sampling is not quantitative and does not yield information on the extent or degree of environmental contamination. Analysts unfamiliar with environmental tape samples may find them difficult to read because environmental samples often contain extraneous material not present in tape samples prepared from laboratory cultures. When possible, investigators should also collect scrapings of material suspected of being microbial growth so that a laboratory analyst can prepare specimens of the material in other ways for examination by microscope.

12.2.2.2 Surface-Wash Sampling In the surface-wash method, a swab, filter, or cheesecloth or gauze swatch is used to wipe a specified surface area (Reynolds et al., 1990; Sandholm and Wirtanen, 1993; Morey, 1994). The collection media may be wetted with sterile water or wash solution (e.g., 0.1% peptone water with 0.01% Tween 80) to enhance particle collection. Samples for culture-based analysis must be handled aseptically; for example, by using sterile forceps or touching only the bare end of a swab stick. A swab can be used to inoculate a culture plate immediately, or swabs, filters, and swatches can be shipped to a laboratory for analysis. Samples may be transferred to a laboratory dry in individual sterile containers or in a test tube with a sterile transport medium. Wipe samples can be processed similarly to dust samples (e.g., mechanically agitated in a sterile wash solution followed by culture-plate or cell-culture inoculation or other appropriate assay).

12.2.3 Sample Analysis

Chapter 6 and the individual chapters in Part III describe methods to analyze biological agents in surface samples. These methods include detection of culturable microorganisms and morphologically distinctive particles as well as bioassays and chemical assays (AIHA, 1996). The concentration of biological material in surface samples is expressed in terms of CFU, particles, or other unit of measurement per area sampled. Sections 13.2.2 and 13.2.3 describe the calculation of the concentration of biological agents in surface samples.

12.2.3.1 Culture-Based Analyses Culture-based analyses involve growing microorganisms from surface samples. For example, contact plates can be incubated directly, and swab samples and other surface washings can be inoculated onto agar-based culture media or cell cultures. The type of growth medium or cells used and the incubation conditions (e.g., temperature, humidity, atmosphere, and duration) influence what bacteria, fungi, viruses, or amebae can be isolated. In addition, factors related to the sampled environment, the microbial agents, and the methods of sample collection and handling determine how representative culture results are of the types and relative proportions of microorganisms present on surfaces.

12.2.3.2 Non-Culture Analyses The primary non-culture analysis performed on surface samples is examination of collected material by microscope [see 12.1.3.2].

12.2.4 Interpreting Sampling Results

Investigators need experience to interpret surface sampling results. The information expected from contact surface samples is often simple confirmation that the collected material is biological in nature or that biological growth can be ruled out. Failure to detect a target material does not necessarily mean it was not present, but simply that the collec-

tion and analytical methods did not detect it. The reliability and relevance of information obtained from culture-based surface samples depends on sample number, where and how samples were collected, and the overall design and purpose of the study. For example, the concentration density of biological agents on the ceilings, walls, and floors of rooms and ventilation ducts may differ substantially. Morey (1994) found significant variations in fungal concentrations in a single diffuser, which suggests that a large number of samples is needed to accurately characterize an area by using surface sampling. Estimating the total surface area assumed to be similar to each sampled section is important because different levels of precaution have been recommended for remediating different degrees of contamination with potentially toxigenic fungi [see 15.2].

The recovery efficiency of various surface sampling methods differ, and some biological materials may be more difficult than others to remove from surfaces. For example, Sandholm and Wirtanen (1993) found that microorganism removal from even relatively smooth surfaces could be difficult and that release of cells to the environment could occur in a random fashion.

Porous or fleecy surfaces (e.g., duct insulation, upholstered furniture, and carpets) are even more difficult to sample adequately than smooth surfaces (e.g., wood, tile, or vinyl floors; table and desk tops; and metal surfaces). Debris other than microbial fragments and spores is sometimes so abundant on environmental surfaces that it obscures accurate recognition and identification of microbial contaminants. This may be especially problematic when collecting samples from fabrics or friable surfaces such as plaster or concrete. Investigators are advised to take multiple samples to improve the chances that at least one will be satisfactory. They also may collect more than one sample from the same area on the chance that the first will remove the grosser particles, leaving the finer material to be removed subsequently.

Several attempts have been made to identify surface concentrations of biological materials that indicate unhealthy conditions. No currently available guidelines have been generally accepted due to the large variability in surface sampling results and poor correlations with inhalation exposure. In general, immediate mitigation is needed for microbial growth found on materials that are in direct contact with indoor air or subject to disturbance that might release biological particles, as well as for materials that building occupants may contact directly [see 14.1.2 and 14.1.3]. However, information on cause–effect relationships between surface concentrations of biological materials and illness is not currently available. Therefore, investigators must consider surface sampling data in conjunction with all available medical and environmental information to decide if the data support the hypotheses under evaluation and if the evidence warrants a recommendation for source removal, cleaning, or repair.

12.3 References

AIHA: Viable Microorganisms. In: Biosafety Reference Manual, pp. 3-29. P.A. Heinsohn, R.R. Jacobs, B.A. Concoby, Eds. American Industrial Hygiene Association, Fairfax, VA (1995).

AIHA: Viable Fungi and Bacteria in Air, Bulk, and Surface Samples. In: Field Guide for the Determination of Biological Contaminants in Environmental Samples, pp. 37-74. H.K. Dillon, P.A. Heinsohn, and J.D. Miller, Eds. American Industrial Hygiene Association, Fairfax, VA (1996).

APHA: Collection and Preservation of Samples. In: Standard Methods for the Examination of Water and Wastewater, pp. 1-18 - 1-24. A.D. Eaton, L.S. Clesceri, and A.E. Greenberg, Eds. American Public Health Association, Washington, DC (1995).

ASM: Manual of Environmental Microbiology. C.J. Hurst, G.R. Knudsen, and M.J. McInerney, et al., Eds. American Society for Microbiology, Washington, DC (1997).

ASTM: Standard Practice for Sampling for Surface Particulate Contamination by Tape Lift. Method E 1216-87 (Reapproved 1992). American Society for Testing and Materials, West Conshohocken, PA (1992).

ASTM: Standard Practices for Collection of Dust from Carpeted Floors for Chemical Analysis. Method D 5438-94. American Society for Testing and Materials, West Conshohocken, PA (1994).

ASTM: Standard Test Method for MICROVACUUM Sampling and Indirect Analysis of Dust by Transmission Electron Microscopy for Asbestos Structure Number Concentrations. Method D 5455-95. American Society for Testing and Materials, West Conshohocken, PA (1997).

Becker, R: Fungal Disfigurement of Constructions — Analysis of the Effects of Various Factors. In: Health Implications of Fungi in Indoor Environments, pp. 361-380. R.A. Samson, B. Flannigan, M.E. Flannigan, et al., Eds. Elsevier, New York, NY (1994).

Burge, H.A.; Solomon, W.R.: Sampling and Analysis of Biological Aerosols. Atmos. Environ. 21(2) 451-456 (1987).

Chew, G.L.; Muilenberg, M.L.; Gold, D.; et al.: Is Dust Sampling a Good Surrogate for Exposure to Airborne Fungi. J. Allergy Clin. Immunol. (1996).

Cole, E.C.; Foarde, K.K.; Leese, K.E.; et al.: Assessment of Fungi in Carpeted Environments. In: Health Implications of Fungi in Indoor Environments, pp. 103-128. R.A. Samson, B. Flannigan, M.E. Flannigan, et al., Eds. Elsevier, New York, NY (1994).

Corrigan, C.A.; Blehm, K.D.: A Comparison of Tape Sampling and Microvacuum Procedures for the Collection of Surface Glass Fiber Contamination. Appl. Occup. Environ. Hyg. 12:751-755 (1997).

Farfel, M.R.; Lees, P.S.J.; Rhode, C.A.; et al.: Comparison of a Wipe and a Vacuum Collection Method for the Determination of Lead in Residential Dusts. Environ. Res. 65:291-301 (1994a).

Farfel, M.R.; Lees, P.S.J.; Bannon, D.; et al.: Comparison of Two Cyclone-Based Collection Devices for the Evaluation of Lead-Containing Residential Dusts. Appl. Occup. Environ. Hyg. 9:212-217 (1994b).

Flannigan, B.; McCabe, E.M.; Jupe, S.V.; et al.: Mycological and Acaralogical Investigation of Complaint and Non-Complaint Houses in Scotland. In: Indoor Air Quality and Climate, Proceedings of the 6th Intern. Conf. Indoor Air Quality and Climate, July 4-8, 1993, Helsinki, Finland. Vol 4, pp 143-148 (1993).

Flannigan, B.; Vicars, S.; Pasensen, A.; et al.: Bioaerosols from Housedust. In: Health Implications of Fungi in Indoor Environments, pp. 65-74. R.A. Samson, B. Flannigan, M.E. Flannigan, et al., Eds. Elsevier, New York, NY (1994).

Fox, A.; Rosario, R.M.T.: Quantification of Muramic Acid, a Marker for Bacterial Peptidoglycan, in Dust Collected from Hospital and Home Air-Conditioning Filters Using Gas Chromatography - Mass Spectrometry. Indoor Air. 4:239-247 (1994).

Freeman, N.C.G.; Wainman, T.; Lioy, P.J.: Field Testing of the LWW Dust Sampler and Association of Observed Household Factors with Dust Loadings. Appl. Occup. Environ. Hyg. 11:476-483 (1996).

Gravesen, S.; Larsen, L.; Gyntelberg, F.; et al.: Demonstration of Microorganisms and Dust in Schools and Offices: An Observational Study of Non-Industrial Buildings. Allergy 41: 520-525 (1986).

Gyntelberg, F.; Suadicani, P.; Nielsen, J.W.; et al.: Dust and the Sick Building Syndrome. Indoor Air 4:223-238 (1994).

Hamilton, R.G.; Chapman, M.D.; Platts-Mills, T.A.E.; et al.: House Dust Aeroallergen Measurements in Clinical Practice: A Guide to Allergen-Free Home and Work Environments. Immunol. Allergy Practice. 15:96(9)-112(25) (1992).

Hoekstra, E.S.; Samson, R.A.; Verheoff, A.P.: Fungal Propagules in House Dust: A Qualitative Analysis. In: Health Implications of Fungi in Indoor Environments, pp. 169-177. R.A. Samson, B. Flannigan, M.E. Flannigan, et al., Eds. Elsevier, New York, NY (1994)

Hyvärinen, A.; Reponen, T.; Husman, T.; et al.: Composition of Fungal Flora in Mold Problem Houses Determined with Four Different Methods. In: Proc. 6th International Conference on Indoor Air Quality and Climate, July 4-8, 1993, Helsinki, Finland. Vol 4., pp. 273-278 (1993).

Leese, K.E.; Cole, E.C.; Hall, R.M.; et al.: Measurement of Airborne and Floor Dusts in a Nonproblem Building. Am. Ind. Hyg. Assoc. J. 58:432-438 (1997).

Lindstrom, A.B.; Beck, M.A.; Henry, M.M.; et al.: A New House Dust Collection System and Its Use in a Study of Asthma in Dust Mite Sensitive Children in Raleigh, North Carolina. In: Proc. 6th International Conference on Indoor Air Quality and Climate, July 4-8, 1993, Helsinki, Finland. Vol 4., pp 279-283 (1993).

Linter, T.J.; Maki, C.L.; Brame, K.A.; et al.: Sampling Dust from Human Dwellings to Estimate the Prevalence of *Dermatophagoides* Mite and Cat Allergens. Aerobiologica 10:23-30 (1994).

Lioy, P.J.; Wainman, T.; Weisel, C.: A Wipe Sampler for the Quantitative Measurement of Dust on Smooth Surfaces: Laboratory Performance Studies. J. Exp. Analysis Environ. Eng. 3:315-330 (1993).

Macher, J.M.; Huang, F.; Flores, M.: A Two-Year Study of Microbiological Indoor Air Quality in a New Apartment. Arch. of Env. Hlth. 46(1):25-29 (1991).

Morey, P.R.: Studies on Fungi in Air-Conditioned Buildings in a Humid Climate. Proc. International Conf. of Fungi and Bacteria in Indoor Air Environ., pp. 79-92. Eastern New York Occupational Health Program. New York, NY (1994).

Nolard, N.; Symoens, F.; Beguin, H.: Mycological Survey in Dwellings and Factories: Application to Diagnosis of Extrinsic Allergic Alveolits. In: Health Implications of Fungi in Indoor Environments, pp. 201-209. R.A. Samson, B. Flannigan, M.E. Flannigan, et al. Elsevier, New York, NY (1994).

O'Rourke, M.K.; Fiorentino, L.; Clark, D.; et al.: Building Characteristics and Importance of House Dust Mite Exposure in the Sonoran Desert, USA. Proc. 6th International Conference on Indoor Air Quality and Climate, July 4-8, 1993, Helsinki, Finland. Vol 4, pp. 155-167 (1993).

OSHA: Sampling for Surface Contamination. OSHA Technical Manual, 4th Ed. U.S. Department of Labor, Occupational Safety and Health Administration (1996).

Pasanen, AL.; Heinonen-Taski, H.; Kalliokoski, P.; et al.: Fungal Microcolonies on Indoor Surfaces — An Explanation for the Base-Level Fungal Spore Counts in Indoor Air. Atmos. Environ. 26B:117-120 (1992).

Ren, J.; Guo, R.R.; Wang, X.: Indoor Airborne Bacteria Concentrations and Respiratory Disease in Old and New Rural Dwelling. Proc. 6th International Conference on Indoor Air Quality and Climate, July 4-8, 1993, Helsinki, Finland. Vol 4, pp. 207-212. Finland (1993).

Reynolds, S.J.; Streifel, A.J.; McJilton, C.E.: Elevated Airborne Concentrations of Fungi in Residential and Office Environments. Am. Ind. Hyg. Assoc. J. 51(11):601-604 (1990).

Reynolds, S.J.; Etre, L.; Thorne, P.S.; et al.: Laboratory Comparison of Vacuum, OSHA, and HUD Sampling Methods for Lead in Household Dust. Am. Ind. Hyg. Assoc. J. 58:439-446 (1997).

Roberts, J.W.; Budd, W.T.; Ruby, M.G.; et al.: Development and Field Testing of a High Volume Sampler for Pesticides and Toxics in Dust. J. of Exp. Anal. and Environ. Epid. 1:43-155 (1991).

Samson, R.A.; Flannigan, B.; Flannigan, M.E.; et al., Eds.: Recommendations. In: Health Implications of Fungi in Indoor Environments, pp. 531-538. Elsevier, New York, NY (1994).

Samson, R.A.; Flannigan, B.; Flannigan, M.E.; et al., Eds.: Media. In: Health Implications of Fungi in Indoor Environments, pp. 589-592. Elsevier, New York, NY (1994).

Sandholm, T.; Wirtanen, G.: Biofilm Growth and Development — Problems in Hygiene Monitoring. Proc. 6th International Conference on Indoor Air Quality and Climate, July 4-8, 1993, Helsinki, Finland. Vol 4., pp. 225-230. Finland (1993).

Schneider, T.; Petersen, O.H.; Nielsen, A.A.; et al.: A Geostatistical Approach to Indoor Surface Sampling Strategies. J. Aerosol Sci. 21:555-567 (1990).

Takatori, K.; Lee, H.J.; Ohta, T.; et al.: Comparison of the House Dust Mycoflora in Japanese Houses. In: Health Implications of Fungi in Indoor Environments, pp. 93-101. R.A. Samson, B. Flannigan, M.E. Flannigan, et al., Eds. Elsevier, New York, NY (1994).

Thorne, P.S.; Lange, J.F.; Bloebaum, P.; et al.: Bioaerosol Sampling in Field Studies: Can Samples be Expressed Mailed? Am. Ind. Hyg. Assoc. J. 55:1072-1079 (1994).

Verhoeff, A.P.; van Wijnen, J.H.; van Reenen-Hoekstra, E.S.; et al.: Fungal Propagules in House Dust in Relation to Home Dampness and Other Residential Characteristics. In: Proc. 6th International Conf. on Indoor Air Quality and Climate. Helsinki, Finland, 1993, Vol 4. pp 213-218. Finland. (1993).

Verhoeff, A.P.; Hoekstra, E.S.; Samson, R.A.; et al.: Fungal Propagules in House Dust: Comparison of Analytical Methods. In: Health Implications of Fungi in Indoor Environments, pp. 49-63. R.A. Samson, B. Flannigan, M.E. Flannigan, et al., Eds. Elsevier, New York, NY (1994).

Wang, E.; Rhoads, G.G.; Wainman, T.; et al.: Effects of Environmental and Carpet Variables on Vacuum Sampler Collection Efficiency. Appl. Occup. Environ. Hyg. 10:111-119 (1995).

Wickman, M.; Gravesen, S.; Nordvall, S.L.; et al.: Indoor Viable Dust-Bound Microfungi in Relation to Residential Characteristics, Living Habits, and Symptoms in Atopic and Control Children. J. Allergy Clin. Immunol. 89:752-759 (1992).

Yang, C.S.; Dougherty, F.J.; Lewis, F.A.; et al.: Microscopic Characterization of Settled Dusts Collected in Office Buildings in the Mid-Atlantic Region. In: Environments for People. American Society for Heating, Refrigeration and Air-Conditioning Engineers. IAQ '92:191-196, Atlanta, GA (1992).

Chapter 13

Data Analysis

Janet M. Macher

13.1 Introduction
13.2 Calculation of Biological Agent Concentrations
    13.2.1 *Agar Impaction Samples*
    13.2.2 *Liquid Samples*
    13.2.3 *Surface and Bulk Material Samples*
    13.2.4 *Samples for Microscopic Examination*
13.3 Data Distributions for Concentration Measurements of Biological Agents
    13.3.1 *Normal Distributions*
    13.3.2 *Skewed Distributions*
    13.3.3 *Comparing Distributions*
13.4 Avoiding Errors in Collection and Analysis of Samples for Biological Agents
    13.4.1 *Detecting Measurement Error in Bioaerosol Samples*
    13.4.2 *Dealing with Measurement Uncertainty*
13.5 Statistical Considerations
13.6 Data Summary and Statistical Analyses
    13.6.1 *Data Summary*
    13.6.2 *Agreement Ratios*
    13.6.3 *Comparing Components of Biological Communities*
    13.6.4 *Cluster Analysis*
    13.6.5 *Comparing the Components of Two Sites or Sampling Times*
    13.6.6 *Comparing the Components of More than Two Sites or Sampling Times*
    13.6.7 *Comparing Multiple Samples for a Single Biological Agent*
        13.6.7.1 Two Sites or Sampling Times
        13.6.7.2 More than Two Sites or Sampling Times
13.7 Summary
13.8 References

13.1 Introduction

This chapter provides practical information on assessing the quality of data (primarily air sampling data) and the performance of simple calculations on such data. Related to this discussion is the material in Chapter 7, which summarizes key points about handling data on bioaerosol exposures and the interpretation of data on individual biological agents. Chapter 14 provides background information on how cause–effect relationships between bioaerosol exposures and health responses are assessed and how investigators use such information to evaluate typical data collected during building investigations.

13.2 Calculation of Biological Agent Concentrations

Concentrations of culturable fungi and bacteria in air and source samples are perhaps the most frequent type of data collected in studies of biological agents in indoor and outdoor environments. Section 6.3.1 describes how typical environmental samples for biological agents are handled in the laboratory. The following section describes how the concentration of biological agents in such samples is calculated. The examples illustrate calculation of the concentration of culturable microorganisms because many investigators are familiar with this type of data. Concentrations of biological agents other than culturable cells in air, liquid, or material samples are calculated in a manner similar to that described below.

13.2.1 Agar Impaction Samples

Samplers 1 to 7 and 10 in Table 11.1 collect particles directly onto an agar surface on which culturable microorganisms grow into CFUs. Assuming that multiple samples were collected for each test location or period, air concentrations of culturable microorganisms are calculated from the colony counts on the plates that have optimal colony densities [see 5.2.3.1]. Air concentrations of individual genera or species are usually expressed in CFU/m^3, that is, colony counts are divided by the volume of air sampled, which is determined by multiplying the sampling rate (typically given in L/min) by the sample collection time and converting units of measurement as required (Equation 13-1).

$$\left(\frac{(plate\ count)}{(sampling\ rate)\ (sampling\ time)} \right)\left(\begin{array}{c} conversion \\ factor \end{array} \right)$$

(13-1)

$$= \left(\frac{CFU}{(\frac{L}{min})\ (min)} \right)\left(\frac{10^3\ L}{m^3} \right) = \frac{CFU}{m^3}$$

13.2.2 Liquid Samples

Samplers 11 and 13 to 15 in Table 11.1 (wetted cyclones, the tangential impactor, and impingers) collect bioaerosols in a liquid and yield bacteria or fungi in suspensions for analysis. Liquid samples may be plated directly on the culture media of choice or may be diluted to achieve suitable colony densities when plated. The concentrations of individual genera or species in an original liquid sample are calculated from the colony counts on the plates with the most appropriate colony densities [see 5.2.3.1]. Colony counts for each genus or species are divided by the dilution factor for the plates that were read and the volume of liquid that was plated (typically 0.1 to 0.5 mL) to give concentration in CFU/mL (Equation 13-2).

$$\frac{(plate\ count)}{(dilution\ factor)\ (volume\ plated)} = \frac{CFU}{(10^{-x})\ (mL)} = \frac{CFU}{mL} \quad \textbf{(13-2)}$$

For air samples, the suspension concentration is converted to air concentration as shown in Equation 13-3. For simplicity, air sampling rate is expressed here in m^3/min.

$$\frac{(suspension\ concentration)\ (total\ liquid\ volume)}{(sampling\ rate)\ (sampling\ time)}$$

(13-3)

$$= \frac{(\frac{CFU}{mL})\ (mL)}{(\frac{m^3}{min})\ (min)} = \frac{CFU}{m^3}$$

Filter washes, surface washes, and washings of bulk materials also yield liquid samples for analysis. After calculating the concentration of a biological agent in a wash liquid, the concentration for a total filter, surface area, or mass of material is calculated as described in Section 13.2.3.

13.2.3 Surface and Bulk Material Samples

Agar-contact samples are collected by touching agar to a surface and collecting easily removed particles or fibers [see 12.2.2.1]. The resulting colonies of individual genera or species on surface samples collected by direct agar contact are counted and identified. Surface concentrations are calculated by dividing the plate counts by the area sampled (plate count/contact area = CFU/cm^2).

Small portions of dust samples (e.g., samples collected with a vacuum device) may be distributed directly onto the surface of agar-based culture media [see 6.3.1.3]. The concentrations of genera or species of culturable fungi or bacteria in direct dust inoculations are calculated by dividing individual plate counts by the mass of dust applied to the culture plate (plate count/mass = CFU/g).

Dust and bulk-material samples may be suspended or washed in liquid and processed as liquid samples. The calculation of biological agent concentration is similar for a bulk-material washing or a dust suspension. The concentrations of culturable microorganisms in the original material (expressed as CFU/g) is calculated from the concentration in the wash suspension (Equation 13-2) multiplied by the total volume of wash liquid and divided by the mass of material tested (Equation 13-4).

$$\frac{(suspension\ concentration)\ (total\ liquid\ volume)}{(mass\ of\ material\ suspended\ or\ washed)}$$

(13-4)

$$= \frac{(\frac{CFU}{mL})\ (mL)}{(g)} = \frac{CFU}{g}$$

Biological agents on surfaces may also be collected by washing a defined area with a swab or gauze swatch or collecting dust on an area basis. Concentrations for surface washings are calculated as for dust or bulk-material suspensions, factoring in the area sampled in place of sample mass and expressing surface density in terms of amount of biological agent per unit area [e.g., CFU of *Bacillus* spp. per square centimeter of surface area (CFU/cm^2)]. Concentrations of dust from sampled surfaces may also be expressed in terms of amount per area. For example, the total concentration of an allergen (µg) in a dust sample would be determined and that amount would be divided by the area of carpet from which the dust was collected to yield the allergen concentration on a per-area basis ($\mu g/m^2$).

13.2.4 Samples for Microscopic Examination

Many different types of samples may be submitted for qualitative examination and identification of materials of biological origin (e.g., adhesive-tape samples or bulk-material specimens). Quantitative estimates of the numbers of distinguishable particles in air samples are typically collected on filters or adhesive-coated glass slides or tapes. Calculation of air concentrations of individual types of particles requires examining multiple microscope fields and determining the average counts for individual types of spores or other biological material. Knowing the area of the specimen examined and the total collection area, the analyst calculates the amount of material collected on the entire sample (Lacey and Venette, 1995; Madelin and Madelin, 1995; AIHA, 1996a). This estimate is divided by the volume of air collected [see 13.2.1] to determine the air concentrations of the different particles.

13.3 Data Distributions for Concentration Measurements of Biological Agents

13.3.1 Normal Distributions

Investigators evaluating measurements of biological agents or other parameters need to understand typical frequency distributions for such data. Samples collected simultaneously or sequentially from an environment in which the bioaerosol concentration is constant show a certain variability (i.e., sampling error or random error). For example, multiple filter samples for airborne endotoxin collected after emissions from a process have reached a steady-state concentration show a certain variability, even though the air concentration throughout the space may be fairly uniform. Likewise, repeated analyses of a

single sample each yield slightly different results (e.g., assays for total culturable bacteria from the same dust sample do not give exactly the same colony counts) (Figure 13.1).

Data from replicate samples or analyses sometimes follow a normal (Gaussian) distribution and are appropriately described by their mean (arithmetic average) and standard deviation, which are used in parametric statistical tests. In Figure 13.1, the shape of a normal curve is fairly obvious, but this plot contains over 50 data points. The mean locates the center of the distribution, and the standard deviation (SD) measures distance from the mean. For two data sets with similar means but different SDs, the data set with the larger SD is more variable. The coefficient of variation (CV) of a data set is calculated by dividing a mean by its SD which, for convenience, is multiplied by 100. For the repeated measurements in Figure 13.1, the sample mean was 833 CFU/mg and the SD was 224 CFU/mg for a CV of 27% [(224 CFU/mg ÷ 833 CFU/mg) (100)]. A CV has no units and is not affected by magnitude. Therefore, this value is useful where the SD tends to increase as measurement values increase. For example, data sets with means 200 and 400 and respective SDs of 20 and 40 would both have CVs of 10% (i.e., the SD is 10% of the mean in each case).

The median is another measure of central tendency and is the value that divides the data into halves. The median is generally used when an underlying population distribution shape is not assumed and is the measure of central tendency used in non-parametric statistics [see 13.6.3]. For data that are normally distributed, the mean and median are approximately the same.

Individual measurements, x_i, can be converted to standard units (Z-scores) by subtracting the mean, x, and dividing by the SD [$(x_i - x)/SD$]. Standard units close to 0 are near the mean; standard units further from 0 are further from the mean. Standard units different than 0 are positive or negative, depending on whether they are above or below the mean. A standard unit of ±1 is one SD from the mean. Calculating standard units provides a scaling that may simplify data assessment for measurements from different sites or times. Standard units identify how many SDs a value is above or below the mean. Roughly 68% of a population is found within one SD of the mean and approximately 95% within two SDs.

13.3.2 Skewed Distributions

For many reasons, measurements of the concentrations of airborne materials vary over time and typically follow skewed distributions (Figure 13.2). Variability is inherent in a sample population and depends on its heterogeneity. For example, outdoor concentrations of some biological agents vary with time of day, wind direction, RH, and other factors. Depending on the source of a biological agent, indoor air concentrations may be affected by

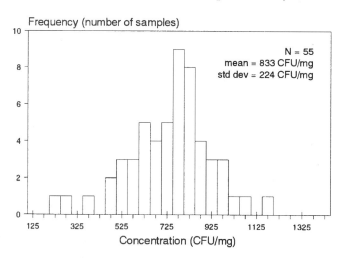

FIGURE 13.1. Example: Frequency distribution of repeated measurements of the concentration of total culturable bacteria from a single dust sample. (Macher, California Department of Health Services, unpublished data)

outdoor air ventilation rate, number of occupants, or occupant activity.

One distribution commonly encountered with biological measurements is the log-normal distribution, in which most data points fall between zero and the mean. Thus, the midpoint of bioaerosol data is often better described by the median (the value above and below which equal numbers of data points occur) or the geometric mean (GM). The GM and geometric standard deviation (GSD) of a data set are obtained by transforming each data point into its logarithm, calculating the mean and standard deviation of these values, and taking the antilogs of the calculated values (Figure 13.3). These values can also be determined from probability plots of the data. The median is the 50% point of the graph. Just as the arithmetic mean and median coincided in the normal distribution, the GM and the median coincide in a log-normal one. The GSD may also be calculated by dividing the point at 84% probability by the GM or by dividing the GM by the 16% probability point. Again, approximately 68% of the data falls within one SD of the center of the distribution (i.e., above the 16% and below the 84% probability points). Exact values for all observations are not needed to calculate medians. Therefore values below LDLs and above UDLs can often be included, provided that fewer than half of the data points fall beyond these limits. However, to calculate GMs, such values must be replaced. Values of zero are often dealt with by adding a small constant (e.g., 0.1) or by using half the LDL. Values above an UDL must also be replaced (e.g., by using twice the UDL). A better alternative for measurements beyond a method's detection limits would be to repeat the samples, adjusting sample volume, or concentrating or diluting samples prior to analysis to obtain data within a measurable range.

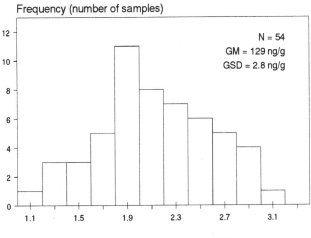

FIGURE 13.3. Example: Histogram of data from Figure 13.2 after logarithmic transformation.

Other types of skewed distributions than the one shown in Figure 13.2 may also be seen with measurements of biological agents or other data. For example, count data are often transformed by taking the square root of the counts to produce an approximately normal distribution and allow use of parametric statistics. Consultation with a statistician is the best way investigators can ensure that they are handling their data appropriately.

13.3.3 Comparing Distributions

Investigators often wish to compare the concentration of biological agents in two environments to better understand what exposures are associated with health responses. The concentrations of specific biological agents (e.g., endotoxin, an allergen, or an indicator compound such as ergosterol) in two populations may have distributions similar to those shown in Figure 13.4. These frequency distributions are of two hypothetical populations, one of measurements from "non-problem environments," the other of measurements from "problem environments." Such distributions might be seen for separate measurements of biological agents in the environments of workers who have been diagnosed with a particular BRI (e.g., asthma or HP) or BRS (e.g., eye, nose, and throat irritation) and matched workers without the BRI or BRS. The distribution for the problem environments is the same as the other, but all values were multiplied by ten. The area under the curves is the same.

The mean values of these hypothetical populations are shown with dotted lines, and the upper 5% of the distribution of the non-problem environments is shaded. Figure 13.5 shows these same data after logarithmic transformation. Again, the upper 5% of the distribution of the non-problem environments is shaded and is seen to overlap with the mean of the log-concentration distribu-

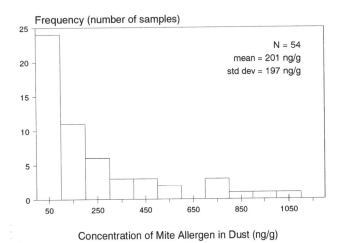

FIGURE 13.2. Example: Frequency distribution of mite allergen concentration in dust samples, illustrating skewing to the right. (Kapadia and Wehrmeister, California Department of Health Services, unpublished data)

Frequency (number or percentage of test sites)

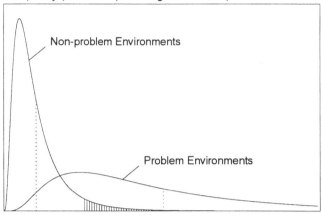

Concentration of Biological Agent

FIGURE 13.4. Frequency distributions of biological agent concentrations in hypothetical non-problem and problem environments. Dotted lines indicate the means of the distributions. The upper 5% of the population of non-problem environments is shaded.

tion from the problem environments (i.e., the GM when transformed back to the arithmetic scale).

These examples are provided to illustrate how transformation can normalize data and how a sufficient number of measurements of a specific biological agent can illustrate differences in frequency distributions for separate sets of environments. It should also be apparent that a single measurement or a small sample (a small number of measurements) may not represent a data distribution well. For example, in Figure 13.5, approximately a quarter of the data overlap (i.e., one-fourth of the measurements could be observed in either of the two sets of environments). Further, approximately 1% of the measurements from the non-problem environments exceed the GM of the problem environments and the same percentage of measurements from the problem environments are below the GM of the other distribution. Real distributions may be more clearly separated and overlap much less. However, unless differences are great and variability is small, it may not be possible to distinguish differences between two populations with small samples (especially single measurements) [see 5.2.4].

13.4 Avoiding Errors in Collection and Analysis of Samples for Biological Agents

Ideally, measurements of airborne biological agents should be accurate and precise, that is, correct (or close to the truth) and repeatable or reproducible. NIOSH has defined a laboratory method as accurate if 95% of the time it provides a result that is within 25% of the correct value (±25%) (i.e., not less than 25% below or more than 25% above the true value) (Taylor et al., 1977). Measurement error is the deviation of a given measurement from the

true value and is composed of chance error (random or sampling error) and bias (systematic or nonsampling error). Replicate samples or repeated analyses of samples are used to assess the former [see 13.3.1]. The SD of a series of replicate measurements or repeated analyses estimates the magnitude of chance error in a single measurement. Bias is a systematic tendency for an estimate to be too high or too low. Comparison with an external standard or reference is needed to detect bias [see 13.4.1]. A sample may be considered representative if it is reasonably accurate and unbiased. Readers should note that the discussion in this section has considered just the variability of individual measurements from the truth, not the variability of the concentration of a biological agent over time or within a space [see 13.3.2 and 5.2.4].

13.4.1 Detecting Measurement Error in Bioaerosol Samples

Bias in aerosol measurements may arise from the sampling device or analytical method used. For example, an air sampler may not collect particles of all sizes with 100% efficiency, leading to underrepresentation of a fraction of an aerosol. Likewise, an analytical method may be less accurate near its detection limits than in midrange. People may also influence test results. For example, laboratory analysts may not treat samples equally if they know which came from case and control locations or from sampling sites near symptomatic and reference workers. Such knowledge may knowingly or unknowingly introduce bias. Misclassification (incorrect identification or assignment) of sampling locations or workers as to test or control category can also seriously confuse a study's outcome. Field and laboratory personnel make errors through lack of training or experience, failure to follow directions, and

Frequency (number or percentage of test sites)

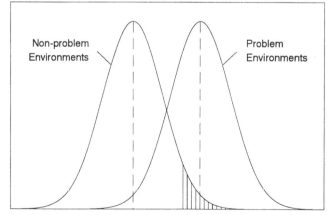

Log Concentration of Biological Agent

FIGURE 13.5. Frequency distributions of log-transformed biological agent concentrations in the hypothetical non-problem and problem environments in Figure 13.4. Dashed lines indicate the mean log-concentrations of the distributions. The upper 5% of the population of non-problem environments is shaded.

fallibility in reading and transcribing information. Such errors tend to affect the precision of measurement data. Table 13.1 outlines some of the measures used to detect and correct potential errors in bioaerosol sample collection and analysis.

A sampling plan should include a minimum of one field and laboratory blank for each sampling event or procedure [see also 5.2.4]. To test a laboratory's analytical performance, field blanks as well as duplicate, split, and spiked samples may be submitted blind (i.e., without identification that would inform an analyst that they are control samples). A laboratory may process some types of samples as received, as well as after dilution and spiking with a known amount of the target biological agent, to assess possible assay inhibition and to measure recovery efficiency.

13.4.2 Dealing with Measurement Uncertainty
The control samples outlined in Table 13.1 can help investigators estimate the reliability of sampling data. Correct measurement of other parameters (e.g., airflow rate)

is also critical for accurate determination of air sample volume and correct calculation of air concentrations. It is generally possible to estimate the variability in the measurement of sampling rate from repeated calibration measurements (i.e., the SD or CV of the mean flow rate) [see 11.4]. Likewise, investigators should attempt to assess the uncertainty related to other aspects of bioaerosol sampling (e.g., measurement of sample collection time, collection efficiency for the targeted particle size, and recovery efficiency for the targeted biological agent). To minimize the overall error in a measurement, investigators should identify and concentrate their efforts on reducing the largest component errors.

Cumulative error for a sample, E_e, is the sum of all individual errors, $E_{1...n}$:

$$E_e = [E_1^2 + E_2^2 + E_3^2 + \cdots + E_n^2]^{1/2}. \qquad \textbf{(13-5)}$$

Investigators can use Equation 13-5 to estimate the overall error in their measurements and evaluate the effect of lowering individual errors. For example, for endotoxin

TABLE 13.1. Samples Used to Assess Errors in Bioaerosol Sample Collection and Analysis

Sample	Description	Use
Sample Collection		
Field blank	Handled as a sample (e.g., placed in an air sampler but no air collected except as required for airflow calibration; or opened and closed in the field but no material collected) and analyzed with samples	To detect contamination during sample collection
Duplicate sample site and intended to be identical	Sample collected simultaneously with another at a	To measure the precision of a collection method
Sample Analysis		
Laboratory blank	Laboratory reagent, unused filter, or other component (free of target organisms or substances) processed with samples	To detect contamination during sample analysis
Positive control sample	Sample known to indicate the presence of a target organism or substance	To confirm proper sample analysis
Negative control sample	Sample known to indicate the absence of a target organism or substance	To confirm negative samples (see also field and laboratory blanks)
Replicate sample	Repeated analysis on a portion of a sample	To measure the precision of an analytical method
Split sample	Portion of sample sent to another laboratory for independent analysis	To assess laboratory performance relative to a reference or other laboratory
Spiked sample	Sample with a known number of target organisms or amount of target substance added	To determine recovery efficiency; to generate calibration curves (see also positive control samples); to identify inhibition or enhancement

collection using a filter cassette, the sample collection time of 240 min may be known to within 5 min (±2%), the desired airflow rate of 5 L/min may be correct to within 0.5 L/min (±10%), particle collection efficiency may be close to 100% (for an error rate of 0), and endotoxin recovery from the filter may be 80% (±20%). In this case, the investigators would gain very little by being more careful in their measurement of collection time and could most improve their measurements by reducing the variability in the recovery efficiency. As is, the cumulative error is ~22%. Reducing the variability of recovery efficiency to 10% would lower the cumulative error to ~14%. Collection and recovery efficiencies can be reported along with results (e.g., concentration: x g/m^3, estimated recovery efficiency 90%) or efficiencies can be factored into the final measurements (e.g., concentration corrected for recovery efficiency: $1.1x$ g/m^3).

In many cases, the inherent variability of the concentration of a biological agent over time or within a space far outweighs any errors associated with measurement of that concentration [see 13.3.2, 13.4, and 13.4.1]. Investigators have observed variations in air concentrations of culturable microorganisms over time and space of three to four orders of magnitude, with GSDs of 3 to 7 (AIHA, 1996b). Investigators should attempt to learn what variability is associated with the measurements they intend to make to decide how critical it is to minimize various measurement errors. Related to this is the issue of what magnitude of differences investigators may be able to detect between test and control environments given the inherent variability of bioaerosol concentrations in the various sampling locations [see also 5.2.4]. This information may be obtained from discussions with other investigators, review of published reports of investigations in environments similar to that currently under study, and by extensive sampling at a study site.

13.5 Statistical Considerations

Occupational and environmental health professionals are familiar with statistical tools for workplace compliance monitoring but may be uncertain how to apply statistical principles to bioaerosol data. Investigators use standard statistical tests to compare mean concentrations for measurements of biological agents from multiple locations or times. For example, investigators may wish to know if the separate distributions of concentrations of a specific biological agent in two environments are sufficiently distinct to conclude that they differ. Typical distributions of bioaerosol concentration data have been discussed because many decisions in planning an investigation and interpreting the results hinge on knowing the underlying distributions of the measurements. Among these decisions is determination of the sample size necessary to have confidence in any conclusions drawn from the data [see 5.2.4].

Eudey et al. (1995) have written a useful discussion of summarizing and presenting bioaerosol data and calculating summary measurements (e.g., central tendencies, variability, relative frequencies or proportions, and correlations). These authors also discuss hypothesis testing and illustrate the application of descriptive and inferential statistics using bioaerosol data as their examples. Atlas and Bartha (1993a) have done the same for studies of microbial ecology. Heber (1995) has discussed bioaerosol particle-size distributions and the statistical analysis of such information. These examples are not repeated here. The AIHA (1996b) has described bias and precision in the determination of viable fungal and bacterial concentrations and has also illustrated the use of statistical tests to evaluate bioaerosol data. Atlas and Bartha (1993b) and AIHA (1996b) have described methods to assess biodiversity (i.e., the range of types of organisms found in an environment). Section 13.6.3 describes some of these methods.

Many statistical tests used to compare contaminant air concentrations at two or more sites or times require that investigators assume their measurements are *independent, normally distributed*, and have *homogeneous variance*. Once transformed to a normal distribution (e.g., by taking the log of, taking the square root of, or squaring the measurements), skewed data may better meet the normality requirement (Figure 13.3). Generally, the variances of transformed data will also become more homogeneous, and investigators can apply parametric statistical procedures to transformed data with less likelihood of violating the normality assumption. Eudey et al. (1995) wrote that while it is often not possible to conduct bioaerosol investigations at a level where formal statistical analysis is possible, it is necessary to consider statistical principles in drawing conclusions from even the smallest data sets.

In the following sections, several simple ways of summarizing and evaluating data are presented. These examples represent tests that investigators have used to assess bioaerosol data and are not necessarily the best or only valid approaches that data analysts may take. For studies requiring more sophisticated data analysis than presented here or covered in the referenced texts, investigators should consult statisticians or standard books on statistical methods. Investigators are advised to seek advice before collecting samples to determine (a) what assumptions they can make about their data, (b) what statistical tests may be appropriate, (c) how to collect sufficient numbers of representative samples, and (d) how to apply suitable statistical procedures to data to test their hypotheses.

13.6 Data Summary and Statistical Analyses
13.6.1 Data Summary
Most readers of this book will be familiar with standard presentations of measurement and survey data in tables

and figures, examples of which appear in other chapters. Often, summation and presentation of data in such forms is all that is needed in a report. Other times, simple summary statistics (e.g., calculation of means, medians, GMs and SDs, GSDs, or confidence intervals) help condense data and make it easier to observe relationships between or among different sampling locations, times, or other data categories.

There are many biases in bioaerosol sampling that make determination of the kinds and concentrations of individual genera and species of microorganisms in air or source samples difficult. Therefore, rather than list concentration measurements for individual taxa, investigators may list each agent as a percentage of the total for that sample. Among samples, counts or concentrations may differ, but the relative percentages of the individual components of the samples may be similar. For example, indoor air that contains primarily bioaerosols from outdoors would have a composition similar to that of outdoor air, but the indoor concentration of each agent may be somewhat lower. Thus, the individual counts or concentrations would differ indoors and outdoors, but the percentages of the agents would be similar in the two environments.

Table 13.2 lists indoor and outdoor samples that illustrate similarities and differences. If possible, the isolates were identified to genus and species and are listed in alphabetical order to simplify comparisons (Morey, 1997). Room A had no apparent water damage; Room B had a history of chronic dampness, rotted wood, and musty odor but no visible fungal growth; and Room C showed both chronic wet conditions and ~20 m² (~200 ft²) of visible fungal growth on interior surfaces. The investigators concluded that *Cladosporium cladosporioides* was the dominant fungus in all samples, but that in Room C, *Paecilomyces variotti* and *Penicillium brevicompactum* (accounting for 14% and 24% of the isolates, respectively) was out of balance relative to the outdoor air (in which the former fungus was not detected and the later comprised 1% of the total).

Another use of percentages is to show relative frequencies, that is, how often particular agents are found in samples from various sites or in repeated samples from single sites. Researchers may use this type of summary to condense data on large numbers of isolates, samples, different sampling sites, or multiple sampling times. Table 13.3 shows an example of this way of summarizing air sampling data. The samples were collected in randomly selected houses in the County of Avon in England (Hunter and Lea, 1994). The isolates were grouped by genus and are listed in order by relative frequency for the

TABLE 13.2 Airborne Fungi Collected on Cellulose Agar Indoors and Outdoors (adapted from Morey, 1997)

Sampling location Total concentration (CFU/m³) Isolate (Percentage of total)	Outdoors 220 (%)	Room A 160 (%)	Room B 190 (%)	Room C 210 (%)
Acremonium strictum	2	–	–	–
Alternaria alternata	–	–	1	–
Aspergillus fumigatus	–	3	1	3
Aspergillus niger	–	1	–	–
Aspergillus ochraceus	–	–	–	4
Aspergillus sydowii	1	–	–	–
Cladosporium chlorocephalum	–	–	2	–
Cladosporium cladosporioides	58	60	49	42
Cladosporium herbarum	5	–	14	–
Mucor heimalis	–	–	2	–
Paecilomyces variotti	–	–	–	14
Penicillium aurantiogriseum	–	–	1	–
Penicillium brevicompactum	1	–	7	24
Penicillium chrysogenum	1	–	2	–
Penicillium citrinum	–	2	–	–
Penicillium commune	–	2	–	–
Penicillium corylophilum	4	5	3	–
Penicillium decumbens	1	2	–	–
Penicillium glabrum	4	–	–	–
Penicillium raistrickii	1	–	–	–
Penicillium simplicissum	–	5	–	–
Stemphylium solani	1	–	–	–
Ulocladium chartarum	7	3	1	–
Non-sporulating fungi	14	15	17	12
Yeast	–	–	–	1
Unidentified	–	2	–	–

entire period. The investigators concluded that *Penicillium* and *Cladosporium* spp., non-sporulating fungi, and yeasts were present in approximately half of the samples year round. *Aspergillus* spp., *Aureobasidium pullulans*, and *Trichoderma viride* showed marked peaks in winter, whereas *Cladosporium* spp. and other leaf-inhabiting fungi (e.g., *Epicoccum*, *Torula*, and *Verticillium* spp.) appeared most frequently in summer.

13.6.2 Agreement Ratios

A very simple means of comparing the types of data presented in Tables 13.2 and 13.3 is calculation of an agreement ratio, R. An agreement ratio reflects the number of identical species isolated from both of two samples (shared species) relative to the total number of species identified in both samples (sum of all species) (Equation 13-6).

$$R = \frac{2\,W}{A + B} \tag{13-6}$$

where, W = number of species both samples have in common

   A = total number of species in Sample 1

   B = total number of species in Sample 2.

Rs of ≥ 0.8 are considered very high. Such a comparison was used to assess the reproducibility of duplicate house dust samples in an assessment of the types of fungi that could be cultured on different media and the types of fungi found in dust samples collected six weeks apart (Verhoeff et al., 1994).

Figure 13.6 shows a correlation matrix or similarity triangle for the data in Table 13.2. When compared with itself, the agreement ratio for each site is 1.0. Sites showing high agreement can be considered to fall within the same relational group. The highest agreement in this example (0.56) was between the outdoors and Room B (where 7 isolates were found in both environments and the respective total numbers of isolates were 13 and 12). The lowest agreement (0.30) was that between the outdoors and Room C (where 3 isolates were found in both environments and the respective total numbers of isolates were 13 and 7).

13.6.3 Comparing Components of Biological Communities

To characterize and compare multiple locations or times, investigators may consider the kinds and relative concentrations of microorganisms that are isolated from air

TABLE 13.3 Percentage of Air Samples Yielding Viable Fungi During a Twelve-Month Period
(adapted from Hunter and Lea, 1994).

Period Number Samples Isolates	Overall (256)	Winter (36)	Spring (68)	Summer (90)	Autumn (58)
Penicillium	91	100	91	82	98
Cladosporium	88	61	82	96	97
Non-sporulating fungi	78	86	78	81	66
Yeasts	57	53	46	63	60
Aspergillus	34	75	26	17	43
Sistotrema	28	58	35	18	12
Basidiomycetes	26	42	37	23	7
Epicoccum	12	3	1	28	5
Botrytis	11	–	7	24	2
Geomyces	9	6	12	11	2
Aureobasidium	8	17	6	8	7
Mucor	7	17	16	1	–
Trichderma	7	25	1	6	3
Verticillium	7	–	–	7	19
Fusarium	6	–	1	10	10
Phoma	6	–	7	10	2
Alternaria	5	–	–	9	7
Torula	5	3	–	10	7
Helminthosporium	4	–	–	–	19
Wallemia	4	–	3	6	3
Scopulariopsis	3	3	1	3	5
Ulocladium	3	6	1	2	2
Oidiodendron	2	3	–	2	2
Sporothrix	1	–	–	–	–
Acremonium	1	–	1	–	2

or source samples. For example, comparisons are often made between indoor and outdoor bioaerosol measurements or between indoor samples from problem and control areas of buildings. If the outdoor environment is the primary source for airborne fungal spores found indoors, the types and relative concentrations of different fungal species should be similar in indoor and outdoor air. However, if indoor fungal growth is releasing spores, it would follow that the types and relative concentrations of fungi indoors and outdoors might differ measurably. Similar comparisons could be made for the kinds of fungal spores on spore-trap samples or the various VOCs collected in indoor air, some of which may be MVOCs arising from microbial contamination.

Non-parametric (distribution-free) tests are used if the assumptions necessary for parametric analyses are not met, the measurement scale that was used was weak, the scale was ordinal (e.g., – –, –, 0, +, ++), samples were ranked rather than measured on a numerical scale, or for simplicity and speed in carrying out the calculations. It is necessary to use non-parametric tests in cases where it is known that data do not fit the assumptions necessary for parametric statistics or where it is desirable to be conservative and not to make any assumptions about a data distribution. While these features make non-parametric analyses attractive, investigators should remember that these procedures are less powerful than parametric statistical tests. Power has to do with the probability of rejecting a null hypothesis when, indeed, it is false. In addition, the use of a non-parametric procedure with data that could be handled with a parametric test results in a waste of data (e.g., if concentration ranks are used rather than concentration measurements).

A chi-square test (χ^2-test) is an example of a non-parametric test that could be performed on the data in Table 13.2 to evaluate the statistical significance of the observed differences. Standard statistics texts describe

the χ^2-test and other non-parametric tests, of which the application of a few examples follows. Parallel parametric tests are available for the non-parametric tests described here and should be used when a data distribution is known to be normal or has been transformed to a normal distribution.

Ecologists use other tests in addition to those described here to assess the diversity of microorganisms in biological communities (i.e., the heterogeneity or variety of different types of organisms occurring together in an environment) (Ludwig and Reynolds, 1988; Atlas and Bartha, 1993a,b). Such tests may help investigators decide if an environment is dominated by a few species of microorganisms or if many different species are present. In some types of environmental studies, measurements of species diversity are used to judge a microbial community's maturity, stability, or flexibility. There are many valid methods for studying environmental data, depending on how the data were collected, the nature of the measurements or observations, the characteristics of the environment from which the data were collected, and the investigation or research goals. For example, factor analysis and principal-component analysis have been used to assess correlations between types or concentrations of biological agents and various environmental parameters that may allow researchers to predict what agents will be found in particular types of environments (Su et al., 1992; Atlas and Bartha, 1993a,b). Cluster analysis has been used to focus a building evaluation on a likely physical source and general characteristics of suspect etiological agents of BRI (Linz et al., 1998).

13.6.4 Cluster Analysis

Cluster analysis is an extension of correlation analysis widely used in the study of microbial ecology. Cluster analysis methods permit grouping of variables according to the magnitudes and interrelationships of their correlations or similarity coefficients (Atlas and Bartha, 1993a). Such methods are used in numerical taxonomy, where the individuals are separate microbial strains the similarity of which an investigator wishes to determine. Biological agents from different sampling sites could also be compared by cluster analysis as well as particular features of sites or organisms [see also 18.4.1.3].

Semi-automated identification systems for bacteria and yeasts use the results of biochemical reactions to calculate a *similarity index* that expresses how closely an unknown isolate matches others (0 = no match, 1 = perfect match). These values are similar to the agreement ratios discussed in Section 13.6.2. However, because the number of test reactions involved is often large, calculation of similarity indices is best done with a computer program and is not described here. For the example in Table 13.4, the unknown GNB most closely matches the pattern for *Pseudomonas fluorescens* because the index of 0.956 is closer to 1.0 than the indices for other possible matches. Per-

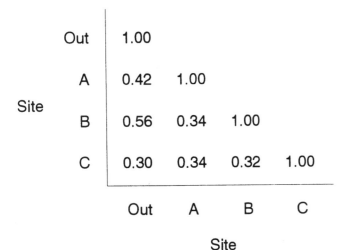

FIGURE 13.6. Similarity triangle depicting relatedness among sampling sites

fect matches for microorganism identification are unlikely due to individual variation among isolates, even if a species matches that in a database. Small variations in test conditions may produce only a slight shift from a perfect match, whereas larger variations may lower an index below that which is acceptable for making an identification. Similar variations are seen when cluster analyses are used for other purposes.

Various methods are available for calculating association coefficients, based on comparing positive and negative matches (e.g., the species of fungi, bacteria, or dust mites found in both, neither, and only one sample from different environments). Clustering techniques may be hierarchical or nonhierarchical, depending on whether they retain rankings as subsidiary clusters are formed. Clustering may be agglomerative, beginning with all separate variables and forming larger and larger groups until a single set contains all components. Divisive techniques work oppositely, beginning with one large set and subdividing it into finer and finer subsets. Some methods even allow overlapping (i.e., inclusion in one group does not preclude inclusion in another). The choice of a particular method depends on the scale in which the data are recorded and the particular reason for performing the cluster analysis (Atlas and Bartha, 1993a).

13.6.5 Comparing the Components of Two Sites or Sampling Times

The microbial species identified in samples from several locations or times are often compared informally to determine if there are striking differences between samples (Tables 13.2 and 13.3). The Spearman rank correlation is a non-parametric statistical test for comparing two samples. A parallel parametric test would be a *t*-test of two sample means. Investigators might use the Spearman rank correlation to assess the similarity of the genera and species of culturable fungi, bacteria, or amebae in air or source samples, the types of fungal spores in spore-trap samples, the species of dust mites or cockroaches from two environments, or the individual types of MVOCs found in different environments.

The first step in such comparisons is to list all categories of agents that were identified and the concentration of each that was measured at the locations or times of interest. Rather than concentrations, some investigators list each agent as a percentage of the total for that sample. Within samples, ranks would be the same whether based on counts, concentrations, or percentages. Between samples, counts or concentrations may differ, but the relative percentages of individual components may be similar.

Table 13.5 shows fictitious data as an example of how different species of organisms would be listed and their concentrations ranked. Some investigators prefer to analyze bacteria and fungi separately; others combine all microorganisms when listing isolates. Ranks are determined from highest to lowest with ties given the average of the corresponding ranks. In the example, the data for a reference site will be compared with two test sites. The concentrations reported for the individual isolates could be the concentrations determined from single samples collected at each location but, preferably, these would be the mean concentrations of multiple samples from each sampling site.

Samples from separate sites must be comparable in all relevant details, for example, collected at approximately the same time with the same or equivalent samplers, analyzed by the same method and, if possible, by the same laboratory personnel. The number of samples in each set should also be comparable. For example, it would be questionable to compare a single sample from one location with the average of multiple samples from another. Investigators must also consider the respective volumes of air collected. A minor component of air may be missed if the sample volume is greatly reduced in one of two test environments (e.g., because one or more bioaerosols were present in large numbers). The minor component may be more readily apparent in a second environment where larger air volumes could be collected because the concentrations of other bioaerosols were also lower.

A null hypothesis (H_0) investigators may wish to test for the data in Table 13.5 would be that the rankings for Samples 1 and 2 are independent from those in the reference sample (i.e., the ranks are not related). The alternative hypothesis (H_1) would be that the rankings are not independent and that there is correspondence between the reference and test samples. The Spearman correlation coefficient, r_S, is computed as follows (Daniel, 1995; AIHA, 1996b):

$$r_S = 1 - \frac{6 \Sigma d_i^2}{n (n^2 - 1)} \qquad \text{(13-7)}$$

where, d_i = rank difference for isolates 1, ..., n
         n = number of isolates.

It should be obvious that if two samples are identical, their summed difference, Σd_i will be zero and r_s will equal 1. Thus, similar samples will have r_s values close to 1. As the value of r_s decreases, the chances that two samples are different increases. Table 13.6 shows d, d^2,

TABLE 13.4. Example: Similarity Index for an Unknown GNB from an Air Sample

Pseudomonas fluorescens	0.956
Pseudomonas putida	0.553
Pseudomonas aeruginosa	0.265

and the sums of d^2 for the example in Table 13.5. If the number of ties is large, a different equation for calculating r_s should be used (Daniel, 1995). Investigators uncertain if substituting averages for ties will affect a test outcome can perform both calculations and compare the results. However, unless the number of ties is excessive, the correction makes very little difference in the value of r_s (Daniel, 1995).

Calculated r_s values can be compared to the critical values listed in Table 13.7. For seven isolates, the respective critical values for one-sided significance levels of 0.05 and 0.01 are 0.6786 and 0.8571. The respective values of r_s calculated for Sites 1 and 2 are 0.96 and 0.40. Therefore, a comparison of Site 1 with the reference site would reject the null hypothesis, and an investigator would conclude that the mix of isolates at the two sites was not independent (i.e., the populations appear to be related or the samples could have been drawn from the same environment). However, a comparison of Site 2 with the reference site would fail to reject (would accept) the null hypothesis that the populations are independent (i.e., the mix of isolates at the two sites does not appear to be related).

While useful for comparing the types of biological agents in samples, the rank correlation test does not take into consideration the magnitude of the values compared (i.e., the actual counts, concentrations, or percentages), just their ranks. For example, the samples in Table 13.8 have similar composition, therefore, their rankings would be determined not to differ by the Spearman test. However, persons in these two settings clearly would have very different exposures to isolate A.

Like other statistical analyses, the rank-order test indicates the confidence with which an investigator can say that two samples differ. However, as for other tests, the results do not assess the importance of any differences that may be determined to be statistically significant. For example, the types of bacteria isolated from indoor and outdoor air may be found to differ significantly. However, if human-source bacteria predominate indoors and environmental bacteria predominate outdoors, the difference may be statistically significant but expected and not a cause for concern [see 7.4.2.1, 18.1.4.2, and 18.5.2]. In addition, rank differences based on very low concentrations may be judged significant but, in fact, may be due to chance variation and unimportant. To determine the true significance of such findings, investigators may need to consult a mycologist, bacteriologist, allergist, or environmental health physician, or may need to conduct further tests using more sensitive sampling and analytical methods.

13.6.6 Comparing the Components of More than Two Sites or Sampling Times

A method to compare multiple biological agents at more than two sites or sampling times is to look at similar categories in the samples to be compared rather than ranking all items together within each location or sampling period. Differences within pairs of observations or among groups of objects are easier to detect if the items being compared are as similar and homogeneous as possible. The Friedman procedure is used to analyze data that is organized into roughly homogeneous blocks (e.g., the rows of isolates in the previous example). Thus, the rank of each isolate is determined relative to that for the same isolate at the other sites, not relative to other isolates found at the same site. To illustrate, the sites from Table 13.5 are treated in Table 13.9 as three locations that an investigator wishes to compare. To avoid confusion, the sites are relabeled I, II, and III. Again, ties in ranks would be assigned the average of the ranks involved. The parallel parametric test would be analysis of variance with two-way classification.

The null hypothesis is that no difference exists among the sampling sites (H_0: the occurrence of the various isolates is the same at all sites, in which case, the sums of the respective ranks would be similar any difference being due to chance factors). The alternative hypothesis is that there is a difference (H_1: the occurrence of the various isolates is different at one or more sites, in which case, the sums of the ranks differ for reasons other than chance). To test the null hypothesis, a chi-square statistic is used, χ_r^2 (Brownlee, 1965; Daniel, 1995):

TABLE 13.5. Rank-Order Comparison: Concentrations and Ranks

Isolate	Reference Site Concentration	Rank	Site 1 Concentration	Rank	Site 2 Concentration	Rank
A	300	1	200	1	170	2
B	100	2	50	3	30	4
C	60	3	75	2	45	3
D	40	4	30	4	20	5
E	10	5	20	5	nd	6.5
F	4	6	6	6	450	1
G	2	7	1	7	nd	6.5

nd = not detected

$$\chi_r^2 = \frac{12}{n\,k\,(k+1)} \sum_{j=1}^{k} (R_j)^2 - 3n\,(k+1) \qquad \text{(13-8)}$$

where,　　n　=　number of rows (isolates)
　　　　　k　=　number of columns (sites)
　　　　　R_j　=　sum of ranks.

Table 13.10 gives the critical values for various degrees of freedom, f, (where, $f = k - 1$). For three sites ($k = 3$), seven isolates ($n = 7$), and a significance level of 0.05, the critical value for $\chi_{.95}^2 (f = 2)$ is 5.991. Thus, the probability of obtaining a χ_r^2 value as large as 5.99 by chance alone, when the null hypothesis is true, is 0.05. The χ_r^2 from the example is 5.43, which is near but below the

TABLE 13.6. Rank-Order Comparison: Ranks and Squared Rank Sums

Isolate	Reference Site Rank	Site 1 Rank	d_i	d_i^2	Site 2 Rank	d_i	d_i^2
A	1	1	0	0	2	-1	1
B	2	3	-1	1	4	-2	4
C	3	2	1	1	3	0	0
D	4	4	0	0	5	-1	1
E	5	5	0	0	6.5	-1.5	2.25
F	6	6	0	0	1	5	25
G	7	7	0	0	6.5	-0.5	0.25
Σd_i^2				2			33.5

TABLE 13.7. Critical Values of Spearman's Rank Correlation Coefficient. Approximate Upper-Tail Critical Values. (Daniel, 1995)

n	$P = 0.10$	$P = 0.05$	$P = 0.01$
5	0.7000	0.8000	0.9000
6	0.6000	0.7714	0.8857
7	0.5357	0.6786	0.8571
8	0.5000	0.6190	0.8095
9	0.4667	0.5833	0.7667
10	0.4424	0.5515	0.7333
11	0.4182	0.5273	0.7000
12	0.3986	0.4965	0.6713
13	0.3791	0.4780	0.6429
14	0.3626	0.4593	0.6220
15	0.3500	0.4429	0.6000
16	0.3382	0.4265	0.5824
17	0.3260	0.4118	0.5637
18	0.3148	0.3994	0.5480
19	0.3070	0.3895	0.5333
20	0.2977	0.3789	0.5203
21	0.2909	0.3688	0.5078
22	0.2829	0.3597	0.4963
23	0.2767	0.3518	0.4852
24	0.2704	0.3435	0.4748
25	0.2646	0.3362	0.4654
26	0.2588	0.3299	0.4564
27	0.2540	0.3236	0.4481
28	0.2490	0.3175	0.4401
29	0.2443	0.3113	0.4320
30	0.2400	0.3059	0.4251

critical value. An investigator would fail to reject (would accept) the H_0 and conclude that the three sites do not differ significantly. The investigator would have accepted the H_0 if, before conducting the test, a significance level of 0.10 (for which the critical value is 4.61) had been chosen. The relatively high χ^2_r might lead the investigator to look closely at the data and consider conducting further source and air sampling specifically to study isolate F for which the difference among samples was greatest.

13.6.7 Comparing Multiple Samples for a Single Biological Agent

13.6.7.1 Two Sites or Sampling Times Multiple independent measurements can be assessed by the two-sample Wilcoxon test, which can be found in many statistics books [see also 13.6.7.2, Equation 13-9]. For example, multiple measurements of a particular bioaerosol may be available from different locations or sampling times. The parallel parametric test would be a comparison of independent means. The two-sample Wilcoxon test allows comparison of samples with different numbers of

components (i.e., the sample size for multiple sites or times need not be equal to compare them). In the Wilcoxon test, all measurements are combined and ranked from lowest to highest in one list (smallest value = rank 1, and so forth up to rank $m + n$, where m is the number of items for the first sample and n that for the second sample). The quantity used to measure possible differences between the two samples is the mean of the rankings for each site. The null hypothesis in this test is that the two populations are identical; the alternative hypothesis is that they differ. If there is no difference between the samples, the means of the ranks should be roughly equal (if the samples are of the same size) or proportional to m and n (if the samples differ in size). If there is a difference, the means will be larger or smaller than expected.

13.6.7.2 More than Two Sites or Sampling Times The H test (Kruskal-Wallis procedure) is a generalization of the two-sample Wilcoxon test and is used to investigate possible differences among independent samples from k populations, where k is greater than two. The parallel parametric test would be analysis of variance with one-way classification. The number of items in each sample is designated by n_i. As in the two-sample test, the entire set of observations are ranked (smallest = rank 1, ..., largest = rank N, where N is the sum of n_i). Mean ranks, $\overline{R_i}$, are calculated for each group based on the sum of the rankings for a sample and the size of that sample. H has approximately a $\chi^2(k - 1)$ distribution. H is shown in Equation 13-9 in a form convenient for computation.

$$H = \frac{12}{N(N+1)} \sum_{i}^{k} \left(\frac{R_i^2}{n_i} \right) - 3(N+1) \tag{13-9}$$

where, N = total number of ranks, sum of n_i
R_i = sum of ranks for population i.

TABLE 13.8. Example: a Rank-Order Comparison that Does not Identify an Apparent Difference

Isolate	Environment 1 Concentration	Rank	Environment 2 Concentration	Rank
A	100,000	1	1000	1
B	500	2	400	2
C	200	3	250	3
D	50	4	60	4
E	10	5	30	5

TABLE 13.9. Two-Way Analysis of Variance: Concentrations and Ranks

Isolate	Site I Concentration	Rank	Site II Concentration	Rank	Site III Concentration	Rank
A	300	1	200	2	170	3
B	100	1	50	2	30	3
C	60	2	75	1	45	3
D	40	1	30	2	20	3
E	10	2	20	1	nd	3
F	4	3	6	2	450	1
G	2	1	1	2	nd	3
R_j		11		12		19

If $k = 2$, this test is identical to the normal approximation for the Wilcoxon test, which has approximately a χ^2 distribution with 1 degree of freedom. If ties occur and midranks are used, H should be divided by

$$1 - \frac{\sum T}{N^3 - N} \qquad \text{(13-10)}$$

where, $\quad T = (t-1)t(t+1)$
$\qquad t = $ number of ties in a group.

T is calculated and included in the sum for each group of ties.

Table 13.11 shows data that might be tested with the Kruskal-Wallis procedure. In this example, investigators collected multiple samples for a particular bioaerosol (in arbitrary units) each day for six consecutive days.

Note that there are three measurements for Day 1 but four measurements for all other days. Also note that

there are several groups of ties. The data are shown in Table 13.12 with their ranks.

The three measurements of 27 (days 3, 4, and 6) receive the mean rank of 5 [(4 + 5 + 6)/3 = 5]; other ties are adjusted similarly. The calculated value of H is 8.753, which needs to be corrected for ties according to Equation 13-8. There are two groups of three tied ranks and three groups of two tied ranks. For each of the triplets, the correction, T, is 24, for the doublets, $T = 6$, and the overall correction is 0.99457. Therefore, H corrected for ties is 8.753/0.99457 = 8.801 with 6 – 1 = 5 degrees of freedom.

The critical value for $\chi^2_{.95}$ ($f = 5$) from Table 13.10 is 11.070, which exceeds the calculated H value of 8.801. Therefore, the investigators would fail to reject (would accept) the null hypothesis and conclude that measurements on different days are identical. While H is distributed asymptotically as $\chi^2(k-1)$ for large k and n_i, in small samples the approximation is not very good. Investigators wanting to analyze small data sets should check with a statistician to determine which test statistic and tables to use for their data. If measurements on the multiple days were also stratified by time of day (e.g., if samples were always collected at 10 a.m., 1 p.m., and 4 p.m.), the two-way analysis of variance described in 13.6.6 or other appropriate test should be used.

13.7 Summary

There are many useful and valid approaches to summarizing and analyzing data from studies of biological agents in indoor and outdoor environments. Reference

TABLE 13.10. Critical Values of the χ^2 Distribution (Adapted from Daniel, 1995)

f	$P = 0.10$	$P = 0.05$	$P = 0.01$
1	2.706	3.841	6.635
2	4.605	5.991	9.210
3	6.251	7.815	11.345
4	7.779	9.488	13.277
5	9.236	11.070	15.086
6	10.645	12.592	16.812
7	12.017	14.067	18.475
8	13.362	15.507	20.090
9	14.684	16.919	21.666
10	15.987	18.307	23.209
11	17.275	19.675	24.725
12	18.549	21.026	26.217
13	19.812	22.362	27.688
14	21.064	23.685	29.141
15	22.307	24.996	30.578
16	23.542	26.296	32.000
17	24.769	27.587	33.409
18	25.989	28.869	34.805
19	27.204	30.144	36.191
20	28.412	31.410	37.566
21	29.615	32.671	38.932
22	30.813	33.924	40.289
23	32.007	35.172	41.638
24	33.196	36.415	42.980
25	34.382	37.652	44.314
26	35.563	38.885	45.642
27	36.741	40.113	46.963
28	37.916	41.337	48.278
29	39.087	42.557	49.588
30	40.256	43.773	50.892

TABLE 13.11. Example: Multiple Bioaerosol Measurements on Multiple Days

	Day 1 Obs.	Day 2 Obs.	Day 3 Obs.	Day 4 Obs.	Day 5 Obs.	Day 6 Obs.
	49	42	46	11	38	24
	52	43	52	27	50	40
	43	42	27	71	41	27
	–	43	51	23	41	37
n_i	3	4	4	4	4	4

TABLE 13.12. Example: Data from Table 13.11 Showing Rankings

	Day 1		Day 2		Day 3		Day 4		Day 5		Day 6	
	Obs.	Rank	Obs.	Rank	Obs.	Rank	Obs.	Rank	Obs.	Rank	Obs.	Rank
	49	18	42	12.5	46	17	11	1	38	8	24	3
	52	21.5	43	15	52	21.5	27	5	50	19	40	9
	43	15	42	12.5	27	5	71	23	41	10.5	27	5
	–	–	43	15	51	20	23	2	41	10.5	37	7
n_i	3		4		4		4		4		4	
$\bar{R}_i$		54.5		55		63.5		31		48		24

texts, published literature, and consultation with experts can help investigators determine how best to analyze data. Preferably, such deliberations take place before investigators collect samples, make measurements, or conduct surveys, rather than afterwards.

13.8 References

AIHA: Total (Viable and Nonviable) Fungi and Substances Derived from Fungi in Air, Bulk, and Surface Samples. In: Field Guide for the Determination of Biological Contaminants in Environmental Samples, pp. 119-130. H.K. Dillon, P.A. Heinsohn, and J.D. Miller, Eds. American Industrial Hygiene Association, Fairfax, VA (1996a).

AIHA: Viable Fungi and Bacteria in Air, Bulk, and Surface Samples. In: Field Guide for the Determination of Biological Contaminants in Environmental Samples, pp. 37-74. H.K. Dillon, P.A. Heinsohn, and J.D. Miller, Eds. American Industrial Hygiene Association, Fairfax, VA (1996b).

Atlas, R.M.; Bartha, R.: Statistics in Microbial Ecology. In: Microbial Ecology: Fundamentals and Applications, pp. 485-505. The Benjamin/Cummings Publishing Company, Inc., Redwood City, CA (1993a).

Atlas, R.M.; Bartha, R.: Microbial Communities and Ecosystems. pp. 130-162. In: Microbial Ecology: Fundamentals and Applications, pp. 130-162. The Benjamin/Cummings Publishing Company, Inc., Redwood City, CA (1993b).

Brownlee, K.A.: Some Nonparametric Tests. In: Statistical Theory and Methodology in Science and Engineering, 2nd Ed., pp. 241-270. Wiley, New York, NY (1965).

Daniel, W.W.: Nonparametric and Distribution-Free Statistics. In: Biostatistics: A Foundation for Analysis in the Health Sciences, 6th Ed. Wiley, New York, NY (1995).

Eudey, L.; Su, H.J.; Burge, H.A.: Biostatistics and Bioaerosols. In: Bioaerosols, pp. 269-307. H.A. Burge, Ed. Lewis Publishers, Boca Raton, FL (1995).

Heber, A.J.: Bioaerosol Particle Statistics. In: Bioaerosols Handbook, pp. 55-75. C.S. Cox and C.M. Wathes, Eds. Lewis Publishers, Boca Raton, FL (1995).

Hunter, C.A.; Lea, R.G.: The Airborne Fungal Population of Representative British Homes. In: Health Implications of Fungi in Indoor Environments, pp. 141-153. R.A. Samson, B. Flannigan, M.E. Flannigan, et al., Eds. Elsevier, New York, NY (1994).

Lacey, J.; Venette, J.: Outdoor Air Sampling Techniques. In: Bioaerosols Handbook, pp. 407-471. C.S. Cox and C.M. Wathes, Eds. Lewis Publishers, Boca Raton, FL (1995).

Linz, D.H.; Pinney, S.M.; Keller, J.D.; et al.: Cluster Analysis Applied to Building-Related Illness. J. Occ. Environ. Med. 40:165-171 (1998).

Ludwig, J.A.; Reynolds, J.F.: Statistical Ecology. Wiley, New York, NY (1998).

Madelin, T.M.; Madelin, M.F.: Biological Analysis of Fungi and Associated Molds. In: Bioaerosols Handbook, pp. 361-386. C.S. Cox and C.M. Wathes, Eds. Lewis Publishers, Boca Raton, FL (1995).

Morey, P.R.: Fungi and Microbial VOCs in Indoor Air. What Does the Data Mean? How Much Mold is too Much? In: Proceedings of Hot and Humid Indoor Environments. Houston, 1997, pp. 1-13. Houston, Texas (1997).

Su, J.J.; Rotnitzky, A.; Burge, H.A.; et al.: Examination of Fungi in Domestic Interiors by Factor Analysis: Correlations and Associations with Home Factors. Appl. Environ. Microbiol. 58:181-186 (1992).

Taylor, D.G.; Kupel, R.E.; Bryant, J.M.: Documentation of the NIOSH Validation Tests. US DHEW Publ. (NIOSH) 77-185. U.S. Government Printing Office, Washington, DC (1977).

Verhoeff, A.P.; Hoekstra, E.S.; Samson, R.A., et al.: Fungal Propagules in House Dust: Comparison of Analytical Methods. In: Health Implications of Fungi in Indoor Environments, pp. 49-63. R.A. Samson, B. Flannigan, M.E. Flannigan, et al., Eds. Elsevier, New York, NY (1994).

Chapter 14

Data Evaluation

Harriet A. Burge with Janet M. Macher, Donald K. Milton, and Harriet M. Ammann

14.1   Types of Epidemiological and Environmental Data
    14.1.1   *Seeking Causal Inferences from Data*
    14.1.2   *Data Indicative of Environmental Contamination*
    14.1.3   *Data Indicative of Exposure*
        14.1.3.1   Environmental Measurements
        14.1.3.2   Biomarkers
    14.1.4   *Data Indicative of Health Risk*
        14.1.4.1   Experimental Dose–Response Data
        14.1.4.2   Epidemiological Dose–Response Data
        14.1.4.3   Using Environmental Data to Indicate Health Risk
        14.1.4.4   Using Biomarkers to Indicate Health Response
        14.1.4.5   Acceptable Risk
14.2   Evaluation of Environmental Data
    14.2.1   *Observational Data*
    14.2.2   *Quantitative Characteristics of Bioaerosol Data*
    14.2.3   *Measurement Data — Concentrations and Kinds of Biological Agents from Single Environments or Sampling Events*
        14.2.3.1   Quantitative Database Comparisons
        14.2.3.2   Indoor/Outdoor Comparisons
        14.2.3.3   Qualitative Database Comparisons
    14.2.4   *Measurement Data — Case-Control Studies*
        14.2.4.1   Activity/Non-Activity Air Sampling Data
        14.2.4.2   Sampling in Complaint and Non-Complaint Environments
    14.2.5   *The Negative Case*
    14.2.6   *Comparison with Existing Standards or Guidelines*
    14.2.7   *Scientific Evidence in the Courtroom*
14.3   Summary
14.4   References

14.1 Types of Epidemiological and Environmental Data

14.1.1 Seeking Causal Inferences from Data

All scientific disciplines share a similar approach to gaining knowledge. The scientific method involves observing empirical evidence and drawing inferences from it. Scientists study what circumstances, events, or combinations thereof lead to particular outcomes. One goal is to identify and distinguish (a) what factors are necessary for an outcome, (b) what, if any, factors may contribute to an outcome but alone are not necessary or sufficient for an outcome, and (c) what, if any, factors have merely non-causal associations with outcomes. This chapter cannot explore this topic in depth but summarizes some of the main points related to the evaluation of epidemio-logical, toxicological, and environmental data. Some investigators speak of using data to "prove" or "disprove" associations such as those between causes and effects. One school of thought holds that testing can only disprove a theory or hypothesis and that investigators can never prove a cause–effect relationship. However, theories about associations that withstand repeated attempts to falsify them become accepted. An alternative understanding of the basis of scientific knowledge is that all investigators begin with a set of prior beliefs about causal relationships. Under this theory of knowledge, progress is made as testable hypotheses are formed based on prior beliefs. Investigators collect data which either support or refute their hypotheses and the underlying beliefs about causality.

Both of these theories of knowledge suggest that the only means of increasing the likelihood of making correct causal inferences is to formulate testable hypotheses and to collect appropriate data. However, it is frequently necessary to make decisions in the face of incomplete data. Therefore, criteria, such as those in Table 14.1, have been proposed to judge possible cause–effect relationships when it is necessary to make public health decisions in the face of uncertainty. These criteria were originally proposed to evaluate the evidence for health effects that might be caused by exposure to tobacco smoke. In brief, an association is considered likely to be causal if it is strong, consistent, specific, plausible, follows a logical time sequence, and shows a dose–response gradient. While these criteria can be useful in decision making, their application should not be confused with the scientific method. Readers should remember that none of these criteria are necessary (except for a temporal relationship) and none are sufficient evidence for causality (Rothman and Greenland, 1998).

To this list could be added the criterion that other causes of the same disease should be excluded or, if present, the contribution of each exposure assessed. Also, elimination or modification of the putative cause or a vehicle carrying it, as well as protection of persons against it, should decrease the incidence of the disease. Data on these points may be collected from epidemiological studies or controlled laboratory experiments.

Epidemiological data consist of measurements of (a) health status, (b) exposure to the agent of interest, and (c) other factors that may be associated with or modify the relationship between exposure and health. The subjects may be cases (persons with a condition) and controls (persons without the condition) whose prior exposure to the agent under study is then assessed. Alternatively, subjects may be known to be exposed (test subjects) or not exposed (control subjects) who are followed to determine the incidence of an adverse health outcome in the two groups over time. Epidemiological studies may be either observational or experimental [see also 3.7]. For an example of an experimental epidemiological investigation see Menzies et al. (1993). In either case, epidemiological investigations can be used to test hypotheses and thus to advance the state of knowledge (Rothman and Greenland, 1998).

Toxicological data from *in-vivo* and *in-vitro* experiments can provide information on health effects and other biological responses under controlled conditions such that causality can sometimes be more easily tested and mechanisms of disease can be identified. However, the results of experimental studies may be difficult to extrapolate to real-world exposures due to differences in dose levels, dose rates, and selection of human subjects or animal species.

TABLE 14.1. Criteria for Determining Whether an Association Between an Exposure and an Apparent Response Means Causation (adapted from USPHS, 1964; Hill, 1965; Evans, 1993)

Criterion	Explanation
Strength of an association	Degree of difference in rates of response for different levels of exposure
Consistency of an association	Similarity of findings from different investigations and different techniques; similar findings by one set of investigators on more than one occasion, by multiple investigators on separate occasions, or through the use of different techniques lends plausibility to an association
Specificity of an association	Narrowness of type of exposure and response
Temporal relationship	Exposure should precede response; the time interval between exposure and response must match known patterns
Biological gradient	A spectrum of responses (none, mild, severe) should be seen in relationship with the degree of exposure (none, moderate, high); responses may be related to intensity, duration, and timing of exposures
Coherence	Similar types and magnitudes of responses should be observed for similar exposures
Biological plausibility of an association	There must be a believable biological basis for the relationship between the exposure and the response
Experimental evidence	Responses should be reproducible experimentally under controlled conditions of exposure and response evaluation
Analogy	Similarity with established exposures and responses may suggest the possibility of such a relationship for another agent

In addition, extrapolation from simple biological phenomena to complex responses of whole organisms may not be valid.

Investigators should be familiar with the types of data used to provide evidence of a potential for exposure to biological agents and adverse health outcomes as a consequence of exposure. Among the most common data collected to evaluate BRIs and BRSs are (a) data indicative of environmental contamination, (b) data indicative of human exposure to contaminants, and (c) data indicative of an adverse response as a result of exposure, as discussed in Sections 14.1.2, 14.1.3, and 14.1.4.

14.1.2 Data Indicative of Environmental Contamination

One approach to interpreting bioaerosol data is to determine whether an unusual exposure exists in a particular environment without establishing a causal link to any specific health effects. Study protocols may emphasize characterization of the environment itself and comparison of the study environment with similar sites or the outdoors. Statistical criteria can then be used to determine if the environment is "unusual." By looking for reservoirs of biological agents, a study may aim to determine whether conditions in a space indicate that sources of biological agents are likely to be present [see 14.2.1]. Various authors have recommended that microbial growth in occupied interiors, in HVAC systems, and on building materials and furnishings, especially if extensive, should be avoided and that any contamination that exists should be removed and further contamination should be prevented.

It is well established that bioaerosols cause infectious and hypersensitivity diseases and that bioaerosols in the indoor environment may cause toxic effects, although data on inhalation exposure is limited. Therefore, it is reasonable to use indicators of environmental contamination as a basis for evaluating the need for improved maintenance or remediation in a preventive context. It is also reasonable to use this approach as a means of making decisions in response to outbreaks of BRIs and BRSs. However, caution should be taken in ascribing specific causal links, as described in Section 14.1.4.3. There is no scientific basis for applying specific exposure limits for total bioaerosol concentration or, at the time of this publication, the concentration of any specific culturable or countable bioaerosol [see 1.2].

14.1.3 Data Indicative of Exposure

14.1.3.1 Environmental Measurements Ideally, researchers would like to know how much of an agent is actually presented to the cells that will mediate a health effect (i.e., dose). However, the dose of a given bioaerosol can rarely be measured and exposure is used as a surrogate for dose. In bioaerosol studies, investigators usually estimate exposures from the concentration of a biological agent in the indoor air or at an environmental source. Approximations can be made of exposure to particles of known size if bioaerosol concentrations are appropriately measured over a reasonable time period and the time people spend in the space is known. This approximation can be further refined if the amount of biological agent per particle can be estimated and some information is available on agent release from particles.

The process of assessing exposure from environmental data involves estimating the "representativeness" of the data that has been or will be collected. The following kinds of environmental sampling data represent, in descending order of accuracy, individual exposure. Unfortunately, most bioaerosol data fall into the last three categories and poorly represent individual exposure. The weakest evidence is that in number six—assumption of exposure based only on the visible presence of microbial growth or other contamination.

1. Personal air sampling over a time period that covers a representative range of potential exposures to a biological agent.
2. Ambient air sampling in the immediate vicinity of a worker over a time period that covers a representative range of potential exposures to a biological agent.
3. Personal or ambient air sampling during worst-case conditions.
4. Personal or ambient grab air sampling.
5. Source sampling.
6. Observation of environmental contamination.

14.1.3.2 Biomarkers Biomarkers are direct markers of dose or of a response to an exposure. For bioaerosols, biomarkers include such analytical measures as the antibody response that follows exposure to allergens and detection of biological toxins or their metabolites in body fluids [see 21.4.4 and 25.2.3.1]. Biomarkers have the enormous advantages of documenting actual exposure to and estimating the received dose of a specific biological agent. However, evidence of contact with an agent is not sufficient to claim a cause–effect association for that agent and a particular health outcome [see 14.1.4.4]. Nevertheless, when linked with environmental data on the air concentration of a biological agent and exposure conditions, biomarkers can provide information to establish guidelines for interpreting environmental concentration measurements.

> Evidence of exposure = Evidence of a potential for a health response
> Evidence of exposure ≠ Evidence of causation of a health response

14.1.4 Data Indicative of Health Risk

This section provides a framework for evaluating the types of scientific data available to investigators for assessing exposures and possible responses to biological agents. The chapters in Part III on individual source organisms provide specific information on evaluating exposure data for different biological agents.

14.1.4.1 Experimental Dose–Response Data

Studies of experimental exposures can provide dose–response data against which investigators can consider data from specific environmental or building studies. Controlled human studies have the advantage that the particular health endpoints of interest can be assessed if they are mild and transient. For example, by exposing panels of subjects in a controlled environment, Castellan et al. (1987) were able to demonstrate that only the endotoxin component of cotton dust, was associated with acute changes in lung function. Such studies can also be designed to identify the responses of sensitive subpopulations to measured exposures.

When human dose–response data are unavailable or such studies are difficult to conduct, experimental exposure of animals may be a useful alternative. Animals can be exposed to varying concentrations of a material such that the actual dose is known and responses can be carefully measured. Researchers then attempt to extrapolate laboratory exposure data to the human populations that may be exposed to the same agent. Such laboratory tests are usually conducted on small numbers of animals (e.g., 10 animals per dose group). To observe statistically significant responses with small numbers of subjects, the dose of the agent under study must often be several orders of magnitude greater than that experienced by the humans whose response is being predicted. Experimental animals may be more or less sensitive than people depending on the particular agent and health outcome.

Several assumptions underlie dose–response experiments in animals. Among these are that (a) the toxic endpoints of an agent are known and measurable, (b) test animals and humans respond similarly to the agent, and (c) high and low doses of the agent affect animals and humans in the same way. Among the reasons for assuming that animals may not predict human responses well is that, unlike human populations, laboratory animals are homogeneous, usually of one sex and age, in optimal health and nutritional status, and unexposed to other contaminants or stressors. However, exposed humans may vary in age, sex, and health status. Further, humans always encounter mixtures of agents via inhalation and, presumably, also by other exposure routes. Mixed exposures are seldom studied experimentally although they are the norm in the real world of human exposures. However, some experimental data exist to demonstrate the importance of potential synergistic effects from mixtures of biological agents (Fogelmark et al., 1994). Researchers generally assume that test animals are less sensitive than humans and add a protection factor into their predictions of human responses.

14.1.4.2 Epidemiological Dose–Response Data

Given the expense and difficulty of conducting human and animal exposure experiments, much of the available information about cause–effect associations for biological agents is gained from epidemiological studies. Epidemiological data may provide evidence for what exposure conditions increase the risk of adverse health effects or conditions. For example, for a one-log increase in CFU/m^3 of *Cladosporium* spp. in a residential environment, Su et al. (1992) predicted that the chance of a child having hay fever symptoms increased by ~20%. Eventually, similar studies may provide sufficient epidemiological data to recommend exposure limits for specific biological agents. If there is a well-recognized dose–response relationship for a particular biological agent, investigators can extrapolate back to the minimal exposure that is likely to result in disease or lead to a significant risk of disease. However, too few studies have yet been conducted to provide an adequate basis for such recommendations. Epidemiological findings are highly method dependent and experimental data to support observed exposure–response relationships are seldom available.

Limited epidemiological data exist for some biological agents and hypersensitivity disease. For example, dust-mite allergen concentrations above 2 μg/g of house dust have been associated with an increased risk of developing dust-mite sensitivity (Platts-Mills and Chapman, 1987). However, a causal link in large prospective studies has yet to be confirmed. Ongoing studies may soon provide the epidemiological basis for confirming or rejecting standards for dust mite and other allergens (Platts-Mills et al., 1995).

> Epidemiological association = Evidence for disease causation
> Epidemiological association ≠ Proof of disease causation

Exposure–response relationships for many infectious agents are well established [see Chapter 9]. Calculations of air concentrations of biological agents that have resulted in clinical illness have been made for a number of infectious diseases, among them tuberculosis [see 9.5]. Unfortunately, exposure–response data are not available for the majority of airborne biological agents. No TLVs have been proposed even for infectious agents for which infectious doses have been determined, primarily because environmental sampling to measure exposure to these agents is difficult and rarely useful except under research conditions.

The risk of contracting an infection is a function of more than the concentration of organisms in the aerosol. Investigators must also consider the likelihood that a sufficient number of the particles are virulent (i.e., contain living cells that are able to cause infection), that the agents reach an appropriate site in the body, and that the organisms overcome the immune system's defenses and survive to multiply. For some infections (e.g., influenza and tuberculosis), it appears that disease may develop if even a single organism reaches an optimum site in the respiratory tract and multiplies. Thus, a single, virulent influenza virus or *Mycobacterium tuberculosis* cell in the air of a space occupied by a susceptible person presents some risk for disease development, even if that risk is small.

14.1.4.3 Using Environmental Data to Indicate Health Risk

After observational data, the most common kind of data from bioaerosol investigations are environmental measurements, often from air samples [see 14.1.2 and 14.1.3.1]. To relate such data to health risk, investigators must consider all the components outlined in Figure 7.1, at least as a thought process. Thus investigators must weigh the evidence for each of the following (a) a source of an agent, (b) exposure of a worker to the agent, (c) inhalation of a sufficient dose of the agent to produce a response, and (d) an adverse response to the exposure. A mistake some investigators make is to assume that a health risk exists based only on the presence of bioaerosols without considering the likelihood of or evidence for exposure of building occupants to the biological agents.

14.1.4.4 Using Biomarkers to Indicate Health Response

When available, biomarkers can provide evidence that exposure has occurred [14.1.3.2]. However, these clinical assays do not necessarily indicate a relationship to either existing or potential disease. For example, a commonly used biomarker is the presence of precipitating antibodies ("precipitins") to specific antigens. The mechanism by which such antibodies are formed is well-known, as is the fact that they do not develop without exposure [see 25.2.2 and 25.2.3.1]. However, the precipitating antibodies themselves are not considered to play a role in disease development. Thus, their presence cannot be used to directly associate disease with a specific exposure. However, the presence of precipitins may indicate an exposure–disease relationship if antibodies are clearly elevated in symptomatic workers as compared to matched control workers. Likewise, the development of antibodies or positive skin-test reactions indicate work-related exposure in persons for whom pre-employment tests were negative and non-occupational sources of exposure have been ruled out.

14.1.4.5 Acceptable Risk

It is important to remember that everyone is exposed to bioaerosols throughout their lives, because bioaerosols of one kind or another are ubiquitous in the indoor and outdoor environment. This is especially true for fungal and bacterial aerosols [see 9.3.4, 18.2, and 19.2]. Many bacteria are essential to human health, for example, the commensal organisms that occupy human skin, the respiratory tract, gut, and so forth. Early exposure to microbial aerosols may be necessary to stimulate normal development of the immune system. Above all, exposures to many bioaerosols are virtually inevitable in everyday life. Thus, such exposures can be considered "unavoidable," "tolerable," or "acceptable" for the majority of healthy persons. This is not to say that employers or building owners or operators should consider that microbial growth in occupied interiors, in HVAC systems, and on building materials and furnishings is normal or allowable. Such contamination is inappropriate and appropriate action should be taken, as discussed elsewhere in this book [see 14.1.2].

Society must weigh the risks associated with bioaerosol exposures for the majority of healthy persons and for individuals who may be more sensitive or susceptible (e.g., very young or elderly persons, immunocom-promised individuals, pregnant women, and persons with certain medical conditions). Under the category of deciding what is reasonable accommodation for persons with special needs falls the considerations surrounding the provision of special environments for unusually sensitive individuals. Complete freedom from risk is an unattainable goal. An important factor in worker reaction to risks is the degree of control they can exercise over the risks associated with a job or work environment. The safety and wholesomeness of a workplace are related to the level of risk that society and individuals regard as reasonable in the context of, and in comparison with, other risks in everyday life. The difficulty with evaluating exposures to many biological agents is that the associated risks have not been determined.

14.2 Evaluation of Environmental Data

Data on kinds and levels of culturable organisms can be used in several ways. An investigator might evaluate the kinds of organisms in the indoor environment and (based on experience or the literature) decide that unusual organisms are present or that the relative concentrations of one or more organisms are unusual. It is often useful to compare the kinds and levels of bioaerosols in different environments (indoors/outdoors or complaint/non-complaint areas) or under different conditions (quiescent/active).

The methods described in Chapter 13 are only a few examples of ways investigators can summarize and evaluate environmental measurement data. Other approaches may be equally useful or more appropriate depending on the data. Readers should be aware that the usefulness of any statistical test depends on the representativeness of the samples from which the data were derived. For sampling and analysis of any biological agent, data can be collected (intentionally or unknowingly) that would falsely be judged significant or not. Section 13.4 discusses measures investigators can take to avoid or detect such biases.

14.2.1 Observational Data

Clearly, if biological growth is visible, there is no doubt it exists [see 14.1.2]. However, such an observation does not mean that occupants are necessarily exposed to biological agents from areas of visible contamination or that health problems workers may be experiencing are due to the biological contamination. To evaluate health risk, investigators must consider the biological agents that may be present in an environmental source and the pathway by which workers could be exposed to the agents. An investigator can use observational data to estimate the likelihood that bioaerosols would be generated, that workers would inhale the material, and that exposure would be sufficient to cause an adverse health effect or predispose a person to one.

An example of using observational data to predict exposure is the finding of a poorly maintained spray humidifier in an office. The water reservoirs in humidifiers can provide a suitable environment for microbial multiplication, and spraying or misting of this water can disseminate potentially allergenic, toxic, or infectious aerosols. Although an investigator could conclude that occupants are likely to be exposed to bioaerosols from the humidifier, the mere presence of this device is not sufficient evidence to predict with certainty that exposure and disease will occur or that a BRI is caused by biological agents from this source. Host factors are inherent in the hypersensitivity, toxic, and infectious disease processes and play a role in which exposed building occupants develop disease and what level of exposure is required for disease initiation. However, presence of a potential source does indicate increased risk of exposure and adverse health effects. A second example is evaluation of the risk from obvious fungal growth on an interior wall. Investigators cannot assume that the presence of such contamination necessarily indicates exposure to spores with their associated allergens or toxins. The fungus may not be producing spores, or the spores may be of a type that do not readily become airborne. Nevertheless, microbial growth in occupied interiors, in HVAC systems, and on building materials and furnishings should not be allowed and such contamination should be removed and further contamination should be prevented (Samson et al., 1994; Health Canada, 1995; Maroni et al., 1995; ISIAQ, 1996).

> Evidence of environmental contamination = Evidence of potential exposure
> Evidence of environmental contamination ≠ Evidence of received dose

The observation of environmental conditions that are known to promote biological growth is less reliable as an indicator of potential bioaerosol exposure than observation of actual contamination. For example, water stains may reflect episodic or chronic excess moisture but not necessarily current or future microbial growth. The environmental conditions may be transient or may not be supporting microbial growth of a kind that could lead to occupant exposure or to disease.

> Growth-promoting environmental conditions
> = Potential for biological contamination
> Growth-promoting environmental conditions
> ≠ Biological contamination

14.2.2 Quantitative Characteristics of Bioaerosol Data

Bioaerosol data derived from visual inspections are usually categorical. That is, the data take on dichotomous values that represent presence or absence (0 or 1) or ordinal values such as "none, a little, an average amount, a lot, or a great deal" (respectively, 0, 1, 2, 3, or 4). Often, these data are not evaluated quantitatively but qualitatively. Data on the concentration of biological agents from environmental source or air samples are nearly always considered in a quantitative manner. The statement "over three sampling days, the average air concentration of *Aspergillus fumigatus* was 1 ± 0.5 CFU/m^3 indoors and 10 ± 5 CFU/m^3 outdoors" is a quantitative statement, the significance of which an investigator must decide. Thus, it is important for investigators to understand a little about the quantitative nature of bioaerosol data.

Bioaerosol sampling data may take on any positive value from 0 to the maximum number obtainable using a particular sampling device and analytical method. A value of 0 may be obtained if a biological agent is absent, but also if its concentration is below the sampling method's lower limit of detection (LDL). For example, the estimated air concentration for a 1-min sample at 28 L/min that yielded 1 CFU would be reported as ~35 CFU/m^3. A similar sample that yielded no growth would be reported as less than 35 CFU/m^3 or between 0 and 35 CFU/m^3. Likewise, a sample at a method's upper limit of detection (UDL) is reported as above the limit value. For example, the plate count for a 200-hole impactor with 200 CFU is reported as >1176 CFU when the positive-hole correction is applied. Various methods are available to handle values below a LDL or above an UDL.

Quantitative data on concentrations of biological agents in environmental samples is usually plotted on a $\log_{10}$ scale with 10-fold increments (i.e., 1, 10, 100, 1000, and so forth) (Figure 14.1). On a logarithmic scale, the difference between 100 and 200 (logs 2.0 and 2.3) is the same as that between 500 and 1000 (logs 2.7 and 3.0) (Sites a-A and b-B, respectively, in Figure 14.1). The log differences are the same in this example even though the arithmetic difference in the latter case is greater (100 at Site a-A; 500 at Site b-B). Although the magnitude of the latter difference is greater, the relative differences in the two examples are the same.

Many replicate samples are required to document small concentration differences with statistical significance. This essentially means that unless investigators have collected a great deal of data, they cannot say that there is a quantitative difference between 100 and 200 or between 500 and 1000. Rather, differences that can be detected with manageable sample sizes are likely to be in 10-fold multiplicative steps (e.g., 100 versus 1000 or 500 versus 5000; respectively, logs 2.0 versus 3.0 and 2.7 versus 3.7) (Eudey et al., 1995). Section 5.2.4 on sample number and Chapter 13 on data analysis discuss related issues.

14.2.3 Measurement Data — Concentrations and Kinds of Biological Agents from Single Environments or Sampling Events

14.2.3.1 Quantitative Database Comparisons As discussed in Section 5.2.1, bioaerosol sampling may attempt to assess (a) background, baseline, or best-case conditions, (b) typical or average conditions, or (c) worst-case or highest exposure conditions. Data from individual sampling episodes is often interpreted with respect to baseline data from other environments or the same environment under anticipated low-exposure conditions. The method most appropriate for interpreting such data depends on the nature of the database. If the baseline data are collected in theoretically low-exposure environments (e.g., the non-problem environment in Section 13.3.3), investigators might decide that the new data must exceed the 90th or 95th percentile of the baseline data to be considered indicative of a potential for harm (Figures 13.4 and 13.5). If the data were collected from a random selection of environments, investigators could assume that both problem and non-problem environments were included in proportion to their actual numbers.

Figure 14.2 uses the hypothetical environmental populations from Section 13.3.3 (Figure 13.5) along with data sets containing hypothetical mixed populations. To illustrate how data sets of mixed populations might look, the figure shows examples of data sets with 80%/20% and 20%/80% combinations of non-problem and problem environments, respectively. While it may be difficult for a layperson to separate the data from two distinct populations when the measurements are mixed, a statistician may be able to do so. The proportions of two populations can often be determined and a randomly selected sample of test sites can be used to deduce the original distributions. However, most data sets are too flawed for such purposes; for example, the sampling sites were not selected randomly, samples were not collected uniformly at all sites, or the sample size was too small. Further, a data set from one region may not be relevant in another climate. Therefore, sites to be compared are generally chosen to be as clearly distinct as possible so that differences in biological agents in the two environments will be distinguishable if they exist.

14.2.3.2 Indoor/Outdoor Comparisons Investigators often use outdoor air as a baseline measurement against which to compare what is found in indoor air. Indoor/outdoor comparisons are commonly used to document the presence or infer the absence of indoor, biologically derived contamination. While this may seem a straightforward comparison, for several reasons it is not. First, an investigator cannot make this comparison for microorganisms unless the genera and species found indoors and outdoors have been identified. For example, comparing the concentration of *Cladosporium cladosporioides* outdoors with that of *Cladosporium sphaerospermum* in-

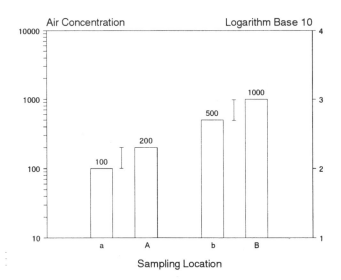

FIGURE 14.1. Logarithmic plot of bioaerosol concentration at paired sites.

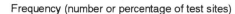
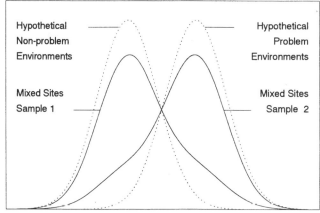

FIGURE 14.2. Frequency distributions of log-transformed biological agent concentrations in the hypothetical non-problem and problem environments in Figure 13.5. Sample 1: mixed distribution; 80% non-problem and 20% problem environments. Sample 2: mixed distribution; 20% non-problem and 80% problem environments.

doors clearly is not meaningful. However, this is what an investigator may unknowingly be doing if the isolates were not identified to species level. Therefore, comparisons of indoor and outdoor bioaerosol concentrations can only be made for like categories of biological agents, for example, particular species of bacteria or fungi, allergens assayed with reagents from the same supplier, or endotoxin measurements from a single laboratory. A comparison of total fungal or bacterial concentrations may be used as a preliminary indicator of a possible difference in two environments, but not as evidence for similarity. Investigators may also use total microbial counts to gauge air concentrations prior to initiating a sampling plan or to identify what microorganisms may be present to select the best culture media and incubation conditions.

Secondly, no single pair of indoor/outdoor samples can be interpreted by themselves. To make quantitative comparisons, at a minimum, representative, replicate samples are essential for each side of a ratio [see 5.2.4, 13.6, and 14.2.3.1].

Thirdly, the magnitude of an indoor/outdoor difference may be important as well as the ratio of two concentrations. For example, if respective indoor and outdoor concentrations in some arbitrary units are 100 and 50, the indoor/outdoor ratio is 2 (Figure 14.3, Site C). It could be concluded that building occupants would be exposed to twice the concentration of the agent as persons outdoors. However, the likelihood of this ratio occurring by chance is high given the large variability in the populations of measurements to be found in indoor and outdoor air and the small difference between 50 and 100 on a logarithmic scale. Further, even if the difference was real, both exposures could still be far below that associated with adverse reactions.

An investigator might conclude that there is essentially no difference between two environments if the variability of the measurements being compared is large. For example, the respective indoor and outdoor concentrations at Site D in Figure 14.3 were estimated to be 100 ± 30 and 50 ± 40, and the two confidence intervals overlap. If outdoor fungal concentrations are very high, indoor/outdoor concentration ratios may appear to be low, even in the presence of substantial indoor growth. For example, the indoor/outdoor concentration ratio at Site E in Figure 14.3 is 0.5 (i.e., the indoor concentration is only half of the outdoor concentration). However, the contribution of outdoor bioaerosols might be low were this a mechanically ventilated building, and the high indoor bioaerosol concentration might instead be due to growth of a potentially hazardous biological agent within the building. In such a situation, exposure for the occupants might be inappropriate even though the indoor/outdoor concentration ratio was low. These examples again illustrate why investigators must be careful to collect data of high quality and be familiar with findings from other

studies before attempting to interpret sampling data and draw conclusions about the potential hazards of the environments from which the samples were taken [see 7.2.2].

Comparing Indoor/Outdoor Fungal Air Concentrations. There is a notable difference in the use of indoor/outdoor concentration ratios for fungi and bacteria. All fungi involved in BRIs and BRSs are ultimately derived from the outdoor environment. Theoretically, overall fungal concentrations should be higher outdoors, except where indoor environments are contaminated with fungi and during periods of snow cover when outdoor concentrations fall to very low levels. This relationship is readily apparent in data collected in clean environments over long time periods (e.g., days or weeks). However, the relationship may not hold for any single pair of measurements or even any single, short-term sample. There is not adequate information on minute-to-minute variability in indoor or outdoor fungal concentrations to allow accurate use of comparisons of fungal concentration ratios for limited numbers of short term samples. Data is also lacking on the rate at which airborne fungi from outdoor sources penetrate buildings and are removed by deposition and other decay mechanisms. There may be greater indoor/outdoor differences for large than small spores. For example, the relative concentrations of large spores may be lower in indoor air than the relative concentrations of smaller spores because of greater losses of large particles as the air infiltrates a building. Therefore, investigators cannot view single, paired, short-term indoor and outdoor samples as sufficiently accurate measures of fungal concentrations to allow meaningful comparisons.

Comparing Indoor/Outdoor Bacterial Air Concentrations. Unlike fungi, bacteria have natural reservoirs both indoors and outdoors, (i.e., indoors: primarily humans; and outdoors: primarily plant surfaces, soil, water, and

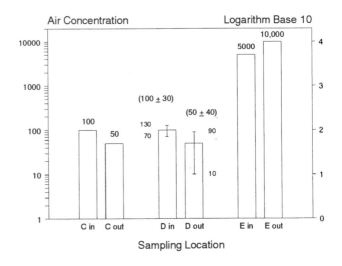

FIGURE 14.3. Logarithmic plot of indoor and outdoor bioaerosol concentrations at three sites.

animals). In fact, in occupied spaces—even when the only occupants present are those persons collecting samples—total bacterial concentrations are often higher than those found outdoors. Most of these are bacteria released from body surfaces and are not considered harmful to other people [see 18.1.4.2 and 18.5.2]. Thus, a strict comparison of indoor and outdoor concentrations for total bacteria is rarely useful. Instead, investigators need to determine the specific taxa of bacteria that are present in the two environments and exclude those of known human origin before concluding that there are building-related sources of bacteria [see 13.6.3].

14.2.3.3 Qualitative Database Comparisons Sometimes, by knowing the nature of a recovered organism, an investigator can determine if it likely entered a building from outdoors or grew indoors. For example, basidiospores are abundant in outdoor air, and few basidiomycetes colonize indoor environments. Thus, the presence of even relatively high concentrations of such spores (e.g., *Coprinus* spp.) in an indoor environment is very unlikely to represent indoor growth. However, such a finding might reflect greater than normal penetration of outdoor air into a mechanically ventilated building [see 10.5.4].

The opposite case is less straightforward, that is, based on indoor samples alone, investigators cannot conclude that the presence of a specific organism in indoor air clearly reflects indoor growth. All microorganisms that grow indoors can also be found outdoors. To document growth in the indoor environment, when visible contamination is not obvious, an investigator must prove that the kinds and relative concentrations of biological agents found in the indoor air differ substantially from those outdoors. Even potentially toxigenic fungi are common outdoors and are occasionally recovered in air samples. Therefore, the presence of these fungi in an indoor air sample is not proof of indoor growth and subsequent increased occupant exposure [see 14.1.3 and 14.1.4.3]. A sufficient number of samples in both environments is needed to allow a meaningful comparison of air concentrations.

14.2.4 Measurement Data — Case-Control Studies

14.2.4.1 Activity/Non-Activity Air Sampling Data Collection of air samples before and after disturbance of a source of a biological agent may provide useful information on the potential of the source to contribute to the bioaerosol burden in the space. This has been termed "aggressive sampling" and may be an example of assessing worst-case exposures [see 5.2.1]. To be realistic, the artificial disturbance must be consistent with what might actually occur in the environment (e.g., dusting of books in a library, foot traffic on potentially contaminated carpet, or the simulated brushing of equipment or persons against contaminated surfaces). A contaminated item can be considered a potential source for exposure if air concentra-

tions of a biological agent identified in an environment increase significantly when the contaminated material is disturbed (i.e., if the variability measured before disturbance does not overlap with that measured during disturbance). The number of samples needed to show a difference depends on the magnitude of the before–after differences and the expected variability in each data set. Because before–after differences may be large with this kind of sampling, relatively few samples may suffice. Note that such data represent the potential contribution of a source when disturbed, not exposure conditions in the space at other times.

14.2.4.2 Sampling in Complaint and Non-Complaint Environments Measuring average concentrations of biological agents in complaint and non-complaint environments may help investigators determine if bioaerosol exposures differ for two groups of workers. Comparison of data from complaint and non-complaint environments can be used to document exposure–complaint relationships if a sufficient number of samples are collected to document the variability for each environment. Here again, investigators can reach this conclusion if the variability measured in the control environment does not overlap that measured in complaint or test areas. Many samples may be required if the difference is small or the variability is large [see 5.2.4].

14.2.5 The Negative Case

Failure to find a biological agent or related environmental conditions is not absolute assurance of their absence nor of the absence of exposure and risk. However, such findings may make absence more probable than presence and may be used to support the assumption that the environment presents conditions of acceptable risk [see 14.1.1 and 14.1.4.5]. Investigators can never definitively conclude or prove that an environment is "safe" and presents no risk of exposure to biological agents. Data can be collected that document the apparent absence of specific hazards (i.e., the relative safety of an environment), but the requirements for data quality to reach this conclusion are stringent.

Frequently, there is some degree of doubt about the reliability of environmental observations. Therefore, it is important to remember that absence of evidence is not evidence of absence of exposure or risk. In part, this means that if investigators have not looked for a biological agent, they cannot say it is not there. It also means that, unless the investigators have used the correct procedures to sample an environment for a biological agent, they cannot say it is not present. Thus, even if testing was conducted specifically for thermophilic actinomycetes (a common cause for HP) and samples were collected appropriately, a laboratory cannot say that these bacteria were not detected, unless the samples were grown on appropriate culture medium and incubated at 50° to 55°C.

> Absence of evidence ≠ Evidence of absence
> No evidence of an effort ≠ Evidence of no effect

Finally, even if investigators can reliably say that a particular agent was not found at the time of sampling, they usually cannot say that the agent was never present in the past. Note that diseases such as asthma and HP result after long periods of exposure (i.e., weeks, months, or even years). Most investigations begin only after disease has developed—and may be fairly advanced—and exposure may no longer be occurring or no longer be occurring at the original level.

14.2.6 Comparison with Existing Standards or Guidelines
By far, comparison of data with an existing standard is the simplest method to interpret environmental measurements. Providing that data are collected in the manner that was used to establish a standard, findings can be compared directly and an immediate determination can be made as to whether the measurements exceed the standard. However, there are no standards that specify acceptable concentrations for airborne materials of biological origin other than those listed in the ACGIH TLV booklet (that are associated with manufacturing environments) (ACGIH, 1998) [see 1.2].

14.2.7 Scientific Evidence in the Courtroom
Uncertainty is an inherent part of any investigation, as discussed in Section 14.1.1. Investigators learn to reach logical conclusions from available evidence and devise recommendations on what needs to be done while acknowledging the uncertainties and possible sources of errors involved in their conclusions. When dealing with uncertainty in legal disputes, investigators and researchers may face different challenges than those they face when making decisions in a purely scientific context.

Researchers use high standards when conducting statistical tests on data. For example, P values of 0.05 or 0.01 are typical (i.e., a probability of error—the likelihood of a positive result being due to chance—of only 1:20 or 1:100). These tests also require that studies have reasonable power to detect an association if one exists (e.g., 80%, an 8 in 10 chance). Scientific judgements are restricted to examining the validity of hypotheses about causal pathways for the population in general rather than for specific individuals. With complex diseases, such as asthma, it is often the case that more than one causal pathway exists. For example, asthma may develop as a result of exposure to dust mites, fungi such as *Alternaria* spp., or rye flour. In addition, a given factor may be a component of more than one causal pathway. For example, low birth weight and exposure to environmental tobacco smoke are risk factors for childhood development of asthma with sensitivity to either cockroaches or dust mites.

Legal decisions may be reached on a preponderance of the available evidence (i.e., that presented in court) or on the decision that a cause "more likely than not" resulted or will result in the effect claimed. The decision in this case is based on a 50-plus% test (i.e., that there is more than a 50% chance that a claimed cause is responsible for a particular effect). This is sometimes referred to as the "balance of possibilities." Further, the persuasiveness of an argument or a witness may have more influence on a judge or jury than the scientific or objective evidence presented. Other references discuss these and related issues of judging causation in law and how this may differ from how the scientific community handles such evidence (Evans, 1993; Faigman, et al., 1997; Rothman and Greenland, 1998).

Either side in a legal dispute can take advantage of the uncertainties associated with the cause–effect relationships of possible or confirmed exposures to biological agents. Establishing the presence of potentially hazardous biological agents in source samples may be sufficient to convince a judge or jury that an environment is unsafe. Scientists may consider source samples imprecise indicators of potential exposure, preferring air samples. However, detecting biological agents in air samples may be difficult. Conversely, medical experts may be able to describe known relationships between biological agents and certain health effects based on experimental or epidemiological data. However, the relevance of this data to the case of a particular worker and workplace may not be convincing without environmental measurements to demonstrate exposure and clinical signs or symptoms of a health response to that exposure.

Expert testimony may be required in legal cases to explain or interpret information that is beyond the common knowledge of lay persons. For example, expert witnesses may testify on the standard of care in building construction and maintenance. Other experts may be needed to describe standard practices and appropriate scientific methods for conducting investigations of BRIs and BRSs. A plaintiff usually must demonstrate (with evidence and expert opinion) that it was more probable than not that a defendant's actions or inactions were a substantial factor in causing a particular outcome. "Substantial" is not well defined and it may be sufficient that a factor is deemed "not insignificant."

Two well-qualified scientific or medical experts can view the same data and reach different conclusions, basing their opinions on their knowledge of available scientific evidence as well as their different experiences and beliefs. Both explanations may be plausible and, given available information and measurement techniques, it may not be possible to determine which, if either, interpretation is correct. Where health-based data is sparse, physicians, toxicologists, or epidemiologists may differ in their interpretations of environmental data with respect to the likelihood or severity of health risks. However, while experts may differ on some points, they may be able to agree on what practical steps should be taken to protect workers from potentially harmful exposures to biological agents.

14.3 Summary

Data analysts consider many criteria when weighing evidence of cause–effect associations. Investigators examining buildings and studying building occupants for evidence of existing or potential adverse effects from exposure to biological agents may apply these same criteria when deciding what data to collect and when evaluating and interpreting available data. Investigators base explanations for possible cause–effect relationships on a series of suppositions they have derived about the environment, the biological agents that may be present, and the effects exposure to these agents may have on humans. The investigators may propose a chain of events and then evaluate the plausibility of each link in the chain individually as well as the strength of the overall connec-

tions. Investigators use a variety of criteria to judge if the explanation they have formed is convincing (Tables 14.1). Table 14.2 summarizes key points in data evaluation.

14.4 References

ACGIH: Biologically Derived Airborne Contaminants. In: 1998 TLVs and BEIs. Threshold Limit Values for Chemical Substances and Physical Agents. Biological Exposure Indices, pp. 11-14. American Conference of Governmental Industrial Hygienists, Cincinnati, OH (1998).

Castellan, R.M.; Olenchock, S.A.; Kinsley, K.B.; et al.: Inhaled Endotoxin and Decreased Spirometric Values, an Exposure–Response Relation for Cotton Dust. N. Engl. J. Med. 317:605-10 (1987).

Eudey L.; Su, H.J.; Burge, H.A.: Biostatistics and Bioaerosols. In: Bioaerosols, pp. 269-307. H.A. Burge, Ed. Lewis Publishers, Boca Raton, FL (1995).

Evans, A.S.: Causation and Occupational Disease. In: Causation and Disease: A Chronological Journey, pp. 177-205. Plenum Medical Book Company, New York, NY (1993).

Faigman, D.L.; Kay, D.H.; Saks, M.J.; et al.: Modern Scientific Evidence: The Law and Science of Expert Testimony. West Publishing Co., St. Paul, MN (1997).

Fogelmark, B.; Sjostrand, M.; Rylander, R.: Pulmonary Inflammation Induced by Repeated Inhalations of Beta(1,3)-D-glucan and Endotoxin. Int. J. Exp. Path. 75:85-90 (1994).

Health Canada: Fungal Contamination in Public Buildings: A Guide to Recognition and Management. Federal-Provincial Committee on Environmental and Occupational Health. Environmental Health Directorate, Ontario, Canada (1995).

Hill, A.B.: The Environment and Disease: Association or Causation? Proc. R. Soc. Med. 56:295-300 (1965).

ISIAQ: Control of Moisture Problems Affecting Biological Indoor Air Quality. TFI-1996. International Society of Indoor Air Quality and Climate, Ottawa, Ontario, Canada (1996).

Maroni, M.; Axelrad, R.; Bacaloni, A.: NATO's Efforts to Set Indoor Air Quality Guidelines and Standards. Am. Ind. Hyg. Assoc. J. 56:499-508 (1995).

Menzies, R.; Tamblyn, R.; Farant, J.P.; et al.: The Effect of Varying Levels of Outdoor-Air Supply on the Symptoms of Sick Building Syndrome. N. Engl. J. Med. 328:821-7 (1993).

Platts-Mills, T.A.E.; Chapman, M.D.: Dust Mites: Immunology, Allergic Disease, and Environmental Control. J. Allergy Clin. Immunol. 80:755-775 (1987).

Platts-Mills, T.A.E.; Sporik, R.B.; Wheatley, L.M.; et al.: Is There a Dose–Response Relationship between Exposure to Indoor Allergens and Symptoms of Asthma? J. Allergy Clin. Immunol. 96:435-440 (1995).

Rothman, K.J.; Greenland, S.: Causation and Causal Inference. In: Modern Epidemiology, 2nd Ed., pp. 7-28. K.J. Rothman and S. Greenland, Eds. Lippincott-Raven, Philadelphia, PA (1998).

Samson, R.A.; Flannigan, B.; Flannigan, M.E.; et al., Eds.: Health Implications of Fungi in Indoor Environments, p. 532-533. Elsevier, New York, NY (1994).

Su, J.J.; Rotnitzky, A.; Burge, H.A.; et al.: Examination of Fungi in Domestic Interiors by Factor Analysis: Correlations and Associations with Home Factors. Appl. Environ. Microbiol. 58:181-186 (1992).

USPHS: Smoking and health: report of the Advisory Committee to the Surgeon General of the Public Health Service. U.S. Department of Health, Education, and Welfare (PHS publication no. 1103), New York, NY (1964).

TABLE 14.2 Key Points in Evaluating Data on Biologically Derived Contaminants

- At this time, ACGIH does not support any numerical criteria for interpreting data on biological agents from source or air samples [except as published in the TLV booklet (e.g., grain dust)] because such criteria are arbitrary given the limited environmental and health–effects data available and because there are more reliable methods of identifying environments in need of intervention [see 1.1, 7.3.2, and 14.2.6].

- The quality of data collected during investigations is the primary factor that allows accurate interpretation—findings are only as reliable as the methods used to reach them [see 7.2.2].

- The negative case [see 14.2.5]:

   True negative – The absence of bioaerosol-related illness or biologically derived contamination can be supported but not proven.

   False negative – If an investigator fails to collect appropriate data on biological contamination or associated illness, the results of the investigation are not proof of the absence of either contamination or illness.

- The positive case (true or false positive): When biological contamination is visible or measurable, it exists and must be addressed. However, presence of biologically derived contamination is not proof of a causal relationship with a health effect [see 14.2.1].

- The investigator, in making decisions to remediate biological contamination, must be aware that if not well considered and properly done, an intervention may create more problems than it solves [see 14.1.4.5].

Chapter 15

Remediation Of Microbial Contamination

Richard J. Shaughnessy and Philip R. Morey

15.1   Introduction
15.2   Removing Existing Contamination
    15.2.1   *Source Containment*
    15.2.2   *Local Containment*
    15.2.3   *Fullscale Containment*
        15.2.3.1   Critical Barrier
        15.2.3.2   Negative Pressure
        15.2.3.3   Decontamination Unit
        15.2.3.4   Containment-Unit Cleaning
        15.2.3.5   Area Cleaning
        15.2.3.6   Uncertainties
15.3   HVAC System Remediation
15.4   Biocide Use
15.5   Judging Remediation Effectiveness
    15.5.1   *Non-Porous and Semi-Porous Materials*
    15.5.2   *Porous Materials*
15.6   Precautions for Potential Exposure to Infectious Agents
15.7   Procedure for Cleaning Cooling Towers and Related Equipment
15.8   Long-Term Prevention Plans
15.9   References

15.1 Introduction

Prevention of microbial growth indoors is only possible if the factors that may allow it are identified and controlled. When prevention has failed and visible microbial growth has occurred in a building, restoration requires (a) removal of porous materials showing extensive microbial growth, (b) physical removal of surface microbial growth on non-porous materials to typical background levels, and (c) reduction of moisture to levels that do not support microbial growth. Preventing water intrusion during the remediation process is advised to prevent further microbial growth. However, Foarde et al. (1997) found that lowering indoor RH triggered spore release from fungal contamination on duct material. Therefore, remediators should consider whether contaminated materials should be removed before measures to thoroughly dry the environment are undertaken. Identification of the conditions that contributed to microbial proliferation in a building is the most important step in remediation. No effective control strategy can be implemented without a clear understanding of the events or building dynamics responsible for microbial growth [see 10.1].

15.2 Removing Existing Contamination

Growth that has occurred in a surface layer of condensation on painted walls or non-porous surfaces (including wood) can usually be removed by (a) vacuuming using equipment with high-efficiency filters or direct air exhaust to the outdoors, (b) washing with a dilute solution of biocide and detergent, or (c) cleaning, thorough drying, and repainting. Porous materials that have sustained extensive microbial growth must often be removed. Examples of porous materials are ceiling tiles, installed carpeting, upholstered furnishings, and wallboard. Extensive microbial growth refers not only to the extent of the area affected but also the degree to which microorganisms have degraded a material for use as a food source. "Extensive" visible fungal growth has been defined as surface areas greater than 3 m² (32 ft²) (NYCDH, 1993; Health Canada, 1995; ISIAQ, 1996). Carpeting and drapes that can be removed for thorough cleaning and drying may be salvageable. Valuable books and papers can sometimes be rescued by fumigation, followed by freeze-drying and vacuum removal of residual particles.

The removal and cleaning of contaminated materials must not be undertaken without proper precautions, because disturbance of contaminated materials can result in bioaerosol release (Flannigan, 1992). Disturbance of microbial growth in air-handling systems may lead to the dissemination of bioaerosols throughout a building (Morey and Williams, 1991). Concentrations of airborne spores indoors during material disturbance and removal may approach levels characteristic of dusty agricultural environments (Hunter et al., 1988; Rautiala et al., 1996). Resulting exposures to biological agents may compromise the health of remediation workers and building occupants.

When visible contamination is extensive, containment procedures similar to those used to handle hazardous wastes (e.g., asbestos) are required to safely remove contaminated materials (Morey, 1994). Remediators can consider using the recommendations others have developed to handle removal of materials visibly contaminated with potentially toxigenic fungi (NYCDH, 1993; Morey, 1992, 1994, 1996; Health Canada, 1995; Morey and Ansari, 1996). Recommended removal methods take into consideration both the nature and extent of contamination, that is, the particular microorganisms present and the amount of material or area affected. Such work should be conducted while buildings are unoccupied (NYCDH, 1993; Weber and Martinez, 1996). Investigators should decide if uncontaminated items in an area to be remediated should be removed to protect them during the cleanup process or if covering the items will provide enough protection.

With appropriate PPE, local maintenance personnel should be able to remediate visibly contaminated areas of less than 3 m^2 (NYCDH, 1993). However, many factors must be considered in deciding what level of precaution is appropriate and how contaminated materials should be contained and removed (IICRC, 1995; ISIAQ, 1996; Morey and Ansari, 1996). In general, the removal and containment precautions required for toxigenic fungi should be used for remediating any visible fungal contamination because virtually all fungi can cause allergy (in sensitized individuals) and many fungi produce toxins (Morey, 1996; NYCDH, 1993) (Table 15.1).

15.2.1 *Source Containment*

Table 15.1 refers to three containment requirements for varying degrees of contaminant removal. Source containment may be as simple as placing a moldy ceiling tile

TABLE 15.1. Guidance for Removing Visible Fungal Growth

CAVEAT: This table is presented as a general guide only. Factors besides those listed may need to be considered when deciding what levels of environmental and personal protection are appropriate during remediation activities (e.g., the contaminating agent, the nature of the contaminated material, and the location of the site requiring remediation). Categorizing the extent of contamination requires professional judgement.

Visible Fungal Growth[A]	Recommendations to Prevent Dust or Spore Dispersion[B]	Suggested Minimal PPE
Minimal	Source containment: material removal with minimum dispersal of dust and spores	N-95 respirator; gloves
Moderate	Local containment: enclosure and negative pressurization to prevent dispersion of dust and spores	N-95 respirator;[B] eye protection; full-body covering[C]
Extensive	Full containment: critical barriers and negative pressurization to contain dust and spores; personnel trained to handle hazardous wastes	As above

(adapted from NYCDH, 1993; Morey, 1992, 1994, 1996; Health Canada, 1995; Morey and Ansari, 1996)

[A]   "Visible contamination" means that fungi are readily observable on surfaces. The presence of hyphae and mycelia on or in materials, as seen by direct microscopic examination, verifies that visible contamination is of fungal origin. [Categorization of the extent of microbial contamination and determination of the required level of containment and appropriate PPE involve professional judgement.]

[B]   Higher levels of respiratory protection [e.g., half- or full-face respirators or full-face powered air purifying respirators (PAPRs) with HEPA cartridges] may be considered necessary for some remediation work. Investigators should seek the opinions of occupational physicians, toxicologists, respiratory protection experts, or health and safety professionals to select appropriate PPE.

[C]   "Full-body covering" is defined as the collective use of full-body disposable coveralls, head covering, eye protection, gloves, and shoe covers.

in a plastic bag, sealing the bag, and removing the sealed bag from a building. This level of containment is considered adequate to prevent spore dissemination from minimal areas of contamination.

15.2.2 Local Containment

Local containment of contaminants may be achieved by constructing an enclosure from two layers of polyethylene film supported on a wood-stud frame. A HEPA vacuum nozzle is used to create negative pressure within the enclosure. Note: The vacuum canister is located outside the enclosure. The negative pressure must be sufficient to ensure containment of bioaerosols [see 15.2.3.2].

Source containment (as in 15.2.1) is used for contaminated material, which is double bagged in 6-mil polyethylene. The bags should be discarded as described in Section 15.2.3.3. Remaining building surfaces and materials should be damp wiped to remove adherent dust. This final cleaning should be performed using minimal water to avoid wetting the material to the point that any residual microbial contamination could regrow [see also 15.4].

15.2.3 Fullscale Containment

A fullscale containment commensurate with an asbestos-abatement program is recommended for removing materials that are extensively contaminated with visible fungal growth (Morey, 1994; Weber and Martinez, 1996).

15.2.3.1 Critical Barrier Two layers of polyethylene sheeting are used to create a critical barrier to isolate a contaminated area from clean or occupied building zones. Critical barriers must block all openings, fixtures, and HVAC system components to prevent the spread of dirt and spores beyond the containment area. The barriers must be constructed without disturbing contaminated materials.

15.2.3.2 Negative Pressure A negative air pressure differential between the work area and the surrounding space must be created to prevent contaminants from leaving the work zone. An air filtration device (e.g., a negative air machine) with a HEPA filter should be used to negatively pressurize the work area. A pressure differential of ≥ 5 Pa (0.02 in. w.g.), which is recommended for asbestos abatement work (OSHA, 1994), may be adequate to contain dust and spores. However, an even higher pressure difference [e.g., –7 Pa (–0.028 in. w.g.)] may be necessary in some cases (Alevantis et al., 1996).

15.2.3.3 Decontamination Unit A decontamination unit should be constructed for entry into and exit from a remediation area. The unit may consist of a single or multiple chambers depending on the size of the operation. A multiple-chamber unit typically has a work room, equipment room, and air lock. At present, there is no scientific evidence to justify the use of showers in the decontamination system.

Contaminated debris should be double bagged (6-mil polyethylene bags) and passed through the decontamination unit. The bag surfaces are HEPA vacuumed before transport into uncontaminated parts of a building. Bags are removed by the most direct exit route. Direct transport of sealed bags to the outdoors is preferable (e.g., through a window or door connected to the decontamination unit). Bagged debris can be disposed of in a landfill as if the contents were moldy compost. Workers disposing of these bags should be aware of their contents, take measures to avoid bag rupture, and be trained in how to deal with such an event.

15.2.3.4 Containment-Unit Cleaning A combination of HEPA vacuuming and damp wiping (with minimum water) should be used to remove settled dust prior to disassembly of critical barriers [see also 15.4]. Water sprays can be used to reduce dust aerosolization but must not wet surfaces excessively. A final inspection of a containment area should be made to ensure that all dust and visible debris have been removed. Air sampling by spore trap or other means may also be conducted to verify that air concentrations of fungal spores and indicator materials (e.g., glucan or ergosterol) in the containment zone are qualitatively and quantitatively similar to ambient outdoor air. Use of surface sampling (e.g., adhesive tape sampling) is advisable to determine that only background concentrations and types of fungi are present on porous surfaces [see 15.5].

15.2.3.5 Area Cleaning Dust that may have settled on surfaces and materials outside a remediation enclosure should be removed by HEPA vacuuming or damp wiping followed by thorough drying.

15.2.3.6 Uncertainties Few data are available on exposures that have occurred during fungal remediation (Morey and Hunt, 1995; Ansari and Morey, 1996; Rautiala et al., 1996; Weber and Martinez, 1996). Decisions on what PPE will adequately protect workers performing remediation of microbial contamination requires experience and professional judgement, which occupational physicians, toxicologists, respiratory protection experts, and health and safety professionals may be able to provide. Individuals employed in the removal of extensive microbial contamination should be informed in writing by a physician of the potential health risks of bioaerosol exposure. This information should include the recommendation that workers immunocompromised for any reason should avoid remediation activities that may expose them to potentially infectious agents [see also 8.4 and 15.6]. NIOSH-approved respirators should be provided, and respirator use must follow a complete respiratory protection program [see also 4.6.2].

The medical personnel providing medical review for respirator users must be aware of the conditions and hazards of the work to be performed. It is highly recommended that the medical personnel have appropriate certification in occupational health. Particular concern should be exercised when workers with a history of asthma are considered for clean-up jobs. Workers should be encouraged to report symptoms while on the job and should be immediately referred for medical evaluation. Follow-up medical examinations after completion of major clean-up operations (at least a symptom questionnaire) would offer both the worker and employer assurance that any untoward effects of exposure are likely to be detected and treated appropriately.

15.3 HVAC System Remediation

The guidelines outlined in Table 15.1 can be applied to remediation of microbial growth on surfaces in HVAC systems. Application of biocides as a substitute for removing microbial growth and settled biological material is not considered acceptable. In the first place, most disinfectants and sanitizers are approved for use on previously cleaned rather than soiled surfaces [see 16.2.4]. Secondly, the allergenicity and toxicity of biological material is not related to microorganism viability (Morey, 1994; ISIAQ, 1996). Contaminated porous materials in HVAC systems must be removed to the bare (underlying) metal and the contaminated materials appropriately discarded. Full-containment procedures should be implemented when removing extensive areas of contaminated porous materials from large HVAC system components (e.g., air-handling plenums) (Morey, 1994). Depending on the extent of visible fungal contamination, removal of porous materials from smaller HVAC system components (e.g., unit ventilators and fan-coil units) requires source or local containment precautions supplemented by HEPA vacuum cleaning.

HVAC system cleaning and decontamination to remove dust and biological growth has expanded into a large industry serving both residential and commercial building markets (Burge, 1995; NADCA, 1992, 1997). The USEPA has a document on duct cleaning, which is available in print or from their website (www.epa.gov/iaq/airduct.html) (USEPA, 1998). Although written for homeowners, much of the information applies to other types of buildings. Cleaning to remove accumulated dust and microbial growth is routinely followed by often inappropriate biocide application [see 15.4 and 16.2.2.2]. To date, no organized effort has been made to evaluate the impact of biocide use in air-conveyance systems on the health of building occupants (Burge, 1995). However, use of a disinfectant in an operating ventilation system has led to building evacuation in at least two cases (Arnow et al., 1978; Sesline et al., 1994).

15.4 Biocide Use

Remediators must carefully consider the necessity and advisability of applying biocides when cleaning microbially contaminated surfaces [see 16.2.3]. The goal of remediation programs should be removal of all microbial growth. This generally can be accomplished by physical removal of materials supporting active growth and thorough cleaning of non-porous materials. Therefore, application of a biocide would serve no purpose that could not be accomplished with a detergent or cleaning agent. Prevention of future microbial contamination should be accomplished by (a) avoiding the conditions that led to past contamination, (b) using materials that are not readily susceptible to biodeterioration, and (c) where necessary, applying compounds designed to suppress vegetative bacterial and fungal growth or using materials treated with such compounds [see 16.3].

15.5 Judging Remediation Effectiveness

The success of a remediation effort is judged in part by the visible degree of contaminant removal that is achieved. Effectiveness may also be confirmed by sampling [see 15.2.3.4]. The ultimate criterion for the adequacy of abatement efforts for treating biological contamination is the ability of people to occupy or re-occupy the space without health complaints or physical discomfort [see 8.6.3]. Cessation of bioaerosol exposure should result in a cessation of bioaerosol-related symptoms. Likewise, mitigation of environmental conditions that led to problems of microbial contamination should result in the absence of microbial growth as long as the control measures continue to be effective. If this is not the case, the investigators did not correctly identify or sufficiently address the underlying causes of the problem [see 10.1].

Following building restoration, the kinds and concentrations of biological agents in air samples should be similar to what is found locally in outdoor air. Concentrations of biological agents in surface samples should be similar to what is found in well-maintained buildings or on construction and finishing building materials (Ström et al., 1990; Nevalainen and Flannigan, 1993; AIHA, 1996). In general, removal of semi-porous items (e.g., enamel-painted wallboard or wood) may not be necessary if washing with a detergent solution returns the material to its original condition (as determined visually). Even then, moisture must be rigorously controlled to prevent future growth. If not, further cleaning may be necessary until no difference is detectable. In some situations, it may be difficult to achieve these levels even with thorough and costly remediation (Ansari and Morey, 1996; Morey, 1996). Investigators may decide to treat a portion of a contaminated area to determine if the remediation process will be satisfactory before proceeding to treat a larger area.

15.5.1 Non-Porous and Semi-Porous Materials

Non-porous materials (e.g., metal ductwork, metal studs, vinyl flooring, and glass) can readily be cleaned and re-used. Slightly porous or semi-porous materials (e.g., wood studs or furniture) that are visibly contaminated may be reusable depending on the depth to which microbial growth has penetrated the substrate [see Figure 19.1]. For example, fungi growing on the surface of wood furniture may be removed by refinishing. However, wood should be discarded if fungal growth has affected its soundness (i.e., if hyphae have penetrated the wood extensively). The National Air Duct Cleaners Association (NADCA, Washington, DC, http://www.nadca.com) has developed an arbitrary industry standard for an acceptable amount of total debris on cleaned ducts of 0.1 g/m² (1.0 mg/100 cm²) (NADCA, 1992). However, this mass limit does not consider the nature of the debris.

15.5.2 Porous Materials

Porous materials from which microbial growth cannot be adequately cleaned must be removed from buildings. In buildings where extensive microbial growth has occurred, porous materials not actively supporting microbial growth may still harbor spores and particles released from other sources. Where appropriate, such materials should be thoroughly cleaned by washing or HEPA vacuuming and should be monitored for residual contamination. Over time, levels of residual contamination should fall. In addition, if porous materials have adsorbed odors, removal of the material may ultimately be necessary to return the building to its pre-contamination state.

15.6 Precautions for Potential Exposure to Infectious Agents

Activities that aerosolize infectious agents from environmental sources may place remediation workers and building occupants at risk [see 8.4 and 9.2.1]. The recommendations in Table 15.1 have been considered appropriate for cleaning internal surfaces contaminated with animal droppings that may contain pathogenic fungi (ISIAQ, 1996). NIOSH has developed work practices and a PPE selection guide for remediation work that may entail exposure to *Histoplasma capsulatum* (Lenhart et al., 1997). This soil fungus can grow in bat and bird droppings, and workers removing such materials from buildings have become infected as a result of inhaling fungal spores. Measures recommended to protect cleaning crews include written notification of the risk of contracting histoplasmosis, work practices that minimize dust generation (e.g., wetting of dry material before sweeping or shoveling), proper waste collection and removal, and PPE. These measures also apply to removal of materials that may be contaminated with other infectious agents, for example, *Cryptococcus neoformans* and *Chlamydia psittaci*.

The AIHA Biosafety Reference Manual (1995) also discusses selection of respiratory protective equipment for potential exposures to infectious aerosols, and the American Thoracic Society has published respiratory protection guidelines that discuss respirator use for protection against bioaerosols (ATS, 1996).

When infection is the primary health concern, disinfectants have been used to treat contaminated soil and accumulations of animal droppings for which removal was impractical or as a precaution before starting a removal process (Weeks, 1984; Lenhart et al., 1997). Formaldehyde solutions (formalin) are the only disinfectants proven effective for decontaminating soil containing *H. capsulatum*. However, because of the potentially serious health hazards associated with exposure to formaldehyde, it should only be handled by persons who know how to apply it safely (Lenhart et al., 1997).

15.7 Procedure for Cleaning Cooling Towers and Related Equipment

The remediation of water-cooled heat-transfer equipment and cooling towers that have been associated with legionellosis outbreaks is sufficiently different than the treatment of other microbial contamination that the following procedure is included. This protocol is reprinted from a CDC publication (CDC, 1997) and is a concise version of a more detailed protocol from the Wisconsin Department of Health and Social Services (WDHSS, 1987).

I.   **Before Chemical Disinfection and Mechanical Cleaning**
   A.   Provide protective equipment to workers who perform the disinfection to prevent their exposure to (a) chemicals used for disinfection, and (b) aerosolized water containing *Legionella* sp. Protective equipment may include full-length protective clothing, boots, gloves, goggles, and a full- or half-face respirator that combines a HEPA filter and chemical cartridges to protect against airborne chlorine levels of up to 10 mg/L.
   B.   Shut off cooling-tower.
      1.   If possible, shut off the heat source.
      2.   Shut off fans, if present, on the cooling tower/evaporative condenser (CT/EC).
      3.   Shut off the system blowdown (i.e., purge) valve. Shut off the automated blowdown controller, if present, and set the system controller to manual.
      4.   Keep make-up water valves open.
      5.   Close building air-intake vents within at least 30 m of the CT/EC until after the cleaning procedure is complete.
      6.   Continue operating pumps for water circulation through the CT/EC.

II. **Chemical Disinfection**
   A.   Add fast-release, chlorine-containing disinfectant in pellet, granular, or liquid form, and follow safety instructions on the product label. Examples of dis-

infectants include sodium hypochlorite (NaOCl) or calcium hypochlorite (Ca[OCl]$_2$), calculated to achieve initial free residual chlorine (FRC) of 50 mg/L (i.e., 3.0 lbs [1.4 kg] industrial grade NaOCl [12%–15% available Cl] per 1,000 gal of CT/EC water; 10.5 lbs [4.8 kg] domestic grade NaOCl [3%–5% available Cl] per 1,000 gal of CT/EC water; or 0.6 lb [0.3 kg] Ca[OCl]$_2$ per 1,000 gal of CT/EC water). If significant biodeposits are present, additional chlorine may be required. If the volume of water in CT/EC is unknown, it can be estimated (in gallons) by multiplying either the recirculation rate in gallons per minute by 10 or the refrigeration capacity in tons by 30. Other appropriate compounds may be suggested by a water-treatment specialist.

   B. Record the type and quality of all chemicals used for disinfection, the exact time the chemicals were added to the system, and the time and results of FRC and pH measurements.

   C. Add dispersant simultaneously with, or within 15 minutes of, adding disinfectant. The dispersant is best added by first dissolving it in water and adding the solution to a turbulent zone in the water system. Automatic-dishwasher compounds are examples of low or nonfoaming, silicate-based dispersants. Dispersants are added at 10–25 lbs (4.5–11.25 kg) per 1,000 gal of CT/EC water.

   D. After adding disinfectant and dispersant, continue circulating the water through the system. Monitor the FRC by using an FRC-measuring device (e.g., a swimming-pool test kit), and measure the pH with a pH meter every 15 minutes for 2 hours. Add chlorine as needed to maintain the FRC at ≥10 mg/L. Because the biocidal effect of chlorine is reduced at a higher pH, adjust the pH to 7.5–8.0. The pH may be lowered by using any acid (e.g., muriatic acid or sulfuric acid used for maintenance of swimming pools) that is compatible with the treatment chemicals.

   E. Two hours after adding disinfectant and dispersant, or after the FRC level is stable at ≥10 mg/L, monitor at 2-hour intervals and maintain the FRC at ≥10 mg/L for 24 hours.

   F. After the FRC level has been maintained at ≥10 mg/L for 24 hours, drain the system. CT/EC water may be drained safely into the sanitary sewer. Municipal water and sewerage authorities should be contacted regarding local regulations. If a sanitary sewer is not available, consult local or state authorities (e.g., Department of Natural Resources) regarding disposal of water. If necessary, the drain-off may be dechlorinated by dissipation or chemical neutralization with sodium bisulfite.

   G. Refill the system with water and repeat the procedure outlined in steps A-F in Section II.

III. Mechanical Cleaning

   A. After water from the second chemical disinfection has been drained, shut down the CT/EC.

   B. Inspect all water-contact areas for sediment, sludge, and scale. Using brushes and/or a low-pressure water hose, thoroughly clean all CT/EC water-contact areas, including the basin, sump, fill, spray nozzles, and fittings. Replace components as needed.

   C. If possible, clean CT/EC water-contact areas within the chillers.

IV. After Mechanical Cleaning

   A. Fill the system with water and add chlorine to achieve FRC level of 10 mg/L.

   B. Circulate the water for 1 hour, then open the blowdown valve and flush the entire system until the water is free of turbidity.

   C. Drain the system.

   D. Open any air-intake vents that were closed before cleaning.

   E. Fill the system with water. CT/EC may be put back into service using an effective water-treatment program.

15.8 Long-Term Prevention Plans

Any remediation attempt that does not include long-term plans to maintain systems and prevent recurrence is short-sighted and destined to fail. There is no one-time, complete "cure" to microbial contamination within structures. Rather, continued oversight and attention to conditions that may allow microbial growth must become an integral part of a control plan [see Chapter 10]. Three basic strategies should be followed to maintain building performance and prevent microbial contamination: (a) routine surveillance inspections and prompt response to problems, (b) adequate preventive maintenance of the building structure as well as HVAC and plumbing systems, and (c) adequate housekeeping including an emphasis on proper and routine cleaning [see 10.7.2].

15.9 References

AIHA Biosafety Committee: Control Methods. In: Biosafety Reference Manual, pp. 51-99. P.A. Heinsohn, R.R. Jacobs, B.A. Concoby, Eds. American Industrial Hygiene Association, Fairfax, VA (1995).

AIHA Biosafety Committee: Field Guide for the Determination of Biological Contaminants in Environmental Samples. American Industrial Hygiene Association, Fairfax, Virginia (1996).

Alevantis, L.E.; Offerman, F.J.; Loiselle, S.; et al.: Pressure and Ventilation Requirements of Hospital Isolation Rooms for Tuberculosis (TB) Patients: Existing Guidelines in the United States and a Method for Measuring Room Leakage. In: Ventilation and Indoor Air Quality in Hospitals, pp. 101-116. M. Maroni, Ed. Kluwer Academic Publishers, The Netherlands (1996).

Ansari, O.; Morey, P.R.: Cleaning to Remove Aspergillus versicolor from a Mold-Contaminated Building. In: IAQ'96, Paths to Better Building Environments, pp 30-36. American Society of Heating, Refrigerating and Air-Conditioning Engineers, Atlanta, GA (1996).

Arnow, P. M.; Fink, J.N.; Schlueter, D.P.; et al.: Early Detection of Hypersensitivity Pneumonitis in Office Workers. Am. J. Med. 64:236-242 (1978).

ATS: American Thoracic Society, Medical Section of the American Lung Association. Respiratory Protection Guidelines. Am. J. Respir. Crit. Care Med. 154:1153-1165 (1996).

Burge, H.A.: Bioaerosols in the Residential Environment. In: Bioaerosols Handbook, pp. 579-597. C.S. Cox and C. M. Wathes, Eds. Lewis Publishers, CRC Press, Boca Raton, FL (1995).

CDC: Appendix B — Maintenance Procedures Used to Decrease Survival and Multiplication of *Legionella* sp. in Potable-Water Distribution Systems. In: Guidelines for Prevention of Nosocomial Pneumonia. Public Health Services, Centers for Disease Control and Prevention. CDC — MMWR 46(RR-1) Atlanta, GA (1997).

CDC: Appendix D — Procedure for Cleaning Cooling Towers and Related Equipment. In: Guidelines for Prevention of Nosocomial Pneumonia. Public Health Services,Centers for Disease Control and Prevention. CDC — MMWR 46 (RR-1) Atlanta, GA (1997).

Flannigan, B.: Approaches to Assessment of Microbial Flora of Buildings. In: IAQ '92,Environments for People, pp. 139-145. American Society of Heating, Refrigerating and Air-Conditioning Engineers, Inc., Atlanta, GA (1992).

Foarde, K.K.; VanOsdell, D.W.; Owen, M.K.; et al.: Fungal Emission Rates and their Impact on Indoor Air. In: Proceedings of the Air and Waste Management Association Specialty Conference, Engineering Solutions to Indoor Air Quality Problems, pp. 581-592 (1997).

Health Canada: Fungal Contamination in Public Buildings: A Guide to Recognition and Management. Federal-Provincial Committee on Environmental and Occupational Health. Environmental Health Directorate, Ontario, Canada (1995).

Hunter, C.A.; Grant, C.; Flannigan, B.; et al.: Mould in Buildings: The Air Spora of Domestic Dwellings. International Biodeterioration. 24:81-101 (1988).

IICRC: Standard and Reference Guide for Professional Water Damage Restoration S500-94. Institute of Inspection, Cleaning and Restoration Certification, Vancouver, WA (1995).

ISIAQ: Control of Moisture Problems Affecting Biological Indoor Air Quality. TFI-1996. International Society of Indoor Air Quality and Climate, Ottawa, Ontario, Canada (1996).

Lenhart, S.W.; Schafer, M.P.; Singal, M.; et al.: Histoplasmosis: Protecting Workers at Risk. Publication No. 97-146. Centers for Disease Control and Prevention. National Institutes for Occupational Safety and Health, Cincinnati, OH (1997).

Morey, P.R.; Williams, C.M.: Is Porous Insulation inside a HVAC System Compatible with a Health Building? In: IAQ '91, Healthy Buildings, pp 128-135. ASHRAE, Atlanta,GA (1991).

Morey, P.: Microbiological Contamination in Buildings; Precautions During Remediation Activities. In: Proceedings ASHRAE IAQ '92, Environments for People. Atlanta, 1992, pp 171-177. ASHRAE, Atlanta, GA (1992).

Morey, P.R.: Studies on Fungi in Air Conditioned Buildings in a Humid Climate. In: Proceedings: Conference on Biological Contaminants. Saratoga Springs, NY (1994).

Morey, P.: Mold Growth in Buildings: Removal and Prevention. In: Proceedings of Seventh International Conference on Indoor Air Quality and Climate. Nagoya, Japan (1996).

Morey, P.; Ansari, S.: Mold Remediation Protocol with Emphasis on Earthquake Damaged Buildings. In: Proceedings of Seventh International Conference on Indoor Air Quality and Climate, Nagoya, Japan, Vol 3, pp. 399-404. Japan (1996).

Morey, P.; Hunt, S.: Mold Contamination in an Earthquake Damaged Building. In: IAQ '95, Healthy Buildings, 3:1377-1381 . ASHRAE, Atlanta, GA (1995).

NADCA: Mechanical Cleaning of Non-Porous Air Conveyance System Components. NADCA 1992-01. National Air Duct Cleaners Association, Washington, DC (1992).

NADCA: General Specifications for the Cleaning of Commercial Heating, Ventilating and Air Conditioning Systems. National Air Duct Cleaners Association, Washington, DC (1997).

Nevalainen, A.; Flannigan, B.: Evaluation of Microbial Contamination of Buildings. In: Workshop Summaries, Indoor Air '93. Helsinki, Finland, 1993, pp. 15-16. Finland (1993).

New York City Department of Health, New York City Human Resources Administration, and Mount Sinai-Irving J. Selikoff Occupational Health Clinical Center: Guidelines on Assessment and Remediation of Stachybotrys atra in Indoor Environments. New York, NY (1993).

OSHA: Occupational Exposure to Asbestos; Final Report. Department of Labor. In: Federal Register, 59(153):41083-41084, 29 CFR Part 1915, §1915.1001.Occupational Safety and Health Administration (1994).

Rautiala, S.; Reponen, T.; Hyvärinen, A.; et al.: Exposure to Airborne Microbes During the Repair of Moldy Buildings. Am. Ind. Hyg. Assoc. J. 57:279-284 (1996).

Sesline, D.; Ames, R.G.; Howd, R.A.: Irritative and Systemic Symptoms Following Exposure to Microban Disinfectant through a School Ventilation System. Arch Environ Health. 49:439-444 (1994).

Ström, G.; Palmgren, U.; Wessen, B.: The Sick Building Syndrome: An Effect of Microbial Growth in Building Constructions? In: Proceedings of the 5th International Conference on Indoor Air Quality and Climate-Indoor Air '90. 1:173-178 (1990).

USEPA: Should You Have the Air Ducts in your Home Cleaned? Doc. No. EPA-402-K-97-002. United States Environmental Protection Agency. Government Printing Office, Washington, DC (1998).

Weber, A.M.; Martinez, K.F.: NIOSH Health Hazard Evaluation Report 93-1110-2575. Martin County Courthouse and Constitutional Office Building, Stuart, Florida, May 1996. U.S. Department of Health and Human Services (1996).

Weeks, R.J. Histoplasmosis Sources of Infection and Methods of Control. Centers for Disease Control and Prevention, Atlanta, GA (1984).

WDHSS: Control of Legionella in Cooling Towers — Summary Guidelines. Section of Acute and Communicable Disease Epidemiology, Bureau of Community Health and Prevention, Division of Health, Wisconsin Department of Health and Social Services, Madison, WI (1987).

Chapter 16

Biocides And Antimicrobial Agents

Eugene C. Cole and Karin K. Foarde

16.1   Introduction
    16.1.1   *Biocides*
    16.1.2   *Antimicrobial Agents*
16.2   Biocides
    16.2.1   *Biocide Efficacy*
    16.2.2   *Biocide Label Claims*
        16.2.2.1   Biocide Resistance
        16.2.2.2   Special Biocide Use Claims
        16.2.2.3   Biocide Selection
        16.2.2.4   Biocide Classes
    16.2.3   *Biocide Use and Application*
    16.2.4   *Aqueous Biocides*
        16.2.4.1   Chlorine Dioxide
        16.2.4.2   Hypochlorites
        16.2.4.3   Biocides for Water Treatment
        16.2.4.4   Aqueous Ozone
    16.2.5   *Gas-Phase/Vapor-Phase Biocides*
    16.2.6   *Precautions During Biocide Application*
16.3   Antimicrobial Agents
    16.3.1   *Efficacy of Antimicrobial Agents*
    16.3.2   *Label Claims and Antimicrobial Product Selection*
    16.3.3   *Antimicrobial Agent Use and Application*
    16.3.4   *Precautions During Antimicrobial Agent Application*
16.4   References

16.1 Introduction

Several panels of experts have stated that microbial growth in HVAC systems and on building materials and furnishings is not appropriate and that such contamination should be removed and further growth prevented (Samson et al., 1994; Health Canada, 1995; Maroni et al., 1995; ISIAQ, 1996) [see also 10.1 and 14.1.4.5]. Biocides and antimicrobial agents may play a role in achieving these goals. However, ACGIH and others discourage the use of biocides and antimicrobial agents without appreciating their limitations and potential hazards of their use instead of removing microbial contamination from building environments (Samson et al., 1994). Biocides are chemical or physical agents that kill or inactivate microorganisms, whereas antimicrobial agents are compounds applied to suppress microbial growth. Thus, biocides are used to treat existing microbial growth, whereas antimicrobial agents are used to prevent growth.

Among the limitations associated with biocide use are the small number of agents that are suitable for ap-plication in buildings. Biocides and antimicrobial agents may show promise in laboratory tests, but safe and effective use in building environments may be difficult. Some agents are designed to treat certain groups of microorganisms. However, suppression of one organism may give others an advantage, leading to different control problems. Further, most disinfectants and sanitizers are approved for use on previously cleaned rather than soiled surfaces. For thoroughly cleaned non-porous building materials, biocides may not be needed. Killing microorganisms usually does not destroy their antigenic or toxic properties. Therefore, even microbial growth that has been treated with a biocide should be removed from indoor environments. Antimicrobial agents should not be used in place of moisture control and regular cleaning and maintenance, but antimicrobial agents may protect some materials from microbial growth.

The following information will (a) help investigators understand the roles biocides and antimicrobial agents may play in the control, remediation, and prevention of microbial contamination in buildings, and (b) inform readers about the limitations and potential hazards of these agents.

16.1.1 Biocides

Biocides are toxic chemicals or physical agents capable of killing or inactivating one or more groups of microorganisms, that is, vegetative bacteria, mycobacteria, or bacterial spores; vegetative fungi or fungal spores; parasites; or viruses (Table 16.1). For the purpose of this book, biocides will be addressed only in the context of their chemical classes rather than by trade names. Physical disinfection mechanisms, such as ionizing or thermal radiation, are impractical for treating biological contamination in building environments and, therefore, are not discussed. Section 9.8.6.4 mentions the use of ultraviolet radiation to disinfect indoor air.

Biocides are sometimes used to control extensive microbial contamination (e.g., sewage backflow, as part of a remediation and clean-up process) (IICRC, 1995). Biocides may also play a preventive role when added to fluid reservoirs (e.g., cooling tower water, metal-working fluids). The definition of a disinfectant (Table 16.1) emphasizes that *disinfection* is a process that significantly reduces or eliminates pathogenic microorganisms (3 $\log_{10}$ or greater reduction). In turn, *decontamination* is a relative term and describes the killing or removal of microorganisms with no specific quantitative implication.

16.1.2 Antimicrobial Agents

Antimicrobial agents are chemical formulations incorporated into or applied onto a material or product to suppress vegetative bacterial and fungal growth as it occurs (Table 16.1). Such compounds may be used to retard microbial growth in potential sources. Typically, antimicrobial agents are incorporated into products during manufacture (e.g., carpet material, ceiling tiles, and air filters). Additionally, antimicrobial agents are often included in products (e.g., paints, coatings, and sealants) that are applied to various building and equipment surfaces.

16.2 Biocides

16.2.1 Biocide Efficacy

For almost 150 years, chemical biocides have been used to kill microorganisms that cause human disease (i.e., since the institution of chloride of lime to kill the agent of puerperal fever) (Semmelweiss, 1975). Today, an arsenal of >8000 biocides for environmental use are registered as pesticides with the USEPA under the Federal Insecticide, Fungicide, and Rodenticide Act (FIFRA). Most of these agents are aqueous formulations that function as sanitizers, disinfectants, or sterilants to kill microorganisms to varying degrees (Table 16.1). However, the killing of microbial growth usually does not destroy the contaminant's or contaminated material's antigenic or toxic properties. In some cases, the concentration of toxins may actually increase after cell death [e.g., dead gram-negative bacteria (GNB) may release endotoxin].

TABLE 16.1. Definitions of Biocides and Antimicrobial Agents

Biocide

Chemical or physical agent capable of killing or inactivating one or more of the following groups of microorganisms:

• Vegetative bacteria	• Bacterial spores	• Mycobacteria	• Viruses
• Vegetative fungi	• Fungal spores	• Parasites	

Sanitizer: biocide that significantly reduces the number of vegetative environmental bacteria of public health importance (e.g., *Staphylococcus aureus* or *Escherichia coli*), as determined by standardized laboratory testing.

Disinfectant: biocide that significantly reduces the number of recognized human pathogens (e.g., *Salmonella* or *Mycobacterium* spp.), as determined by standardized laboratory testing. It may kill some bacterial spores.

Sterilant: biocide that can completely kill or inactivate significant challenges of resistant bacterial spores (e.g., *Clostridium* or *Bacillus* spp.), as determined by standardized laboratory testing.

Antimicrobial agent

Chemical formulation incorporated into or applied onto a material or product to suppress or retard the development of vegetative bacterial or fungal growth.

Bacteriostatic agent: chemical agent that suppresses or retards bacterial growth on direct contact with the treated material.

Fungistatic agent: chemical agent that suppresses or retards fungal growth on direct contact with the treated material.

16.2.2 Biocide Label Claims

The labels on USEPA-registered biocides provide useful information, including the biocidal claims the manufacturer makes for a product (Figure 16.1) (see also the National Antimicrobial Information Network (NAIN, 800-447-6349, http://www.ace.orst.edu/info/nain).

Label claims are approved following USEPA review of efficacy data generated using standard methods such as those of the Association of Official Analytical Chemists (AOAC, 1995). This standardized laboratory testing requires specified contact times and temperatures, the use of designated strains of microorganisms, and the preparation of the use-concentration of a biocide as specified on the manufacturer's label. Therefore, the testing is standardized, but claims are not based on in-use evaluations on the variety of materials and surfaces that may be found in buildings. Table 16.2 defines the various biocidal activities manufacturers may claim their products possess.

TABLE 16.2. Definitions of Biocidal Activities in Order of Increasing Range of Effectiveness

Activity	Description
Germicide	Inactivates one or more groups of pathogenic microorganisms.
Bactericide	Inactivates vegetative GNB and GPB, but not *Mycobacterium* or *Legionella* spp. nor any bacterial spores.
Fungicide	Inactivates both vegetative fungi and fungal spores.
Virucide	Inactivates only the viruses listed on the product label. (Is also bactericidal, fungicidal)
Tuberculocide	Inactivates *Mycobacterium* spp. (e.g., *Mycobacterium bovis* and especially *Mycobacterium tuberculosis*); is also bactericidal, fungicidal, and virucidal.
Sporicide	Inactivates environmentally resistant bacterial spores; therefore, is also germicidal, bactericidal, fungicidal, virucidal, and tuberculocidal.

```
Name of Product                 Description of Product
Active Ingredients              Inert Ingredients
  USEPA Registration Number     Biocide Claims (Table 13.2)
  USEPA Establishment Number    Instructions for Use
                                  • Coverage
Precautionary Statements          • Area preparation
  • Hazard to humans and          • Surface preparation
    domestic animals              • Mixing instructions
  • Environmental hazards         • Application
  • Physical and chemical         • Application precautions
    hazards                       • Clean-up
```

FIGURE 16.1. Sample biocide label.

16.2.2.1 Biocide Resistance A solid framework of information on biocide resistance for the major groups of microorganisms has emerged from many years of research (Cole and Robison, 1996). Table 16.2 reflects the resultant general scale of biocide resistance. In increasing order of resistance to biocides are vegetative bacteria, vegetative fungi and fungal spores, viruses, mycobacteria, and bacterial spores. Viruses are divided into two groups with regard to biocide resistance, that is, the less-resistant, enveloped (lipophilic) viruses and the more-resistant, non-enveloped (hydrophilic) viruses (Klein and DeForest, 1983) [see Figure 21.1]. The resistance scale can be useful when selecting biocides for specific applications. For example, a USEPA-registered disinfectant with a fungicidal claim also predicts effectiveness against vegetative bacteria. Similarly, a tuberculocidal claim would predict effectiveness against all categories except bacterial spores. A sporicidal claim would predict the inactivation of all microbial forms. It is also important to realize

that the effectiveness of an environmental biocide can only be ensured when it is used according to label directions, as is required by law. Potential biocide users should understand that it is a violation of Federal law to use a biocide in any manner inconsistent with its label directions.

16.2.2.2 Special Biocide Use Claims Biocide manufacturers can claim applications other than those listed in Table 16.2 if they submit additional documentation specific to the claim prior to USEPA approval. Such claims may include efficacy of a product against organisms other than the standard test strains or suitability for use on designated materials or surfaces. Thus, some products may claim effectiveness against *Legionella pneumophila*, *Aspergillus fumigatus*, *Penicillium chrysogenum*, or the hepatitis viruses. Likewise, products may be approved as sanitizers for use on materials or surfaces such as carpet or ductwork and other components of ventilation systems.

A few products were previously registered with the USEPA specifically for use on the inside of air ducts (USEPA, 1997). However, these products were registered solely for the purpose of sanitizing the smooth surfaces of unlined sheet-metal ducts. The USEPA no longer registers biocides for specific use in HVAC systems. No products are registered as biocides for use on glass-fiber ductboard or glass-fiber-lined ducts. Therefore, it is important to determine if sections of a system contain such materials before permitting the application of any biocide. For the most current information on air duct cleaning, contact the National Air Duct Cleaners Association

(NADCA, Washington, DC) [see 15.3 and 15.5.1]. Specific questions about biocide product formulations and claims can be directed to the Antimicrobials Division, Registration Division, USEPA Office of Pesticide Programs and Toxic Substances, USEPA (Washington, D.C.).

16.2.2.3 Biocide Selection The selection of a chemical biocide for a particular application depends on three factors. First, biocide users must consider the type and extent of microbial contamination. Second, selection among candidate products is based on factors that determine biocide effectiveness and suitability for the specified application. Third, the selection process should identify the hazards of candidate biocides. These factors must be carefully considered as a whole in the biocide selection process (Table 16.3).

16.2.2.4 Biocide Classes Table 16.4 lists the most common biocide classes. These compounds have different modes of action and spectra of microorganism inactivation. In addition, biocides differ in stability, reactivity with materials, and hazard based on their chemical and physical properties (AIHA, 1985, 1995; Block, 1991). Chlorine compounds [e.g., hypochlorites and chlorine dioxide (ClO_2)], are widely used because of their antibacterial and antifungal effectiveness, deodorizing capability, and low cost [see 16.2.4].

16.2.3 Biocide Use and Application

An international workshop concluded that until the health risk of using biocides indoors is more fully understood, the use of biocides should only be considered as a last option for controlling fungal growth in indoor environments (Samson et al., 1994). Biocide use should not be considered if careful and controlled removal of contaminated material is sufficient to address a problem [see 15.4]. However if the remediation or restoration of a microbially contaminated non-porous surface or material warrants biocide use, a trained person should oversee the selection and application of a suitable biocide. Such persons should understand biocide capabilities, limitations, and health and safety hazards. Environmen-

tal disinfectants are routinely and safely used in healthcare facilities and laboratories to promote patient and worker health. Likewise, biocide use may play an important role in the remediation of certain conditions (e.g., microbial contamination from sewage backflow into buildings) (Berry et al., 1994; IICRC, 1995). With biocide use, four basic precautions must be observed (Table 16.5) [see also 15.4].

The last precaution in Table 16.5 (i.e., removal of biocide residues and microbial remnants) is necessary if building occupants may be exposed to residual biocide via inhalation, skin contact, or ingestion (e.g., if the biocide has contaminated food preparation areas). How-

---

Effective remediation of water-damaged or microbially contaminated buildings involves (a) the use of appropriate techniques to promote rapid drying, and (b) complete removal of contaminated materials rather than the application of biocides without these steps.

---

ever, a biocide residue may be desirable, in certain cases, to inhibit future microbial growth on the treated surface. Removal of residual biocide with water should be followed by thorough drying of the cleaned items. The workers doing the cleaning should wear PPE similar to that used during biocide application [see 16.2.6]. Residual microbial growth must be removed because the antigenic and toxic properties of microorganisms and microbially contaminated materials are not eliminated when microorganisms are killed.

16.2.4 Aqueous Biocides

Water-based biocides are the most commonly used treatments. These compounds are typically marketed as concentrated solutions that require mixing with tap water to achieve the recommended use concentration. Aqueous biocides can be used to wipe down contaminated

TABLE 16.4. Important Biocide Classes

Alcohols (ethyl, isopropyl)	Hydrogen peroxide
Aldehydes (formaldehyde, glutaraldehyde)	Phenolic compounds
Halogens (chlorine, iodine, and bromine compounds)	Quaternary ammonium compounds (cationic detergents)

TABLE 16.3. Factors to Consider in Biocide Selection

Type and extent of microbial contamination; biocide effectiveness against specific microbial contaminants
- Efficacy label claims   • Active ingredients

Suitability in use
- Use concentration   • Application or delivery requirements
- Use pH   • Inactivation by organic matter
- Temperature   • Compatibility with materials and surfaces
- Contact time   • Cost

Safety in use
- Health and safety hazards; PPE requirements
- Impact on the general environment

TABLE 16.5. Basic Precautions for Biocide Use

- Never use biocides in occupied buildings.
- Never spray biocides into functioning air-handling units.
- Ensure that biocide applicators are trained and use PPE to prevent respiratory, mucus membrane, and dermal exposures.
- After treatment, remove residual biocide and any remaining surface microbial growth.

surfaces or mop floors or can be sprayed onto materials or fogged into spaces. However, application of water-based biocides introduces additional moisture into the building environment, which may not be desirable. In addition, the USEPA approves most disinfectants and sanitizers for use on previously cleaned, hard rather than soiled or porous surfaces. *Cleaning* is the process of removing all foreign material (e.g., soil, fungal growth, or other organic matter) from hard surfaces. Disinfectant-detergents and sanitizer-detergents clean and inactivate microbial contamination in one step. Studies have shown that cleaning with a detergent may be as effective as cleaning and treatment with a biocide. Again, it is critical that such products are USEPA registered and used according to label directions for the stated applications.

Kapyla (1985) assessed the effectiveness of nine aqueous biocides to inactivate fungal contamination on insulated window frames. Tested products included aldehydes as well as quaternary ammonium chloride and phenolic compounds (with and without alcohols). The best results were obtained by spraying aldehydes or quaternary ammonium chlorides in alcohol.

Biocide application is not recommended in the restoration of water-damaged indoor environments except where they have suffered extensive sewage backup. Widespread pollution from raw sewage presents a significant health risk from a variety of infectious agents, and biocides may help to control and contain these agents during the restoration process (Berry et al., 1994; IICRC, 1995). That process includes the following steps: sewage extraction, application of biocide, removal of organic residuals, cleaning of all surfaces, and secondary application of biocide if deemed necessary. Sewage-contaminated porous materials must be discarded. Likewise, porous materials that, for any reason, remain wet for >24 hours should be discarded. The Institute of Inspection, Cleaning, and Restoration Certification (IICRC, Vancouver, WA) has published guidance for biocide spray application (IICRC, 1995). The standard addresses spray orifices, droplet size, and recommended biocide application rates (ft^2/gal) for floor and subfloor surfaces.

16.2.4.1 Chlorine Dioxide ClO_2 preparations are available with registered claims for a variety of uses, including treatment of air-handling systems. Once again, surfaces to be treated must first be cleaned. Typically, a 500-ppm solution is prepared for spraying or fogging. At the time of use, ClO_2 is generated by mixing sodium chlorite and an organic acid such as citric or lactic acid. ClO_2 has a number of desirable features. It is a broad-spectrum biocide and deodorizer, is effective over a wide pH range, and is less corrosive than hypochlorite [see 16.2.4.2]. ClO_2 is used in a variety of industrial applications, for example, drinking water and wastewater treat-

ment, paper pulp bleaching, and food processing. Residual ClO_2 breaks down to water and sodium chloride.

16.2.4.2 Hypochlorites Hypochlorites are popular biocides because of their broad spectrum effectiveness as disinfectants and sanitizers. Additional attractive features are their deodorizing properties, availability, and low cost. Sodium hypochlorite (household bleach) is commonly used in a 1:10 dilution (5000 ppm) to treat a variety of surfaces and materials (FOH, 1993). However, sodium hypochlorite must be used with extreme caution because it is a strong bleaching agent, corrodes metals, and is inactivated by organic matter. This biocide should never be mixed with ammonia or ammonia-containing products to avoid liberation of toxic chlorine gas.

Increased ventilation is required when using hypochlorite for building remediation. Applicators also need skin and mucus membrane protection as well as respirators with inorganic vapor cartridges. Hypochlorite will bleach many carpet and fabric materials. To prevent corrosion, metal surfaces should be rinsed with water following treatment with hypochlorites. The Canada Mortgage and Housing Corporation and the American Red Cross/Federal Emergency Management Agency have published general guidance on using hypochlorite in residential environments for flood clean up and remediation of microbial contamination (ARC/FEMA, 1992; CMHC, 1993, 1994). Although hypochlorites are frequently used to treat microbial growth in indoor sources, little applied research has been conducted to define an acceptable procedure to maximize their effectiveness.

16.2.4.3 Biocides for Water Treatment Biocides should never be used in humidification systems. Rather, microbial contamination should be controlled in humidifiers through use of clean water, limiting the time water remains stagnant in reservoirs, and periodic cleaning. Biocide use in condensate drain pans in operating air-handling units is discouraged in favor of appropriate drain design, operation, and routine inspection and maintenance [see 10.4.4.3]. However, aqueous biocides are often used to control biological contamination in other water systems (e.g., for control of algae and legionellae in cooling towers).

The importance of selecting a biocide with appropriate label claims was illustrated in a cooling tower that continued to harbor legionellae despite biocide treatment (Hung et al., 1993). It was eventually determined that the "biocide" was an algicide that had no effect on GNB. Not all of the many biocides that have demonstrated efficacy against legionellae in the laboratory have proven as effective under in-use conditions (ASM, 1993). Chlorine at 1 to 3 ppm has been shown to be effective in controlling legionellae, although corrosion and pH maintenance are major concerns with its use (Barbaree, 1991). Section 10.6 describes a CDC-recommended procedure for cleaning cooling towers and related equipment. Section 20.6 describes the use of sodium hypochlorite and hydrogen peroxide to decontaminate sources of amebae. Section 23.6.2 discusses reasons why biocides may not be the best approach for controlling other bacteria.

16.2.4.4 Aqueous Ozone Aqueous ozone (O_3) is recognized as an effective biocide in water systems (Wickramanayake, 1991). However, aqueous ozone is subject to rapid degradation and certainly is not recommended for use on visible fungal growth. Ozone is affected by temperature, sunlight, pH, and organic matter. Therefore, these factors, along with initial concentration, determine ozone's effectiveness. Ozone has been used as a drinking water disinfectant since 1893 and is still used in a variety of applications such as wastewater treatment, preparation of high purity water for the pharmaceutical industry, and treatment of cooling tower water. Generally, relatively low ozone concentrations (≤1 ppm) are effective.

16.2.5 Gas-Phase/Vapor-Phase Biocides

No gas- or vapor-phase biocides can effectively and safely remediate a microbially contaminated building because of problems with biocide delivery, efficacy, and toxicity. First and foremost, there is no means to deliver a gas- or vapor-phase biocide to all spaces and surfaces within a building. Microbial growth will remain viable unless an extremely effective agent reaches every wall and ceiling space, literally every nook and cranny. Even then, non-viable residues of microbial contamination that remain following treatment would continue to pose a possible allergic or toxic hazard until physically removed [see 16.2.3]. The efficacy of any gas-phase biocide is compromised when a sufficient concentration cannot be maintained in a space for the necessary time.

However, several gas-phase biocides work well to decontaminate small, confined areas. For example, formaldehyde gas (HCHO) has long been used to decontaminate high-containment microbiology laboratories and laboratory equipment (e.g., biological safety cabinets) (AIHA, 1985, 1995; NSF, 1992; USDHHS, 1993; WHO, 1993). In such situations, it is possible to completely seal a room or piece of equipment, generate formaldehyde gas, maintain a desired concentration for several hours, neutralize the agent, ventilate the space, and confirm treatment effectiveness with biological indicators. However, such decontamination is seldom possible for an entire building because effective treatment cannot be achieved on a practical, safe, and cost-effective basis, particularly in severely microbially contaminated structures. Even repeated attempts to sterilize a clean, new building with formaldehyde gas were unsuccessful (Taylor, 1974). Another deterrent to formaldehyde gas use is its recognized toxicity and carcinogenicity.

A vaporized hydrogen peroxide (H_2O_2) system was developed as an alternative to formaldehyde gas to fumigate contaminated laboratory equipment (Klapes and Vesley, 1990). However, hydrogen peroxide cannot be used on a larger scale for the reasons discussed above for formaldehyde. Ethylene oxide (ETO) is used to sterilize heat-sensitive medical equipment. However, ETO is also excluded from consideration as a space decontaminant for its recognized toxicity, mutagenicity, and teratogenicity.

While an effective deodorizer, gas-phase ozone is a poor biocide, particularly against bacterial and fungal spores. Ozone is also a strong oxidizing agent, reactive (rubber and electrical wire insulation with unsaturated carbon-carbon bonds), and very unstable. Gas-phase ozone was recognized as a poor biocide for surface contamination over 50 years ago (Elford and van den Ende, 1942). Recent controlled laboratory studies confirmed the ineffectiveness of gas-phase ozone against a variety of fungal and bacterial contaminants (Foarde, et al., 1997). The first phase of the study tested high concentrations of fungal and bacterial spores and yeast cells dried on glass surfaces. The samples were exposed to ozone concentrations in excess of 5 ppm for 23 hours at 30% and 90% RH. Some organisms suffered minimal inactivation at this ozone concentration. To achieve a significant kill for other microorganisms, ozone concentrations of 6 to 10 ppm were required. A second study with fungal spores on a variety of building materials demonstrated no decontamination of any item, even at 9 ppm ozone for 23-hour exposure. Again, even if gas-phase ozone killed all microbial growth, the dead microbial residue would continue to pose a potential allergic or toxic hazard if left in place [see 16.2.3].

16.2.6 Precautions During Biocide Application

Biocides that kill or inactivate microorganisms may also harm humans. With few exceptions, disinfectants are low-vapor-pressure chemicals for which neither threshold limit values nor permissible exposure limits have been established. Chemicals used as biocides for which there are recommended exposure limits are chlorine, chlorine dioxide, ethanol, formaldehyde, glutaraldehyde, hydrogen peroxide, isopropyl alcohol, ozone, and phenol (ACGIH, 1998). Biocides should not be applied in occupied buildings, nor should biocides be used in operating ventilation systems (Sesline et al., 1994).

Data on user exposure during biocide application is limited. Biocide application covers transfer of the product from a shipping container to other containers or reservoirs, dilution and mixing with other reagents, if required, and delivery to the surface to be treated. Biocides may be liquids, loose solids (e.g., powders or flakes), or compact solids (e.g., premeasured packets). Gas-phase biocides are usually generated from liquids. Except for the latter, the airborne route of exposure for workers applying biocides is generally low relative to skin exposure (Popendorf and Selim, 1995). Dermal exposure may occur when users pour, pump, or otherwise transfer liquid or solid biocides from shipping containers to secondary containers for measuring, transportation, or filling of other containers. Splashing may also occur when biocides are diluted with water or mixed with other reagents. Unless skin protection is used, applicators may experience considerable dermal exposure when spraying, mopping, or wiping biocides onto floors, walls, other surfaces, or carpeting. Inhalation exposure may occur during the use of hypochlorites and gas-phase biocides. Appropriate precautions and PPE for biocide applicators should be selected based on the product manufacturer's warnings and recommendations (e.g., goggles or face shield, aprons or other protective clothing, gloves, and respiratory protection) [see also 4.6.2].

16.3 Antimicrobial Agents
16.3.1 Efficacy of Antimicrobial Agents
For the purposes of this book, environmental antimicrobial agents are those chemical formulations used to suppress vegetative bacterial and fungal growth on direct physical contact (Table 16.6). Antimicrobial agents are intended to prevent microbial contamination and are referred to as bacteriostatic or fungistatic agents (see Table 16.1 for definitions of these terms). These compounds have been shown, in laboratory and limited field studies, to suppress or retard initial bacterial or fungal growth rather than to kill established growth, which may also contain spores.

Antimicrobial compounds are now incorporated into a variety of consumer products to retard microbial growth (Table 16.7). In some cases, such agents can also be applied to existing materials (e.g., carpet, upholstery, and similar furnishings) to retard bacterial and fungal growth and resultant odors in humid or damp environments. Antimicrobial agents for environmental applications are distinctly different from, and do not include, antimicrobial products used on the living human body (e.g., antimicrobial handwashes and a variety of antiseptics). These latter products are subject to Food and Drug Administration approval and are not used to address contamination in the inanimate indoor environment.

16.3.2 Label Claims and Antimicrobial Product Selection
As for biocides, manufacturers must register antimicrobial agents with the USEPA and submit efficacy data. Antimicrobial agents are evaluated using standard methods from a variety of organizations [e.g., the American Association of Textile Chemists and Colorists (AATCC, Research Triangle Park, NC) and the American Society for Testing and Materials (ASTM, West Conshohocken, PA)]. Ideally, a suitable antimicrobial product will have label claims for efficacy against a spectrum of environmental microorganisms and will identify materials into which the agent can be incorporated or onto which it can be applied to control these microorganisms.

It is important that an antimicrobial agent have demonstrated activity against the environmental organisms of concern if its use is considered for materials such as carpet or air filters. Because carpet collects fungal and bacterial soil organisms and air filters retain the spores of common outdoor fungi, typical target organisms include *Aspergillus* and *Penicillium* spp., yeasts, and a variety of GNB. Thus, efficacy claims against only innocuous environmental *Bacillus* spp. and human-source bacteria (e.g., *Staphylococcus* or *Micrococcus* spp.) are inappropriate. Claims against the latter do not necessarily signify in-use effectiveness against the spectrum of environmental microorganisms that may multiply on building materials. Much applied research regarding the long-term effectiveness of antimicrobial products in the indoor environment remains to be done.

16.3.3 Antimicrobial Agent Use and Application
Many antimicrobial agents are referred to as "bound" products, because the chemical formulations bind to fibers or material surfaces. Such binding enables an antimicrobial agent to inhibit the growth of bacteria and fungi that come into contact with the treated material. However, a layer of dirt or debris on treated materials and surfaces may act as a physical barrier between a chemical agent and a microorganism and eliminate antimicrobial effectiveness. Thus the cleanability of a treated product becomes a significant issue. Dirt and debris can readily be removed from many hard, cleanable surfaces (e.g., those coated with antimicrobial paints). Interference is a greater concern with textile materials (e.g.,

TABLE 16.6. Representative USEPA-Registered Environmental Antimicrobial Agents

Phosphated quaternary amine complex

Bis (tri-n-butyltin) oxide

Tri-n-butyltin maleate

5-chloro-2-(2,4-dichlorophenoxy) phenol

10-10', oxybisphenoxarsine

3-trimethoxysilylpropyloctadecyldimethyl
   ammonium chloride

TABLE 16.7. Examples of Materials and Products that may Contain or be Treated with Antimicrobial Agents

Drapes	Roofing products
Curtains	Tents
Linens	Concrete reinforcements
Upholstery	Air filters
Vinyl furniture	Packaging
Counter tops	Sealants
Plastic products	Protective coatings
Carpet	Paints
Hard-surface flooring	Stains
Mop heads	Wallpaper

carpet) and becomes a considerable concern for non-cleanable or porous materials (e.g., duct liner). Table 16.7 lists materials and products that may contain antimicrobial agents. A few specific products treated with antimicrobial agents have been evaluated in the peer-reviewed literature (Gettings et al., 1990; Kemper et al., 1990; Price et al., 1993; Simmons and Crow, 1995).

Antimicrobial products may play a role in preventing microbial contamination on some surfaces and finishes, particularly those in environments and microenvironments where moisture control is difficult. However, it should be stressed that the use of antimicrobial products and materials does not eliminate the need for attention to moisture control in conjunction with a dedicated and routine preventive program of building operation, inspection, maintenance, and cleaning.

> Antimicrobial agents should not be used in place of moisture control in buildings or reduction of microbial growth through regular cleaning and maintenance.

16.3.4 Precautions During Antimicrobial Agent Application
Like biocides, antimicrobial agents may harm humans, and exposure may occur during the application process [see 16.2.6]. In one study, the airborne exposure route for workers applying antimicrobial agents was much lower than skin exposure (Popendorf et al., 1995). Appropriate precautions and PPE should be selected based on the product manufacturer's hazard warnings and recommendations [see also 4.6.2].

16.4 References

ACGIH: 1998 TLVs and BEIs. Threshold Limit Values for Chemical Substances and Physical Agents. Biological Exposure Indices. American Conference of Governmental Industrial Hygienists, Cincinnati, OH (1998).

AIHA: Decontamination and disposal. In: Biohazards Reference Manual, pp. 65-77. American Industrial Hygiene Association, Akron, OH (1985).

AIHA: Biohazards Reference Manual, 2nd Ed., pp. 101-109 American Industrial Hygiene Association, Fairfax, VA (1995).

AOAC: Official Methods of Analysis, 16th Ed. Association of Official Analytical Chemists International, McLean, VA (1995).

ARC/FEMA: Repairing Your Flooded Home. American Red Cross, Federal Emergency Management Agency, ARC 4477. FEMA Publications, Washington, DC (1992).

ASM: Legionella. Current Status and Emerging Perspectives. J.M. Barbaree, R.F. Breiman, A.P. Dufour, Eds. American Society for Microbiology, Washington, DC (1993).

Barbaree, J.M.: Controlling Legionella in Cooling Towers. ASHRAE Journal 33:38-42 (1991).

Berry, M.; Bishop, J.; Blackburn, C.; et al.: Suggested Guidelines for Remediation of Damage from Sewage Backflow into Buildings. J. Environ. Health. 57:9-15 (1994).

Block, Seymour S.: Disinfection, Sterilization, and Preservation, 4th Ed. Lea & Febiger, Malvern, PA (1991).

CMHC: Clean-Up Procedures for Mold in Houses, Canada Mortgage and Housing Corporation, Ottawa, Ontario, Canada (1993).

CMHC: Cleaning Up Your House After a Flood, Canada Mortgage and Housing Corporation, Ottawa, Ontario, Canada (1994).

Cole, E.C.; Robison, R.: Test Methodology for Evaluation of Germicides.

In: Handbook of Disinfectants and Antiseptics, pp. 1-16. J.M. Ascenzi, Ed. Marcel Dekker, Inc., New York, NY (1996).

Elford, W.S.; van den Ende, J. Investigation of Ozone as an Aerial Disinfectant. J Hyg. 42:240-265 (1942).

Foarde, K.K.; VanOsdell, D.W.; Steiber, R.S.: Investigation of Gas-Phase Ozone as a Potential Biocide. Appl. Occup. and Environ. Hyg. 12:535-542 (1997).

FOH: Federal Occupational Health Protocol for Controlling Microbial Growth After a Flood. In: Indoor Air Quality Update, Vol.6, No. 11. Cutter Information Corp., Arlington, MA (1993).

Gettings, R.L.; Kemper, R.A.; White, W.C.: Use of an Immobilized Antimicrobial for Intervention of Environmental Sources of Microbial Populations in the Homes of Mold-Sensitive Subjects and Subsequent Monitoring of the Presentation of Allergic Symptoms. Developments in Industrial Microbiology, Vol. 31. J. Indust. Micro., Suppl. No. 5:231-235 (1990).

Health Canada: Fungal Contamination in Public Buildings: A Guide to Recognition and Management. Federal-Provincial Committee on Environmental and Occupational Health. Environmental Health Directorate, Ottawa, Ontario, Canada (1995).

Hung, L.L; Copperthite, D.C.; Yang, C.S.; et al.: Environmental Legionella Assessment in Office Buildings of Continental United States. Indoor Air 3:349-353 (1993).

ISIAQ: Control of Moisture Problems Affecting Biological Indoor Air Quality. TFI-1996. International Society of Indoor Air Quality and Climate, Ottawa, Ontario, Canada (1996).

IICRC: Standard and Reference Guide for Professional Water Damage Restoration S500-94. Institute of Inspection, Cleaning and Restoration Certification, Vancouver, WA (1995).

Kapyla, M.: Frame Fungi on Insulated Windows. Allergy 40:558-564 (1985).

Kemper, R.A.; White, W.C.; Gettings, R.L.: Sustained Aeromicrobiological Reductions Utilizing Silane-Modified Quaternary Amines Applied to Carpeting: Preliminary Data from an Observational Study of Commercial Buildings. Developments in Industrial Microbiology, Vol. 31. J. of Indust. Micro., Suppl. No. 31:237-244 (1990).

Klapes, N.A.; Vesley, D.: Vapor-phase Hydrogen Peroxide as a Surface Decontaminant and Sterilant. Appl. Environ. Microbiol. 56: 503-506.

Klein, M.; DeForest, A.: Principles of Viral Inactivation. In: Disinfection, Sterilization and Preservation, 3rd Ed. S. Block, Ed. Lea & Febiger, Philadelphia, PA (1983).

Maroni, M.; Axelrad, R.; Bacaloni, A.: NATO's Efforts to Set Indoor Air Quality Guidelines and Standards. Am. Ind. Hyg. Assoc. J. 56:499-508 (1995).

National Sanitation Foundation (NSF): NSF 49: Class II (Laminar Flow) Biohazard Cabinetry. Ann Arbor, MI (1992).

Popendorf, W.; Selim, M.: Exposure while Applying Commercial Disinfectants. Am. Ind. Hyg. Assoc. J. 56:1111-1120 (1995).

Popendorf, W.; Selim, M.; Lewis, M.: Exposure while Applying Industrial Antimicrobial Pesticides. Am. Ind. Hyg. Assoc. J. 56:993-1001 (1995).

Price, D.L.; Simmons, R.B.; Ramey, D.L.; et al.: Assessment of Air Filters Treated with a Broad Spectrum Biostatic Agent. Indoor Air '93 Proceedings 6:527-532 (1993).

Samson, R.A.; Flannigan, B.; Flannigan, M.E.; et al., Eds.: Recommendations. In: Health Implications of Fungi in Indoor Environments, pp. 531-538. Elsevier, New York, NY (1994).

Semmelweiss, I.P.: Lecture on the Genesis of Puerperal Fever (Childbed Fever). In: Milestones in Microbiology, pp. 80-82. T. Brock, Ed. American Society for Microbiology, Washington, DC (1975).

Sesline, D.; Ames, R.G.; Howd, R.A.: Irritative and Systemic Symptoms Following Exposure to Microban Disinfectant through a School Ventilation System. Arch Environ Health. 49:439-444 (1994).

Simmons, R.B.; Crow, S.A.: Fungal Colonization of Air Filters for Use in Heating, Ventilating, and Air Conditioning (HVAC) Systems. J. Indust. Micro. 14:41-45 (1995).

Taylor, R.J.: The Sterilization of a New Building Designed for the Breeding of Specific-Pathogen-Free Animals. J. Hyg., Camb. 72:41-45 (1974).

USDHHS: Biosafety in Microbiological and Biomedical Laboratories, 3rd Ed. HHS Publication no. (CDC)93-8395. Public Health Service, Centers for Disease Control and Prevention, and National Institutes of Health. USGPO, Washington, DC (1993).

USEPA: Should You Have the Air Ducts in your Home Cleaned? United States Environmental Protection Agency, EPA-402-97-002. Washington, DC (1997).

WHO: Laboratory Safety Manual. World Health Organization, Geneva (1993).

Wickramanayake, G.B.: Disinfection and Sterilization by Ozone. In: Disinfection, Sterilization, and Preservation, pp. 182-190. S.S. Block, Ed. Lea and Febiger Publishers, Philadelphia, PA (1991).

Chapter 17

Source Organisms — An Overview

Harriet A. Burge and Janet M. Macher

17.1   Biologically Derived Airborne Contaminants
17.2   Classification of Biological Agents
17.3   The Biological Sciences
17.4   Studying Biological Agents
17.5   Organism Lifestyles and Metabolisms
       17.5.1   *Autotrophic and Heterotrophic Organisms*
       17.5.2   *Obligate Parasites*
       17.5.3   *Facultative Parasites*
       17.5.4   *Commensal Microorganisms*
       17.5.5   *Saprobic Microorganisms*
       17.5.6   *Phagotrophic Organisms*
17.6   Controlling Exposure to Bioaerosols

17.1 Biologically Derived Airborne Contaminants

ACGIH has adopted the phrase "biologically derived airborne contaminants" to describe bioaerosols (airborne particles composed of or derived from living organisms) and VOCs released from living organisms. Bioaerosols include whole microorganisms as well as fragments, toxins, and particulate waste products from all varieties of living things (e.g., bacteria, fungi, plants, and animals). This third part of the book describes in detail specific biological agents of concern in indoor environments. For a specific agent or group of agents, each chapter addresses organism characteristics, health effects, sample collection and analysis, and data interpretation. This introductory chapter provides an overall framework for this part of the book and introduces some general terms.

17.2 Classification of Biological Agents

There is no single system for classifying life forms, but Figure 17.1 represents the major kingdoms into which many biologists divide organisms. Biological classification systems are designed both for convenience in communicating about the natural world and for illustrating and understanding relationships among organisms. Classification systems may be artificial or natural. Artificial classifications are based on convenience. However, such systems may result in the grouping together of genetically unrelated organisms simply because they share some readily observed features. Natural systems are based on presumed evolutionary relationships among organisms, which are primarily deduced from their methods of sexual reproduction and, more recently, from their DNA structures.

Section 6.2 discusses classification and naming systems, as do the sections on *Characteristics* in the following chapters. Figure 17.2 gives examples of common bioaerosol sources or components for major groups of organisms of significance in indoor environments. Botanists and mycologists have traditionally used the term "divisions" for the major groups that zoologists have called "phyla."

17.3 The Biological Sciences

Biologists usually specialize in a single discipline. For example, microbiology, mycology, botany, and zoology are the respective studies of microorganisms (viruses, bacteria, and parasites), fungi, plants, and animals. Within these broad disciplines, further specialization is common. Although sometimes grouped as "microbiologists," virologists, bacteriologists, parasitologists, and mycologists receive different training. These and other biological sciences share basic common principles and use many of the same tools but, beyond this, a biologist's skills and knowledge are very specific. Microbiologists may specialize in the medical, manufacturing, or environmental aspects of their disciplines, and specialists may be unfamiliar with other areas of a shared general science. For example, a medical or public health microbiologist may be qualified to answer questions about infectious diseases, their causative agents, and sources of these agents. However, a medical microbiologist is less qualified to comment on the significance of many of the microorganisms found in indoor environments than an environmental bacteriologist or mycologist.

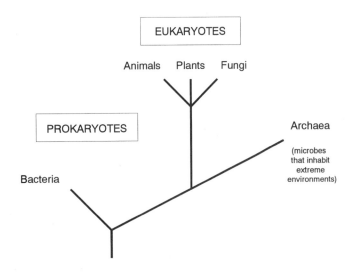

FIGURE 17.1. The biological kingdoms.

17.4 Studying Biological Agents

The conditions and events required to generate a bioaerosol are (a) the presence of a source of a biological agent, (b) multiplication of the original organism and, in some cases, manufacture of cell products, and (c) aerosolization and dissemination of the biological material into indoor air. These ideas echo the concepts of hosts, agents, and environments as well as sources, pathways, and receivers introduced in Sections 2.1.2 and 2.6.

Often, microbiologists must separate the members of a microbial community from each other and their environment to identify and study the individual components. However, it is often the interactions among microorganisms, plants, and animals and between them and their environments that are of interest. These interactions may be impor-

---

CLASSIFICATION
Kingdom > Phylum or Division > Class > Order > Family > **Genus** >**Species**

EXAMPLES	
Bacterium:	***Staphylococcus epidermidis*** (bacterium from healthy human skin)
Fungus:	***Cladosporium sphaerospermum*** (common saprobic fungus)
Plant:	***Ambrosia artemisiifolia*** (ragweed – source of pollen allergens)
Animal:	***Felis domesticus*** (cat – source of cat allergens)

FIGURE 17.2. Classification of life forms with examples
(in the examples, only the kingdoms, genera, and species are named).

---

tant to understand how microorganisms enter buildings, how they survive and multiply indoors, and what effect biological agents may have on humans. Using sampling or analytical methods that focus too narrowly on one or a few biological agents may lead investigators to overlook important ones [see 6.1.1].

Bioaerosols are often complex with respect to the size and nature of their individual components. For example, airborne skin scales may carry bacteria that are part of the normal human flora, and droplets released in coughs and sneezes may carry both bacteria and viruses as well as mucus and saliva from the nasal and oral regions. Water droplets from a contaminated humidifier may contain several types of bacteria and associated cell products, one or more types of fungal spores, mycotoxins, and dissolved antigens. Individual bioaerosols range from those that are submicroscopic (<0.1 µm) to particles greater than 100 µm in diameter. Table 17.1 lists some key characteristics of common bioaerosols and offers examples of organisms with different lifestyles.

17.5 Organism Lifestyles and Metabolisms
17.5.1 Autotrophic and Heterotrophic Organisms

Plants and some bacteria are *autotrophs*, that is, they synthesize complex organic molecules from inorganic raw materials. Plants are important (primarily outdoor) sources of airborne allergenic particles (e.g., pollen grains and plant fragments). Most bacteria, fungi, protozoa, and animals are *heterotrophs*, that is, they acquire organic nutrients from the environment. Bacteria and fungi lack internal digestive systems and feed mainly by absorption. These microorganisms may be saprobic (saprophytic—living and feeding on dead organic matter) or parasitic (living on or in other organisms and feeding on them) [see 17.5.2, 17.5.3, and 17.5.5]. Animals (e.g., protozoa, arthropods, birds, and mammals) are phagotrophs and feed by ingesting energy-rich, carbon-containing compounds from plants or other animals. Thus, organisms can be roughly classified as producers, decomposers, or consumers depending on how they obtain food (Figure 17.3). Organisms can be free living or can live together with others in symbiotic relationships. Symbiotes can both benefit (mutualism), one can benefit while the other receives little benefit or harm (commensalism), or one can benefit at the expense of the other (parasitism).

17.5.2 Obligate Parasites

Obligate parasites cannot reproduce outside a living host. Therefore, the reservoirs for parasitic microorganisms are other living organisms. All viruses, as well as a few bacteria and fungi, are obligate parasites and are only found in association with their hosts. Multiplication of an obligate parasite occurs within its host, from which the parasite is disseminated. For example, people are sources of many viruses that are spread by coughing and sneezing.

TABLE 17.1. Characteristics and Sources of Common Bioaerosols

Source Organism	Airborne Unit	Disease Agent	Health Effects	Organism Lifestyle[A]	Indoor Sources
Viruses	Organisms	Influenza virus	Respiratory infection	Obligately parasitic	Human hosts
Bacteria	Organisms	*Legionella pneumophila*	Pneumonia, inhalation fever	Facultatively parasitic	Cooling towers
		Mycobacterium tuberculosis	Respiratory infection	Obligately parasitic	Human hosts
	Spores	*Thermoactinomyces* allergen	HP	Saprobic	Hot water systems; hot, damp surfaces
	Cell fragments, cell products	Endotoxin	Fever, chills	—	Stagnant water reservoirs
		Allergens	Asthma	—	Industrial processes
Fungi	Organisms	*Sporobolomyces* allergens	HP	Saprobic	Damp environmental surfaces
	Spores	*Histoplasma capsulatum*	Systemic infection	Facultatively parasitic	Bird droppings
		Alternaria allergens	Asthma, rhinitis	Saprobic	Outdoor air, damp surfaces
		Aspergillus toxins: aflatoxin, ochratoxin, sterigmatocystin	Liver cancer	—	Damp surfaces supporting the growth of specific fungi
	Cell fragments, cell products	Allergens, glucans, toxins, MVOCs	Headaches, mucous membrane irritation	—	Damp surfaces
Protozoa	Organisms	*Naegleria fowleri*	CNS infection	Facultatively parasitic	Contaminated water reservoirs
	Cell fragments	*Acanthamoeba* allergens	HP	—	Contaminated water reservoirs
Algae	Organisms, cell fragments	*Chlorococcus* allergens	Asthma, rhinitis	Autotrophic	Outdoor air
Vascular plants	Pollen	*Ambrosia* (ragweed) allergens	Asthma, rhinitis	Autotrophic	Outdoor air
Arthropods	Fecal pellets, body parts	*Dermatophagoides* allergen	Asthma, rhinitis	Phagotrophic	Settled dust, house dust
Mammals	Dried skin scales, urine, saliva	Dogs, rodents, cats	Asthma, rhinitis	Phagotrophic	Animals, animal bedding
Birds	Dried excreta	Pigeon allergens	HP, asthma, rhinitis	Phagotrophic	Animals, nests, droppings

[A] Autotrophic: synthesizes complex organic molecules from simple ones
Facultatively parasitic: can use a living host as well as non-living organic matter (see saprobic)
Phagotrophic: obtains energy-rich, carbon-containing compounds by eating plants or animals
Obligately parasitic: requires a living host
Saprobic: externally digests non-living organic material

17.5.3 Facultative Parasites

Most bacteria and some fungi are facultative parasites and can survive and grow on or in living as well as dead or decaying organic matter [see 17.5.5]. Humans serve as reservoirs for some of these organisms (e.g., *Mycobacterium tuberculosis*). Other facultative parasites are found in environmental reservoirs, which can act as amplifiers and disseminators of biological materials. *Legionella* spp. are facultative parasites that reside in natural streams and soil. These bacteria often find their way into man-made water systems (e.g., cooling towers) where they multiply. People are exposed to these infectious agents by inhaling airborne bacteria disseminated from such sources. Healthy people may resist infection with facultative parasites, but factors such as smoking, chronic lung disease, and immuno-suppression put some persons at higher risk of respiratory infection.

17.5.4 Commensal Microorganisms

Commensal microorganisms have adapted to coexist with a host. Some biologists use this term to designate a symbiotic relationship that does not cause harm. In this book, the term commensalism is used as a distinction between parasitism (where one organism benefits but the other is harmed) and mutualism (where both organisms benefit) [see 18.1 and 19.1.4].

17.5.5 Saprobic Microorganisms

Saprobes use non-living organic material (e.g., dead or decaying plant or animal debris) as nutrient sources. The environmental reservoirs of saprobic microorganisms are usually located outdoors, associated with plants and the upper, aerated layers of soil. Most fungi, as well as many bacteria and protozoa, are saprobes. These organisms are responsible for the natural aerobic decay of organic materials and nutrient recycling.

Saprobes grow in and are disseminated from their outdoor reservoirs and may colonize indoor substrates if they find suitable environmental conditions there. The common fungus *Aspergillus versicolor* is an example of a saprobic organism. The reservoir of *A. versicolor* is in the outdoor environment, but this fungus can also multiply on indoor substrates (e.g., damp walls). Spores and cell fragments may become airborne when an organism or its substrate is disturbed.

17.5.6 Phagotrophic Organisms

Phagotrophs are free-living organisms that ingest food, which they process before absorption. Phagotrophic animals release glycoproteins in their fecal material, urine, skin secretions, saliva, and fragmented body parts. These proteins are often allergens. Protozoa, arthropods, and larger animals (e.g., birds and mammals) act as the reservoirs, amplifiers, and disseminators of their associated bioaerosols. For example, cats groom their fur by licking it, which transfers saliva to the hairs. When a cat scratches itself, it releases fine flakes of dried saliva and associated allergens. Amebae also produce allergens and, in addition, may cause infections. Dust mites, cockroaches, birds, and dogs are also sources of airborne allergens [see Table 25.2].

17.6 Controlling Exposure to Bioaerosols

People may have little control over bioaerosols in the outdoor environment. However, we generally can control bioaerosol presence or concentration indoors. Control may involve preventing entry of outdoor aerosols, avoiding conditions that allow organisms to multiply indoors, and cleaning or removing materials that have become contaminated [see Chapters 10 and 15]. To choose appropriate control and remediation measures, investigators must understand the nature, sources, and health effects of the various types of biological agents. In addition, investigators must know what collection and assay methods are available to identify biological agents and their sources and to measure bioaerosol concentrations. The following chapters provide such information for specific groups of biological agents.

AUTOTROPHS	HETEROTROPHS	
Producers	Decomposers (saprobes)	Consumers (phagotrophs)
Plants	Fungi	Animals
Some bacteria	Bacteria	Amebae

FIGURE 17.3. Lifestyle adaptations.

Chapter 18

Bacteria

James A. Otten and Harriet A. Burge

18.1 Characteristics
    18.1.1 *Morphology*
    18.1.2 *Classification*
    18.1.3 *Lifestyles and Metabolism*
    18.1.4 *Ecology*
        18.1.4.1 Outdoor Environments
        18.1.4.2 Indoor Environments
18.2 Health Effects
    18.2.1 *Infectious Diseases*
    18.2.2 *Hypersensitivity Diseases*
    18.2.3 *Toxic Effects*
    18.2.4 *Volatile Organic Compounds*
    18.2.5 *Beneficial Bacterial Products*
18.3 Sample Collection
    18.3.1 *Sampling Strategies*
    18.3.2 *Source Sampling*
    18.3.3 *Air Sampling*
    18.3.4 *Sample Handling*
18.4 Sample Analysis
    18.4.1 *Culture*
        18.4.1.1 Culture Media
        18.4.1.2 Incubation Conditions
        18.4.1.3 Traditional Culture Identification Methods
        18.4.1.4 Commercial Identification Kits
        18.4.1.5 Semi-Automated Systems for Rapid Bacterial Identification
    18.4.2 *Fluorescence Microscopy*
    18.4.3 *Other Analytical Methods*
18.5 Data Interpretation
    18.5.1 *Bacteria from Outdoor Air*
    18.5.2 *Human-Source Bacteria*
    18.5.3 *Gram-Negative Bacteria*
    18.5.4 *Significant or Substantial Health Risk*
18.6 Remediation and Prevention
18.7 References

18.1 Characteristics

18.1.1 Morphology

Bacteria are single-celled prokaryotic organisms that have cell membranes but not organized nuclei or membrane-bound organelles. Most bacteria take the form of very small spheres—cocci; straight or curved rods—bacilli; spiral rods—spirochetes; or branched filaments—actinomycetes (Figure 18.1). Bacteria are generally 1 to 5 μm in diameter, although some are narrower or wider and some rods are quite long. Bacteria are unicellular and reproduce by simple cell division. After division, bacterial cells may remain grouped in pairs, tetrads, or chains. The actinomycetes (e.g., *Nocardia, Streptomyces, Thermoactinomyces,* and *Faenia* spp.) form long, branching cell chains that mimic fungi but are distinguishable by their smaller size. Some bacteria produce endospores (i.e., internal spores) that are extremely resistant to environmental stresses. A bacterial endospore is a dormant structure capable of surviving harsh conditions for long periods. Endospores can reestablish the vegetative form of a bacterium under appropriate circumstances. Actinomycetes form true spores, which they commonly release into the air. Other bacterial spores (e.g., the endospores of *Bacillus* spp.) may become airborne when contaminated material is disturbed.

18.1.2 Classification

The true bacteria form a kingdom with thousands of known, living species. Bacteria are classified by Gram-stain reaction, cell shape (Figure 18.1), oxygen requirement, spore production, substrate utilization, metabolite production, and ability to cause disease (Holt et al., 1994) [see 6.2 and 18.4]. Table 18.1 lists some bacteria commonly found in indoor air as well as important airborne pathogens, classifying them according to their morphology.

The Gram stain reveals cell shape and size and separates bacteria into two groups based on the structure and chemistry of their cell envelopes (Chapin, 1995). Staining depends on cell-wall thickness, surface pore size, and cell envelope permeability. Two stains are used, one of which only Gram-positive bacteria (GPB) retain (i.e., crystal violet). A second stain, saffranin, is used to stain Gram-negative bacteria (GNB), which otherwise would not be observable. Therefore, GPB appear blue or purple, and GNB appear pink to red. Nearly all of the cocci that are recovered from air are Gram positive (e.g., *Micrococcus* spp.) as are approximately half the rods (e.g., *Bacillus* spp.). Airborne GNB are usually rod shaped (e.g., *Pseudomonas* spp.).

Some bacteria (including *Mycobacterium* spp.) are Gram positive but their cell walls contain large amounts of lipid that prevent the uptake of many stains. The acid-fast stain, an alternative procedure to the Gram stain, is used to identify these organisms (Chapin, 1995). Fluorescent Gram-stain kits are now commercially available, as are reagents that differentially stain live and dead organisms (Hensel and Petzoldt, 1995; Madelin and Madelin, 1995; Morris, 1995; Terzieva et al., 1996; Lawrence et al., 1997). Many bacteria, including the mycobacteria, stain readily with fluorescent dyes that bind with specific surface proteins.

18.1.3 Lifestyles and Metabolism

Most bacteria are free living and contain the necessary genetic information as well as energy-producing and biosynthetic systems for growth and reproduction. Bacteria are able to use diverse food sources, ranging from completely inorganic substrates to complex organic compounds [see 17.5]. Most of the bacteria commonly isolated from air are saprobic and utilize non-living, complex organic matter [see Table 17.1]. A few bacteria are autotrophic and can synthesize organic compounds from inorganic constituents, although not as efficiently as the plants. A few bacteria (e.g., rickettsiae, chlamydiae, and ehrlichiae) are obligate parasites, requiring living hosts for whom they are pathogenic. Some saprobic bacteria are opportunistic pathogens, invading living hosts only under unusual circumstances, that is, when a host's immune system is not functioning well.

Bacteria have adapted to a wide variety of ecological habitats with extremes in temperature, acidity, and oxygen tension. Bacteria metabolize carbon compounds in the presence or absence of oxygen, that is, aerobically or anaerobically, respectively. Many bacteria can grow under a wide range of oxygen tensions and are termed facultative anaerobes.

Bacteria are divided into four groups on the basis of their preferred temperature range (Table 18.2). Although the terms for the bacterial groups are similar to those used to described fungal temperature requirements, the temperature ranges defining the terms differ [see 19.1.5.3]. Most bacteria that are recovered from indoor or outdoor air will grow across

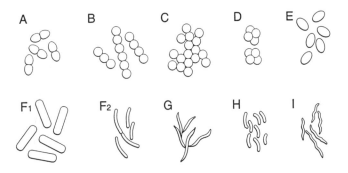

FIGURE 18.1. Cell shapes and arrangements for common bacterial forms.

A. diplococci, B. streptococci, C. staphylococci, D. sarcinae (tetrads), E. coccobacilli, F1. bacilli, F2. slender bacilli, G. filamentous bacilli, H. vibrios, I. spirilla.

a broad temperature range that may span several categories [see 6.3.3.1 and 18.4.1.2]. Human pathogens are generally mesophiles, while thermophilic bacteria inhabit very warm environments (e.g., hot springs and compost piles). Thermophilic organisms are often obligate thermophiles and will only reproduce at high temperatures. This means that air samples collected to assess the presence of obligate thermophiles must be incubated at or above 50°C. Many saprobic bacteria can be collected from the air and other environments and grown in culture in the laboratory—providing the organism's physical and nutritional requirements are met [see 18.4.1].

18.1.4 Ecology

Different physical and chemical conditions in indoor and outdoor environments segregate microorganisms into habitats. Familiarity with natural bacterial habitats helps investigators understand which disease-causing bacteria may be found in particular environments. Bacteria are abundant in air, water, and soil; in and on animals; on plant surfaces; and on man-made surfaces both indoors and outdoors. Bacteria may become airborne from any of these natural reservoirs.

18.1.4.1 Outdoor Environments Most bacteria in the outdoor air were released from the surfaces of living plants. Many of these are rod-shaped GPB and GNB such as species of *Achromobacter, Actinomycetes, Bacillus, Corynebacterium, Flavobacterium, Micrococcus, Pseudomonas, Sarcina,* and *Staphylococcus.* Water may also be a significant source of bacterial aerosols near oceans, lakes, ponds, and streams. The bacteria responsible for Legionnaires' disease (*L. pneumophila* and other species) are found in natural waters and soil. Some industries probably emit bacteria from ventilation exhausts (e.g., operations that use water-based coolants or recycled wash

water for industrial processes). Except for resistant spore stages, airborne bacteria typically survive for only a short time outdoors primarily due to the "open air factor." This effect is considered to be due to a mixture of photo-oxidants in ambient air (Cox, 1995; Marthi, 1994). Section 18.5.1 discusses interpreting the significance of the indoor presence of bacteria that arise from outdoor sources.

18.1.4.2 Indoor Environments The indoor air of occupied buildings typically contains higher concentrations and more types of bacteria than outdoor air. In non-manufacturing environments, the majority of bacteria in air are shed from human skin and respiratory tracts. These human-source or human-commensal bacteria include Gram-positive *Micrococcus, Staphylococcus,* and *Corynebacteria* spp. (from healthy human skin and scalp) along with *Streptococcus* spp. (from oral and nasal secretions).

Talking, coughing, and sneezing release progressively more oral and respiratory droplets, many of which contain bacteria and viruses. Very large droplets travel only a short distance (≤1 m) before falling from the air. A few diseases are transmitted primarily via these large droplets and are considered among those infections contracted via direct contact [see 9.4.1.1]. Smaller droplets released from the nose and mouth as well as from aerosol-generating equipment contain water, salts, proteins, cell debris, and, occasionally, suspended microorganisms or other biological agents. These fine particles dry rapidly—leaving droplet nuclei that may remain airborne long enough to travel throughout an indoor space [see 9.4.1.2]. Section 18.5.2 discusses interpreting the significance of the indoor presence of bacteria from humans.

Humidifiers may also introduce bacteria into the air as may any water spray, especially those that use re-circulated water or water from stagnant reservoirs. Indoors, *Legionella* spp. can colonize warm to hot water

TABLE 18.1. Classification of Bacteria that are Common in Indoor Air or are Agents of Airborne Disease

Organism	Gram Reaction	Cell Shape	Spores	Disease	Occurrence in Air Outdoors	Indoors[A]
Micrococcus spp.	+	Cocci	No	Usually none	Common	Common
Bacillus spp.	+	Rods	Yes (endospores)	HP	Common	Common
Thermoactinomyces spp.	+	Filamentous	Yes	HP	Uncommon	Rare
Corynebacterium spp.	(+)	Rods	No	Usually none	Rare	Occasional
Mycobacterium tuberculosis	(+)	Irregular rods	No	Tuberculosis	Rare[B]	Rare
Pseudomonas spp.	−	Rods	No	Usually none	Common	Rare
Enteric bacteria	−	Rods	No	Various	Rare	Rare
Legionella pneumophila	−	Rods	No	Legionellosis	Rare	Rare

[A] Background or ambient occurrence. Presence in problem buildings may differ.
[B] Environmental *Mycobacterium* spp. may be fairly common in outdoor air.

systems, living in biofilms that develop on surfaces in contact with water. A biofilm is a community of organisms—usually dominated by bacteria but also including protozoa and fungi—that live attached to some surface within a polysaccharide layer the bacteria have secreted (Marshall, 1997). Sections 10.4 and 10.5 discuss microbial growth in indoor environments. Legionellae also live symbiotically with some amebae [see 20.1.4].

18.2 Health Effects

Most naturally occurring bacteria do not cause human illness or other complaints. In fact, many bacteria are essential to human health (i.e., the commensal organisms that occupy human skin, the respiratory tract, gut, and so forth) and to the earth's ecology [see 14.1.4.5]. The risk of illness from environmental bacteria increases only when they enter buildings in inappropriate numbers or multiply indoors [see also 18.5.4].

18.2.1 Infectious Diseases

Bacteria are best known as agents of infectious disease. Bacterial infections are acquired by a variety of mechanisms, including ingestion of food and water, direct (hand) transfer of respiratory secretions or fecal material, inoculation via foreign bodies and animal vectors, as well as direct contact with the skin and mucosa via sprays or splashes. Inhalation is the primary transmission route of interest for bacterial aerosols [see 9.4].

Infectious aerosols may be released from people (e.g., *M. tuberculosis*), from environmental reservoirs (e.g., *Legionella* spp.), or from animals (e.g., the agents of anthrax, brucellosis, psittacosis, Q-fever, and tularemia). For the most part, the bacteria that comprise normal human flora do not cause infectious disease. However, some healthy persons carry potentially pathogenic bacteria (e.g., *Staphylococcus aureus*) that can be transmitted to susceptible persons (i.e., those with some level of immune incompetence). *S. aureus* in the air of houses, offices, and schools is normal and is usually not considered a health risk in these environments. However,

this bacterium may cause severe problems when found in hospitals, especially in surgical suites.

Some bacteria are obligate intracellular parasites and are generally handled in virology laboratories (e.g., chlamydia, rickettsia, and related bacteria). These bacteria are responsible for several important airborne infections (e.g., Q-fever from *Coxiella burnetti*, respiratory disease from *Chlamydia pneumoniae*, and psittacosis from *Chlamydia psittaci*).

18.2.2 Hypersensitivity Diseases

The first cause of farmer's lung disease (a form of HP) to be described was *Faenia rectivirgula*, an actinomycete formerly called *Micropolyspora faeni*. This and other thermophilic actinomycetes, when present in high concentrations, have the potential to cause HP in any exposed person. *Bacillus* spp. have also been implicated in HP outbreaks, and proteases from these bacteria used in detergent manufacture have been linked with several HP outbreaks (Dolovich and Little, 1972; IOM, 1993).

HP associated with exposure to GNB has also been reported, although it may be that endotoxin from these bacteria acted as an adjuvant (a stimulant to the immune system) rather than the GNB itself functioning as the responsible antigen [see Chapter 23]. An association between IgE-mediated immediate hypersensitivity diseases, such as asthma and hay fever, has not been clearly documented, although, when other antigens are present, a GNB may function as an adjuvant.

18.2.3 Toxic Effects

Bacteria produce both exotoxins and endotoxins (Table 18.3). *Exotoxins* are proteins a bacterium produces internally and secretes into its environment. The exotoxin produced by the anaerobic bacterium *Clostridium botulinum* is one of the most powerful poisons known and is responsible for often-fatal food and wound botulism. Exotoxins have not been reported in air.

Endotoxins are lipopolysaccharides that are a part of the cell wall of GNB. Low exposures to endotoxins

TABLE 18.2. Preferred Temperature Ranges for Bacteria Found in the Indoor Environment

Classification	Temperature Range	Example
Psychrophilic bacteria	0 to 10°C	*Bacillus psychrophilis*
Mesophilic bacteria		
Environmental bacteria	18 to 30°C	*Pseudomonas* spp.
Human-source bacteria	35 to 44°C	*Staphylococcus epidermidis*
Thermophilic bacteria	50 to 55°C	*Thermoactinomyces* spp.

stimulate the immune system and are probably essential to its appropriate development. At higher levels, endotoxins act as irritants. Very high inhalation exposure to endotoxins can cause a flu-like illness that includes fever and difficulty breathing. Some cases of humidifier fever have been attributed to inhalation exposure to GNB, although it is not clear whether the disease results from exposure to endotoxins alone or from the adjuvant characteristics of endotoxins acting in conjunction with other allergens (IOM, 1993) [see also 18.2.2]. Chapter 23 discusses these effects as well as possible reactions to exposure to other bacterial cell-wall components (e.g., peptidoglycans).

18.2.4 Volatile Organic Compounds

Bacteria produce VOCs while growing and degrading substrates. Some of these compounds have distinctive odors and low odor thresholds [see 26.2.2.2]. For example, some body odors derive from bacteria that grow on human skin and use perspiration as a substrate. People may associate malodors with wastes or spoilage and find them offensive or annoying. Exposure to such compounds may be responsible for some nonspecific BRSs (Gerber 1979). However, there is little evidence that bacterial VOCs cause specific health effects (Batterman, 1995) [see 26.3].

18.2.5 Beneficial Bacterial Products

Some bacteria produce antibiotics (e.g., streptomycin from *Streptomyces* spp.). In nature, these compounds may convey an advantage for growth or survival. Many antibiotics are selectively toxic to prokaryotic cells, directly killing bacterial cells or preventing their reproduction. Bacteria are also commonly used to make certain foods (e.g., cultured milk products and vinegar) and in biotechnology and agriculture.

18.3 Sample Collection

18.3.1 Sampling Strategies

The most successful sampling strategies are those formulated to answer specific questions. For example, people are the source of most bacteria found in non-

manufacturing, indoor environments, and people's activities affect bioaerosol release. Sampling indoors and outdoors, with and without people present, during periods of high and low occupant activity, and with the ventilation system on and off can identify important patterns in bacterial aerosol concentrations [see 5.2.1]. Sampling potential sources directly followed by active disturbance of a potential source during air sampling can document the contributions of specific sites to the numbers and types of airborne bacteria. The primary assay for identifying and quantifying bacteria in environmental samples remains culture isolation. Therefore, this chapter discusses primarily sample collection to assess air and source concentrations of culturable bacteria. Chapter 23 on endotoxin and other bacterial cell-wall components and Chapter 26 on MVOCs discuss sampling for other bacterial products.

18.3.2 Source Sampling

Bacteria growing in the environment seldom form the discrete, visible colonies seen in the laboratory. However, the presence of slime, biofilm, or foam suggests the presence of bacteria. In addition, fruity, sour, or putrid odors are often indicators of bacterial growth. Source samples (bulk or surface samples) are useful to determine the types of bacteria present. Typical source samples are (a) fluids from water reservoirs (e.g., humidifiers, cooling towers, and poorly drained drip or condensate pans), (b) pieces of water-damaged materials (e.g., carpeting, wall coverings, and other building materials and furnishings), and (c) samples of biofilms and sludges [see Chapter 12].

Clinical specimens (e.g., sputum from patients suspected of having active tuberculosis or throat swabs from patients with pharyngitis) represent another type of source sample. These samples are used for infectious disease diagnosis, to determine if a person represents a potential source of infection for others, and to establish that transmission of infectious bacteria from an environmental source to a person has occurred. Source samples are especially useful for fragile and fastidious

TABLE 18.3. Bacterial Toxins

Characteristics	Exotoxins	Endotoxins
Source bacteria	GPB and GNB	GNB
Toxin secreted or produced as a structural component	Secreted	Structural component
Chemical nature of toxin	Protein	Lipopolysaccharide
Toxin stability to heat	Usually labile	Stable
Biological activity of toxin	Unique to each exotoxin	Similar for all endotoxins

organisms (e.g., *Legionella* spp.) (Breiman et al., 1990; CDC, 1994; AIHA, 1996). Source samples are also preferable for bioaerosols likely to be present in low concentrations (e.g., *M. tuberculosis*) for which air sampling is rarely satisfactory [see 9.6].

18.3.3 Air Sampling

Bacteria vary widely in their ability to survive collection from the air and to grow under laboratory conditions following collection. Therefore, an understanding of a target bacterium's physiological and morphological characteristics is important both in the decision to undertake air sampling and in air sampler selection. In general, GPB (especially spore-formers) are better able than GNB to survive the rigors of the airborne state and of sample collection. Several researchers have studied the effects of collection on bacterial recovery (Juozaitis et al., 1994; Stewart et al., 1995; Terzieva et al., 1996). Methods that expose bacterial cells to drying (e.g., collection on filters or in dry cyclones) are generally not suitable if culture analysis is to be used (Jensen, et al., 1992). However, these methods may be appropriate for other analytical approaches [see 6.4, 6.5, 6.6, and 6.7].

All of the samplers listed in Table 11.1 could theoretically be used to collect bacteria. However, direct-agar impactors are most frequently chosen because of their convenience (AIHA, 1996) [e.g., slit samplers (Sampler 1), multiple-hole impactors (Samplers 2 to 8), and centrifugal samplers (Sampler 10)]. Collection directly onto agar is generally limited to environments where the air concentration of culturable microorganisms is relatively low (e.g., <10^4 organisms/m³). Collection into a liquid provides a sample that is readily diluted for culture and other assays [e.g., collection with a wetted cyclone (Sampler 11), tangential impactor (Sampler 13), or liquid impinger (Samplers 14 and 15)] (AIHA, 1996). Therefore, concentrated aerosols that would overload direct-agar impactors and samples for analyses that require delivery of the test material in a liquid can be successfully sampled with cyclones, impactors or impingers containing appropriate collection fluids.

18.3.4 Sample Handling

Handling begins as soon as a sample is collected and is of particular concern for bacterial analysis. The abrupt change from environmental conditions may cause cells either to die (and thus be undetectable by culture analysis) or to multiply (and thus be over-represented regardless of the analytical method). Collected bacteria may begin to divide following a period of rest and repair (e.g., bacteria collected on agar-based media or in a collection fluid as well as source samples inoculated onto agar-based media in the field immediately after collection). The length of this period depends on the bacterium but

may be as short as a few hours. Growth on agar plates before a sample reaches the laboratory is acceptable if bacteria that multiply at room temperature are of interest. However, if the plan is to incubate plates at 56°C to isolate thermophilic actinomycetes, the growth of mesophilic microorganisms during transport may interfere with subsequent growth of the bacteria of interest. Therefore, sample plates should reach the laboratory within 24 hours of collection and should be protected from extreme temperatures during transit.

Liquid samples (collected either directly from sources or from air) present a more serious transport problem. If individual cells reproduce during transit, there is no way to know how many cells were present in the original sample. For culture analysis, such samples must reach the laboratory within 24 hours of collection, be kept cool (~4°C) during transit, and be processed on the day they reach the laboratory. A biocide may be added to liquids to prevent microbial growth if an analysis does not require that the cells be alive. The chosen biocide must be compatible with any subsequent assays. Three-percent formalin is often used when bacteria will be counted using fluorescence microscopy.

18.4 Sample Analysis

Isolation of bacteria by laboratory culture remains the primary method of environmental sample analysis, although many other approaches are also used (Jensen et al., 1994; Hensel and Petzoldt, 1995; AIHA, 1996; Buttner et al., 1997) [see 6.3]. Because of the complex effects of environmental conditions on bacterial growth, laboratory reports should always specify the methods of sample preparation as well as the culture medium and incubation conditions used so that reviewers can accurately interpret the findings. Section 6.3.1 briefly describes how laboratories process samples and how concentrations of culturable bacteria are calculated for air and source samples.

18.4.1 Culture

Bacterial colonies that grow on the surface of agar-based culture media are clusters of cell progeny resulting from repeated division of one original cell or a small cluster of cells. Because it is seldom known if the original particle was a single cell or consisted of multiple cells, each colony is reported as a colony-forming unit (CFU). If the original particle contained only one type of organism, the resulting colony is a collection of virtually identical cells. However, if more than one microorganism was present, the resulting colony may include more than one type of bacterium (and possibly fungus). Such impurity can make subsequent identification difficult. Bacterial counts are frequently reported as total colony counts or as separate total counts of GPB and GNB. Even when culturable bacteria are identified to the genus and species levels, data may be difficult to interpret for the reasons discussed in Section 6.1.2.

18.4.1.1 Culture Media Culture media for bacteria are selected either to isolate as many culturable types as possible or to select specific bacterial groups. General purpose media are used for source and air samples in routine environmental investigations [e.g., nutrient agar or soybean-casein digest agar (SCDA, also known as tryptic or trypticase soy agar, TSA)] (Table 18.4). Although analyses based on culture always underestimate the actual concentrations of bacteria, some media are better suited for environmental samples than others. For example, a low-nutrient medium, such as R2A, may allow damaged cells to repair themselves, ultimately yielding higher counts than richer media such as nutrient agar and SCDA.

Special purpose media are used to select specific bacteria, usually by preventing the growth of others or by supplying special nutrients [see 6.3.2]. Selective media generally underestimate even the organisms for which they are intended. For example, MacConkey's agar allows growth of GNB and is used to select these and to suppress competition from often more abundant GPB. However, concentrations of GNB on MacConkey's agar are likely to be lower than they would be on a less stressful medium. Samples for estimating legionella concentrations must be plated on a medium with special nutrients (i.e. amino acids as carbon sources rather than sugars). However, the growth of other environmental bacteria, which are often more abundant than the legionellae, must be controlled. Therefore, environmental water samples are often heat or acid treated before plating (CDC, 1994). Such treatment makes estimation of legionella concentrations questionable because an unknown percentage of these bacteria are also likely to be killed. Similar cautions apply to the addition of antibiotics and other growth-restricting compounds to culture media, because these agents may also inhibit the target organisms to some degree.

18.4.1.2 Incubation Conditions Different incubation temperatures are used for bacteria with distinct temperature requirements [see 18.1.3]. Therefore, different bacteria may be recovered from a single sample depending on the incubation conditions used. Clinical microbiology laboratories typically focus on isolation of human pathogens. Therefore, these laboratories use relatively rich media and an incubation temperature of 35° to 37°C, similar to human body temperature. However, many environmental bacteria prefer lower temperatures (e.g., 20° to 30°C). Thermophilic bacteria (including some actinomycetes) require temperatures above 50°C. These temperature requirements can sometimes be used to selectively isolate specific bacteria. For example, thermophilic and facultatively thermophilic bacteria that can tolerate temperatures as high as 40 to 45°C can be detected among large populations of strict mesophiles by incubating samples at an elevated temperature. In addition to culture medium and incubation temperature, the atmosphere provided (i.e., air or 2.5% to 10% CO_2 in air) also affects bacterial growth in the laboratory.

Some bacteria can divide as often as every 20 min. Therefore, bacteria that grow well at 35° to 37°C often form visible colonies within 24 to 48 hours. These include most human-source organisms. Bacteria collected directly from environmental sources and incubated at 20°C grow more slowly but usually form visible colonies within 3 to 5 days. Other bacteria may require many days of incubation. For example, *M. tuberculosis* requires incubation for at least 10 to 14 days at 37°C before visible colonies appear. Laboratory personnel must take precautions to prevent media from drying when holding plates for long periods or at elevated temperatures (e.g., enclosing plates in special plastic bags).

TABLE 18.4. Culture Media for Bacteria (ingredients per liter of distilled water)
(Nash and Krenz, 1991; APHA, 1995)

Nutrient Agar		R2A Agar		
Beef extract	3 g	Yeast extract	0.5	g
Peptone or pancreatic		Proteose peptone or polypeptone	0.5	g
digest of gelatin	5 g	Casamino acids	0.5	g
Agar	15 g	Glucose	0.5	g
Final pH 6.8		Soluble starch	0.5	g
		Dipotassium hydrogen		
Soybean-Casein Digest Agar		phosphate, K_2HPO_4	0.3	g
Pancreatic digest of casein	15 g	Magnesium sulfate		
Papaic digest of soy meal	5 g	heptahydrate, $MgSO_4 \times 7\ H_2O$	0.05	g
Sodium chloride	5 g	Sodium pyruvate	0.3	g
Agar	15 g	Agar	15.0	g
Final pH 7.3		Final pH 7.2		

18.4.1.3 Traditional Culture Identification Methods
Microbiologists use various approaches to identifying isolates, but the most common method involves using a branching key or flowchart, such as the dichotomous one illustrated in Figure 18.2. Microbiologists use other charts to identify other groups of bacteria (e.g., aerobic rods).

Gram and Acid-Fast Stains. A specialized stain (such as the Gram stain) is generally used as a first step in categorizing bacterial isolates (Figure 18.2) [see 18.1.2]. However, performance and interpretation of the Gram stain are somewhat subjective. Some organisms do not stain well or may easily lose a stain, sometimes due to age or the physiological condition of a culture. Mycobacteria and related bacteria require special staining techniques, such as the acid-fast stain.

Isolation of Bacteria in Culture. Isolation and identification of bacteria are time-consuming tasks that require repeated transfers and manipulations of cultures. After Gram-staining, isolates from initial platings are subcultured to a variety of liquid or agar-based media containing nutritional and enzymatic substrates chosen to further separate the members of large bacterial groups. Microbiologists observe and record the reactions of an isolate on selected substrates and compare the results to key characteristics for specific taxonomic groups (Table 18.5). If needed, further tests are performed until a presumptive identification can be made. For example, from the information in Table 18.5, the unknown legionella most closely matches *Legionella micdadei*.

A bacterium in culture (especially one recently isolated from an environmental sample) can show great variability in its reactions. Reactions depend on the composition of the growth medium (which can vary from batch to batch and from manufacturer to manufacturer), inoculum size, atmosphere, aeration, temperature, and incubation time. In addition, judging the nature of a reaction on a given substrate is often subjective and experience in the range of reactions that may be seen is valuable. From Figure 18.2 and Table 18.5, it should be clear that an error on even one test may lead to an incorrect identification or cause confusion in a subsequent series of tests because the chosen set of tests was inappropriate for the organism [see also 13.6.4].

18.4.1.4 Commercial Identification Kits In an effort to standardize some of the more widely used identification tests, commercially prepared kits have been developed and packaged for differentiation within certain bacterial groups. Most of these kits were developed for use in clinical laboratories, which must rapidly screen and identify large numbers of potential pathogens for patient diagnosis. Typically, these kits contain various media that react with bacterial products to show a color change. Examples include the Enterotube for screening enteric bacteria and the Minitek System for differentiating

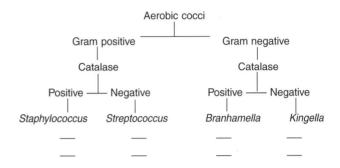

FIGURE 18.2. Sample flowchart for identification of aerobic, coccoid bacteria.

enterics and anaerobes (BBL, Cokeysville, Md.). However, laboratory personnel must remember that kits developed for clinical isolates are not appropriate for some environmental bacteria for the reasons discussed in Section 6.1.2.

18.4.1.5 Semi-Automated Systems for Rapid Bacterial Identification The discipline of bacterial taxonomy has benefited greatly by recent technological developments in the areas of biochemistry and molecular biology (e.g., carbon source and fatty acid analyses). The tedious task of manipulating large numbers of cultures has been somewhat simplified by the development of semi-automated systems for rapid bacterial identification. Several such systems are now commercially available and are finding increasing application in non-clinical laboratories to analyze large numbers of environmental samples.

18.4.2 Fluorescence Microscopy
Total bacterial counts can be made using fluorescence microscopy after cells deposited on filters are stained (APHA, 1989, 1995; Chapin, 1995) [see 6.4]. Such total bacterial counts may allow estimation of the reliability of culture data. Specific bacteria (both living and dead) can often be directly counted after staining with corresponding antibodies labelled with a fluorescent dye. Fluorescent antibody methods have been used to detect *L. pneumophila* in environmental samples as well as to establish serotypes, provided appropriate serotype-specific antibodies are available (Palmer et al., 1995; Fields, 1997).

18.4.3 Other Analytical Methods
Chapter 6 provides information and references on other techniques for identifying and quantifying bacteria from air and source samples [see 6.5, 6.6, and 6.7]. Sections 23.4 and 23.7.3 describe analytical methods for measuring endotoxin and other bacterial cell-wall components.

18.5 Data Interpretation
Data from appropriately designed sampling studies should allow investigators to evaluate hypotheses regarding a

study site. Few guidelines or standards are available to interpret bacterial aerosol data [see 7.3.2 and 14.2.6]. The information obtained may help investigators answer, completely or partially, the following types of questions:

1. Are environmental bacteria multiplying in a building?
2. If so, what is their source?
3. Are human-source bacteria present at concentrations that should cause concern?
4. Do the types and concentrations of bacteria present indicate a risk to health?

18.5.1 Bacteria from Outdoor Air

Some bacteria that are common in outdoor air may penetrate to building interiors and may also grow indoors [see also 7.4.2.1, 10.5.4 and 18.1.4.1]. Decisions on whether identification of bacteria and bacterial agents in indoor air represents simple entry with outdoor air or actual growth in the indoor environment depend on the relative indoor and outdoor concentrations. The dominance of a single species may reflect indoor growth, whereas the presence of mixed bacterial species may reflect intrusion of outdoor air or shedding from building occupants. Actinomycetes (especially thermophilic ones) are rare in buildings and outdoors. Therefore, the presence of actinomycetes indoors may be considered an indication of an indoor source.

18.5.2 Human-Source Bacteria

Human-source bacteria typically dominate indoor air in occupied buildings and are abundant in dust and on surfaces (Burge, 1995) [see 7.4.2.1 and 18.1.4.2]. Air concentrations of these bacteria depend on the number of persons present, their activity levels, the types of clothing they wear, and the ventilation rate to the space. Thus, concentrations of human-source bacteria in a space occupied only by the investigator collecting samples should be quite low ($<100/m^3$). In contrast, air concentrations may be very high ($>5000/m^3$) in a schoolroom where many lightly-clad children are actively playing. Nevertheless, both measurements may be considered "normal." Extremely high concentrations of human-source bacteria in low-activity spaces may indicate overcrowding or inadequate ventilation. Such suspicions can be confirmed by measuring CO_2 concentration and ventilation rate. Recommendations may involve decreasing the number of persons in a space as well as increasing or redistributing ventilation air.

18.5.3 Gram-Negative Bacteria

A predominance of GNB in an indoor environment may suggest the presence of sources that would be considered unusual in clean, properly ventilated buildings. For example, oxidase-negative, glucose-fermenting GNB typical of fecal contamination may be found where plumbing has leaked or toilet exhausts are improperly vented. However, finding a few of these bacteria mixed with normal human-source organisms is not a cause for concern. Oxidase-positive GNB are often found in standing water. These bacteria are usually members of the genus *Pseudomonas* or related genera and are a concern if they produce odors or can become aerosolized. Dominance of these bacteria in air may suggest the presence and aerosolization of standing water.

18.5.4 Significant or Substantial Health Risk

The question of what constitutes a serious health risk is difficult to assess for any bioaerosol [see 18.2]. Some information is available on concentrations of infectious bacteria that have been associated with disease outbreaks. For example, elevated concentrations of *Legionella* spp. in cooling towers have been linked epidemiologically with disease outbreaks (Shelton et al., 1994; Fields, 1997). Likewise, tuberculosis transmission has been related to the parameters discussed in Section 9.5. However, little is known about the potential effects for healthy adults and children of exposure to the majority of bacteria recovered during routine air and source sampling. Often, it is only possible to establish that an unusual exposure situation does or does not exist with respect to a control environment [see 14.2].

TABLE 18.5. Example: Species Identification of an Unknown Legionella from an Environmental Water Sample

	Unknown legionella	L. jordanis	L. bozemanii	L. micdadei
Oxidase	+	+	−	+
Catalase	+	+	+	±
Gelatinase	+	+	+	+
b-lactamase	−	+	+	−
Hippurate hydrolysis	−	−	−	−

18.6 Remediation and Prevention

Readers are directed to (a) Chapter 10 for a description of measures to control indoor bacterial growth [see 10.2, 10.3, 10.4, 10.5, and 10.6], (b) Chapter 15 for a discussion of cleaning and remediation of existing contamination [see 15.2, 15.3, 15.6, and 15.7], and (c) Chapter 23, for recommendations on limiting exposures to endotoxin and other bacterial cell-wall components [see 23.6].

18.7 References

AIHA: Viable Fungi and Bacteria in Air, Bulk, and Surface Samples. In: Field Guide for the Determination of Biological Contaminants in Environmental Samples, pp. 37-74. American Industrial Hygiene Association, Fairfax, VA (1996).

APHA: Standard Methods for the Examination of Water and Wastewater, 17th Ed. American Public Health Association, American Water Works Association and Water Pollution Control Federation, Washington, DC (1989).

Batterman, SA: Sampling and Analysis of Biological Volatile Organic Compounds. In: Bioaerosols, pp. 249-268. H.A. Burge, Ed. Lewis Publishers, Boca Raton, FL (1995).

Breiman, R.F.; Cozen, W.; Fields, B.S.; et al.: Role of Air Sampling in Investigation of an Outbreak of Legionnaires' Disease Associated with Exposure to Aerosols from an Evaporative Condenser. J. Infect. Dis. 161: 1257-1261 (1990).

Burge, H.A.: Bioaerosols in the Residential Environment. In: Bioaerosols Handbook. C.S. Cox and C.M. Wathes, Eds. CRC Press Inc., Boca Raton, FL (1995).

Buttner, M.P.; Willeke, K.; Grinshpun, S.A.: Sampling and Analysis of Airborne Microorganisms. In: Manual of Environmental Microbiology, pp. 629-640. C.J. Hurst, G.R. Knudsen, M.J. McInerney, et al., Eds. American Society for Microbiology Press, Washington, DC (1997).

CDC: Procedures for the Recovery of Legionella from the Environment. Centers for Disease Control and Prevention, Atlanta, GA (1994).

Chapin, K.: Clinical Microscopy. In: Manual of Clinical Microbiology, 6th Ed., pp. 33-51. P.R. Murray, Ed. American Society for Microbiology, Washington, DC (1995).

Cox, C.S.: Stability of Airborne Microbes and Allergens. In: Bioaerosols Handbook, pp. 77-99. C.S. Cox and C.M. Wathes, Eds. Lewis Publishers, Boca Raton, FL (1995).

Dolovich, J.; Little, D.C.: Correlates of Skin Test Reactions to Bacillus subtilis Enzyme Preparations. J. Allergy Clin. Immunol. 49(1):43-53 (1972).

Eaton, A.D.; Clesceri, L.S.; Greenberg, A.E., Eds.: Standard Methods for the Examination of Water and Wastewater, 19th Ed., pp. 9-110 to 9-117. American Public Health Association, Washington, DC (1995).

Fields, B.S.: Legionellae and Legionnaires' Disease. In: Manual of Environmental Microbiology, pp. 666-675. C.J. Hurst, G.R. Knudsen, M.J. McInerney, et al., Eds. American Society for Microbiology Press, Washington, DC (1997).

Gerber, N.N.: Odorous Substances from Actinomycetes. Dev. Ind.

Microbiol. 20:225-238 (1979).

Hensel, A.; Petzoldt, K.: Biological and Biochemical Analysis of Bacteria and Viruses. In: Bioaerosols Handbook, pp. 335-360. C.S. Cox and C.M. Wathes, Eds. Lewis Publishers, Boca Raton, FL (1995).

Holt, J.G.; Krieg, N.R.; Sneath, P.H.A.; et al.: Bergey's Manual of Determinative Bacteriology, 9th Ed. Williams & Wilkins, Baltimore, MD (1994).

IOM: Agents, Sources, Source Controls and Diseases. In: Indoor Allergies: Assessing and Controlling Adverse Health Effects, pp. 86-130. A.M. Pope, R. Patterson, and H.A. Burge, Eds. Institute of Medicine, Washington, DC (1993).

Jensen, P.A.; Todd, W.F.; Davis, G.N.; et al.: Evaluation of Eight Bioaerosol Samplers Challenged with Aerosols of Free Bacteria. Am. Ind. Hyg. Assoc. J. 53:660-667 (1992).

Jensen, P.A.; Lighthart, B.; Mohr, A.J.; et al.: Instrumentation Used with Microbial Bioaerosols. In: Atmospheric Microbial Aerosols: Theory and Applications, pp. 226-284. B. Lighthart and A.J. Mohr, Eds. Chapman and Hall, New York, NY (1994).

Juozaitis, A.; Willeke, K.; Grinshpun, S.A., et al.: Impaction onto a Glass Slide or Agar Versus Impingement into a Liquid for Collection and Recovery of Airborne Microorganisms. Appl. Environ. Microbiol. 60:861-870 (1994).

Lawrence, J.R.; Korber, D.R.; Wolfaardt, G.M.; et al.: Analytical Imaging and Microscopy Techniques. In: Manual of Environmental Microbiology, pp 29-51. C.J. Hurst, G.R. Knudsen, M.J. McInerney, et al., Eds. American Society for Microbiology, Washington, DC (1997).

Madelin, T.M.; Madelin, M.F.: Biological Analysis of Fungi and Assorted Molds. In: Bioaerosols Handbook, pp. 361-386. Lewis Publishers, Boca Raton, FL (1995).

Marshall, K.C.: Colonization, Adhesion and Biofilms. In: Manual of Environmental Microbiology, pp. 358-365. C.J. Hurst, G.R. Knudsen, M.J. McInerney, et al., Eds. American Society for Microbiology Press, Washington, DC (1997).

Marthi, B.: Resuscitation of Microbial Bioaerosols. In: Atmospheric Microbial Aerosols: Theory and Applications. B. Lighthart and A.J. Mohr, Eds. Chapman and Hall, New York, NY (1994).

Morris, K.J.: Modern Microscopic Methods of Bioaerosol Analysis. In: Bioaerosols Handbook, pp. 285-316. Lewis Publishers, Boca Raton, FL (1995).

Nash, P.; Krenz, M.M.: Culture Media. In: Manual of Clinical Microbiology, 5th Ed., pp.1226-1288. A. Balows, Ed. American Society for Microbiology, Washington, DC (1991).

Palmer, C.J.; Bonilla, F.G.; Roll, B.; et al.: Detection of Legionella Species in Reclaimed Water and Air with the EnviroAmp Legionella PCR Kit and Direct Fluorescent Antibody Staining. Appl. Environ. Microbiol. 61(2):407-412 (1995).

Shelton, B.G.; Flanders, W.D.; Morris, G.K.: Legionnaires' Disease Outbreaks and Cooling Towers with Amplified Legionella Concentrations. Curr. Microbiol. 28:359-363 (1994).

Stewart, S.L.; Grinshpun, S.A.; Willeke, K.; et al.: Effect of Impact Stress on Microbial Recovery on an Agar Surface. Appl. Environ. Microbiol. 61:1232-1239 (1995).

Terzleva, S.; Donnelly, J.; Ulevicius, V.; et al.: Comparison of Methods for Detection and Enumeration of Airborne Microorganisms Collected by Liquid Impingement. Appl. Environ. Microbiol.

Chapter 19

Fungi

Harriet A. Burge and James A. Otten

19.1   Characteristics
    19.1.1   *Morphology*
    19.1.2   *Reproduction*
    19.1.3   *Classification*
        19.1.3.1   General Groupings
        19.1.3.2   The Sexual Fungi
        19.1.3.3   The Asexual Fungi
    19.1.4   *Lifestyles and Metabolism*
    19.1.5   *Ecology*
        19.1.5.1   Fungal Responses to Environmental Factors
        19.1.5.2   Moisture
        19.1.5.3   Temperature
        19.1.5.4   Light
19.2   Health Effects
    19.2.1   *Infectious Diseases*
    19.2.2   *Hypersensitivity Diseases*
        19.2.2.1   Immediate-Type Allergic Reactions
        19.2.2.2   Hypersensitivity Pneumonitis
    19.2.3   *Toxic Effects*
        19.2.3.1   Mycotoxins
        19.2.3.2   Glucans
    19.2.4   *Volatile Organic Compounds*
    19.2.5   *Beneficial Fungal Products*
19.3   Sample Collection
    19.3.1   *Sampling Strategies*
    19.3.2   *Source Sampling*
    19.3.3   *Air Sampling*
    19.3.4   *Sample Handling*
19.4   Sample Analysis
    19.4.1   *Culture*
        19.4.1.1   Culture Media
        19.4.1.2   Incubation Conditions
    19.4.2   *Microscopy*
        19.4.2.1   Examination of Fungal Structures
        19.4.2.2   Spore-Trap Air Samples
    19.4.3   *Chemical Analyses*
    19.4.4   *Immunoassays*
19.5   Data Interpretation
    19.5.1   *Other Recommendations*
        19.5.1.1   Health Implications of Fungi in Indoor Environments
        19.5.1.2   Canadian Recommendations
        19.5.1.3   American Industrial Hygiene Association
        19.5.1.4   International Society of Indoor Air Quality and Climate
    19.5.2   *Earlier ACGIH Recommendations*
    19.5.3   *Current ACGIH Recommendations*
        19.5.3.1   Indoor/Outdoor Relationships
        19.5.3.2   Indicator Species
        19.5.3.3   Potentially Pathogenic Fungi
19.6   Remediation and Prevention
19.7   References

19.1 Characteristics

19.1.1 Morphology

The fungi form a kingdom of eukaryotic organisms, without chlorophyll, that have cells bound by rigid walls usually formed of chitin and glucans. Fungi may be unicellular or multicellular. Yeasts are unicellular fungi that reproduce primarily by budding and, in culture, form pasty colonies similar to those of bacteria. The multicellular fungi are formed of microscopic, branched filaments called hyphae (Figure 19.1). A colony is a visible mass of interwoven hyphae that form a mycelium. Fungal colonies may appear cottony, velvety, granular, or leathery and may be white, gray, brown, black, yellow, greenish, or other colors. The mycelial fungi most commonly found growing in the indoor environment are often called molds (moulds) (e.g., species of *Penicillium*, *Aspergillus*, and *Cladosporium*). Mushrooms, puffballs, bracket fungi, cup fungi, morels, and truffles—to name a few—are also mycelial fungi, many of which form typical mold-like colonies in the laboratory. In the wild, these fungi form large, sometimes edible, fruiting bodies.

19.1.2 Reproduction

Fungi have complex life cycles that may include both sexual (teleomorphic) and asexual (anamorphic) stages. Spores—which are often small, light, and easily transported through the air—may be produced during either of these stages. Sexual spores are produced following (a) fusion of two compatible nuclei, and (b) meiosis (reduction division) that results in four daughter nuclei, each containing a combination of genetic information derived from the two original nuclei. These spores are named according to the morphology of their production [see 19.1.3.2]. Asexual spores are produced by mitosis without nuclear fusion, although genetic recombination may occur through other processes. Asexual spores are called *sporangiospores* if they are formed and remain within sporangia (e.g., *Rhizopus* or *Mucor* spp.)

[see 19.1.3.3]. Asexual spores are called *conidia* if they are formed externally to the spore-producing cell (i.e., on a conidiophore).

19.1.3 Classification

19.1.3.1 General Groupings Biological classification systems are designed both for convenience in communicating about the natural world and for illustrating and understanding relationships among organisms [see 17.2]. Classification systems may be artificial or natural. Artificial classifications are generally developed for convenience and may result in the grouping together of genetically unrelated organisms. Natural classification systems are based on presumed evolutionary relationships among fungi.

The simplest classification of the fungi is an artificial one based on observations made by eye. Thus, use of the term "mold" for visible fungal growth is more or less equivalent to a gardeners' use of the term "weed" for a plant growing where it is not wanted. A mold may belong to one of three natural classes of fungi. This term has no taxonomic significance and is used in this book only in a very general sense. "Mildew" is a term a layperson might use to refer to fungi growing on fabrics, window sills, or bathroom tile. Mycologists use the term mildew to refer to certain fungi that cause plant diseases, such as downy mildew and powdery mildew. The term "yeast" refers to fungi that, at least in culture, are unicellular (i.e., do not form mycelia). Yeasts include fungi from three separate natural classes.

Mycologists use sexual lifecycle stages (the teleomorphs) of fungi to classify them (Table 19.1). In addition, the asexual stages (the anamorphs) are often classified separately because these are the stages that appear most often in culture.

19.1.3.2 The Sexual Fungi Sexual reproduction is considered the most stable characteristic in most biological

---

TABLE 19.1. Example: Classification of Some Fungal Genera Found in Indoor and Outdoor Air

Phylum	Class	Sexual Form	Asexual Form
Zygomycota (Zygomycetes)	—	*Mucor* *Rhizopus*	*Mucor* *Rhizopus*
Ascomycota	Saccharomycetes	*Saccharomyces* *Saccharomycopsis* *Dipodascus*	*Torulopsis* *Candida* *Geotrichum*
	Ascomycetes	*Emericella* *Eurotium* *Eupenicillium* *Pleospora* *Mycosphaerella*	*Aspergillus* *Aspergillus* *Penicillium* *Stemphylium* *Cladosporium*
Basidiomycota	Holobasidiomycetes	*Filobasidiella* *Coprinus*	*Cryptococcus* —
	Teliomycetes	*Ustilago*	*Ustilago*

classification systems, and the fungi common in indoor environments are placed in three major groups based on their sexual reproduction processes. These groups are the Zygomycetes, Ascomycetes, and Basidiomycetes (Table 19.1).

Zygomycetes The hyphae of zygomycetes are generally broad, with many nuclei in each cell and few crosswalls (septae) between cells. Sexual spores in the zygomycetes are formed following fusion of two multinucleate hyphal tips. Nuclear pairing occurs and unpaired nuclei degenerate. The paired nuclei fuse, undergo meiosis, and the cell becomes surrounded by a thick, resistant wall. The zygospore thus formed is a resting stage and is rarely found in air. The asexual spores of zygomycetes are produced in sporangia (i.e., they are sporangiospores) and are the dissemination stage. Examples of zygomycetes that are important for air quality investigations are *Mucor* and *Rhizopus* spp. (Table 19.1).

Ascomycetes The ascomycetes produce sexual spores within specialized cells called asci, in which nuclear fusion and meiosis occur. Meiosis is often followed by a single mitotic division resulting in eight nuclei in each ascus. Each nucleus becomes enclosed within a spore (an ascospore). In many ascomycetes, these spores are forcibly squirted from an ascus in response to changing water pressure. Ascospores are produced in abundance and are a major factor in the dissemination of these fungi. Each spore germinates to form narrow, septate hyphae, usually with one nucleus per cell. Conidia are also produced and disseminated in great abundance. Most of the fungi considered important in indoor environments are asexual (conidial) stages of ascomycetes, for example, species of *Alternaria, Aspergillus, Cladosporium,* and *Penicillium* (Table 19.1), as well as many of the common yeasts. Some ascomycetes produce macroscopic fruiting bodies that contain the sexual spores. The delicious morels and truffles are also members of this group, as are *Peziza* spp. and *Pyronema domesticum,* which occasionally are produced on damp materials in buildings.

Basidiomycetes In the basidiomycetes, nuclear fusion and meiosis occur within a cell (the basidium) that is the equivalent of an ascus. However, instead of undergoing a mitotic division, each of the four meiotic nuclei migrate into the swollen ends of special projections. These become basidiospores that are forcibly discharged and disseminated in air. Some basidiomycetes have asexual stages that are classed in the form-genus *Geotrichum.* The pink yeasts *Rhodotorula* and *Sporobolomyces* spp.—which may be common in some indoor environments—as well as *Cryptococcus* spp. belong to the Basidiomycetes.

19.1.3.3 The Asexual Fungi For many fungi, only the anamorphic or asexual stages are known. These fungi have been called "imperfect" fungi or Deuteromycetes. Different mycologists classify the asexual fungi in different ways. One scheme groups the fungi that are common in indoor environments into (a) Blastomycetes: yeasts and yeast-like fungi (e.g., *Rhodotorula* spp.), (b) Hyphomycetes: mycelial fungi that form asexual spores on conidiophores not enclosed in fruiting bodies (e.g., *Aspergillus* spp.), and (c) Coelomycetes: mycelial fungi that form asexual spores inside fruiting bodies (e.g., *Phoma* spp.). Table 19.2 illustrates different asexual spore types for some common hyphomycetes.

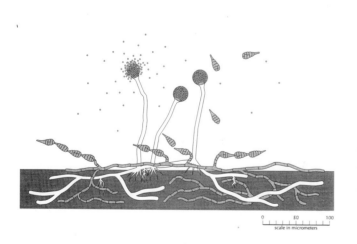 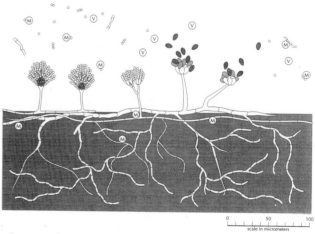

FIGURE 19.1. Example: fungal growth showing aerial and vegetative hyphae and reproductive structures.

19.1.4 Lifestyles and Metabolism

All fungi are heterotrophs, which means they must have external food sources and cannot make their own carbohydrates from water and carbon dioxide as can plants [see 17.5.1]. External sources of carbon may be simple sugars and starches or more complex carbon-containing substances. Common carbon sources in the indoor environment include the starchy pastes used with wallpaper, the cellulose in paper and fabrics (polysaccharides), the keratin in animal skin scales (proteins), and the lignin in wood (complex aromatic polymers). Fungi release enzymes (e.g., cellulase, protease, or ligninase) to digest these complex organic compounds into glucose which they then absorb. Of critical importance to this process is the availability of adequate moisture [see 19.1.5.2]. Fungi also require nitrogen in complex form as well as many elements (e.g., phosphorus, sulfur, and manganese).

Fungi may be saprobic, parasitic, or both (i.e., facultative parasites) or may live as symbiotes with other organisms [see 17.5.2, 17.5.3, and 17.5.5]. Saprobes use non-living organic material. Almost all fungi can grow as saprobes and can be cultured on non-living growth media in the laboratory. The common saprobic fungi are responsible for most natural aerobic decay and nutrient recycling. These fungi are abundant on the surfaces of dead plants and in the upper, aerated soil layers. Most fungi that grow on indoor materials are saprobic (e.g., *Aspergillus* and *Penicillium* spp.).

Most of the parasitic fungi are facultative and survive primarily as saprobes, invading living tissues only when suitable hosts are available. A few fungi (e.g., the plant rusts) are obligate parasites and require specific living hosts in which to grow and reproduce. These organisms do not grow in artificial culture, and they do not occupy indoor sites except when their specific living host plants are also present.

A few fungi are symbiotic, which means that they grow in close association with another living organism, often to the benefit of both. The mycorrhizal fungi (including most mushrooms) grow in partnerships with plants (usually trees), forming mycelial masses around and within a tree's peripheral roots. The fungus provides increased access to water and mineral nutrients for the tree while the tree provides soluble carbohydrates for the fungus. Neither the tree nor the fungus would thrive without the other. Another group of symbiotes, called lichens, are associations between fungi and algae, in which a fungus provides protection against desiccation while an alga provides soluble carbohydrates. Mycorrhizal fungi and lichens usually do not grow indoors, although their spores are often abundant in outdoor air and may penetrate to interior spaces.

In the process of degrading substrates to release the energy contained in their carbohydrates, the fungi produce many metabolic products. The principal metabolites are CO_2 and water. Under some circumstances, most fungi can also produce ethanol, and many fungi produce a variety of other volatile and non-volatile organic compounds [see 19.2.4 and Chapter 26].

Ergosterol is the major sterol in most fungi, but is absent from vascular plants, animals, and most other organisms (Gessner and Newell, 1997). Ergosterol is mainly a membrane component, and membranes are critical in maintaining cell function. Therefore, ergosterol is likely to be indicative of cytoplasm-containing, metabolically active cells. The molecule probably undergoes quite rapid deg-

TABLE 19.2. Example: How Some Common Asexual Fungi are Classified by their Method of Spore Production (von Arx, 1970)

Spore type	Example	Description
Arthrospores	*Geotrichum*	Mycelium fragments into individual cells
Blastospores	*Cladosporium, Monilia*	Spores formed by sequential budding at the ends of hyphae and spore chains, usually on conidiophores; youngest spores distal to the conidiophore
Botryoblastospores	*Botrytis*	Spores formed by budding simultaneously over the conidiophore
Porospores	*Alternaria, Ulocladium*	Spores formed through pores in the conidiophore or the last spore
Phialospores	*Aspergillus, Penicillium, Stachybotrys*	Spores formed internally within a spore-bearing cell (phialide) and extruded; youngest spores proximal to the conidiophore

radiation upon cell death, suggesting that it is basically an indicator of living fungal biomass. Ergosterol assays are becoming popular as a means to assess overall concentrations of fungal growth in indoor environments and to measure exposures to organic dust (AIHA, 1996b; Flannigan, 1997; Saraf et al., 1997).

Fungal cell walls are composed primarily of chitin (an acetyl glucosamine polymer) and glucans (or occasionally mannans). The glucans are under investigation as agents of BRS (see 24.7). Assays for glucans are also used to estimate fungal biomass (see 19.4.4).

19.1.5 Ecology

Few places on earth are completely free of fungi. Among the many environmental factors that influence fungal survival and growth are (a) water, (b) temperature, (c) light, and (d) nutrient concentration and kind. The first three factors are discussed below. Nutrients were discussed in Section 19.1.4.

19.1.5.1 Fungal Responses to Environmental Factors

Fungi respond to their environment in both absolute and relative ways. For example, for every fungus, there are absolute minimum and maximum temperatures below and above which growth will not occur. Near these temperature extremes, growth rate is very slow, increasing gradually as conditions approach the optimum, that is, the temperature at which the growth rate is maximal. The same pattern holds for water availability, nutrient levels and types, light, and atmosphere.

Complicating this pattern is the fact that environmental factors affect fungi interactively. Thus, the effect of water availability on fungal growth must be considered relative to the influences of temperature, nutrient availability, lighting conditions, and so forth. Environmental factors that are physically related (e.g., water availability and temperature) introduce further complications. Both the biological and physical interactions between water availability and temperature must be considered when discussing the effects of either of these parameters on fungal growth. This essentially means that for any fungus there is no single optimal set of environmental conditions.

19.1.5.2 Moisture

Water availability is one of the most critical factors controlling microbial growth in indoor environments. Mycologists discuss water availability in terms of water activity (a_w), which is a measure of the water within a substrate that an organism can use to support growth [see 10.3.3]. Table 19.3 classifies a few fungi according to their water requirements. For most fungi, regardless of whether they are hydrophilic or xerotolerant, the optimum a_w will be above 0.90. For any fungus, a_w range and optimum depend on temperature and the amounts and types of nutrients available. Thus, both the range and the optimum water activity for a specific fun-

gus can differ greatly for growth in different environments (e.g., on cooling coils in a ventilation system and a warm piece of wallboard).

19.1.5.3 Temperature

Temperature affects fungal growth both directly and through its control over water activity. Temperature controls the rate at which the biochemical reactions leading to growth occur. Most organisms have a minimum, maximum, and optimum temperature range for growth. The fungi fall into three categories with respect to temperature tolerance. Most fungi are mesophilic, with an optimum temperature range between 15° and 30°C. Psychrophiles (e.g., *Acremonium psychrophilum*) can grow at temperatures below 0°C, and their growth rate slows above approximately 17°C. Psychrotolerant fungi (e.g., *Cladosporium herbarum*) have temperature minima below 15°C but also grow well above 20°C. Thermophiles (e.g., *Thermomyces* spp.) usually cannot grow below approximately 20°C, and their growth rate is optimal between 35° and 50°C. *Aspergillus fumigatus* is considered a thermotolerant fungus and can grow well over a wide temperature range (i.e., 18° to 45°C). There is some indication that fungi adapt to temperature regimes so that, for example, an isolate of *C. herbarum* from the arctic might be psychrotolerant, while one from a temperate or tropical climate might be strictly mesophilic (Möller and Dreyfuss, 1996).

As mentioned in 19.1.5.2, temperature also controls a_w and, therefore, substrate water availability. Likewise, measures of a_w requirements for specific fungi are highly dependent on the temperatures at which the experiments are conducted.

19.1.5.4 Light

Light primarily influences fungal sporulation rather than mycelial growth. Many fungi require some exposure to light to stimulate spore production (e.g., *Epicoccum* spp.). In general, dark-spored fungi are resistant to ultraviolet radiation—an essential characteristic because their spores travel in the outdoor air during the day. The myth that fungi preferentially grow in dark environments has arisen because dark environments are often also damp. For example, in a cool climate, a dark closet on an exterior wall will often be colder than the rest of a house. Consequently, humidity will be higher and condensation may occur, allowing the growth of fungi. If a light is left on in the closet, the air is warmed, the humidity falls, and both condensation and fungal growth are prevented.

19.2 Health Effects

People are continuously exposed to fungi through both inhalation and ingestion with no apparent ill effects [see 14.1.4.5]. In fact, some fungi are used as foods and others are sources of drugs that control or prevent disease. However, certain fungi and fungal products are important agents of human disease (IOM, 1993; Horner et al., 1995; Levetin, 1995).

19.2.1 Infectious Diseases

Many persons suffer superficial fungal infections of the skin, nails, hair, and mucous membranes. The dermatophytic fungi (i.e., *Trichophyton, Microsporum,* and *Epidermophyton* spp.) can utilize keratin as a nutrient source and are known primarily from human skin infections. These fungi cause ringworm of the head (tinea capitis) and other regions, athletes' foot, and jock-strap itch. *Candida albicans* is another common fungal infectious agent. This yeast-like fungus is a human-commensal organism on skin and mucous membranes (e.g., in the GI tract). However, *C. albicans* may cause superficial mucous membrane infections in persons taking antibiotics that destroy competing bacteria or may cause disseminated disease in severely immunosuppressed persons.

Several fungi that may be common in outdoor reservoirs can cause infections in all previously uninfected persons who are exposed (e.g., *Histoplasma capsulatum, Blastomyces dermatitidis,* and *Coccidioides immitis*) (APHA, 1995; ASM, 1995). These fungi usually cause transient flu-like diseases that only rarely become a serious threat [see also 9.2.1]. *H. capsulatum* and *B. dermatitidis* are soil fungi that occupy wet sites enriched with protein (e.g., from bird droppings). Most persons living in the Mississippi river valley have been exposed to *H. capsulatum* and carry specific antibodies against the organism. *C. immitis* inhabits dry, alkaline soil and is common in southwestern U.S. where coccidioidomycosis (Valley fever) outbreaks often follow dust storms (Kirkland and Fierer, 1996). *Cryptococcus neoformans* is a fungus commonly recovered from pigeon nests and droppings as well as from moist bat droppings and contaminated soil. This fungus is an opportunistic pathogen, although intense exposure may lead to transient pulmonary disease in normal, healthy individuals. The severe forms of cryptococcosis (disseminated infections) are primarily encountered in immunocompromised persons and are especially common in AIDS patients (Ampel, 1996; Dromer et al., 1996).

The majority of fungi commonly encountered in the environment are unable to cause infectious disease unless an exposed person is severely immunodeficient [see 8.4 and 9.2.1]. Even in susceptible persons, only those fungi that are opportunistic human pathogens able to grow at the elevated temperatures and reduced oxygen levels found within the human body can cause infection. The most familiar of these fungi is *A. fumigatus,* a common fungus to which everyone is exposed throughout their lives. Although *A. fumigatus* is an important cause of infection in immunosuppressed persons, even exposure to high concentrations ($>10^6$ spores/m^3 of air) is unlikely to cause infection in a healthy person. Another example of an opportunistic fungus is *Pneumocystis carinii,* to which exposure is probably universal. However, disease is primarily confined to severely immunodeficient persons and is most often diagnosed in HIV-infected patients.

19.2.2 Hypersensitivity Diseases

19.2.2.1 Immediate-Type Allergic Reactions
Most fungi produce highly allergenic proteins or glycoproteins that, under appropriate exposure conditions, can cause hypersensitivity diseases in susceptible hosts. Between 10% and 60% of genetically susceptible (atopic) persons develop immediate hypersensitivity (allergy) to fungi, as demonstrated by skin tests (IOM, 1993). Exposure to some fungi is clearly associated with symptoms of asthma and hay fever (Delfino et al., 1996; Neas et al., 1996). There is also some evidence that sensitization and exposure to fungi increases the chances that an asthma attack will be fatal (O'Holloran et al., 1991; Targonski et al., 1995). Unfortunately, the nature of exposures that stimulate either sensitization or subsequent symptoms remains unknown. Chapter 25 on antigens discusses these reactions and allergic bronchopulmonary mycoses (i.e., fungal colonization of mucous secretions in the central airways of some asthmatics).

TABLE 19.3. Moisture Requirements for Some Common Fungi

Water Requirements	Common Indoor Fungi	Comments
Hydrophilic fungi (minimum a$_w$ >0.90)	*Fusarium, Rhizopus, Stachybotrys* spp.	Colonize continuously wet materials (e.g., soaked wallboard, water reservoirs for humidifiers, drip pans).
Mesophilic fungi (minimum a$_w$ ≥0.8≤0.9, optimum a$_w$ >0.9)	Most mycelial fungi, including *Alternaria, Epicoccum, Ulocladium, Cladosporium* spp., *Aspergillus versicolor*	Colonize continuously damp materials (e.g., damp wallboard, damp fabrics). Xerotolerant fungi can also grow under these conditions.
Xerotolerant fungi (minimum a$_w$ <0.8, optimum a$_w$ >0.8)	*Eurotium (Aspergillus glaucus* group*)*, some *Penicillium* spp.	Colonize relatively dry materials (e.g., house dust at high RH).
Xerophilic fungi (minimum a$_w$ <0.8)	*Aspergillus restrictus*	Colonize very dry materials (e.g., high-sugar foods, some building materials).

19.2.2.2 Hypersensitivity Pneumonitis HP may also result from exposure to fungal antigens. For HP development, it appears that heavy, continuous or repeated exposure to small fungal particles is essential. Genetic factors that may control susceptibility are unknown [see also 3.3.3 and 8.2.1.3].

19.2.3 Toxic Effects

19.2.3.1 Mycotoxins Mycotoxins produce a variety of health effects via ingestion, skin contact, and inhalation [see 24.2]. Depending on the kind of mycotoxin and the nature of the exposure, effects may include mucous membrane irritation, skin rashes, dizziness, nausea, immunosuppression, birth defects, and cancer. Nearly all of the mycotoxin literature focuses on ingestion exposure, although the role of inhaled mycotoxins in human disease is currently under scrutiny. In view of the potential severity of resulting diseases, a conservative approach to limiting exposure to mycotoxins is recommended [see 24.6].

19.2.3.2 Glucans Glucans comprise the bulk of the cell walls of most fungi. Glucans have antitumor activity and modulate the endotoxin-stimulated release of cytokines in Gram-negative bacterial infections (Williams et al., 1996). Glucans have irritant effects similar to—although less potent than—those of endotoxin. Exposure to glucans in dust has been associated with BRSs (Rylander et al., 1992; Rylander, 1995). Whether the glucans, some other fungal agent, or other factors associated with conditions leading to fungal growth actually mediated the effects remains to be investigated [see 24.7.2].

19.2.4 Volatile Organic Compounds

Fungi produce VOCs while growing and degrading substrates [see 19.1.4]. Some of these compounds have distinctive odors and low odor thresholds, and many people find these VOCs offensive or annoying [see 26.2.2.1]. Exposure to such compounds may be responsible for some nonspecific BRSs. However, the role of fungal VOCs in clinically evident disease has not been studied [see 26.3].

19.2.5 Beneficial Fungal Products

Fungi are well-known as important food sources (e.g., mushrooms, truffles, morels, tempeh, soy sauce, yeast bread, cheese, wine, and beer). In addition, many of our most effective antibiotics (e.g., penicillin, griseofulvin, and cephalosporin) are produced by common fungi. Cyclosporin A, an immunosuppressive mycotoxin produced by the fungus *Tolypocladium inflatum*, is one of the drugs that makes organ transplants possible. In addition, many anti-cancer agents are products of fungal metabolism. As mentioned in Section 19.2.3.2, fungal glucans show promise as agents to minimize the effects of Gram-negative bacterial infections.

19.3 Sample Collection

19.3.1 Sampling Strategies

Visible fungal growth on interior surfaces is clear evidence that fungi have colonized an environment. Although the health impacts of surface growth have not been documented, the potential exists for exposure to fungal allergens, toxic metabolic products, and malodorous VOCs. In most cases, the discovery of visible fungal growth warrants a recommendation for cleanup and identification of the underlying reason for the growth [see 10.1 and 14.1.2]. In many cases, further air or source sampling is not necessary. However, investigators may decide to collect air or source samples to document that fungal growth has occurred or to record the kinds of fungi that predominate. The most successful sampling strategies are those formulated to answer specific questions.

19.3.2 Source Sampling

Investigators who suspect exposure to specific allergens or toxins may collect source samples to evaluate the possible contributions of visible fungi or fungi in dust to specific disease processes [see Chapter 12]. Source sampling is also useful to document that discoloration or deposits on surfaces actually represent either fungal growth or spore accumulation. Simple tape sampling, with microscopic analysis, will often confirm the presence of hyphae and spores as well as allow identification of many kinds of fungi. However, culture of surface or bulk samples may be necessary to allow microscopic identification of some fungi (e.g., species of *Aspergillus* or *Penicillium*) [see 19.4].

Bulk dust samples are often cultured to evaluate the kinds of fungi present. In addition, chemical analyses (e.g., measures of ergosterol or glucans) may be used to estimate total fungal biomass. Immunoassays for specific fungal allergens are also under development. Investigations occasionally require collection of bird, bat, or rodent droppings; other animal debris; or soil samples for detection of fungal saprobes that cause human infections (Levitz, 1991; Lenhart, 1994; APHA, 1995; ASM, 1995; Lenhart et al., 1997) [see 19.2.1].

19.3.3 Air Sampling

Air sampling for particles of fungal origin is complicated by their diversity in size, shape, density, and surface features—all of which affect particle behavior while airborne and during collection. Most commonly, airborne fungal spores are collected using impaction onto agar or an adhesive-coated transparent surface (spore trapping) (Samplers 1 to 7 and 10 and Samplers 16 to 20 in Table 11.1). Culture-based analysis tends to underestimate actual fungal concentrations because many spores are either not living or are unable to grow on the culture medium provided [see 19.4]. Spore trapping allows accurate counting of total fungal spores and identification of some spores, but many spores cannot be identified by microscope (e.g., *Penicillium* and *Aspergillus* spp.).

19.3.4 Sample Handling

Samples of living fungi must be handled carefully to preserve viability if cultural analysis is to be used and to prevent growth before the samples reach a laboratory. For microscopy or chemical analysis, it may be possible to treat samples with a biocide or dry them to kill the cells or prevent replication. Stored properly, treated samples can be held indefinitely. Air and dust samples for immunoassay should also be kept dry, but must be analyzed within approximately 48 hours or frozen if longer delays are necessary. Investigators should consult the mycologist who will examine the samples about how to ship them (e.g., the type of container to use and whether to refrigerate samples or keep them at room temperature). Samples to be cultured should reach the laboratory within 24 hours of collection and should be examined and processed on the day received to minimize changes in the kinds and concentrations of fungi or fungal products present.

19.4 Sample Analysis

The primary methods of environmental sample analysis for fungi remain (a) isolation of fungi by laboratory culture, and (b) microscopic examination of fungal cultures and individual fungal spores. However, other approaches are also used (Table 19.4) (Madelin and Madelin, 1995; AIHA, 1996a,b; Buttner et al., 1997). In particular, analysis of bulk samples for ergosterol or glucan concentration as estimates of total fungal biomass are of increasing interest [see 19.1.4 and 19.2.3.2]. Immunoassays are under development for measurement of some specific fungal allergens. Section 6.3.1 briefly describes how laboratories process samples and how concentrations of culturable fungi are calculated for air and source samples. Table 19.4 compares available methods for identifying common fungi.

19.4.1 Culture

Total (non-differential) fungal colony counts can provide investigators some information for evaluating the microbial status of an indoor or outdoor environment. However, it is essential that fungi be identified if investigators wish to compare bulk, surface, or air samples from different sampling locations or times and to interpret the potential health risks of the fungi present. A well-trained environmental mycologist may be able to identify many fungi to the genus level directly on a culture plate using a low-power dissecting microscope. Characteristics used for identification include colony color, size, and texture; the type of spore-bearing structures present; and, in some cases, the arrangement of the spores on these structures. A few fungi are so distinctive they can be identified to group or even species level by this simple method. For example, a good environmental mycologist can readily assign many fungi to the species level (e.g., *Epicoccum nigrum*, *Paecilomyces variotii*, *Paecilomyces lilacinus*, *Stachybotrys chartarum*, *Cladosporium sphaerospermum*, *Pithomyces chartarum*, *Trichoderma viride*, *Botrytis cinerea*, and several *Penicillium* spp.). Some *Aspergillus* isolates can also be identified to the species level, and others to the group level, with further characterization pursued where indicated. For identification, fungi must be grown under carefully controlled conditions. Identification of fungal species generally requires skillful use of a high-power, light microscope and highly specific training and experience. Such experience is available from mycologists working in academic research laboratories and others who have obtained specialized training.

Yeasts are often abundant in dust samples but are usually reported only as total colony counts and are seldom identified to genus or species level. Unlike other fungi, in culture, yeasts produce bacteria-like colonies with few distinguishing features. Physiological and biochemical tests are required for yeast identification. Consequently, little is known about which yeasts are common in indoor environments and the potential health effects of aerosol exposure to them. One exception is *Sporobolomyces* sp., which has been associated with hypersensitivity disease and is readily recognized by its distinctive colony morphology and by its production of forcibly discharged ballistoconidia.

19.4.1.1 Culture Media If investigators wish to compare their sampling results with an existing database, they should adopt the same culture medium used in the studies in the database. Except when specific fungi are of special concern, culture media for fungi from air or source samples should support a wide variety of common taxa. Various formulations of malt extract agar (MEA) have a long history of use for air monitoring (Samson et al., 1994; AIHA, 1996a). The second formulation in Table 19.5 is one also used as a diagnostic medium to identify *Aspergillus* spp. This formulation works well in most non-industrial indoor environments, provided rapid-growing fungi, such as *Rhizopus*, *Mucor*, *Monilia*, or *Trichoderma* spp., are not abundant. If these fungi are present in large numbers, 2% MEA (without glucose or peptone) or a medium such as DG-18 that contains growth inhibitors can be used.

The addition to fungal culture medium of compounds to suppress bacterial growth is generally needed only for samples from areas where very high concentrations of bacteria (>10^5 CFU/m^3) are expected (e.g., agricultural, metal-working, or domestic waste-handling environments). Commonly used bactericides are rose bengal and antibiotics (e.g., chloramphenicol, penicillin, and streptomycin). Rose bengal becomes fungicidal when exposed to light and must be kept covered as much as possible during sample collection and in the laboratory.

Cellulose agar has been used to isolate *S. chartarum*. This fungus grows well on other media (e.g., MEAs, after initial isolation), but its growth is slow and other fungi in environmental samples may outcompete it. However, few fungi other than *S. chartarum* are able to grow on water agar (20 g agar per L of water) overlaid with sterile filter paper. Therefore, this or other formulations in which cellulose is the only carbon source (AIHA, 1996a) may be useful to isolate this fungus from environmental samples.

19.4.1.2 Incubation Conditions Culture plates are incubated at room temperature (18° to 22°C) for isolation of most fungi. If *only* thermotolerant or thermophilic fungi are of concern, plates can be incubated at temperatures up to 45°C. Some near-ultraviolet exposure is necessary to induce sporulation in many dark-spored fungi and can be provided by sunlight or by placing lamps designed for illuminating house plants in an incubator. Fungal cultures should be monitored daily, but typically colonies cannot be accurately counted or identified before at least five days of incubation. Plates should be disturbed as little as possible during incubation because spores are readily dislodged, and dislodged spores may form new colonies, leading to inaccurate counts.

19.4.2 Microscopy

Mycologists use dissecting and light microscopes to examine fungi from air and source samples. Both species identification of fungi in culture and examination of spore-trap air samples should be performed by experienced mycologists familiar with environmental fungi.

19.4.2.1 Examination of Fungal Structures Many common fungi can be identified to genus and sometimes to species based on macroscopic colony morphology (e.g., color and texture) or spore-bearing structures viewed at 10- to 60x magnification (Table 19.2). More exact identifications can be made by examining stained wet mounts.

Fungi that have not sporulated in culture generally cannot be identified. In some cases, these colonies are sterile forms of common fungi, such as species of *Cladosporium*. Although an experienced mycologist can provisionally classify these fungi based on colony morphology, such identification is never entirely reliable. However, most non-sporulating fungi seen in the laboratory are types that are unable to produce spores in culture. Many of these colonies represent growth from sexual spores (ascospores and basidiospores) of fungi that do not produce asexual stages.

19.4.2.2 Spore-Trap Air Samples Many kinds of individual fungal spores are identifiable—at least to general category—when examined at 1000X magnification. Experienced aerobiologists can identify some fungal genera and species from microscopic examination of single spores (e.g., *E. nigrum* and *S. chartarum*). In other cases, the genus can be identified from a spore, but not the species (e.g., *Alternaria* and *Cladosporium* spp.). The spores of some genera are similar (e.g., *Aspergillus* and *Penicillium* spp.), and culture is needed to identify these fungi to the species level.

19.4.3 Chemical Analyses

Although culture and microscopy are still the most commonly used methods for analyzing samples for fungal content, chemical analyses for indicators of fungal presence are of increasing interest (e.g., assays for ergosterol and glucans) (AIHA, 1996b; Flannigan, 1997; Saraf et al., 1997) [see 6.7 and 24.7.3]. Requirements for air sample collection for these assay methods differ from the requirements for either culture or microscopy in needing much larger air volumes to overcome the methods' relatively low sensitivities. These methods may prove most useful in highly contaminated environments, such as agricultural settings or wastewater treatment plants.

TABLE 19.4. Relative Usefulness of Available Methods for Analysis of Fungal Samples

Fungal Category	Microscopy	Culture	Ergosterol	Glucan	Immunoassay
Total fungi	++[A]	+	++[B]	++[B]	–
Alternaria spp.	+++	+	–	–	++
Cladosporium spp.	+++	++	–	–	(++)
Stachybotrys chartarum	+++	+	–	–	
Penicillium and *Aspergillus* spp.	++	++	–	–	–
Penicillium viridicatum	–	++	–	–	
Aspergillus fumigatus	–	++	–	–	(++)
Aspergillus niger	++	++	–	–	–
Chaetomium globosum	+++	+	–	–	–
Basidiospores	+++	–	–	–	–

[A] most useful for air samples	– not useful	++ useful
[B] most useful for bulk samples	+ somewhat useful	+++ very useful

19.4.4 Immunoassays

Immunoassays are available for detecting a few fungal allergens (e.g., mixtures of allergens from *Alternaria alternata*, *A. fumigatus*, and *C. herbarum*) (Horner et al., 1995) and for a few mycotoxins (e.g., aflatoxin and T-2 toxin) [see 24.4.2 and 25.4]. None of these assays has yet been used extensively in air quality investigations. Large-volume air samples are required for immunoassays to overcome the methods' relatively low sensitivity.

19.5 Data Interpretation

Data from appropriately designed sampling studies should allow investigators to evaluate hypotheses regarding possible exposure to fungal aerosols at a study site. Few health-based guidelines or standards are available to assist investigators in the interpretation of fungal aerosol data, and only limited exposure–dose or dose–response data are available on which to base guidelines. Therefore, the interpretation of data on fungi in air and source samples generally focuses on the fungal genera and species identified, comparisons among different environments, and the potential susceptibilities of exposed populations to various fungal agents.

19.5.1 Other Recommendations

Various groups have offered guidance on the interpretation of environmental air and source samples for fungi. These recommendations are meant to be applied by properly trained and experienced persons who understand the

strengths and limitations of the available methods for environmental sampling for fungi.

19.5.1.1 Health Implications of Fungi in Indoor Environments A 1992 workshop in the Netherlands led to the recommendation that certain fungi should be considered indicator organisms that may signal moisture presence or a potential for health problems if above a baseline level (to be established) in air or surface samples (Samson et al., 1994). The following fungi were named as indicator species (an asterisk indicates those fungi the authors considered to be important toxigenic taxa): (a) materials with a_w >0.90 to 0.95: *A. fumigatus* and species of *Trichoderma*, *Exophiala*, *Stachybotrys**, *Phialophora*, *Fusarium**, *Ulocladium*, and yeasts (*Rhodotorula*), (b) materials with a_w 0.85 to 0.90: *A. versicolor**, and (c) materials with $a_w \leq 0.85$: *A. versicolor**, species of *Eurotium* and *Wallemia*, and species of *Penicillium* (e.g., *Penicillium chrysogenum* and *Penicillium aurantiogriseum*).

19.5.1.2 Canadian Recommendations Health Canada (1993, 1995) has recommended the following guidelines for recognizing and managing fungal contamination in public buildings. Bird or bat droppings may contain pathogenic fungi (e.g., *C. neoformans* and *Histoplasma* spp.) and toxigenic *A. fumigatus*. Appropriate action (described in the document) should be taken for the safe removal of materials that may contain these fungi. The persistent presence, demonstrated on repeated sampling, of toxigenic

---

TABLE 19.5. Culture Media for Fungi (ingredients per liter of distilled water)

Category	Medium	Ingredients		
Most saprobic fungi[A]	2% MEA	20	g malt extract	
		20	g agar	
Most saprobic fungi[B]	MEA	20	g malt extract	
		20	g dextrose	
		1	g peptone	
		15	g agar	final pH 4.5-5.0
Xerotolerant fungi (including fungi cultured from settled dust samples)	DG-18 a_w = 0.955	10	g glucose	
		5	g peptone	
		1	g KH_2PO_4	
		0.5	g $MgSO_4 \cdot 7H_2O$	
		0.1	g chloramphenicol	
		15	g agar	
		220	g glycerol	
		2	mg dichloran	final pH 5.6
Antibacterial medium for fungi	MEA with:[c] chloramphenicol rose bengal, penicillin, or streptomycin	0.1	g/L	final pH 5.6
		35	mg/L	
		20	units/mL	
		40	units/mL	

[A] environments in which rapid-growing fungi are abundant
[B] environments in which rapid-growing fungi are not abundant

[c] Except for chloramphenicol, add anti-bacterial agent after autoclaving.

fungi (e.g., *S. chartarum* and species of *Aspergillus, Penicillium,* and *Fusarium*) indicates that further investigation and appropriate action (described in the document) should be taken. The confirmed presence of one or more fungal species seen as a significant percentage of an indoor sample but not similarly present in concurrent outdoor samples is considered evidence of a fungal amplifier, and appropriate action (described in the document) should be taken. Fungi in indoor air should be qualitatively similar to and quantitatively lower than what is found in outdoor air, but factors such as sampling technique, season, and weather affect what fungi are isolated from outdoor air. Numeric criteria are suggested for single fungal species (other than *Cladosporium* or *Alternaria* spp.), mixed species reflective of those typically found in outdoor air, and fungi primarily from plants in summer.

19.5.1.3 American Industrial Hygiene Association The AIHA (1996a,b) has offered guidelines for interpreting culture results for air and source samples for fungi. These guidelines state that genera such as *Cladosporium, Alternaria,* and *Epicoccum* as well as Basidiomycetes are present in outdoor air on a seasonal basis. However, in mechanically ventilated buildings with air filtration, the concentrations of these typically outdoor fungi should be lower than concentrations measured at the OAI. Dominance in indoor air of fungal species not predominant in outdoor air indicates that these fungi are growing in a building and that the air quality is degraded. The confirmed presence of *S. chartarum, A. versicolor, A. flavus, A. fumigatus,* or *Fusarium moniliforme* requires that urgent risk management decisions be made (references provided). "Confirmed presence" is defined as colonies in several samples, many colonies in any sample, or, where a single colony was found in a single sample, evidence of the growth of these fungi on building materials by visual inspection or source sampling. The AIHA guide states that certain pathogenic fungi may be problems if present in indoor air (e.g., *A. fumigatus, A. flavus,* and other species that may cause aspergillosis as well as *F. moniliforme, H. capsulatum,* and *C. neoformans*). Because the latter two fungi are difficult to culture and are seldom detectable in air samples, a default assumption is that they may be present in any bird or bat droppings and that disinfection and removal of such material is required.

19.5.1.4 International Society of Indoor Air Quality and Climate The International Society of Indoor Air Quality and Climate (ISIAQ) (1996) has proposed guidelines for interpreting environmental samples for fungi. These guidelines state that in naturally ventilated, non-problem buildings, the relative abundance of different fungi in indoor air tends to follow the pattern found in outdoor air, although the numbers are usually smaller. When air-conditioning or mechanical ventilation with filtration is used, indoor fungal concentrations in non-problem buildings may be even lower than in naturally ventilated buildings. When windows are closed or when snow cover reduces outdoor sources of fungi, indoor sources of *Penicillium* spp. and other soil fungi may be more obvious. While a diversity of fungi is usually found in non-problem buildings, one or two fungal species may dominate the indoor air in buildings with persistent moisture problems. The presence or dominance of toxigenic or allergenic species indicates a problem that may cause deterioration of the quality of the indoor air.

19.5.2 Earlier ACGIH Recommendations
The last ACGIH (1989) guidelines stated that the taxa of fungi isolated from indoor and outdoor air should be similar and that the concentration of airborne fungi should be lower indoors than outdoors, the degree of difference varying with the type of building ventilation. This statement is still correct, with the understanding that in regions with winter snow cover, outdoor fungal air concentrations may be lower than indoor concentrations even in non-problem buildings. It is assumed that the samples used to make such comparisons were collected appropriately and that the numbers of samples for the indoor and outdoor locations are sufficient in number to allow meaningful comparison.

Based on the authors' experience when the 1989 guidelines were written, they stated that outdoor fungal air concentrations exceeding 1000 CFU/m^3 were routine and that concentrations near 10,000 CFU/m^3 were not uncommon in summer months. Except in specialized environments where immunosuppressed persons are routinely present, levels of any saprophytic fungus below 100 CFU/m^3 were reported as not of concern. Concentration data collected since 1989 continue to support these statements about concentration ranges for total culturable or countable fungi. The earlier authors also wrote that in buildings with mechanical ventilation and minimal air filtration, indoor fungal concentrations typically are less than half of outdoor levels. Currently, it is understood that comparing the ratio of indoor and outdoor fungal concentrations requires consideration of many factors. The authors never intended the above statements to be interpreted that indoor fungal air concentrations of 50 CFU/m^3, 500 CFU/m^3, or 5000 CFU/m^3 were to be used as criteria for judging indoor air quality in various settings or different seasons. The following sections summarize ACGIH's current recommendations for evaluating fungi in indoor environments [see also Chapters 1 and 14].

19.5.3 Current ACGIH Recommendations
Rather than focusing on specific kinds of fungi or on quantitative measures of fungal prevalence, the ACGIH approach has been to emphasize that active fungal growth in indoor environments is inappropriate and may lead to

exposure and adverse health effects. Evidence that active growth is occurring is most often sensory (visual identification or odor perception) confirmed by judicious source sampling. If air sampling is to be used, ACGIH guidance has emphasized the importance of well-designed sampling protocols and reliance on carefully collected baseline data for comparison. Following is a summary of guidelines for assessing fungal problems in non-industrial indoor environments:

1. The presence of visible fungal growth confirmed by source sampling in occupied indoor environments is strong evidence that exposure may occur. The conditions leading to such growth should be corrected and the growth removed, using appropriate precautions.

2. The presence of moldy odors in occupied indoor environments is strong evidence that fungal growth is occurring. Such growth should be located and confirmed by source sampling. The conditions leading to the growth should be corrected and the growth removed, using appropriate precautions.

3. The persistent presence of water in indoor environments (except in places designed for the carriage or storage of water) is likely to lead to fungal growth. The conditions allowing such water to accumulate should be corrected.

4. The presence of accumulations of organic debris, especially bird or animal droppings, is presumptive evidence of the presence of fungal contamination. The conditions allowing the accumulation of such debris should be corrected and the debris removed, using appropriate precautions.

5. Interpretation of source or air sampling data in the absence of any of the above conditions requires a sufficient number of samples (including controls) to ensure that results are not due to random chance. If these data requirements are met, an investigator may consider sampling results in light of the following discussions.

19.5.3.1 Indoor/Outdoor Relationships Indoor/outdoor relationships are assessed both by comparing concentrations and species composition of comparably collected samples. In non-problem environments, the concentration of fungi in indoor air typically is similar to or lower than the concentration seen outdoors, except when outdoor air concentrations are near zero (e.g., during periods of snow cover). If fungal concentrations indoors are consistently higher than those outdoors, then indoor sources are indicated. However, indoor fungal growth may also be present in situations where indoor concentrations of airborne fungi are equal to or lower than those outdoors, and interpretation of data depends on a knowledge of the kinds of fungi present in the two environments.

Note that exposure to fungi actively growing indoors may present unusual health risks even when total fungal concentrations are higher outdoors. If the variability of the data is high (which is common), it may be difficult to establish that two locations differ with respect to concentrations of airborne fungi [see 14.2.3.2].

The species of fungi found in indoor and outdoor air typically are similar if outdoor air is the primary source for the fungi in indoor air (Burge et al., 1977; Strachan et al., 1990; Targonski et al., 1995; AAAAI, 1996; Delfino et al., 1996; Neas et al., 1996). Comparisons of the species compositions of indoor and outdoor populations requires accurate identification of fungal species, not simply identification to the genus level. For example, a report stating that *Cladosporium* spp. predominated in both indoor and outdoor air samples may lead investigators to conclude that the indoor environment did not present a particular problem. In fact, *C. herbarum* may have been dominant outdoors and *C. sphaerospermum* dominant indoors; the former arising from outdoor sources and the latter released from areas of indoor growth. However, without identification of the fungal isolates to the species level, this possible cause of occupant complaints would be missed.

19.5.3.2 Indicator Species Fungi whose presence indicate excessive moisture or a health hazard have been termed indicator organisms. Interpreting the presence or absence of an indicator species (e.g., a recognized toxigenic fungus that is uncommon in outdoor air) requires the ability to identify fungi to the species level and a knowledge of the prevalence of various fungal species in indoor and outdoor environments. The mere presence of a few CFUs or spores of an indicator species should be interpreted with caution. Identification of the presence of a particular fungus in an indoor environment does not allow investigators to conclude that building occupants are exposed to antigenic or toxic agents. Investigators should also recognize that fungi named as indicator species are not the only fungi of significance. Many fungi other than those specifically listed by various groups may cause problems for building occupants exposed through inhalation of fungal aerosols or via other contact.

19.5.3.3 Potentially Pathogenic Fungi Some fungal pathogens should be assumed to be present when materials known to support their growth are found (e.g., *H. capsulatum* and *C. neoformans* in bird and bat droppings) [see 19.2.1]. Removal of such materials should be conducted as if they contained pathogenic fungi. Disturbance of soil or other material that may contain fungal pathogens (e.g., compost containing *Aspergillus* spp. or soil containing *H. capsulatum*, *B. dermatitidis*, or *C. immitis*) should be conducted with consideration that occupants of neighboring buildings may be exposed if airborne fungal spores enter the buildings.

19.6 Remediation and Prevention

Readers are directed to (a) Chapter 10 for a description of measures to control indoor fungal growth [see 10.2, 10.3, 10.4, and 10.5], (b) Chapter 15 for a discussion of cleaning and remediation of existing contamination [see 15.2, 15.3, and 15.6], and (c) Chapter 24 for recommendations on limiting exposures to mycotoxins [see 24.6].

19.7 References

AAAAI: Aeroallergen Monitoring Network, Annual Report. American Academy of Allergy, Asthma, and Immunology, Milwaukee, WI (1996).

ACGIH: Guidelines for the Assessment of Bioaerosols in the Indoor Environment. American Conference of Governmental Industrial Hygienists, Cincinnati, OH (1989).

AIHA: Viable Fungi and Bacteria in Air, Bulk, and Surface Samples. In: Field Guide for the Determination of Biological Contaminants in Environmental Samples, pp. 37-74. American Industrial Hygiene Association, Fairfax, VA (1996a).

AIHA: Total Fungi and Substances Derived from Fungi. In: Field Guide for the Determination of Biological Contaminants in Environmental Samples, pp. 119-130. American Industrial Hygiene Association, Fairfax, VA (1996b).

Ampel, N.M.: Emerging Disease Issues and Fungal Pathogens Associated with HIV Infection. Emerg. Infect. Dis. 2:109-116 (1996).

APHA: Control of Communicable Diseases Manual. A.S. Benenson, Ed. American Public Health Association, Washington, DC (1995).

ASM: Manual of Clinical Microbiology, 6th Ed. P.R. Murray, Ed. American Society for Microbiology, Washington, DC (1995).

Burge, H.A.; Boise, J.R.; Rutherford, J.A.; et al.: Comparative Recoveries of Airborne Fungus Spores by Viable and Nonviable Modes of Volumetric Collection. Mycopathologia. 61:27-33 (1977).

Buttner, M.P.; Willeke, K.; Grinshpun, S.A.: Sampling and Analysis of Airborne Microorganisms. In: Manual of Environmental Microbiology, pp. 629-640. C.J. Hurst, Ed. American Society for Microbiology, Washington, DC (1997).

Delfino, R.J.; Coate, B.D.; Zeiger, R.S.; et al.: Daily Asthma Severity in Relation to Personal Ozone Exposure and Outdoor Fungal Spores. Am. J. Respir. Crit. Care Med. 154:633-641 (1996).

Dromer, F.; Mathoulin, S.; Dupont, B.; et al.: Epidemiology of Cryptococcosis in France: A 9-year Survey (1985-1993). French Cryptococcosis Study Group. Clinical Infectious Diseases. 23:83-90 (1996).

Flannigan, B.: Air Sampling for Fungi in Indoor Environments. J. Aerosol Sci. 28:381-392 (1997).

Gessner, M.O.; Newell, S.Y.: Bulk Quantitative Methods for the Examination of Eukaryotic Organoosmotrophs in Plant Litter. In: Environmental Microbiology, pp. 295-308. C.J. Hurst, G.R. Knudsen, M.J. McInerney, et al., Eds. ASM Press, Washington, DC (1997).

Health Canada: Indoor Air Quality in Office Buildings: A Technical Guide. Federal-Provincial Advisory Committee on Environmental and Occupational Health. Environmental Health Directorate, Ottawa, Ontario, Canada (1993).

Health Canada: Fungal Contamination in Public Buildings: A Guide to Recognition and Management. Federal-Provincial Committee on Environmental and Occupational Health. Environmental Health Directorate, Ottawa, Ontario, Canada (1995).

Horner, W.E.; Helbling, A.; Salvaggio, J.E.; et al.: Fungal Allergens. Clin. Microbiol. Rev. 8:161-179 (1995).

Institute of Medicine (IOM): Indoor Allergens: Assessing and Controlling Adverse Health Effects. National Academy Press, Washington, DC (1993).

ISIAQ: Control of Moisture Problems Affecting Biological Indoor Air Quality. TFI-1996. International Society of Indoor Air Quality and Climate, Ottawa, Ontario, Canada (1996).

Kirkland, T.N.; Fierer, J.: Coccidioidomycosis: A Reemerging Infectious Disease. Emerg. Infect. Dis. 2:192-199 (1996).

Lenhart, S.W.: Recommendations for Protecting Workers from Histoplasma capsulatum Exposure during Bat Guano Removal from a Church's Attic. Appl. Occup. Environ. Hyg. 9:230-236 (1994).

Lenhart, S.W.; Schafer, M.P.; Singal, M.; et al.: Histoplasmosis: Protecting Workers at Risk. Publication No. 97-146. Centers for Disease Control and Prevention. National Institutes for Occupational Safety and Health (NIOSH), Cincinnati, OH (1997).

Levetin, E.: Fungi. In: Bioaerosols, pp. 87-120. H. Burge, Ed. Lewis Publishers, Boca Raton, FL (1995).

Levitz, S.M.: The Ecology of Cryptococcus neoformans and the Epidemiology of Cryptococcosis. Reviews of Infectious Diseases. 13:1163-1169 (1991).

Madelin, TM.; Madelin, M.F.: Biological Analysis of Fungi and Associated Molds. In: Bioaerosols Handbook, pp. 361-386. C.S. Cox and C.M. Wathes, Eds. Lewis Publishers, Boca Raton, FL, (1995).

Möller, C.; Dreyfuss, M.M.: Microfungi from Antarctic Liches, Mosses, and Vascular Plants. Mycologia. 88:922-933 (1996).

Neas, L.M.; Dockery, D.W.; Burge, H.; et al.: Fungus Spores, Air Pollutants, and Other Determinants of Peak Expiratory Flow Rate in Children. Am. J. Epidemiology. 143:797-807 (1996).

O'Hollaren, M.T.; Yunginger, J.W.; Offord, K.P.; et al.: Exposure to an Aeroallergen as a Possible Precipitating Factor in Respiratory Arrest in Young Patients with Asthma. New England J. Med. 324:359-363 (1991).

Rylander, R.; Persson, K.; Goto, H.; et al.: Airborne beta-1,5-glucan May be Related to Symptoms in Sick Buildings. Indoor Environ 1:263-267 (1992).

Rylander, R.: Office and Domestic Environments. In: Organic Dusts: Exposure, Effects, and Prevention, pp. 247-255 (1995).

Samson, R.A.; Flannigan, B.; Flannigan, M.E.; et al., Eds.: Recommendations. In: Health Implications of Fungi in Indoor Environments, pp. 531-538. Elsevier, New York, NY (1994).

Samson, R.A.; Flannigan, B.; Flannigan, M.E.; et al., Eds.: Media. In: Health Implications of Fungi in Indoor Environments, pp. 589-592. Elsevier, New York, NY (1994).

Saraf, A.; Larsson, L.; Burge, H.; et al.: Quantification of Ergosterol and 3-hydroxy Fatty Acids in Settled House Dust by Gas Chromatography-Mass Spectrometry: Comparison with Fungal Culture and Determination of Endotoxin by a Limulus Amebocyte Lysate Assay. Appl. Environ. Microbiol. 63:2554-2559 (1997).

Strachan, D.P.; Flannigan, B.; McCabe, E.M.; et al.: Quantification of Airborne Moulds in the Homes of Children with and without Wheeze. Thorax. 45:382-387 (1990).

Targonski, P.V.; Persky, V.W.; Ramekrishnan, V.: Effect of Environmental Molds on Risk of Death from Asthma during the Pollen Season. J Allergy Clin Immunol. 1:955-961 (1995).

von Arx, J.A.: The Genera of Fungi Sporulating in Pure Culture. Lubrecht & Cramer (1981)

Williams, D.L.; Mueller, A.; Browder, W.: Glucan-Based Macrophage Stimulators: A Review of their Anti-infective Potential. Clin Immunother. 5:392-399 (1996).

Chapter 20

Amebae

James A. Otten and Harriet A. Burge

20.1 Characteristics
    20.1.1 *Morphology*
        20.1.1.1 *Naegleria* spp.
        20.1.1.2 *Acanthamoeba* spp.
    20.1.2 *Classification*
    20.1.3 *Lifestyles and Metabolism*
    20.1.4 *Ecology*
20.2 Health Effects
    20.2.1 *Infectious Diseases*
        20.2.1.1 *Naegleria fowleri* Infections
        20.2.1.2 *Acanthamoeba* and *Balamuthia* spp. Infections
    20.2.2 *Inhalation Fever*
20.3 Sample Collection
    20.3.1 *Source Sampling*
    20.3.2 *Air Sampling*
        20.3.2.1 Liquid Impingers
        20.3.2.2 Multiple-Hole Impactors
        20.3.2.3 Electrostatic Precipitators
    20.3.3 *Sample Handling*
20.4 Sample Analysis
    20.4.1 *Culture*
    20.4.2 *Most Probable Number*
    20.4.3 *Microscopy*
    20.4.4 *Pathogenicity Tests*
    20.4.5 *Immunoassays*
20.5 Data Interpretation
20.6 Remediation and Prevention
20.7 References

20.1 Characteristics

20.1.1 Morphology

Free-living amebae are relatively small, unicellular eukaryotic organisms with asymmetric bodies that constantly change shape through the formation and retraction of pseudopodia (pseudopods). Pseudopodia are cytoplasmic projections with blunt or rounded ends used in locomotion and feeding. Although multinucleate forms are occasionally seen, amebae generally contain a single nucleus and divide by simple binary fission. Fixed and stained amebae reveal at least one contractile vacuole (for expelling excess water and wastes), a granular cytoplasm, and a double-layered cell membrane (Rondanelli, et al., 1987; APHA, 1995a).

20.1.1.1 Naegleria spp. *Naegleria* spp. are slug-like in the active, feeding (trophozoite) stage, 10 to 35 mm long, with the anterior end broader than the posterior (Tyndall and Ironside, 1990) (Figure 20.1a). Movement is through formation of pseudopodia. In addition to the infectious, trophozoite form, *Naegleria* spp. can exist in flagellate and cyst forms. Transformation from the trophozoite to the pear-shaped flagellate form is transitory (i.e., the flagellar form may revert to a trophozoite or may encyst) and most often occurs when the supporting medium is diluted with water. The rapid motility of the flagellate form is by means of two to four anterior flagellae, which are long, hairlike structures (Figure 20.1b). When nutrients become sparse, *Naegleria* spp. can also form smooth-walled, spherical cysts (7 to 15 μm in diameter) (Visvesvara, 1995) (Figure 20.1c). Return from the encysted to the trophozoite or flagellate state occurs when suitable food sources become available.

20.1.1.2 *Acanthamoeba* spp. The trophozoites of *Acanthamoeba* spp. are somewhat larger (15 to 45 μm) than those of *Naegleria* spp. Acanthamoeba trophozoites are characterized by spike-like cytoplasmic projections called acanthopodia (Figure 20.1). Acanthamoebae have no flagellate stage and produce double-layered cysts, 10 to 25 μm long, with wrinkled outer walls (Figure 20.2).

20.1.2 Classification
The amebae are classified in the subkingdom *Protozoa*, phylum *Sarcomastigophora*, subphylum *Sarcodina*. Classification of free-living amebae based solely on morphology is not practical because of their amorphous nature (Tyndall and Vass, 1995). Nevertheless, amebae can be identified from their microscopic appearance using the features mentioned in Section 20.1.1. Serological identification methods avoid the ambiguities of morphological and cytological systems and provide more reliable identification.

20.1.3 Lifestyles and Metabolism
Amebae are phagotrophic and usually digest food particles in vacuoles [see 17.5.1]. Free-living amebae feed by phagocytosis, that is, engulfing algae, bacteria, and other small organisms and absorbing dissolved organic material. Engulfed GNB, including *Legionella* spp., may remain alive (and infectious) within amebae [see 20.1.4].

20.1.4 Ecology
Free-living amebae are widely distributed in the environment (Caron, 1997) and are found on vegetation and in soil. Amebae are also found in water, especially warm waters (e.g., industrial process fluids, swimming pools, natural hot springs, and even laboratory eyewash stations) (Bowman et al., 1996). *Acanthamoeba* spp. have been found in home humidifiers (Tyndall et al., 1995). *Acanthamoeba, Echinamoeba, Hartmannella, Naegleria,* and *Valkampfia* spp. are known to harbor *Legionella* and *Mycobacteria* spp. symbiotically in the environment, which contributes to the amebae's public health significance (Fields, 1993; Visvesvara, 1995). In fact, amebae—especially *Hartmannella* spp.—may play a major role in supporting the growth and survival of legionellae in environmental waters (Fields, 1993; States et al., 1993; Tyndall and Vass, 1995; Fields, 1997; Steinert et al., 1997). Within amebae, bacteria are protected from environmental stresses, including contact with biocides.

20.2 Health Effects
Pathogenic protozoa cause serious health problems worldwide, especially in tropical and Third World countries. However, the majority of protozoa cause gastrointestinal diseases transmitted via ingestion (e.g., amebiasis and giardiasis) or hematogenous diseases transmitted by insect vectors (e.g., malaria) and are not of concern in indoor air. Protozoa that may be transmitted by air and cause infectious disease or inhalation fever are free-living amebae in the genera *Naegleria* and *Acanthamoeba*. Two other genera of amebae, *Hartmannella* and *Balamuthia*, have been confused with *Acanthamoeba* spp. While *Hartmannella* spp. are not known to cause human disease, *Balamuthia* spp. have recently been associated with encephalitis similar to that caused by *Acanthamoeba* spp. (Visvesvara, 1995). This discussion addresses infectious amebae possibly associated with airborne transmission (i.e., *Naegleria fowleri* and *Acanthamoeba* spp.) and amebic antigens that may be associated with hypersensitivity diseases.

20.2.1 Infectious Diseases
Free-living amebae, primarily in the genera *Naegleria* and *Acanthamoeba*, can cause fatal central nervous system (CNS) infections (Visvesvara, 1995). Risk factors for susceptibility are unknown. Apparently healthy people have developed naegleria infections, while immunosuppressed people may be more susceptible to acanthamoeba and probably balamuthia infections (APHA, 1995b).

20.2.1.1 *Naegleria fowleri* Infections Only one species of *Naegleria* (i.e., *N. fowleri*) is known to infect humans, causing a primary amebic meningoencephalitis (PAM) that is nearly always fatal. Over 160 cases of PAM have been reported worldwide (70 in the U.S.) as of January 1994 (Visvesvara, 1995). Human *N. fowleri* infections generally occur in healthy children and young adults who have bathed in spas or engaged in active water sports such as swimming and diving. Waterborne amebae enter the nasal passages where they traverse the nasal mucosa and migrate to the olfactory lobes and cerebral cortex. The ensuing infection results in high fever, severe headache, erratic behavior, and death (Tyndall and Ironside, 1990). At least one report indicated that *Naegleria* encephalitis occurred after exposure to airborne amebic cysts from soil (Lawande et al., 1979).

20.2.1.2 *Acanthamoeba* and *Balamuthia* spp. Infections Several species of *Acanthamoeba* can cause a usually fatal granulomatous amebic encephalitis (GAE). Amebae in the genus *Balamuthia* produce an encephalitis with a clinical and histological picture similar to GAE. Most GAE patients have no history of contact with fresh water. It is believed that penetration of the CNS is via the blood from a primary focus in the lower respiratory tract. Initial entry of amebae may be via inhalation of airborne agents or via skin lesion or other trauma (Visvesvara, 1995). GAE is usually chronic, characterized by an insidious onset and prolonged course. Over 110 cases of GAE have occurred worldwide (~64 in the U.S.) as of January 1994 (Visvesvara, 1995). Acanthamoebae introduced into the eyes have caused severe infections of the cornea, especially in contact lens wearers (Hirst et al., 1984; Tyndall and Vass, 1995). Acanthamoeba eye infections require prompt antibiotic treatment.

20.2.2 Inhalation Fever
Exposure to amebae can lead to inhalation fever. For example, humidifier fever in office and industrial workers has been attributed to aerosolized *Naegleria* antigens (Edwards, et al., 1976; Medical Research Council, 1977; Edwards, 1980) [see 3.4.1 and 8.2.2.1]. However, amebae are better known as agents of infection.

20.3 Sample Collection
20.3.1 Source Sampling
Testing samples of suspect water or soil for amebae is usually more appropriate than attempting to isolate the organisms from the air. Several hundred grams of soil or large volumes of water (≥4 L) should be collected to allow concentration of samples (APHA, 1995a; Caron, 1997). Schaeffer (1997) gives detailed directions for proper water sample collection to recover protozoa.

20.3.2 Air Sampling
Because of their relatively large size, amebic trophozoites and cysts can travel only short distances and remain airborne only briefly. Consequently, exposure is more similar to large-droplet contact transmission than true airborne infection [see 9.4.1]. Relatively large-volume air samples should be collected near points of aerosol generation. Investigators collecting samples should take precautions to avoid mucous membrane contact with splashing water and inhalation of water aerosols.

Kingston and Warhurst (1969) isolated amebae from hospital and outdoor air using a slit sampler, and Lawande (1983) recovered amebae from outdoor air using settle plates. Tyndall and Ironside (1990) described the use of three types of air samplers to collect amebae, that is, liquid impingers, multiple-hole impactors, and high-volume electrostatic precipitators [see 11.3]. However, air sampling studies have recovered few amebae even near known positive water supplies (Tyndall and Vass, 1995).

20.3.2.1 Liquid Impingers To improve collection efficiency, two liquid impingers are often used in series, and their collection fluids are combined for analysis. An impinger with a high airflow rate (e.g., a Greenberg-Smith impinger: flow rate, 28 L/min; sampling time, 30 to 60 min; fluid volume, 150 mL) may be more suitable than an impinger operated at a lower flowrate (e.g., a three-jet, tangential impinger or an all-glass impinger (AGI): flow rate, 12.5 L/min; Samplers 13 and 14, in Table 11.1). A wetted cyclone sampler may also be suitable for collecting airborne amebae (Sampler 11).

20.3.2.2 Multiple-Hole Impactors Multiple-hole impactors have been used to collect samples of amebae (flow rate, 28 L/min; sampling time, 10 to 30 min; Samplers 5 to 7). Impactors should be located to avoid direct splashes but sufficiently close to collect airborne droplets. The agar plates used in these samplers are coated with GNB as described in Section 20.4.1.

20.3.2.3 Electrostatic Precipitators Tyndall and Ironside (1990) used an electrostatic precipitator (ESP) to collect amebae (flow rate, ≤10^3 L/min; sampling time, ≤3 hours; collection fluid volume, ~300 mL). Collection fluid in an ESP circulates over a charged plate onto which particles deposit after being oppositely charged as they enter the sampler [see 11.3]. The tubing for the collection fluid must be decontaminated (e.g., with 70% ethanol) and rinsed to prevent carry over of amebae from one test to another.

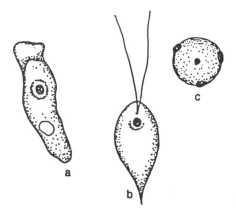

FIGURE 20.1. *Naegleria* **spp. (a) ameboid stage, (b) flagellated stage, (c) cyst stage** (reproduced with permission, APHA, 1995).

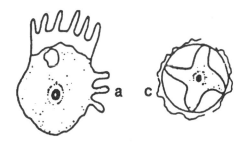

FIGURE 20.2. *Acanthamoeba* **spp. (a) ameboid stage, (c) cyst stage** (reproduced with permission, APHA, 1995).

20.3.3 Sample Handling

Water and air samples for amebae should be kept at room temperature (24° to 28°C) and may be stored at 4°C for up to 24 hours but should not be frozen (Visvesvara, 1995; Schaeffer, 1997).

20.4 Sample Analysis

20.4.1 Culture

Water concentrates and sediments can be examined directly by microscope and can be plated on non-nutrient agar. The agar is covered with a lawn of bacteria isolated from the test water (on the assumption that the amebae were feeding on these in the environment) or with laboratory cultures of *Escherichia coli* or *Klebsiella aerogenes*. Once isolated, *Acanthamoeba* and *Naegleria* spp. can also be cultured on many nutrient media and grown in cell culture (APHA, 1995a; Visvesvara, 1995; Warren et al., 1997).

Culture plates are incubated at 25°, 37°, or 45°C for 7 days or until amebic growth is observed at 10X magnification. The ability of pathogenic *Acanthamoebae* and *Naegleria* spp. to grow at 45°C has proven useful as a semiselective technique for isolating pathogenic free-living amebae from environmental samples (Tyndall and Ironside, 1990). However, some amebae involved in eye infections have neither grown at 37°C nor infected laboratory animals (Bier and Sawyer, 1990). Culture samples are examined daily and amebae belonging to the genera *Naegleria* and *Acanthamoeba* are identified by the morphological characteristics of their trophozoite and cyst forms [see 20.1.1].

20.4.2 Most Probable Number

Analysis of water samples for amebae can be quantitative or qualitative. Tyndall and Ironside (1990) described a quantitative, most-probable-number method using five replicates each of sample volumes of 0.01, 0.1, 1.0, 10, and 100 mL. The 100-mL samples are filtered through 1.2-μm pore cellulose filters, which are then quartered or halved and inverted onto culture plates with bacteria. The 10-mL samples are centrifuged and the pellets are resuspended and plated. The smaller water samples are inoculated directly onto culture plates. Soil samples can be suspended in sterile water, allowed to settle, and treated as water samples. Laboratories should always run known positive and negative control cultures of amebae at the same time as environmental samples.

20.4.3 Microscopy

Water samples concentrated by filtration or centrifugation can be examined in wet-mount preparations at 10X and 40X magnification, preferably using phase-contrast optics (APHA, 1995a; Visvesvara, 1995; Schaeffer, 1997). Amebae are detected by their active directional movement and are identified based on their (a) cell size, shape, and morphological features (Figure 20.1), (b) characteristic pat-

terns of locomotion, and (c) ability to form flagellates (Caron, 1997). Amebic mounts can be stained and preserved as permanent specimens.

Naegleria-like organisms that grow at 45°C can be tested for ability to enflagellate by harvesting trophozoites into sterile distilled water and incubating the suspensions at 35°C. The samples are examined for flagellates at 1, 2, and 3 hours incubation (Tyndall and Ironside, 1990). Up to 50% of *N. fowleri* trophozoites typically transform to the pear-shaped biflagellated form (Visvesvara, 1995) (Figure 20.1b).

20.4.4 Pathogenicity Tests

Naegleria spp. that grow at 45°C and demonstrate flagellation can be instilled intranasally into two-week-old mice to test the amebae's infectivity and to identify *N. fowleri* (Visvesvara, 1995). Mice infected with *N. fowleri* show signs of encephalitis and die within 5 to 7 days. Moribund mice are sacrificed, and their brain tissues are cultured or are stained for microscopic examination.

The presence and concentration of *Acanthamoeba* spp. can be detected by allowing trophozoites on culture plates to encyst and by observing the characteristic double-walled morphology of *Acanthamoeba* cysts (Figure 20.2c). Mouse pathogenicity tests for encephalitic *Acanthamoeba* spp. are carried out as described above. Mice infected with *Acanthamoeba* spp. die of acute disease in 5 to 7 days and of chronic disease after several weeks (Visvesvara, 1995).

20.4.5 Immunoassays

While routinely used, plating methods and mouse pathogenicity tests are labor intensive and cumbersome. Monoclonal antibodies specific for *N. fowleri* and pathogenic *Acanthamoeba* spp. allow rapid and specific detection of pathogenic amebae. If available, flow cytometry can be used to detect and count cells to which fluorescein-tagged antibodies have attached. Fluorescence microscopy is only used to detect amebae concentrated by centrifugation or filtration. Techniques for detecting amebic antigens are analogous to those for detecting other microbial antigens [see 6.5.2].

20.5 Data Interpretation

Investigators should be aware that amebae can cause inhalation fever (e.g., humidifier fever), severe eye and wound infections, and fatal encephalitis (Edwards et al., 1976; Edwards, 1980; Tyndall and Vass, 1995; Visvesvara, 1995). Fortunately, the size of amebic trophozoites and cysts causes them to fall rapidly from air and greatly diminishes the risk of infection when they are discharged as aerosols. Physicians should consider the presence of pathogenic amebae when a patient's infection does not respond to traditional antibiotic treatment. Potential environmental sources of amebae should be tested for pathogenic types if infections are identified. These tests should be conducted by laboratories with experience in the assay of pathogenic amebae. Matching of amebae from clinical specimens and environmental samples may implicate a water system as the source of an infection.

20.6 Remediation and Prevention

Tyndall and Ironside (1990) recommend that decontamination of sources of free-living amebae include removal of biofilm and scale, followed by treatment with an oxidizing biocide such as sodium hypochlorite or hydrogen peroxide. Hydrogen peroxide at 300 to 3000 ppm may act more slowly than bleach, but may be preferred because it is less corrosive and safer to use [see 16.2.4.2]. It has been recommended that eyewash stations be flushed for several minutes at least weekly or monthly to control microbial populations (Bier and Sawyer, 1990; Bowman et al., 1996).

20.7 References

APHA: Pathogenic Protozoa. In: Methods for the Examination of Water and Wastewater, 19th Ed., pp. 9-110-9-117. American Public Health Association, American Water Works Association and Water Pollution Control Federation, Washington, DC (1995).

Benenson, A.S., Ed. Naegleriasis and Acanthamebiasis. In: Control of Communicable Diseases Manual, 16th Ed., pp. 321-323. American Public Health Association, Washington, DC. (1995)

Bier, J.W.; Sawyer, T.K.: Amoebae Isolated from Laboratory Eyewash Stations. Current Microbiol. 20:349-350 (1990).

Bowman, E.K.; Vass, A.A.; Mackowski, R.; et al.: Quantitation of Free-Living Amoebae and Bacterial Populations in Eyewash Stations Relative to Flushing Frequency. Am. Ind. Hyg. J. 57:626-633 (1996).

Caron, D.A.: Protistan Community Structure. In: Manual of Environmental Microbiology, pp. 284-294. C.J. Hurst, G.R. Knudsen, M.J. McInerney, et al., Eds. American Society for Microbiology Press,Washington, DC (1997).

Eaton, A.D.; Clesceri, L.S.; Greenberg, A.E., Eds.: Standard Methods for the Examination of Water and Wastewater, 19th Ed., pp. 9-110 to 9-117. American Public Health Association, Washington, DC (1995).

Edwards, J.H.; Griffiths, A.J.; Mullins, A.J.: Protozoa asSsources of Antigen in 'Humidifier Fever.' Nature, 264:438-439 (1976).

Edwards, J..H.: Microbial and Immunological Investigations and Remedial Action after an Outbreak of Humidifier Fever. Br. J. Ind. Med. 37:55-62 (1980).

Fields, B.S.: Legionella and Protozoa: Interaction of a Pathogen and its Natural Host. In Legionella: Current Status and Emerging Perspectives, pp. 129-136. J.M. Barbaree, R.F. Breiman, and A.P. Dufour, Eds. American Society for Microbiology, Washington, DC (1993).

Fields, B.S.: Legionellae and Legionnaires' Disease. In: Manual of Environmental Microbiology, pp. 666-675. C.J. Hurst, G.R. Knudsen,

M.J. McInerney, et al., Eds. American Society for Microbiology Press, Washington, DC(1997).

Hirst, L.W.; Green, W.R.; Merz, W.; et al.: Management of Acanthamoeba Keratitis: A Case Report and Review of the Literature. Ophthalmology 91:1105-1111 (1984).

Kingston, D.; Warhurst, D.C.: Isolation of Amoebae from the Air. J Med Microbiol. 2:27-36 (1969).

Lawande, R.V.; Abraham, S.N.; John, I.; et al.: Recovery of Soil Amebas from Nasal Passages of Children during the Dusty Harmattan Period in Zaria. Am. J. Clin. Path. 71:201-203 (1979).

Lawande, R.V.: Recovery of Soil Amoebae from the Air During the Harmattan in Zaria, Nigeria. Ann. Trop. Med. Parasitol. 77:45-49 (1983).

Medical Research Council Symposium: Humidifier Fever. Thorax, 32:653-663 (1977).

Page, F.C.: An Illustrated Key to Freshwater and Soil Amoebae. In: Freshwater Biological Association Scientific Publication No. 34, pp. 1-155 (1976).

Rondanelli, E.G.; Carosi, G.;Lanzarini, P.; et al.: Ultrastructure of Acanthamoeba-Naegleria Free-living Amoebae. In: Infectious Diseases. Color Atlas Monographs. I. Amphizoic Amoebae Human Pathology. E.G. Rondanelli, Ed. Piccin, Padua, Italy (1987).

Schaeffer, F.W.:. Detection of Protozoan Parasites in Source and Finished Drinking Waters. In: Manual of Environmental Microbiology, pp. 153-167. C.J. Hurst, G.R. Knudsen, M.J. McInerney, et al., Eds. American Society for Microbiology Press, Washington, DC (1997).

States, S.J.; Podorski, J.A.; Conley, L.F.; et al.: Temperature and the Survival and Multiplication of Legionella pneumophila Associated with Hartmannella vermiformis. In Legionella: Current Status and Emerging Perspectives, pp. 147-149. J.M. Barbaree, R.F. Breiman, and A.P. Dufour, Eds. American Society for Microbiology, Washington, DC (1993).

Steinert, M.; Emödy, L.; Amann, R.; et al.: Resuscitation of Viable but Nonculturable Legionella pneumophila Philadelphia JR32 by Acanthameoba castellanii. Appl. Environ. Microbiol. 63:2047-2053 (1997).

Tyndall, R.L.; Ironside, K.S.:. Free-living Amoebae: Health Concerns in the Indoor Environment. In Biological Contaminants in Indoor Environments, pp. 163-175. P.R. Morey, J.C. Feeley, and J.A. Otten, Eds. American Society for Testing and Materials, Philadelphia, PA (1990).

Tyndall, R.L.; Vass, A.A.:. The Potential Impact on Human Health from Free-Living Amoebae in the Indoor Environment. In Bioaerosols, pp. 121-132. H.A. Burge, Eds. Lewis Publishers, Boca Raton, FL (1995).

Tyndall, R.L.; Leman, E.S.; Bowman, E.K.; et al.: Home Humidifiers as a Potential Source of Exposure to Microbial Pathogens, Endotoxins, and Allergens. Indoor Air 5:171-178 (1995).

Visvesvara, G.S.:. Pathogenic and Opportunistic Free-Living Amebae. In: Clinical Manual for Microbiology, 6th Ed., pp. 1196-1203. P.R. Murray, E.J. Baron, M.A. Pfaller, et al., Eds, American Society for Microbiology, Washington, DC (1995).

Warren, A.; Day, J.G.; Brown, S.: Cultivation of Algae and Protozoa. In: Manual of Environmental Microbiology, pp. 61-71. C.J. Hurst, G.R. Knudsen, M.J. McInerney, et al., Eds. American Society for

Chapter 21

Viruses

James A. Otten and Harriet A. Burge

21.1   Characteristics
    21.1.1   *Morphology*
    21.1.2   *Classification*
    21.1.3   *Lifestyles and Metabolism*
    21.1.4   *Ecology*
21.2   Health Effects
21.3   Sample Collection
    21.3.1   *Collection Methods*
    21.3.2   *Sample Handling*
21.4   Sample Analysis
21.5   Data Interpretation
21.6   Remediation and Prevention
    21.6.1   *Host Immunization and Treatment*
    21.6.2   *Engineering Controls*
21.7   References

21.1 Characteristics

Characteristics that distinguish viruses from cellular microbial forms are (a) viruses contain only one type of nucleic acid (DNA or RNA), and (b) viruses are obligate parasites and depend completely on their hosts for reproduction (i.e., the virus provides the instructions for making new viruses, but the host cell manufactures them). The science of *virology* emerged during the last decade of the nineteenth century with the discovery that some plant and animal diseases (e.g., tobacco mosaic as well as foot and mouth disease) were caused by infectious agents smaller than bacteria or fungi. The 1911 discovery of a virus that caused malignant tumors in chickens began the study of tumor virology, and today oncoviruses are known to cause many animal tumors. In the late 1940s, researchers developed techniques for growing live cells in culture, which allowed the study of virus and cell interactions outside normal hosts.

21.1.1 Morphology

Viruses are replicating, non-cellular particles that constitute the smallest and simplest known infectious agents. Viruses were first described as "filterable agents" because they could pass through filters that retained bacteria and other cells. Viruses are a heterogenous class of agents varying in chemical composition, host range, effect on a host, and size and morphology (Figure 21.1).

The size of animal viruses ranges from ~20 to 300 nm. In the air, it is unusual to find single virus particles; viruses typically travel attached to other particles. The simplest viruses consist of nucleic acid (RNA or DNA, which is either single or double stranded and either linear or circular) enclosed in a protein coat called a capsid. A capsid typically consists of a single layer of protein molecules arranged in an icosahedral shell or a helical tube. In addition, some viruses have a lipoprotein envelope that surrounds the nucleocapsid. Viral envelopes are derived from the host cell's membrane and from virus-specific proteins synthesized within the host.

21.1.2 Classification

Viruses are not easy to classify among living organisms because they are not truly "alive," except in association with a living cell. The key properties on which the 73 viral families were established are the kind and strandedness of the viral nucleic acid and the presence or absence of a lipoprotein envelope (Melnick, 1995) (Figure 21.1). Classification within seven clusters is based on viral size, shape, structure, and mode of replication.

Viruses can also be divided into animal, plant, and bacterial viruses (bacteriophages). A particular virus can typically infect only a certain type of cell, although some viruses can infect more than one host or cell type, for example, both humans and other animals or both kidney

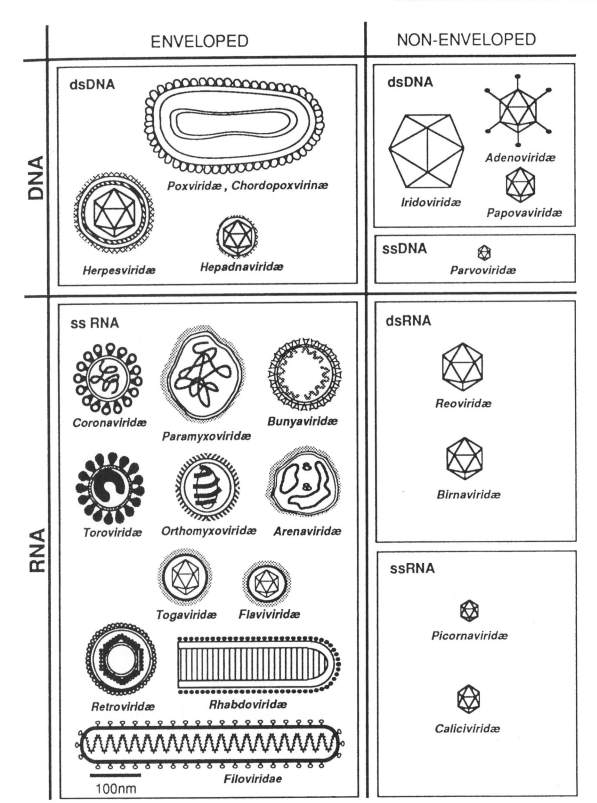

FIGURE 21.1. Drawings of viruses in the common virus families of vertebrates (Melnick, 1995; reproduced with permission, ASM).

and lung cells. The host range of a virus is determined primarily by the specificity of the viral attachment to a cell membrane, which depends on properties of the virus coat and cell surface receptors.

21.1.3 Lifestyles and Metabolism

Two types of interactions are common between viruses and cells (a) host cell destruction or lysis, and (b) nonlytic infection that may alter the host cell and include integration of viral genetic material with host DNA. Viral multiplication is divided into distinct events. First, a virus adsorbs to a cell surface, usually at a specific site. Next, the virus (either the entire virus or only the genetic material) penetrates the cell membrane and the capsid separates or uncoats from the viral nucleic acid. Uncoating begins the eclipse phase of viral multiplication, during which viral genetic material replicates and viral proteins are synthesized. Formation of complete viruses signals the end of the eclipse phase. Viruses that integrate with their host's genome may not complete all of these steps. Finally, the complete viruses are assembled, and the progeny are released. Release may involve lysis of the host cell, resulting in its death, or individual viruses may bud through the host's cell membrane without damaging the cell. The entire process of viral multiplication can take from 6 to 48 hours.

21.1.4 Ecology

Although viruses are obligate parasites, they are found in a wide variety of environments. Viruses do not survive well outside their hosts, and enveloped viruses die more quickly in the environment than non-enveloped viruses (Hierholzer, 1990). Environmental factors that de-termine whether an airborne virus will survive outdoors include season, moisture content and temperature of the air, wind conditions, sunlight, and presence of atmospheric pollutants (Cox, 1995). Indoors, RH and temperature are important factors controlling the infectivity of airborne viruses.

Some viruses undergo continual evolutionary change, for example, the antigenic makeup of the influenza virus changes over time. Viruses are of interest to public health authorities as agents of wide-spread, common infections (e.g., colds and influenza) as well as emerging infectious diseases [e.g., acquired immunodeficiency syndrome (AIDS) and Hantavirus pulmonary syndrome (HPS)]. Human immunodeficiency virus (HIV) is not transmitted through the air, but Sin Nombre virus [SNV, or the Four Corners virus (FCV)] may be, as well as by contact. Natural and manmade environmental changes may affect a virus' geographical and host range and its impact on people (Lederberg, et al., 1992).

21.2 Health Effects

As far as is known, viruses cause only infectious disease [see 3.5.2 and 8.2.3.3]. Table 21.1 lists examples of the many areas of the human body in which airborne viruses can cause infection. Viral infection may also cause birth defects, cancer, and behavioral changes.

In addition to the respiratory tract, viruses enter human hosts through the mucous membranes of the nose, mouth, and urogenital area; the conjunctiva of the eyes; the gastrointestinal tract; and the skin via, insect or animal bites or through open wounds. Airborne transmission of viral diseases has been known since the World War I era. The majority of common viral infections are spread through

TABLE 21.1 Health Effects of Selected Airborne and Droplet-Borne Viruses

Symptom Site	Disease	Virus
Skin	Papular-vesicular rash: varicella (chicken pox), shingles	Varicella-zoster virus
	Macular-erythematous rash: rubella (German measles)	Rubella virus
	Macular-hemorrhagic rash: measles (rubeola)	Measles virus
Central nervous system	Encephalomyelitis, viral meningitis	Measles virus, arenaviruses, mumps virus
Salivary glands	Infectious parotitis (mumps)	Mumps virus
Upper respiratory tract	Rhinitis, pharyngitis, common cold	Rhinovirus, enterovirus, adenovirus, parainfluenza virus, coronavirus
Middle and lower respiratory tract	Tracheitis, bronchitis, pneumonia	Parainfluenza virus, respiratory syncytial virus, influenza virus
	Hantavirus pulmonary syndrome	Hantavirus (Sin Nombre or Four Corners virus)

the air (e.g., influenza, measles, chicken pox, and some acute viral respiratory diseases) (Jennings et al., 1988). Large-droplet transmission plays a role in the spread of common cold and mumps viruses [see 9.4.1]. Some enteric viruses may also be transmitted through the air (Sawyer et al., 1988; Caul, 1994). Rarely, viral diseases normally transmitted by an animal bite or direct, intimate contact (e.g., respectively, rabies or hepatitis) may be transmitted by air if viruses become airborne in sufficient quantity.

Nearly all airborne viral diseases are transmitted directly from one person to another although transmission from animals to humans may also occur. Some viral infections show strong seasonal variation associated with periods when people spend more time together indoors and buildings are sealed most tightly. Riley et al. (1978) demonstrated the effects of crowding on measles transmission, and Brundage et al. (1988) observed rates of acute respiratory infections due to adenoviruses, among other agents. The latter study followed recent army recruits housed in older, drafty barracks and in newer, tighter buildings. Soldiers in the newer barracks, where there was a lower supply rate of outdoor air, shared more indoor air and the infectious viruses it contained [see 3.5.2 and 9.8.6.3].

In general, disease risk associated with exposure to viral aerosols depends on individual susceptibility, viral virulence and concentration, and particle size. The deposition site for an airborne virus depends on the anatomy of the host's respiratory tract and on the size of the particle carrying a virus (Knight, 1980). Very large particles are trapped by nasal hairs, particles larger than 10-µm deposit on the nasal epithelium, and particles below 5 µm reach the deeper portions of the lungs [see 9.3.4].

Viruses cannot initiate infection unless they survive the body's natural defense mechanisms and attach to cells that will support their replication. The mechanisms and determinants of viral pathogenicity are still poorly understood. Although cytopathic viruses kill cells, a virus itself usually plays a passive role in cell damage. There is no convincing evidence that viruses produce toxins—as do some bacteria and fungi; although viruses have themselves been viewed as "molecular toxins." Successful viral infection may go undetected or may cause mild, severe, or fatal disease.

21.3 Sample Collection
21.3.1 Collection Methods
Researchers may attempt environmental sampling to better understand how viral diseases are transmitted and to verify that airborne transmission of a particular virus is possible (Hurst, 1997a,b; Sattar and Ijaz, 1997). Viruses associated with airborne infections generally do not remain infective very long in the environment (either in air or in inanimate reservoirs) nor do viruses multiply alone in

water or on organic substrates as do saprobic bacteria and fungi and phagotrophic amebae [see 21.1.4].

Air sampling for infectious viruses is not conducted routinely as part of IEQ investigations. Confirmed infection in one or more persons may be considered sufficient evidence that a particular viral agent was present. However, for research purposes, air sampling has been used to document transmission of airborne viral diseases. Techniques that have been used to collect air samples for viral analysis have included filtration, liquid impingement, impaction, electrostatic precipitation, and centrifugal scrubbing (Sattar and Ijaz 1997). The technique chosen to collect viral aerosols depends on the virus itself, the environmental conditions, and the assay methods available [see 21.4].

21.3.2 Sample Handling
Samples for viral culture need careful preservation after collection; for example, sample containers must be sterile and free of interfering compounds and samples must be protected from acids, heat, and drying. Many viruses are unstable at an acid pH or at temperatures above 4°C. Materials should be refrigerated or placed on ice if delivery time is longer than an hour. Most viruses lose infectivity rapidly when kept at –15 to –20°C, but can be frozen above –70°C for long-term storage. Drying can be prevented by storing samples in gelatin, serum, or a buffered salt solution. Small amounts of penicillin (50 units/mL), amphotericin (1 µg/mL), and streptomycin (5 µg/mL) can be added to storage liquid to prevent fungal and bacterial growth.

21.4 Sample Analysis
The concentration of active viruses in environmental and clinical samples have traditionally been determined by (a) introducing test material into cultured cells, (b) injecting test material into an animal (e.g., a newborn or suckling mammal or a chick embryo), or (c) virus counting using electron microscopy, measurements of optical density of viral suspensions, or hemagglutination assays (Gerba and Goyal, 1982; Hierholzer, 1990; Sawyer et al., 1994; Hensel and Petzoldt, 1995; Jones Brando, 1995; Hurst, 1997a,b; Payment, 1997). However, modern molecular methods, such as PCR, have reduced the necessity for viral cultivation [see 6.2.3 and 6.6]. Specifically designed molecular probes now permit the screening of environmental samples for some viral agents, primarily in water and soil samples. Clinical specimens from exposed workers can also be tested for biological indicators of viral infection using assays such as immunoprecipitation, radioimmunoassays, countercurrent immunoelectrophoresis, detection of viral nucleic acids in body fluids, and tissue staining (Herrmann, 1995; Podzorski and Persing, 1995).

The type of collection medium and host cell chosen for viral culture depends on the virus under study, and the type of culture medium used depends on the chosen host cells (Hensel and Petzoldt, 1995; Payment, 1997). Viral culture yields information about the number of infectious viral units in a sample, which are identified by a virus' ability to infect host cells, multiply, and produce progeny. The sample used to infect host cells is diluted (titrated) to determine the endpoint at which infection still occurs. Infected cells are recognized by changes in their appearance, attachment of fluorescent-labeled antibodies specific for viral proteins, or degenerative changes (cytopathology). Culturing of viruses from environmental samples is complicated by the high rate of cell culture contamination by other microorganisms and the possibility of toxic interference from chemicals in environmental samples (Payment, 1997).

21.5 Data Interpretation

The number of virus particles required to cause infection in a susceptible individual is not known for most viruses. However, evidence suggests that for some viruses one particle may be sufficient to initiate infection (Goodlow and Leonard, 1961). The data retrieved from environmental sampling is usually the number of viruses in a sample that are able to infect cultured cells or laboratory animals. This number is not necessarily the same as that needed to infect a human host. Therefore, data from such samples may indicate only (a) the presence and concentration of the virus, or (b) the probability of its absence. Investigators must establish their own criteria for interpreting such data with respect to risks for humans.

21.6 Remediation and Prevention

21.6.1 Host Immunization and Treatment

The body can mount effective immune responses to viruses following either natural infection or artificial immunization. The Chinese practiced smallpox immunization centuries before anyone understood what caused the disease. Jenner introduced vaccination to prevent smallpox in 1798, and Pasteur demonstrated the effectiveness of the rabies vaccine in 1877. Vaccination continues to be the best means of preventing viral diseases [see 8.2.3.3 and 8.2.3.4]. In fact, smallpox has been eradicated as a result of worldwide immunization programs. Unfortunately, vaccines are not available for all viral agents.

Personal hygiene also plays an important role in preventing viral transmission. Therefore, people should wash their hands frequently, cover sneezes and coughs, and remain at home when ill with a communicable viral infection [see 9.8].

Viral infections do not respond to antibiotics as do bacterial infections, but antibiotics are used to treat the secondary bacterial infections that may occur in weakened hosts. Viral infections can be treated with a number of antiviral chemotherapeutic drugs that block various stages of viral replication. However, many such drugs work efficiently in cultured cells, but either inhibit viral replication poorly in the body or are highly toxic, producing serious side effects.

21.6.2 Engineering Controls

Three engineering control measures are used to reduce the spread of viral aerosols indoors: ventilation, filtration, and air disinfection [see 9.8, 10.5.2.1, 10.5.3, and 10.5.6].

21.7 References

Brundage, J.F.; Scott, R. McN.; Lednar, W.M.; et al.: Building-Associated Risk of Febrile Acute Respiratory Diseases in Army Trainees. J. Am. Medical Assoc. 259:2108-2112 (1988).

Caul, E.O.: Small Round Structured Viruses: Airborne Transmission and Hospital Control. Lancet. 343:1240-1242 (1994).

Cox, C.S.: Stability of Airborne Microbes and Allergens. In: Bioaerosols Handbook, pp. 77-99. C.S. Cox and C.M. Wathes, Eds. Lewis Publishers, Boca Raton, FL (1995).

Gerba, C.P.; Goyal, S.M.: Methods in Environmental Virology. Marcel Dekker, Inc. New York, NY (1982).

Goodlow, R.J.; Leonard, F.A.: Viability and Infectivity of Microorganisms in Experimental Airborne Infection. Bacteriol. Rev. 25:182-187 (1961).

Hensel, A.; Petzoldt, K.: Biological and Biochemical Analysis of Bacteria and Viruses. In Bioaerosols Handbook, pp. 335-360. C.S. Cox and C.M. Wathes, Eds. Lewis Publishers, Boca Raton, FL (1995).

Herrmann, J.E.: Immunoassays for the Diagnosis of Infectious Diseases. In: Manual of Clinical Microbiology, 6th Ed., pp. 110-122. P.R. Murray, Ed. American Society for Microbiology, Washington, DC (1995).

Hierholzer, J.C.: Viruses, Mycoplasmas and Pathogenic Contaminants in Indoor Environments. In Biological Contaminants in Indoor Environments, pp. 21-49. P.R. Morey, J.C. Feeley, and J.A. Otten, Eds. American Society for Testing and Materials, Philadelphia, PA. (1990).

Hurst, C.J.: Detection of Viruses in Environmental Waters, Sewage, and Sewage Sludge. In: Manual of Environmental Microbiology, pp. 168-175. C.J. Hurst, G.R. Knudsen, M.J. McInerney, et al., eds. American Society for Microbiology Press, Washington, DC (1997a).

Hurst, C.J.: Sampling Viruses from Soil. In: Manual of Environmental Microbiology, pp. 400-405. C.J. Hurst, G.R. Knudsen, M.J. McInerney, et al., Eds. American Society for Microbiology Press, Washington, DC (1997b).

Jennings, L.C.; Dick, E.C.; Mink, K.A.; et al.: Near Disappearance of Rhinovirus Along a Fomite Transmission Chain. Journal of Infectious Diseases 158(4):888-892 (1988).

Jones Brando, L.V.: Cell Culture Systems. In: Manual of Clinical Microbiology, 6th Ed., pp. 158-165. P.R. Murray, Ed. American Society for Microbiology, Washington, DC (1995).

Knight, V.: Viruses as Agents of Airborne Contagion. Annuals New York Acad Sciences. Part V, 147-156 (1980).

Lederberg, J.: Shope, R.E.; Oaks, S.C., Eds.: Emerging Infections: Microbial Threats to Health in the United States. National Academy Press, Washington, DC (1992).

Melnick, J.L.: Taxonomy of Viruses. In: Clinical Manual for Microbiology, 6th Ed., pp. 859-867. P.R. Murray, E.J. Baron, M.A. Pfaller, et al., Eds. American Society for Microbiology, Washington, DC, (1995).

Payment, P.: Cultivation and Assay of Viruses. In: Manual of Environmental Microbiology, pp. 72-78. C.J. Hurst, G.R. Knudsen, M.J. McInerney, et al., Eds. American Society for Microbiology Press, Washington, DC (1997).

Podzorski, R.P.; Persing, D.H.: Molecular Detection and Identification of Microorganisms. In: Manual of Clinical Microbiology, 6th Ed., pp. 130-157. P.R. Murray, Ed. American Society for Microbiology, Washington, DC (1995).

Riley, E.C.; Murphy, G.; Riley, R.L.: Airborne Spread of Measles in a Suburban Elementary School. Am. J. Epidemiol. 107:421-432 (1978).

Sattar, S.A.; Ijaz, M.K.:. Airborne Viruses. In: Manual of Environmental Microbiology, pp. 682-692. C.J. Hurst, G.R. Knudsen, M.J. McInerney, et al., Eds. American Society for Microbiology Press, Washington, DC (1997).

Sawyer, L.A.; Murphy, J.J.; Kaplan, J.E.; et al.: Twenty-five to 30-nm Virus Particle Associated with a Hospital Outbreak of Acute Gastroenteritis with Evidence for Airborne Transmission. Am. J. Epidemiol. 127:1261-1271 (1988).

Sawyer, M.H.; Chamberlin, C.J.; Wu, Y.N.; et al.: Detection of Varicella-Zoster Virus DNA in Air Samples from Hospital Rooms. J. Inf. Dis. 169:91-94 (1994).

Chapter 22

House Dust Mites

Larry G. Arlian

22.1 Characteristics
    22.1.1   *Morphology and Size*
    22.1.2   *Classification*
        22.1.2.1   Relationship of Dust Mites to Other Acari Subclasses
        22.1.2.2   Mites Commonly Found in House Dust
    22.1.3   *Lifestyles and Metabolism*
        22.1.3.1   Feeding
        22.1.3.2   Dependence of Mite Feces (Allergen) Production on RH and Mite Feeding
        22.1.3.3   Life Cycle
        22.1.3.4   Fecundity
    22.1.4   *Ecology*
        22.1.4.1   Importance of Ambient RH on Dust-Mite Prevalence
        22.1.4.2   Mechanism for Active Uptake of Water Vapor
        22.1.4.3   Dependence of Mite Survival on RH Rather Than Absolute Water in Air
22.2 Health Effects
22.3 Sample Collection
    22.3.1   *Source Sampling*
        22.3.1.1   Vacuum Samples
        22.3.1.2   Heat Escape and Passive Transfer Samples
    22.3.2   *Air Sampling*
    22.3.3   *Sample Handling*
22.4 Sample Analysis
    22.4.1   *Processing Dust Samples*
        22.4.1.1   Suspension Extraction Separation
        22.4.1.2   Flotation Separation
    22.4.2   *Mounting Mites and Species Determination*
        22.4.2.1   Temporary and Permanent Mounts
        22.4.2.2   Mite Identification
    22.4.3   *Mite Allergen Assay*
    22.4.4   *Guanine Assay*
22.5 Data Interpretation
22.6 Remediation and Prevention
    22.6.1   *Acaricides*
    22.6.2   *Environmental Controls*
22.7 References

22.1 Characteristics

22.1.1 Morphology and Size

The phylum Arthropoda includes insects, crustaceans, and arachnids (e.g., spiders, ticks, and mites) [see 22.1.2]. Characteristics of this phylum are an exoskeleton, jointed appendages, and a hemocoel—a blood-filled body cavity. The Arthropoda contain two major subphyla of living groups of organisms, that is, the Mandibulata and the Chelicerata. House dust mites belong to the latter subphylum and are more closely related to spiders than to insects, which belong to the former.

1. **Mandibulata** — arthropods with mouthparts consisting of mandibles, labrum, maxillae, and labium. These structures may be modified for chewing (e.g., jaw-like, as in the grasshopper) or for piercing and sucking (e.g., stylet-like, as in the mosquito, louse, and flea). Representative groups include the (a) Insecta — insects, (b) Crustaceans — lobsters and crabs, (c) Diplopoda — millipedes, and (d) Chilopoda — centipedes.

2. **Chelicerata** — organisms, including mites, with mouthparts consisting of chelicerae and pedipalps. The chelicerae vary in structure but may consist of fixed and movable digits that are pincer-like (as in dust mites), modified for piercing (e.g., stylet-like, as in spider mites), or modified for cutting (e.g., sickle-like, as in ticks). The Chelicerata are divided into three classes (a) Merostomata — horseshoe crabs, (b) Pycnogonida — sea spiders, and (c) Arachnida — spiders, mites, ticks, scorpions, harvestman spiders (daddy-longlegs), and so forth.

Organisms in the class Arachnida, including dust mites, have four pairs of legs, numbered I to IV from the anterior (forward) section backward. The class Arachnida is divided into 11 subclasses based on the number of body regions and type of abdominal segmentation. Mites belong to the subclass Acari. Mites differ from other arachnids in having only one body region (an idiosoma, Figure 22.1) while other arachnids have two body regions.

Although present in many residences and other buildings, dust mites are unnoticed because of their small size. Female *Dermatophagoides farinae* measure ~360 to 425 μm in length and weigh (fresh) between 10 and 16 μg.

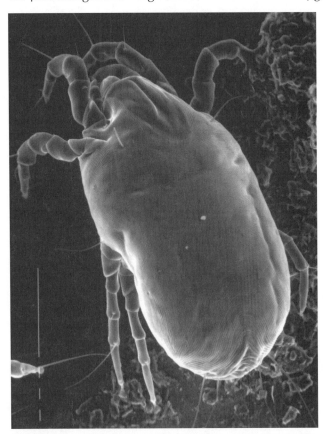

FIGURE 22.1. Adult female *D. farinae.* (magnification: 234X; the long bar in the lower left corner is ~90 μm, the shorter bars are ~12 μm)

Female *Dermatophagoides pteronyssinus* are only ~300 to 350 μm long. By comparison, *Euroglyphus maynei* is very small, with females measuring 225 to 280 μm. *Blomia tropicalis* is a large mite compared to the other three, with females measuring 440 to 520 μm. Males are smaller than females for each species. Likewise, developmental stages are smaller than adults, with larvae being the smallest [see 22.1.3.3].

22.1.2 Classification

Section 17.2 outlines the taxonomic organization of life forms. Dust mites belong to the kingdom of animals; phylum, Arthropoda (subphylum, Chelicerata); class, Arachnida (subclass, Acari); and group, Astigmata; with three genera (*Dermatophagoides*, *Euroglyphus*, and *Blomia*) important for humans indoors.

It is important to keep in mind that many mite species, besides house dust mites, may be found indoors. Therefore, some acarological expertise is necessary to recognize house dust mites when analyzing dust samples to identify what mite species are present. Some mites are parasitic on humans and other animals (e.g., *Sarcoptes scabiei*, the scabies mite) and others may transmit infectious diseases (e.g., rickettsial disease transmitted from mice to humans by *Liponyssoides sanguineus*). However, these diseases are acquired through skin penetration rather than inhalation of mite-related particles. Follicle mites are common in the hair follicles and sebaceous glands of humans. These mites occasionally induce inflammation that could be confused with atopic dermatitis or an inflammation from an infestation with scabies mites.

Occasionally, reference is made to "paper mites" occurring in buildings. It should be pointed out that there is no such creature. However, people may see predaceous mites, spider mites (which are plant parasites), and book lice walking across the pages of books and papers, which may be the source of this myth.

22.1.2.1 Relationship of Dust Mites to Other Acari Subclasses

The subclass Acari is divided into seven groups of mites and ticks based on the location of the external opening (stigmata) to the respiratory tract. The allergy-causing house dust mites and other domestic mites found indoors belong to the Astigmata (no stigmata) group. All groups are listed below because ticks and mites other than allergy-causing dust mites may also be found in dwellings and workplaces, and investigators may need to differentiate among these groups.

1. **Mesostigmata.** As the name "meso" (middle) implies, the mesostigmata are characterized by paired stigmata in the middle of the body lateral to legs III and IV. These mites are avian and mammalian parasites, and several predaceous species may be found in house dust, where they feed on dust mites.

2. **Prostigmata.** The stigmatal openings of the prostigmata are located on the anterior margin of the body. Examples include chiggers, follicle mites, straw itch mites, and cheyletid mites. Several species of predaceous cheyletid mites are found in house dust where they feed on dust mites.

3. **Metastigmata.** The respiratory openings for these mites are located behind legs IV. This group contains the hard and soft ticks; for example, *Dermacentor andersoni* (Rocky Mountain wood ticks), *Dermacentor variabilis* (American dog tick), *Ixodes scapularis* (Lyme disease tick in eastern to midwestern U.S.), and *Ixodes pacificus* (Lyme disease tick on the U.S. west coast).

4. **Astigmata.** Mites of this group have no organized respiratory tract and external stigmata. Most mites found in human dwellings belong to this group, for example, house dust mites (Table 22.1), storage (domestic) mites (Table 22.2), the scabies mite (*Sarcoptes scabiei*), and mange mites (*Psoroptes* spp.).

TABLE 22.1. Common House Dust Mites

Blomia tropicalis	*Dermatophagoides farinae*
Euroglyphus maynei	*Dermatophagoides microceras*
	Dermatophagoides pteronyssinus

TABLE 22.2. Common Storage (Domestic) Mites

Acarus siro	*Glycyphagus domesticus*
Aleuroglyphus ovatus	*Lepidoglyphus destructor*
Chortoglyphus arcuatus	*Tyrophagus putrescentiae*

5. **Cryptostigmata.** The stigmata on these mites (commonly known as soil mites) are hidden at the base of the legs. These mites are heavily sclerotized (hardened and rigid), beetle-like, and usually brown in color as adults. They may be carried indoors on clothing or in pet hair.

6. **Notostigmata.** The stigmata on these mites are located on the dorsal (back) idiosoma.

7. **Tetrastigmata.** The idiosoma of this group bears four pairs of stigmatal openings.

22.1.2.2 Mites Commonly Found in House Dust The mites most commonly found in house dust in homes worldwide are *D. farinae*, *D. pteronyssinus*, *E. maynei*, and *B. tropicalis* (Arlian, 1989). In the U.S., all of the dust mites in Table 22.1 may be found indoors (Arlian et al., 1992), but *D. farinae* and *D. pteronyssinus* are found most frequently

and are most widely distributed geographically. Persons analyzing dust samples should be aware that *D. microceras* may occur, but its prevalence in U.S. homes is not clear because this species is morphologically similar to and may be confused with *D. farinae* (Griffiths and Cunnington, 1971). *Blomia* spp. and *E. maynei* occur in some homes in the southern U.S. (Fernandez-Caldas et al., 1990; Arlian et al., 1992). Both *D. farinae* and *D. pteronyssinus* co-inhabit U.S. homes, with one species usually predominating (Arlian et al., 1992). The more prevalent species in co-inhabited homes generally constitutes >70% of the total mite population, but which species predominates varies among the residences in a geographical area.

22.1.3 Lifestyles and Metabolism

22.1.3.1 Feeding All arthropods are phagotrophs and feed by ingestion. Respectively, house dust and storage mites feed on detritus of animal or vegetable origin and live in close association with their food sources (Squillace, 1995). Dust mites are most prevalent in high-use areas (e.g., living rooms, bedrooms, and areas with carpets and upholstered furniture) where shed skin scales collect and serve as food. The genus name, *Dermatophagoides* (from "derma," skin, and "phagein," to eat), illustrates that human skin scales and animal dander are important food sources for dust mites. It has been estimated that the skin one human sheds each day (~1 g) (Hamilton et al., 1992) could feed several thousand mites for up to three months (Lundblad, 1991).

22.1.3.2 Dependence of Mite Feces (Allergen) Production on RH and Mite Feeding The major allergens responsible for hypersensitivity reactions to dust mites are found in their feces and dried body fragments. During their life span, mites have been reported to produce feces weighing

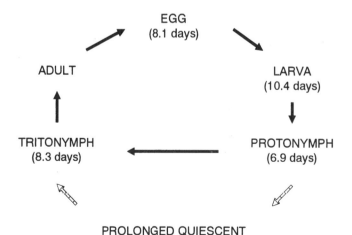

FIGURE 22.2. Life Cycle of *D. pteronyssinus* at 23°C, 75% RH.

up to 200 times their body weight (Hamilton et al., 1992). Fecal pellets, containing partially digested food and digestive enzymes, are ~10 to 35 μm in diameter (IOM, 1993). Ambient RH may influence the rate at which feeding mites produce allergen and its accumulation in dust (Arlian, 1977, 1992). Experiments have shown that mites feed sparingly at RHs < 70%, producing little fecal material and associated allergen. Mites consumed 75% less food (with a corresponding reduction in fecal pellet production) at 75% RH than at 85%. Therefore, simply lowering RH from 85% to 75% may result in a significant drop in the production of fecal allergen even though mites continue to survive [see 22.1.4.1].

22.1.3.3 Life Cycle The life cycle of house dust mites consists of five stages: egg, larva, protonymph, tritonymph, and adult (Figure 22.2). Each stage has an active feeding period followed by a short, non-feeding, quiescent period before the next stage emerges from the old exoskeleton.

22.1.3.4 Fecundity The desiccation-resistant, prolonged, quiescent protonymphal stage of dust mites (Figure 22.2) allows them to survive long dry periods (e.g., the heating season) (Ellingsen, 1975; Arlian et al., 1983). The duration of the protonymphal stage may be several months under unfavorable climatic conditions. The protonymph remains glued to a substrate (e.g., a carpet fiber) so that it is not removed by vacuuming. This stage serves to initiate the development of a mite population when indoor conditions again become optimal (e.g., spring in temperate zones). Dust mites have no visual organs (i.e., they are sightless), and males probably locate females by detecting a female pheromone (Furumizo, 1975).

The duration of the life cycle of a dust mite depends directly on temperature (Arlian et al., 1990; Arlian and Dippold, 1996). The microhabitats in which mites are found indoors vary in temperature (cf., mattresses and floors), and temperatures can fluctuate rapidly within a microhabitat. Therefore, mite development in low-temperature locations (e.g., floors) is slower than in warmer locations (e.g., mattresses or couches) (Colloff, 1987; Hart and Fain, 1988; Arlian et al., 1990; Arlian and Dippold, 1996). Table 22.3 illustrates the influence of temperature on mite life cycle length. Table 22.4 provides information on mite fecundity for *D. farinae* and *D. pteronyssinus* and shows the effect of temperature on the latter's reproductive rate.

22.1.4 Ecology

22.1.4.1 Importance of Ambient RH on Dust-Mite Prevalence Both food and water are critical for mite survival. However, ambient RH determines mite prevalence because adequate food (e.g., human skin scales) is available in most homes and other occupied spaces regardless of climate (Arlian, 1992). In humid regions, practically all homes and many other buildings contain breeding populations of dust mites. In dry climates (e.g., Colorado), fewer homes contain dust mites, and mite densi-

---

TABLE 22.3. Length of Life Cycle (Egg to Adult) for *Dermatophagoides* spp. at Different Temperatures (75% RH)

Temperature (°C)	D. farinae[A] (mean days ± SD)	D. pteronyssinus[B] (mean days ± SD)
16	140.1 ± 14.7[C]	122.8 ± 14.5
23	35.6 ± 4.4	34.0 ± 5.9
30	17.5 ± 1.2	19.3 ± 2.5
35	no data[D]	15.0 ± 2.0

[A] Data from Arlian and Dippold (1996).
[B] Data from Arlian et al. (1990).
[C] <9% of eggs completed the life cycle.
[D] <2% of eggs completed the life cycle.

TABLE 22.4. Reproductive Statistics for *Dermatophagoides* spp. Reared at 23° or 35°C (75% RH)

Parameter	D. farinae[A] 23°C (mean ± SD)	D. pteronyssinus[B] 23°C (mean ± SD)	D. pteronyssinus[B] 35°C (mean ± SD)
Female longevity (days)	100.4 ± 59.8	31.2 ± 11.1	15.5 ± 9.6
Reproductive period (days)	34.0 ± 10.7	26.2 ± 10.7	11.6 ± 6.4
Number egg-laying days	31.3 ± 8.6	23.3 ± 10.8	10.8 ± 6.0
Total number eggs/female	65.5 ± 17.4	68.4 ± 30.4	48.0 ± 29.6
Number eggs/day/female	2.0 ± 0.4	2.5 ± 0.7	3.3 ± 1.3
Number eggs/egg-laying day/female	2.1 ± 0.3	2.8 ± 0.5	4.3 ± 1.0
Eggs/day/female (range)	0 – 5	0 – 8	0 – 12

[A] Data from Arlian and Dippold (1996).
[B] Data from Arlian et al. (1990).

ties in buildings that have them are low. However, altering the indoor environment by raising indoor humidity through the use of evaporative coolers may lead to conditions that facilitate mite survival (Ellingson et al., 1995). Mite densities exhibit a seasonal cycle that parallels RH changes, with the highest mite concentrations occurring during periods of high RH (Bronswijk and Sinha, 1971; Charlet et al., 1978; Furumizo, 1978; Lang and Mulla, 1978; Murray and Zuk, 1979; Arlian et al., 1982; Colloff, 1991). Mites die from desiccation during long dry periods (e.g., the heating season and summers in some regions). However, depending on housekeeping practices, a graveyard of dead mites may remain for some time after all mites die.

22.1.4.2 Mechanism for Active Uptake of Water Vapor

The microenvironments in which dust mites live (e.g., carpets, upholstered furnishings, and bedding) have no liquid water available. At RHs above the critical humidity for mite survival, active mite stages extract sufficient water from unsaturated air to compensate for water lost by evaporation, excretion, defecation, egg production, and other avenues (Arlian and Wharton, 1974). Paired glands opening above the most anterior legs are involved in the active uptake of ambient water vapor (Wharton and Furumizo, 1977). The glands open into a canal or trough that delivers secretions from these and the salivary glands to a cavity near the mouth. The secretion is high in sodium chloride, potassium chloride, and other materials. It has been proposed that the secretion is hygroscopic and absorbs water as it flows along the canal to the mouth, where the mite ingests the vapor-enriched solution. In the gut, the water and salts pass into the hemolymph, from which the glands absorb the salts and recycle them.

For fasting *D. farinae* and *D. pteronyssinus*, respectively, laboratory studies have shown that the critical RH at 25°C is ~70% and 73% (Larson, 1969; Arlian, 1975). However, field surveys and population growth studies indicate that feeding, active mite stages may survive at 20° to 22°C and re-produce at an ambient RH as low as 50% (Arlian et al., 1998). Laboratory studies have shown that the critical humidity is temperature dependent (Arlian and Veselica, 1981). Lowering indoor humidity can reduce mite population density because mites gradually dehydrate and die below 50% RH (Arlian, 1977, 1992; Arlian et al., 1998). Males and immature mites are more susceptible to desiccation than females. *D. pteronyssinus* is more susceptible to desiccation than *D. farinae* at humidities at or below the critical humidity.

22.1.4.3 Dependence of Mite Survival on RH Rather Than Absolute Water in Air

The maintenance of water balance and mite survival have been found to be functions of RH and not the absolute amount of water in air. All mites gradually dehydrate and die when held for more than 11 days at humidities below 50%. Approximately half of test populations of adult *D. farinae* and *D. pteronyssinus* dehydrated and died in 2 to 3 days at 40% to 50% RH and 28° to 34°C (Arlian, 1975; Brandt and Arlian, 1976). However, some female mites may survive 6 to 11 days in these conditions. Even though the absolute amount of water in the air (13.6 g/m³) was the same in both cases, mites survived at 20°C and 79% RH but died from dehydration at 27°C and 56% RH (Table 22.5) (Arlian, 1992). This occurred because water uptake is driven by chemical potential, which is related to temperature as described in the Arrhenius equation (Arlian and Veselica, 1982).

22.2 Health Effects

Hypersensitivity diseases caused by allergens from mites that live indoors constitute a major health problem in the U.S. and elsewhere. The mite species most often the source of these indoor allergens are *D. farinae*, *D. pteronyssinus*, *E. maynei*, and *B. tropicalis*. House dust mites are primarily a concern in human dwellings, but dust mites and mite allergen have been identified in office buildings in association with and without health com-

TABLE 22.5. RHs for Selected Conditions of Absolute Water Content of Air and Temperature (data from Arlian, 1992)

Temperature (°C)	Absolute Water (g H₂O/m³ air)		
	9.4	13.6	17.3
	Resultant RH (%)		
30	31	45	57
27	37	56	70
24	45	66	82
20	54	79	100
18	63	89	
16	69	100	
13	85		
10	100		

plaints (Lundblad, 1991; Hung et al., 1993; Menzies et al., 1993; Janko et al., 1995; Arlian, unpublished). Storage (domestic) mites and/or mite allergen (*A. siro* and *L. destructor*) have also been found in human dwellings (Table 22.2). Wickman et al. (1993) found exposure to storage mites a risk factor for sensitization to dust mites.

Clinical reactions to these mite species include chronic or perennial rhinitis, allergic asthma, atopic dermatitis, and keratoconjunctivitis. In western Europe, ~5% to 10% of the population is estimated to have active dust-mite allergy (Lundblad, 1991). Another 5% to 20% of people are estimated to be atopic (i.e., predisposed to develop mite allergy) but not sensitized and currently allergic.

22.3 Sample Collection

22.3.1 Source Sampling

The International Association of Allergology and Clinical Immunology (IAACI, Milwaukee, Wisconsin) has recommended collecting settled dust to evaluate exposure to dust mites and their allergens (Platts-Mills, et al., 1992). House dust mites are most prevalent in high-use areas (e.g., mattresses, bedding, carpets, sofas, and couches) where shed human skin scales collect and serve as food. It is important to sample multiple indoor locations to obtain a reasonable determination of mite prevalence.

22.3.1.1 Vacuum Samples Dust samples can be obtained from major mite foci by sampling a 1-m^2 area for 2 min using a specially designed dust trap attached to the end of a vacuum hose (Arlian et al., 1982, 1992). Dust is collected on tight-mesh filters constructed from bed sheets (180 to 200 thread count) or cellulose-ester filters (0.8-μm pore) in cassette holders (Janko et al., 1995). Once dust is collected, sheet filters must be folded tightly and secured and filter cassettes must be sealed to confine live mites. Collected dust is analyzed for the density of live or total mites [see 22.4.1] or the concentration of mite allergen [see 22.4.3].

22.3.1.2 Heat Escape and Passive Transfer Samples Carpets may be directly sampled for live mites using a heat escape or passive transfer method (Bischoff et al., 1992). For the heat escape method, a large piece of clear tape is placed on the surface of the test substrate (e.g., a carpet) and a heat source is placed beneath it. To avoid the heat, mites migrate to the surface where they adhere to the tape. For the passive transfer or mobility test method, a piece of tape is placed on the surface of an unheated substrate. Mites migrate to the surface during normal activity, and some adhere to the tape. With either method, the tape containing captured mites is removed after several hours (usually overnight) and examined under a microscope to determine the number of live mites per surface area.

22.3.2 Air Sampling

Only limited measurements of mite allergen concentrations in air have been made, even though that is the assumed exposure route. Volumetric sampling has repeatedly demonstrated that mite allergen concentrations are low to undetectable under normal circumstances (Tovey et al., 1981; Swanson et al., 1985; Platts-Mills et al., 1986a; Sakaguchi et al., 1989; deBlay et al., 1991). Exposure measurements are difficult to obtain because dust-mite fecal and body particles are large (>10 μm) and settle rapidly. Therefore, mite antigen may not be detected in air even where dust mites are known to be present (IOM, 1993).

22.3.3 Sample Handling

Dust samples can be stored at 75% RH and 5° to 8°C for 2 to 3 days without altering live mites (Arlian et al., 1982). Samples can be frozen for later analysis if only total mite counts or allergen concentration will be determined.

22.4 Sample Analysis

Training and experience are required to detect whole dust mites, identify their genera and species, and measure their density (number per area sampled or per gram of sample) [see 22.4.2]. Detection and measurement of dust-mite antigen (allergen) is performed using an immunoassay [see 22.4.3]. Laboratory personnel can conduct large numbers of immunoassays with a modest degree of training.

22.4.1 Processing Dust Samples

It is almost impossible to find and count mites in freshly vacuumed dust samples without processing the material to reduce extraneous debris and increase contrast between mites and other matter. Once dust has been collected, the mites can be separated from other material by a suspension or flotation method (Bischoff et al., 1992).

22.4.1.1 Suspension Extraction Separation There are several variations of the suspension extraction method using either a lactic acid, ethanol, or sodium chloride solution. Arlian et al.'s (1982, 1992) modified suspension method uses duplicate subsamples of dust (each ~100 mg) or the total dust collected from a 1-m^2 area if the amount of sample is small. The samples are each suspended in 25 mL of saturated sodium chloride solution containing 2 drops of liquid detergent. The mixture is vibrated thoroughly to break apart the material and create a uniform suspension of mites and dust. The suspension is rinsed on a 235-mesh sieve (sieve size, 45 μm) to remove fine dust. The mites and dust retained on the sieve are stained by passing a crystal violet solution through the sieve. After the excess stain is rinsed off, the material remaining on the sieve is washed into a gridded Petri dish and the number of live and dead mites are counted at ~25X magnification with the aid of a stereomicroscope. Dust components other than mites take up the violet stain providing a good contrast for recognizing mites, which appear white or clear against a stained background. Live mites will be moving or plump, whereas dead mites will be immobile and shriveled.

22.4.1.2 Flotation Separation The flotation method to separate mites in dust samples involves sieving dry samples to remove coarse material and retain the fine fraction with mites (Feldman-Muhsam et al., 1985). The fine dust is suspended in a liquid, such as dichloromethane, in which mites float but heavier particles sink. The supernatant containing mites is filtered, cleared, and refiltered (preferably on black filter paper) for viewing with a microscope [see 22.4.2 and 22.4.1.1]. This method is not as widely used as the suspension extraction method because mites are killed and, therefore, immobile. In addition, mites may more easily be lost in the process and it is difficult to see them on the filter paper.

22.4.2 Mounting Mites and Species Determination

22.4.2.1 Temporary and Permanent Mounts Temporary mounts for identifying mite species can be made using two to three drops of lactic acid on a microscope slide. Mites are removed from processed dust samples using a fine needle flattened at the tip and are placed in the lactic acid. A cover slip is added, and the slides are placed in an oven (~60°C) for 2 to 4 days until the mites appear clear. It is very difficult to identify mites if specimens are not completely clear.

Permanent mounts for identifying mite species can be made using Hoyer's medium: distilled water, 50 g; gum Arabic, 30 g; chloral hydrate, 200 g; and glycerin, 20 g. Mites must also be cleared before preparing permanent mounts, for example, by placing them in a small dish or vial with lactic acid. The container is warmed for 2 to 4 days (see above) or until mites appear clear. Cleared specimens are rinsed with 70% ethanol to remove the lactic acid, which otherwise will form crystals that may obscure key morphological characteristics. Cleared and rinsed specimens are placed in 1 to 2 drops of Hoyer's medium on a microscope slide. After adding a cover slip, the preparation is warmed in an oven for several days to dry the mounting medium. The cover slip on dry preparations is ringed with nail polish if the slides are to be stored for a year or longer.

22.4.2.2 Mite Identification Once clear, temporary and permanent mite mounts can be examined at 125X to 250X magnification using a compound microscope. Mite identification is based on the distinguishing characteristics listed and illustrated in reference publications (Fain, 1965, 1966; Arlian, 1989; Colloff, 1992). Researchers may record mite genera and species in dust samples and the ratios of male to female, mature to immature, and live to dead mites [see also 22.4.1].

22.4.3 Mite Allergen Assay

An ELISA is used to quantify mite allergen concentration in dust samples (Chapman et al., 1984, 1987a,b; Heymann et al., 1986; Platts-Mills et al., 1986b; Chapman, 1988) [see 6.5.2]. The capture antibodies used for the ELISA are specific for either *D. farinae* or *D. pteronyssinus* group-1 antigens (abbreviated *Der f* 1 and *Der p* 1, respectively) [see Step 1, Figure 6.1]. Bound allergen is detected with a second enzyme-conjugated monoclonal antibody that recognizes a cross-reacting site common to both *Der p* 1 and *Der f* 1 [Step 3 in Figure 6.1]. The concentration of *Der p* 1 and *Der f* 1 antigens are obtained by comparing the optical density reading for a dust sample with a standard curve of optical densities generated from allergen standards of known concentration. It is beneficial to determine allergen concentration for the same dust samples for which mite densities were determined [see 22.4.1].

22.4.4 Guanine Assay

Guanine ($C_5H_5N_5O$) is a major component of mite feces and body parts and has been measured as a surrogate for dust-mite antigen (Janko et al., 1995). Liquid chromatography (LC) can be used to detect guanine (Lundblad, 1991). A commercial kit for semi-quantitative colorimetric guanine analysis has been used (Lundblad, 1991). However, interferences have been encountered when sampling chairs and carpets (Janko et al., 1995).

22.5 Data Interpretation

The IAACI has proposed dust-mite allergen limits for residential dwellings (Platts-Mills et al., 1992). Exposure to dust containing 2 µg/g of group-1 mite allergen (roughly equivalent to 100 mites/g) (Hamilton et al., 1992) or 0.6 mg/g of guanine is considered to increase the risk of dust-mite sensitization as well as the development of asthma and bronchial hyperreactivity in affected persons. Exposure to dust containing 10 µg/g of group-1 mite allergen or >2.5 mg/g of guanine represents a higher risk level and an increased chance of acute asthma attacks (IOM, 1993). Epidemiological studies have found better correlations between symptoms and ELISA measurements of mite allergen than mite counts. However, measurement of dust allergen content on a per-gram basis may be misleading if different substrates are compared because the density of dust from various sources may differ markedly (cf. dust from bedding or upholstered furnishing, which contains lint and fine feathers, and carpet dust, which contains soil, sand, and coarse fibers).

22.6 Remediation and Prevention

Lundblad (1991) suggested that hypersensitivity to dust mites may be considered a possible explanation for IEQ complaints when (a) affected persons exhibit symptoms of mite-related allergy (principally rhinitis that ceases when away from an implicated environment), (b) a diagnosis of mite sensitivity has been made for affected persons, (c) environmental conditions (i.e., temperature, humidity, and potential sources) are compatible with mite infestation, and (d) other explanations for symptoms have been ruled out. Confirming dust-mite presence and comparing mite or mite-allergen concentrations with recommended guidelines are relatively simple [see 22.4 and 22.5]. Below are suggestions on how to minimize mite reproduction and allergen release following confirmation that dust mites are present indoors.

22.6.1 Acaricides

Laboratory studies have shown that several chemicals are potent acaricides able to kill house dust mites, but few chemicals have been suitably formulated and approved for domestic use (Colloff et al., 1992; Squillace, 1995). Therefore, it remains difficult to reduce mite concentrations in mattresses, carpets, and fabric-covered sofas and chairs by chemical means, in part because of difficulty applying acaricides. Safety issues must be addressed if acaricides (e.g., pyrethroids, natamycin—an antifungal agent, pirimiphos methyl, or benzyl benzoate) are used in occupied spaces (Lundblad, 1991; IOM, 1993) [see also 16.1 and 16.2.6]. A 1% to 3% solution of tannic acid has been recommended to denature mite allergens. However, this treatment does not kill mites, and the effect is temporary, lasting ~6 to 8 weeks (IOM, 1993).

TABLE 22.6. Environmental Control Measures to Minimize Dust-Mite Populations and Allergen Production

---

- Where mites may breed, maintain RH below 50% (Arlian, 1975, 1992; Brandt and Arlian, 1976). Regular use of air conditioners may lower RH and temperatures sufficiently to reduce the rates of mite feeding, reproduction, and allergen production, although this remains to be quantified (Arlian, 1977, 1992).

- Remove carpets to eliminate a major mite-breeding substrate.

- Vacuum regularly and discard vacuum bags immediately so collected mites do not breed in them, producing more allergen.

- Replace fabric-upholstered furniture with vinyl or leather-covered pieces or wooden chairs and benches to eliminate mite-breeding areas.

- In homes, encase mattresses and pillows in covers impermeable to mites and their allergens to prevent skin scales and moisture from accumulating on these items and serving as food and water for mites. This procedure also isolates mites and their allergens from contact with a sleeping individual. Wash all bedding weekly in hot water (54°C) to kill mites and remove allergenic material.

- Air cleaners with HEPA filters can remove suspended allergenic particles from room air. However, this intervention may be of limited benefit because allergenic mite particles are large (10 to 35 μm in diameter) and settle rapidly (IOM, 1993).

22.6.2 Environmental Controls

Table 22.6 lists engineering controls recommended to reduce dust-mite numbers and allergen concentrations [see also 10.5.2.2 and 10.5.6].

22.7 References

Arlian, L.G.: Dehydration and Survival of the European House Dust Mite, Dermatophagoides pteronyssinus. J. Med. Entomol. 12:437-442 (1975).

Arlian, L.G.: Humidity as a Factor Regulating Feeding and Water Balance of House Dust Mites, Dermatophagoides farinae and D. pteronyssinus (Acari: Pyroglyphidae). J. Med. Entomol., 14:484-488 (1977).

Arlian, L.G.: Biology and Ecology of House Dust Mites, Dermatophagoides spp. and Euroglyphus spp. Immunol. Allergy Clin. N. Am. 9:339-356 (1989).

Arlian, L.G.: Water Balance and Humidity Requirements of House Dust Mites. Exp. Appl. Acarol., 16:15-35 (1992).

Arlian, L.G.; Dippold, J.S.: Development and Fecundity of Dermatophagoides farinae (Acari: Pyroglyphidae). J. Med. Entomol., 33:257-260 (1996).

Arlian, L.G.; Veselica, M.M.: Re-Evaluation of the Humidity Requirements of the House Dust Mite Dermatophagoides farinae (Acari: Pyroglyphidae). J. Med. Entomol., 18:351-352 (1981).

Arlian, L.G.; Veselica, M.M.: Relationship between Transpiration Rate and Temperature in the Mite Dermatophagoides farinae. Physiol. Zool. 55:344-354 (1982).

Arlian, L.G.; Wharton, G.W.: Kinetics of Active and Passive Components of Water Exchange between the Air and a Mite, Dermatophagoides farinae. J. Insect. Physiol. 20:1063-1077 (1974).

Arlian, L.G.; Bernstein, I.L.; Gallagher, J.S.: The Prevalence of House Dust Mites, Dermatophagoides spp., and Associated Environmental Conditions in Homes in Ohio. JACI, 69:527-532 (1982).

Arlian, L.G.; Woodford, P.J.; Bernstein, I.L.; et al.: Seasonal Population Structure of House Dust Mites, Dermatophagoides spp. (Acari: Pyroglyphidae). J. Med. Entomol. 20:99-102 (1983).

Arlian, L.G.; Rapp, C.M.; Ahmed, S.G.: Development of Dermatophagoides pteronyssinus (Acari: Pyroglyphidae). J. Med. Entomol. 27:1035-1040 (1990).

Arlian, L.G.; Bernstein, D.; Bernstein, I.L.; et al.: Prevalence of Dust Mites in the Homes of People with Asthma Living in Eight Different Geographical Areas of the United States. JACI, 90:292-300 (1992).

Arlian, L.G.; Confer, P.D.; Rapp, C.M.; et al.: Population Dynamics of the House Dust Mites Dermatophagoides farinae, D. pteronyssinus, and Eurogylphus maynei (Acari: Pyroglyphidae) at Specific Relative Humidities. J. Med. Entomol. 35:46-53 (1998).

Bischoff, E.R.C.; Fischer, A.; Liebenberg, B.: Assessment of Mite Numbers: New Methods and Results. Exp. Appl. Acarol. 16:1-14 (1992).

Brandt, R.L.; Arlian, L.G.: Mortality of House Dust Mites, Dermatophagoides farinae and D. pteronyssinus, Exposed to Dehydrating Conditions or Selected Pesticides. J. Med. Entomol., 13:327-331 (1976).

Chapman, M.D.: Allergen Specific Monoclonal Antibodies: New Tools for the Management of Allergic Disease. Allergy 43:7-14 (1988).

Chapman, M.D.; Sutherland, W.M.; Platts-Mills, T.A.E.: Recognition of Two Dermatophagoides pteronyssinus Specific Epitopes on Antigen P1 using Monoclonal Antibodies: Binding to Each Epitope can be Inhibited by Sera from Mite Allergic Patients. J. Immunol. 133:2488-2495 (1984).

Chapman, M.D.; Heymann, P.W.; Wilkins, S.R.; et al.: Monoclonal Immunoassays for Major Dust Mite (Dermatophagoides) Allergens, Der p I and Der f I, and Quantitative Analysis of the Allergen Content of Mite and House Dust Extracts. JACI, 80:184-194 (1987a).

Chapman, M.D.; Heymann, P.W.; Platts-Mills, T.A.E.: Epitope Mapping of Two Major Inhalant Allergens, Der p I and Der f I, from Mites of the genus Dermatophagoides. J. Immunol., 139:1479-1484 (1987b).

Charlet, L.D.; Mulla, M.S.; Sanchez-Medina, M.: Domestic Acari of Colombia: Population Trends of House Dust Mites (Acari: Pyroglyphidae) in Homes in Bogota, Colombia. Intl. J. Acarol., 4:23-31 (1978).

Colloff, M.J.: Effects of Temperature and Relative Humidity on Development Times and Mortality of Eggs from Laboratory and Wild Populations of European House-Dust Mite Dermatophagoides pteronyssinus (Acari: Pyroglyphidae). Exp. Appl. Acarol. 3:279-289 (1987).

Colloff, M.J.: Practical and Theoretical Aspects of the Ecology of House Dust Mites (Acari: Pyroglyphidae) in Relation to the Study of Mite Mediated Allergy. Rev. Med. Vet. Ent., 79:611-630 (1991).

Colloff, M.J.: Pictorial Keys for the Identification of Domestic Mites. Clin. Exp. Allergy 22:823-830 (1992).

Colloff, M.J.; Ayres, J.; Carswell, F.; et al.: The Control of Allergens of Dust Mites and Domestic Pets: A Position Paper. Clin. Exp. Allergy 22:1-28 (1992).

deBlay, F.; Heymann, P.W.; Chapman, M.D.; et al.: Airborne Dust Mite Allergens: Comparison of Group II Allergens with Group I Allergen and Cat-Allergen Fel d I. J. Allergy Clin. Immunol. 88:919-926 (1991).

Ellingsen, I.J.: Permeability to Water in Different Adaptive Phases of the Same Instar in the
American House Dust Mite. Acarologia, 17:734-744 (1975).

Ellingson, A.R.; LeDoux, R.A.; Vedanthan, P.K.; et al.: The Prevalence of Dermatophagoides Mite Allergen in Colorado Homes Utilizing Central Evaporative Coolers. J. Allergy Clin. Immunol. 96:473-479 (1995).

Fain, A.: Les acariens nidicoles et detriticoles de la famille Pyroglyphidae Cunliffe (Sarcoptiformes). Rev. Zool. Bot. Afr. 72:257-288 (1965).

Fain A.: Nouvelle Description de Dermatophagoides pteronyssinus (Trouessart, 1897) Importance de cet Acarien en pathologie humaine (Psoroptidae: Sarcoptiformes). Acarologia, 8:302-327 (1966).

Feldman-Muhsam, T.; Mumcuoglu, Y.; Osterovich, T.: A Survey of House Dust Mites (Acari: Pyroglyphidae and Cheyletidae) in Israel. J. Med. Entomol. 22:663-669 (1985).

Fernandez-Caldas, E.; Fox, R.W.; Bucholtz, G.A.; et al.: House Dust Mite Allergy in Tampa, Florida. Allergy Proc. 11:263-267 (1990).

Furumizo, R.T.: Laboratory Observations on the Life History and Biology of the American House Dust Mite Dermatophagoides farinae (Acarina: Pyroglyphidae). Calif. Vector Views 22:49-60 (1975).

Furumizo, R.T.: Seasonal Abundance of Dermatophagoides farinae Hughes 1961 (Acarina: Pyroglyphidae) in House Dust in Southern California. Calif. Vector Views 25:13-19 (1978).

Griffiths, D.A.; Cunnington, A.M.: Dermatophagoides microceras sp. n.: A Description and Comparison with its Sibling Species, D. farinae Hughes, 1961. J. Stored Prod. Res. 7:1-14 (1971).

Hamilton, R.G.; Chapman, M.D.; Platts-Mills, T.A.E.; et al.: House Dust Aeroallergen Measurements in Clinical Practice: A Guide to Allergen-free Home and Work Environments. Immunol. Allergy Practice. 15:96/9-112/25 (1992).

Hart, B.J.; Fain, A.: Morphological and Biological Studies of Medically Important House-Dust Mites. Acarologia, 19:285-295 (1988).

Heymann, P.W.; Chapman, M.D.; Platts-Mills, T.A.E.: Antigen Der f I from the Dust Mite Dermatophagoides farinae: Structural Comparison with Der p I from D. pteronyssinus and Epitope Specificity of murine IgG and Human IgE Antibodies. J. Immunol. 137:2841-2847 (1986).

Hung L.L.; Lewis F.A.; Yang C.S.; et al.: Dust Mite and Cat Dander Allergens in Office Buildings in the Mid-Atlantic Region. In: Environments for People — Proceedings of IAQ92 Conference, San Francisco, California, pp. 163-170. American Society of Heating, Refrigerating and Air-Conditioning Engineers, Inc., Atlanta, GA

(1993).

Institute of Medicine (IOM): Agents, Sources, Source Controls, and Disease. In: Indoor Allergens: Assessing and Controlling Adverse Health Effects, pp. 86-130. A.M. Pope, R. Patterson, and H.A. Burge, Eds. National Academy Press, Washington, DC (1993).

Institute of Medicine (IOM): Assessing Exposure and Risk. In: Indoor Allergens: Assessing and Controlling Adverse Health Effects, pp. 185-205. A.M. Pope, R. Patterson, and H.A. Burge, Eds. National Academy Press, Washington, DC (1993).

Janko, M.; Gould, D.C.; Vance, L.; et al.: Dust Mite Allergens in the Office Environment. Am. Ind. Hyg. Assoc. J. 56:1133-1140 (1995).

Krieger, R.I.; Dinoff, T.M.; Peterson, J.: Human disodium octaborate tetrahydrate Exposure Following Carpet Flea Treatment is not Associated with Significant Dermal Absorption. J. Exposure Anal. Environ. Epidem. 6:279-288 (1996).

Lang, J.D.; Mulla, M.S.: Seasonal Dynamics of House Dust Mites, Dermatophagoides spp., in Homes in Southern California. Environ. Ent. 7:281-286 (1978).

Larson, D.G.: The Critical Equilibrium Activity of Adult Females of the House Dust Mite, Dermatophagoides farinae Hughes. Ph.D. Thesis, Ohio State University, Columbus, OH (1969).

Lundblad, F.P.: House Dust Mite Allergy in an Office Building. Appl. Occup. Environ. Hyg. 6:94-96 (1991).

Menzies, R.; Tamblyn, R.; Comtois, P.; et al.: Case-Control Study of Microenvironmental Exposures to Aero-allergens as a Cause of Respiratory Symptoms — Part of the Sick Building Syndrome (SBS) Symptom Complex. In: Environments for People. Proceedings of IAQ92 Conference, pp. 201-210. San Francisco, California. American Society of Heating, Refrigerating and Air-Conditioning Engineers, Inc., Atlanta, GA (1993).

Murray, A.B.; Zuk, P.: The Seasonal Variation in a Population of House Dust Mites in a North American City. JACI, 64:266-269 (1979).

Platts-Mills, T.A.E.; Heymann, P.W.; Longbottom, J.L.; et al.: Airborne Allergens Associated with Asthma: Particle Size Carrying Dust Mite and Rat Allergens Measured with a Cascade Impactor. J. Allergy Clin. Immunol. 77:850-857 (1986a).

Platts-Mills, T.A.E.; Heymann, P.W.; Chapman, M.D.; et al.: Cross-Reacting and Species Specific Determinants on a Major Allergen from Dermatophagoides pteronyssinus and D. farinae: Development of a Radioimmunoassay for 'Antigen P1 Equivalent in House Dust and Dust Mite Extracts. JACI, 78:398-407 (1986b).

Platts-Mills, T.A.E.; Thomas, W.R.; Aalberse, R.C.; et al.: Dust Mite Allergens and Asthma: Report of a Second International Workshop. J. Allergy Clin. Immunol. 89:1046-1060 (1992).

Sakaguchi, M.; Inouye, S.; Yasueda, H.; et al.: Measurement of Allergens Associated with Dust Mite Allergy II. Concentrations of Airborne Mite Allergens (Der I and Der II) in the House. Int. Arch. Allergy Appl. Immunol. 90:190-193 (1989).

Squillace, S.P.: Allergens of Arthropods and Birds. In: Bioaerosols, pp. 133-148. H. Burge, Ed. Lewis Publishers, Boca Raton, FL (1995).

Swanson, M.C.; Agarwall, M.K.; Reed, C.E.: An Immunochemical Approach to Indoor Aeroallergen Quanitation with a New Volumetric Air Sampler: Studies with Mite, Roach, Cat, Mouse, and Guinea Pig Antigens. J. Allergy Clin. Immunol. 76:721-729 (1985).

Tovey, E.R.; Chapman, M.D.; Wells, C.W.; et al.: The Distribution of Dust Mite Allergen in the Houses of Patients with Asthma. Am. Rev. Respir. Dis. 124:630-635 (1981).

van Bronswijk, J.E.; Sinha, R.N.: Pyroglyphid Mites (Acari) and House Dust Allergy. J. Allergy 47:31-52 (1971).

Wharton, G.W.; Furumizo, R.T.: Supracoxal Gland Secretions as a Source of Fresh Water for Acaridei. Acarologia 19:112-116 (1977).

Wickman, M.; Lennarat Nordvall, S.; Pershagen, G.; et al.: Sensitization to Domestic Mites in a Cold Temperate Region. Am. Rev. Respir. Dis. 148:58-62 (1993).

Chapter 23

Endotoxin And Other Bacterial Cell-Wall Components

Donald K. Milton

23.1   Characteristics
    23.1.1   *Chemical Composition*
    23.1.2   *Units of Measurement*
23.2   Health Effects
    23.2.1   *Experimental Exposures*
    23.2.2   *Epidemiological Studies*
        23.2.2.1   Acute Effects
        23.2.2.2   Chronic Effects
        23.2.2.3   Low Exposures and BRS
    23.2.3 *Endotoxin Monitoring*
23.3   Sample Collection
    23.3.1   *Filter Type*
    23.3.2   *Handling and Storage*
23.4   Sample Analysis
    23.4.1   *Extraction Methods*
    23.4.2   *Limulus Amebocyte Lysate Assay for Endotoxin*
        23.4.2.1   Quantitative LAL Assays
        23.4.2.2   Parallel-Line LAL Analysis
        23.4.2.3   Interferences with LAL Assays
        23.4.2.4   Variability in LAL Reagents
    23.4.3   *Chemical Assays for LPS*
23.5   Data Interpretation
23.6   Remediation and Prevention
    23.6.1   *Recommendations — Relative Limit Values*
    23.6.2   *Remedial Actions*
23.7   Bacterial Cell-Wall Components
    23.7.1   *Characteristics*
    23.7.2   *Health Effects*
    23.7.3   *Sampling*
    23.7.4   *Data Interpretation*
23.8   References

23.1 Characteristics

23.1.1 Chemical Composition

Endotoxin is a term for the toxin characteristic of the outer membrane of GNB (Rietschel and Brade, 1992). Thus, as the name implies, the toxin is an intrinsic part of these bacteria. When purified from the outer membrane, endotoxin consists of a family of molecules called lipopolysaccharides (LPS). Lipid A, the lipid portion of LPS, consists of a 1,4′ diphosphoryl β-(1,6)-*D*-glucosamine disaccharide backbone with two ester- and two amide-linked 3-hydroxy fatty acids. The 3-hydroxy fatty acids usually carry secondary, non-hydroxylated fatty acids esterified to their hydroxyl groups (Sonesson et al., 1994). Lipid A is chemically distinct from other lipids in biological membranes and is primarily responsible for the molecule's characteristic toxicity. While the family of LPS molecules with toxic activity have certain structural similarities, there is significant heterogeneity of lipid A structure among bacterial species and of polysaccharide among serotypes within species (Sonesson et al., 1994). Even within one organism, the length of the polysaccharide portion and the number of non-hydroxylated fatty acids in each LPS molecule may vary. LPS is fairly heat stable and, therefore, it is not eliminated by autoclaving. Endotoxin is encountered in the environment as part of whole cells or membrane fragments. In the environmental literature, the term *LPS* is frequently reserved for purified preparations while *endotoxin* is used to denote the naturally occurring material (Jacobs, 1989). This convention will be followed here, except that the term LPS will also be used when referring to the chemical characteristics or physical amount of the molecule in a sample, and the term endotoxin when referring to the toxic activity or potency of an environmental sample.

LPS refers to the chemical characteristics or physical amount of the molecule.
Endotoxin refers to the characteristic toxic activity or potency.

23.1.2 Units of Measurement

Endotoxin potency is measured by determining the biological activity of a sample in a specified assay system. The results are then reported in Endotoxin Units (EU) with reference to the biological activity in the same test system of a standard LPS preparation. By convention, 1 EU is equivalent to 0.1 ng of the reference standard endotoxin. The current U.S. reference standard is an *Escherichia coli* O113:H10:K LPS preparation known as EC6 [United States Pharmacopoeial (USP) Convention, Rockville, MD, lot G]. Prior to 1995, EC5 (USP, lot F) was the reference standard. Lots F and G—and thus EUs referenced to EC5 and EC6—are equivalent. Results may be consistently expressed as ng of EC6 or as EU. However, older data were often expressed in the units of the various LPS standards used in the individual assays and may not be consistent with current units. Whatever the units, endotoxin measurements are based on a biological assay for biological activity or potency and do not represent a measure of the concentration of a single chemical substance.

On the other hand, the concentration of LPS can be measured by chemical methods, such as detection of 3-hydroxy fatty acids by GC–MS. Using the structural information that each LPS molecule contains four 3-hydroxy fatty acids, the number of moles of LPS in a sample can be obtained directly from the 3-hydroxy fatty acid measurements. To compute the mass of LPS in a sample requires assumptions about the average molecular weight of the LPS—approximately 8000 daltons based on SDS-polyacrylamide gel analysis of environmental samples (Milton, unpublished observations).

Airborne endotoxin is ubiquitous except during times of snow cover. Outdoor endotoxin levels range up to ~3 EU/m³ (0.3 ng/m³) during the growing season due to aerosolization of GNB from leaves (Andrews and Hirano, 1992). Airborne endotoxin levels in homes and offices can be higher than those outdoors, for example, in the presence of contaminated humidifiers, after severe water damage, and under other as yet poorly defined conditions (Milton, 1996). High airborne endotoxin levels (1 x 10⁴ to 27 x 10⁴ EU/m³) are associated with a variety of industrial work environments. In particular, airborne endotoxin may be found where organic dusts are present (e.g., cotton mills, swine barns, and grain handling) or

where recirculated industrial washwater or other water-based fluids [e.g., metal-working fluid (MWF) used in machining] are aerosolized (Rylander and Morey, 1982; Olenchock, 1994; Milton et al., 1995; Jacobs, 1997).

Airborne endotoxin is ubiquitous except during times of snow cover.

23.2 Health Effects

Endotoxin is highly toxic. During infection with GNB and after experimental intravenous injection, endotoxin causes fever and malaise, changes in white blood cell counts, respiratory distress, shock, and even death (Galanos et al., 1990; Kotani and Takada, 1990). Endotoxin is also a powerful, nonspecific stimulant to the immune system, resulting in beneficial effects (e.g., reduced cancer risk and enhanced vaccine effectiveness) but also in adverse effects (e.g., potentially increased risk of hypersensitivity disease) (Enterline et al., 1985; Loppnow et al., 1990; Rylander, 1990; Engelhardt et al., 1991; Fogelmark et al., 1991; Baker et al., 1992; Rose et al, 1998). Advances in cell biology have shown that endotoxin exerts its effects through specific binding to a complex set of serum and cell-surface proteins (Ulevitch, 1993).

23.2.1 Experimental Exposures

High-concentration inhalation exposure of experimental animals and humans to pure LPS produces airway and alveolar inflammation as well as chest tightness, fever, and malaise (Pernis et al., 1961; Cavagna et al., 1969; Hudson et al., 1977; Gordon, 1992; Sandström et al., 1992; Schwartz et al., 1994). Inhalation of 20 µg of nebulized LPS caused airflow obstruction and increased bronchial hyperreactivity in asthmatics, and higher doses (200 µg) caused symptoms in nonasthmatics (Pernis et al., 1961; Cavagna et al., 1969; Michel et al., 1992; Clapp et al., 1994).

Experimental human studies of cotton dust exposure demonstrated a strong dose–response relationship between endotoxin and acute airflow obstruction with thresholds at 90 to 330 EU/m³ (Rylander et al., 1985; Castellan et al., 1987) (Figure 23.1). These threshold levels were lower than would be expected based on human experimental studies with pure LPS. Differences in dose-rate and the presence of particulate matter and other irritating substances may affect responses at low endotoxin levels. There are also some data to suggest that the biological methods employed to measure endotoxin in environmental samples may underestimate the amount of LPS present [see 23.4.2]. Thus, responses may appear to have been produced after low exposures when, in fact, the exposures were approximately three times higher (Rylander et al., 1989a; Sonesson et al., 1990).

Experimental human studies of cotton dust exposure have demonstrated a strong dose–response relationship between endotoxin and acute airflow obstruction with thresholds at 90 to 330 EU/m³.

23.2.2 Epidemiological Studies

23.2.2.1 Acute Effects Some epidemiological studies have observed acute effects at relatively high endotoxin exposures (Thelin et al., 1984; Donham et al., 1989) but, in at least one study, the healthy worker effect appears to have blunted acute effects at high exposures (Kennedy et al., 1987). Two studies observed acute airflow obstruction at endotoxin levels in the expected ranges based on experimental cotton dust studies (i.e., 300 to 400 EU/m³) (Smid et al., 1994) and 45 to 150 EU/m³ (Milton et al., 1996). In the latter study, analysis of air samples by chemical as well as biological methods found no evidence that the biological methods underestimated LPS in the samples (Walters et al., 1994). The lowest endotoxin exposure associated with acute airflow obstruction (personal GM, 64 pg/m³) was reported among workers exposed to mist from a cold-water humidification system in a synthetic textile yarn-manufacturing plant (Kateman et al., 1990). Workers in areas with

no humidification or with steam humidification had GM exposures of 18 and 19 pg/m³. The exposed group's endotoxin levels were well below those associated with health effects in other studies published by the same investigators (Smid et al., 1994) and suggest that either there were transient measurement problems (possibly due to reagent lots) [see 23.4.2.4] or that another factor was responsible for the acute airflow obstruction and history of humidifier fever in the exposed workers.

23.2.2.2 Chronic Effects Chronic effects of endotoxin on pulmonary function and respiratory symptoms have been suggested by cross-sectional studies (Thelin et al., 1984; Kennedy et al., 1987; Donham et al., 1989; Smid et al., 1992a; Milton et al., 1995; Schwartz et al., 1995b). The apparent threshold endotoxin exposure for chronic effects was 1 µg/m³ (~10,000 EU/m³) in the earliest study from which quantitative exposure–response relationships can be deter-

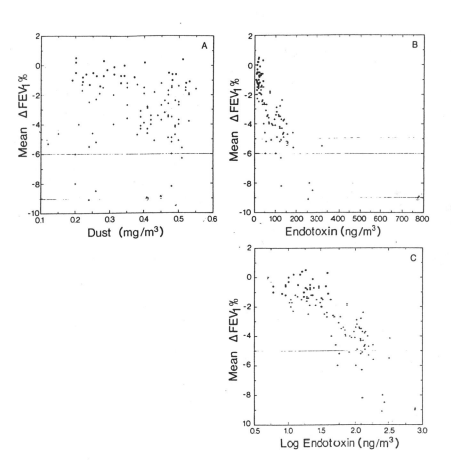

FIGURE 23.1. Group mean percentage change in forced expiratory volume in one second (rFEV₁%) versus airborne dust concentration (Panel A), airborne endotoxin concentration (B), and log airborne endotoxin concentration (C). Data are from 108 sessions of exposure to cotton dust, each involving between 24 and 35 subjects. Group mean values for rFEV₁% are plotted as open circles if not different from zero and as crosses if different from zero (P<0.05 by two-tailed t-test). Exposure concentrations were determined from air samples collected by vertical elutriators (Castellan et al., 1987).

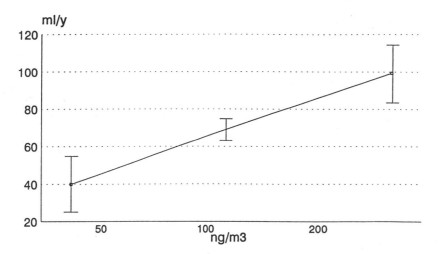

FIGURE 23.2. Predicted association between logtransformed endotoxin exposure and annual rate of decline of FEV₁ (with standard error); corrected for age, baseline FEV₁, and pack-years of smoking. Based on 171 pig farmers (Vogelzang et al., 1998).

mined (Thelin et al., 1984). In later studies, the thresholds were 10 to 400 EU/m³ (Kennedy et al., 1987; Smid et al., 1992a) and any exposure above background but ≤100 EU/m³ (Schwartz et al., 1995b). However, some of these studies found apparent chronic effects on lung function in the absence of significant acute airflow obstruction and vice versa (Kennedy et al., 1987; Schwartz et al., 1995b).

Four prospective longitudinal studies of endotoxin exposure and lung function have been reported. Two studies observed an association of endotoxin with accelerated loss of lung function (Schwartz et al., 1995a; Vogelzang et al., 1998), and two did not (Christiani et al., 1994; Heederik et al., 1994). The lack of standardized measurement methods and reagents may have contributed to exposure misclassification, particularly in one of the longitudinal studies where exposure assessment depended on repeated measurement of area endotoxin over many years (Christiani et al., 1994), resulting in a bias toward the null (i.e., an inability to detect health effects because workers' exposures are not accurately or consistently measured) [see 23.3 and 23.4]. Both of the positive studies were among pig farmers, although one suggested that the effect of endotoxin was present in both pig farmers and other control farmers (Schwartz et al., 1995a). A log-linear exposure–response relationship was demonstrated in the most recent report (Vogelzang et al., 1998). A doubling of the endotoxin level was associated with an increase of approximately 20 mL/yr. in the rate of FEV₁ decline (Figure 23.2). However, even in the lowest exposure group (400 EU/m³), the rate was in excess of the rate expected in the normal population.

As the ability to reproducibly measure endotoxin improves, so may the evidence for chronic effects from longitudinal studies. Agriculture-based exposures are highly complex and other agents in organic dust may be important contributors to observed acute and chronic health effects (Malmberg and Larsson, 1993; Zhiping et al., 1996).

Therefore, longitudinal studies of endotoxin's health effects in other environments (e.g., ones involving exposure to contaminated washwater mists) are also needed.

> A doubling in endotoxin level was associated with an increased rate of decline in FEV1 of ~20 mL/yr. Even the rate of decline in the lowest exposure group (400 EU/m3) exceeded the rate expected in the normal population.

23.2.2.3 Low Exposures and BRS The preceding discussion concerned exposures occurring in industrial and agricultural environments where exposure levels exceeded 45 EU/m³. Low-level endotoxin exposures, only slightly in excess of the normal outdoor background level, have been reported in association with increased severity of asthma and BRSs (Rylander et al., 1989b; Michel et al., 1991; Gyntelberg et al., 1994; Teeuw et al., 1994). Michel et al. (1991) found that endotoxin above the median level in house dust (1.1 ng/mg) was associated with increased asthma severity among adults. Endotoxin level was not associated with dust-mite concentrations. In a subsequent study, Michel et al. (1996) found that among asthmatics exposed to high levels of house dust mite antigen, severity of the disease was associated with the endotoxin level detected in house dust samples. Overall, concentrations of dust-mite antigen were associated with having asthma, but not with its severity. Rylander et al. (1989b) also studied residences by sampling for endotoxin and β-(1→3)-D-glucan. However, this group looked at the prevalence of BRSs in apartment dwellers. Endotoxin was associated with fatigue and mucosal irritation when present above 0.2 ng/m³; the maximum endotoxin level reported was 18 ng/m³.

Two large epidemiological studies suggest that endotoxin may be related to BRSs. Teeuw et al. (1994) studied airborne endotoxin in 19 Dutch government buildings and interviewed 1355 office workers. The investigators reported higher airborne endotoxin levels in mechanically ventilated problem buildings than in mechanically or naturally ventilated non-problem buildings. Levels of airborne culturable GNB followed a similar pattern. However, the levels of endotoxin reported in the naturally ventilated buildings (mean, 35 ng/m³) were 100-fold higher than expected based on data from other laboratories. The levels in problem buildings (200 ng/m³) were similar to those that others from the same region reported for dusty agricultural environments (Smid et al., 1992b). The authors reported using some of the appropriate controls for endotoxin measurement methods. However, these researchers did not investigate the possibility that the high overall results represented enhancement of the reaction with endotoxin due to the presence of β-(1→3)-D-glucan—a well known feature of the type of *Limulus* amebocyte lysate (LAL) used (Roslansky and Novitsky, 1991).

The second large epidemiological study (Gyntelberg et al., 1994) selected 12 buildings from an earlier Danish Town Hall Study to include a range of symptom prevalences. The study examined the association of symptoms with a broad range of parameters measured in carpet dust collected from each building. The strongest associations with symptoms were found for culturable GNB. In particular, fatigue, heavy-headedness, and throat irritation were strongly associated with GNB concentrations. However, there were no associations with endotoxin levels in the dust. The authors speculated that the airborne endotoxin levels were more strongly correlated with culturable GNB than with endotoxin levels in dust. However, the reported mean endotoxin level, 20 EU/g (2 ng/g), was 500-fold lower than levels Michel et al. (1991) reported in dust from homes. Michel's data are in general agreement with those reported by other investigations of house dust (Douwes et al., 1995; Milton et al., 1997), and based on the recent data from Hines et al. (1997), it seems likely that levels in offices are similar to those in homes.

> Low-level endotoxin exposures, only slightly in excess of the normal outdoor background level, have been associated with increased severity of asthma and BRSs.

23.2.3 Endotoxin Monitoring

While some intriguing data suggest that there may be health effects at low-level endotoxin exposures, definitive statements regarding causality, establishment of a TLV, or recommendations for routine monitoring in low exposure environments would be premature. Even in the higher-exposure environments, where evidence for endotoxin health effects is relatively well established, a TLV or recommendations for routine monitoring are premature due to the lack of standardized methods for sample collection and analysis and orders-of-magnitude measure-

ment differences among laboratories [see 23.3 and 23.4]. In addition, the health effects of endotoxin may vary with the type of endotoxin and the presence of other factors so that environment-specific endotoxin exposure limits may be required. Therefore, sampling for endotoxin remains primarily a research technique due to the problems with reagent standardization and data interpretation [see 23.4.2.4 and 23.5]. However, sampling may be useful to confirm suspected exposure in agricultural and related industrial environments and where aerosols are generated from recirculated water in either manufacturing operations (e.g., washwaters or MWFs) or buildings with humidification systems.

> A TLV for endotoxin or recommendations for routine monitoring are premature. Sampling may be useful to confirm suspected exposure in agricultural and related industrial environments and where aerosols are generated from recirculated water in manufacturing operations or buildings with humidification systems.

23.3 Sample Collection

Endotoxin aerosols are ordinarily collected on filter media as these are easy to use and allow long sampling times. Collection with all-glass impingers has occasionally been reported. However, this method may underestimate endotoxin levels due to its low collection efficiency for the submicrometer particles in which endotoxin is present (Olenchock et al., 1983).

23.3.1 Filter Type

Several authors have reported attempts to examine the effect of filter medium selection on endotoxin measurements made with LAL assays (Milton et al., 1990; Sonesson et al., 1990; Gordon et al., 1992; Douwes et al., 1995). Milton et al. (1990) reported that cellulose mixed ester (CE), two types of Teflon, and polyvinyl chloride (PVC) filters inactivated or adsorbed endotoxin activity from pure LPS solutions and suggested that a desirable medium would not have this property. The apparent irreversible adsorption of endotoxin by filter media suggested that measured endotoxin levels would be affected by filter type, endotoxin loading, and extraction time. PVC filters produced the largest reduction in endotoxin activity (90%); the other tested media produced ~60% reductions. A wider range of materials have been tested since that publication. The only filters that have not reduced the endotoxin activity of LPS and whole GNB suspensions were polycarbonate (PC) capillary-pore membrane filters (Milton and Johnson, 1995). These findings agreed with results of a chemical method for LPS analysis by GC–MS that suggested that samples collected on PVC filters and assayed for endotoxin activity levels underestimated by orders of magnitude the LPS concentrations in samples of poultry barn aerosols (Sonesson et al., 1990).

Gordon et al. (1992) reported on aerosol-chamber experiments with LPS/saline solutions and cotton dust as well as MWF spiked with LPS and MWF that had been used and contained GNB. The authors found that the amount of endotoxin activity detected in filter extracts depended on the type of aerosol. Extracting the filters at 68°C by incubation in water without agitation, the highest activity levels from the LPS/saline aerosols were obtained with glass fiber (GF), cellulose acetate (CA), and PVC filters. The GF filters gave 3-fold higher endotoxin activities than did CA, the second best filter medium. Virtually no endotoxin activity was recovered from PC filters. Further, relative to GF filters, the PC filters yielded approximately half the chloride concentration from the saline aerosol. In experiments with cotton dust, the CA and PVC filters gave 2-fold higher endotoxin activities than the GF filters. In experiments with LPS-spiked MWF and used MWF, the CA filters gave ~2-fold higher endotoxin estimates than either PVC or GF filters. Unfortunately, the authors did not report the expected airborne levels of endotoxin, chloride, or particulate matter calculated from the amounts of material aerosolized and chamber air flow. Thus, endotoxin activity, chloride, and particle collection and extraction efficiencies cannot be determined from these data. The authors provided no explanation for the poor yields of saline from the PC filters.

Douwes et al. (1995) found that GF, Teflon, and PC filters gave similar estimates of airborne endotoxin level from replicate samples of a water mist. CE filters produced lower estimates. Walters et al. (1994) used GC–MS analysis to compare LPS concentration in aqueous extracts of PC filters with total LPS concentration on simultaneously collected filters as determined by direct methanolysis of the filter medium (Walters et al., 1994). These data suggest that, on average, extraction with aqueous buffer and sonication released 100% of the LPS from the PC filters. These results are in direct contrast with the earlier report that endotoxin could not be extracted from PC filters (Gordon et al., 1992). However, the GC–MS results agree with the finding that PC filters did not bind or inactivate LPS solutions and with the results from Douwes et al. (1995).

Air sample collection on PC filters and extraction by sonication in 0.05-M potassium phosphate, 0.01% triethylamine, pH 7.5 buffer should be considered the standard method for environmental studies until similar validations, using GC–MS to compare endotoxin yields after aqueous extraction, document the suitability of other methods [see also 23.4].

23.3.2 Handling and Storage

In general, environmental samples for endotoxin analysis should not be frozen, particularly once they have been extracted. In some cases, even unextracted samples should not be frozen; however, the literature on these points is conflicting (Olenchock et al., 1989; Douwes et al., 1995; Milton and Johnson, 1995; Milton et al., 1997). Olenchock et al. (1989) reported that freezing did not affect endotoxin activity in extracts of grain dust. However, Douwes et al. (1995) and Milton et al. (1997) found that house-dust extracts lost 25% to 86% of their activity with freezing. The endotoxin activity in house-dust samples stored desiccated at –20°C without extraction was stable for at least several months (Milton et al., 1997). In contrast, the endotoxin activity of bulk MWF samples dropped—sometimes by several orders of magnitude—when frozen (Milton and Johnson, 1995). Thus, the stability of endotoxin activity during sample storage may vary depending on the sample matrix, and stability should be established for each type of sample to be stored.

While bulk dust samples can be desiccated and stored below 0°C to prevent bacterial growth during shipping and storage, liquid samples present a greater problem. Storage of MWFs at 4°C in the presence of formaldehyde preserved endotoxin activity. However, storage with formaldehyde may result in significant activity losses when endotoxin levels are low. Walters et al. (1994) reported that preserving endotoxin air samples during shipping and storage by attaching a container of desiccant directly to a sampling cassette resulted in stable endotoxin estimates for 6 to 8 weeks from samples stored at 4°C.

Endotoxin is ubiquitous in the laboratory environment. Therefore, to destroy background endotoxin, all glassware, forceps, and metal cassettes should be baked using an appropriate protocol (e.g., 210°C for 1 hour or 270°C for 30 min) (Tsuji and Harrison, 1978; Gail and Jensch, 1987). Plasticware should be new material that was machine-packaged in sealed containers. Plastic sampling cassettes should also be new material or may be cleaned by sonication in endotoxin-free 1% triethylamine. Successful decontamination must be demonstrated. Polystyrene is the preferred polymer. Polypropylene should be avoided due to the tendency of some lots of this polymer to irreversibly adsorb large amounts of LPS (Novitsky et al., 1986). The inhibitory effect of polypropylene has not always been evident (Douwes et al., 1995), possibly because of variation in this property from lot to lot of test tubes (Milton, unpublished observation).

Suggested standard method for studying environmental endotoxin: collect air samples on PC filters and extract by sonication in 0.05-M potassium phosphate, 0.01% triethylamine, pH 7.5 buffer.

The stability of endotoxin activity during sample storage may vary depending on the sample matrix. Stability should be established for each type of sample to be stored.

23.4 Sample Analysis
23.4.1 Extraction Methods

Although a part of sample analysis, the extraction process is closely coupled with the choice of filters for sample collection [see 23.3.1]. In this section, aqueous extraction methods in preparation for LAL assay will be discussed. Douwes et al. (1995) compared endotoxin activity measurements using a device to collect parallel air samples from a water mist in a potato-processing plant. In an incomplete factorial design, the authors tested two extraction media (water and 0.05% Tween 20), two extraction temperatures (room temperature and 68°C), two agitation methods (vigorous and quiet), and four filter types (GF, Teflon, PC, and CE). The most significant factor was extraction medium (e.g., Tween 20 yielded a 7-fold higher activity level). The CE filters gave ~2-fold lower endotoxin estimates than did the other media. There were no significant differences between the GF, Teflon, and PC filters.

An earlier attempt to compare extraction media used 1% Tween 20 to extract organic dust and found that the non-ionic detergent produced drastic changes in the endotoxin standard curve (Olenchock et al., 1989). When endotoxin activity was corrected for these changes, extraction in Tween 20 produced lower endotoxin estimates for bulk dust samples than did extraction in water. However, Douwes et al. (1995) did not find important changes in the standard curve with 0.05% Tween 20. The earlier study used an endpoint and the latter a kinetic assay, but both studies used chromogenic LAL from the same manufacturer. In attempts to replicate these experiments, it was observed that 0.05% Tween 20 strongly inhibited the activity of the U.S. reference standard EC6 in some, but not all, lots of kinetic chromogenic LAL from the same manufacturer (Milton, unpublished data). More troubling was the finding that, depending on the LAL lot used, Tween 20 only slightly inhibited the activity of other LPS preparations, at the same time that it inhibited the activity of the reference standard, resulting in up to 15-fold differences in the estimated potency of a single preparation of E. coli O55:B5 LPS. Therefore, unless the LAL reagent is formulated for use with this detergent, even dilute Tween 20 cannot be recommended for use with LAL assays. Furthermore, these data show that—when comparing different extraction methods using the LAL assay—higher endotoxin activity levels may only indicate interference with the assay and not increased recovery of material from a filter medium [see 23.4.3].

The major difficulty interpreting results from the LAL assay, especially in comparing results with different filter media and extraction methods, derives from the LAL assay being a comparative and not an analytical bioassay (Finney, 1978) (i.e., changes in other factors than the concentration of LPS can result in changes in measured endotoxin activity levels) [see 23.4.2]. Thus, there is a need for chemical methods to validate extraction efficiencies [see 23.4.3]. Such methods would avoid the confusion generated when an extraction method causes even pure LPS preparations to appear more or less potent in the LAL assay without changing the concentration of LPS. Chemical methods suffer from lower sensitivity but results are not enhanced or inhibited by other factors.

> The LAL assay is a comparative, not an analytical, bioassay. Factors other than actual LPS concentration can change the measured endotoxin activity level. This makes it difficult to interpret results from the LAL assay, especially when comparing samples collected on different filter media and extracted and assayed in different solutions.

23.4.2 Limulus Amebocyte Lysate Assay for Endotoxin

Environmental endotoxin is usually measured using a test based on the reaction of LAL with LPS because this test is easy, fast, and sensitive (Levin, 1987). LAL tests are in-vitro biological assays made from a crude lysate of the amebocyte (blood cell) of the horseshoe crab (Limulus polyphemus). The lysate contains a serine protease specifically activated by endotoxin, which in turn activates a cascade of serine proteases. The amplification produced by the cascade is responsible for the ease with which low endotoxin levels can be detected. An alternate pathway exists in LAL, allowing activation by β-$(1\rightarrow3)$-D-glucan. The alternate pathway can be inhibited by addition of a specific detergent (zwittergent), or removed by chromatographic purification, followed by recombining only the desired LAL components (Obayashi et al., 1985; Roslansky and Novitsky, 1991).

LAL assays respond to heterogeneous LPS molecules with varying effectiveness, depending on LPS aggregation and polysaccharide content (Morrison et al., 1987). The sensitivity of LAL to LPS varies with the type of bacterium from which the LPS was derived, presumably because of differences in lipid A structures. Comparison with fever thresholds in the rabbit pyrogen model suggest that LAL over-represents the biological activity of environmental endotoxin (Pearson et al., 1982). However, human inhalation experiments suggest that LAL reactivity may under-represent the biological potency of crude preparations of whole bacteria for causing airways and alveolar inflammation (Rylander et al., 1989a). An assumption made for the purposes of environmental studies is that variations in LAL sensitivity to environmental endotoxins correspond to variations in the toxic potency of inhaled endotoxin, even though the absolute amounts of LPS cannot be determined from LAL data. However, whether this assumption holds may depend on the type of toxic effect as well as the structure and form of the inhaled LPS. Therefore, it is important to keep in mind that the LAL test is a comparative method providing an estimate of the relative toxicity of a broad class of molecules rather than an analytical method that measures the concentration of a single chemical moiety.

23.4.2.1 Quantitative LAL Assays Quantitative LAL methods based on chromogenic and turbidimetric assays are preferred for environmental work. Early studies were performed with gel-clot methods, but these have a large potential for error and do not lend themselves to internal validation. LAL preparations that do not respond to β-(1→3)-D-glucans (from fungi and plants) should be used when the objective is to measure endotoxin. However, glucan-sensitive and -insensitive LAL gave similar results in environments with very low concentrations of fungal and plant materials (Walters et al., 1994). At present, relatively glucan-insensitive LAL containing zwittergent is available from BioWhittaker (Walkersville, Md.) in two chromogenic formulations (i.e., one for an endpoint assay and one for a kinetic assay) and a kinetic turbidimetric formulation. A kinetic chromogenic LAL, purified to remove glucan-activated proteases, is produced in Japan and may be available from Seikagaku America (Ijamsville, Md.) or Associates of Cape Cod (Falmouth, Mass.). The kinetic assays offer the significant advantages of a much wider dynamic range and suitability for parallel-line analysis [see 23.4.2.2]. Commercially available LAL kits are designed and FDA approved for use in the pharmaceutical industry. Unfortunately, these kits are neither intended nor certified for analysis of environmental samples collected to measure exposure to airborne endotoxin.

23.4.2.2 Parallel-Line LAL Analysis Because LAL tests are comparative dilution assays, they must be rigorously standardized if any generalizations or comparisons with other observations are to be made. Results of a comparative method may not be valid under conditions other than those of a given assay because factors other than differences in LPS concentration can affect the results (Finney, 1978). The standard method of endotoxin extraction and analysis should control pH and ionic strength throughout the extraction, dilution, and assay steps because the method is sensitive to extraneous factors such as the type of glassware used to store and dilute samples as well as interference from other constituents of environmental samples (Milton et al., 1992; Hollander et al., 1993; Douwes et al.; 1995, Milton et al., 1997). Both an endotoxin standard and an environmental sample must be tested over a range of dilutions to ensure parallel dose–response relationships. Significant deviations from parallelism indicate interference with the LAL reaction and invalid results. Appropriate statistical methods for analysis of parallel-line biological assays should be employed to calculate the endotoxin activity (relative potency) of a test sample and to give an estimate of its intra-assay variance (Finney, 1978).

23.4.2.3 Interferences with LAL Assays Routine application of parallel-line methods has shown that there are two distinct types of interference with the LAL assay, that is, dilution-dependent and dilution-independent interference (Milton et al., 1992; Milton et al., 1997) (Figure 23.3). Dilution-dependent interference (DDI) occurs when the effect of an interfering compound diminishes as the sample is diluted. In this case, the standard and unknown dose–response curves become parallel after sufficient dilution of a sample, and a valid estimate of endotoxin activity can be obtained from the parallel portions of the curves. Some authors have argued that it may be sufficient to employ the parallel-line analysis on a limited number of samples to determine the appropriate dilution for all subsequent assays (Hollander et al., 1993). However, DDI may vary from sample to sample in a single environment, making this approach prone to increased measurement error (Milton et al., 1992; Milton et al., 1997). Dilution-independent interference (DII) also occurs sporadically and will be an uncontrolled source of error if parallel-line methods are not routinely used (Milton et al., 1992; Milton et al., 1997). DII may result from either chemical interference (e.g., pH effects on LPS micelle conformation) (Galanos et al., 1979; Coughlin et al., 1985) or random propagation of dilution error (Higgins et al., 1996).

23.4.2.4 Variability in LAL Reagents There is another serious problem with LAL assay of environmental samples that falls outside the control of individual laboratories — the available LAL reagents themselves vary widely in their sensitivity to environmental endotoxins. Saraf et al. (1997) found significant differences between endotoxin measurements of five bacterial supernatants (*E. coli*, *Branhamella catharralis*, *Pseudomonas aeruginosa*, *Pseudomonas cepacia*, and *Helicobacter pylori*) made with kinetic chromogenic LAL preparations from three manufacturers. Two of the lysates produced similar results, but the third manufacturer's lysate consistently gave 100- to 1000-fold lower estimates of endotoxin potency. Inconsistency among LAL preparations may not be restricted to lysates from different manufacturers. Assay of house dust samples stored at –20°C using lysate manufactured in different years resulted in an average 3.5-fold increase in endotoxin potency (range 0.6 to 8.5) over the course of storage. Most of this change in potency was a result of difference in responsiveness to environmental endotoxin between two lots of lysate (Milton et al., 1997).

The source of this problem appears to be that current manufacturing practice calls for testing each LAL lot with a variety of pharmaceutical products and for control of each lot's formulation to ensure consistent sensitivity to a standard LPS preparation before a lot is released. However, there has been no quality control for the sensitivity of LAL to environmental endotoxins. It is expected that this situation will improve (BioWhittaker, Inc.,

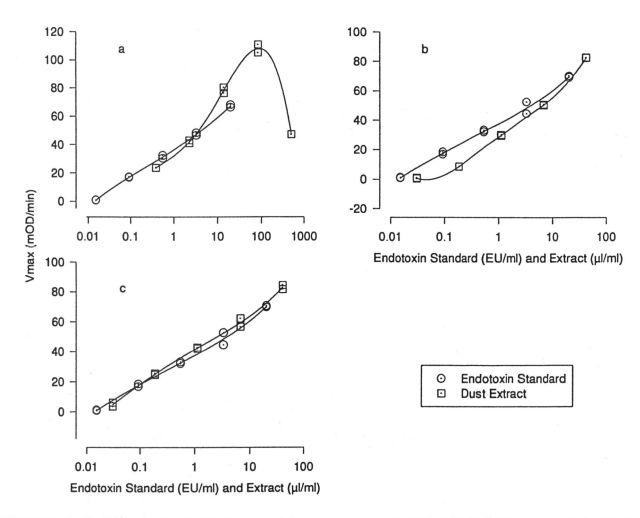

FIGURE 23.3. Dose–response curves for three house dust extracts showing (a) DDI with both dilution-dependent inhibition and enhancement, (b) DII, and (c) no evidence for interference. Regression lines fit to the data with fourth order polynomials are shown to illustrate the shape of the dose–response curves (Milton et al., 1997).

personal communication). In the past, however, lot-to-lot and year-to-year differences in LAL sensitivity to environmental endotoxin may account for some unexpected results that have been observed (e.g., the increase of airborne endotoxin reported in a cotton mill while the elutriated dust concentration declined over a five-year period) (Olenchock et al., 1990). Thus, failure to find exposure–response relationships in longitudinal studies and interlaboratory differences in endotoxin measurements may be largely attributable to differences in LAL lots.

> Available LAL reagents vary widely in their sensitivity to environmental endotoxins.

23.4.3 Chemical Assays for LPS

Chemical analysis for LPS in environmental samples has been based on detection of unique constituents including 2-keto-3-deoxyoctanoic acid, dideoxyheptose, and 3-hydroxy fatty acids. The 3-hydroxy fatty acids have been most extensively employed in environmental studies. Analysis of 3-hydroxy fatty acids by HPLC (Morris et al., 1988) has been reported, but the majority of studies used the more specific GC–MS methods (Maitra et al., 1978; Parker et al., 1982; Sonesson et al., 1990; Mielniczuk et al., 1993). Chemical methods have advantages in that they are not subject to issues of variable biological sensitivity, they enable characterization of environmental LPS, and they make possible identification of the organisms present in a sample. However, 3-hydroxy fatty acid methods are ~1000-fold less sensitive than the LAL assay, and these methods cannot determine whether the detected LPS existed in a biologically active form in the original sample.

Thus, the amounts of LPS detected may not correspond directly with biological activity either because of error in the biological measurement or because of inherent differences in endotoxin potency of LPS from different species. The types of 3-hydroxy fatty acids in the sample may be related to the biological potency of the

LPS (Saraf et al., 1997). Thus, it may be possible to estimate biological potency (endotoxin exposure) from LPS measurements. Also, the types of 3-hydroxy fatty acids in a sample are characteristic of the bacterial species present and so may be used to investigate the microbial ecology and sources of endotoxin in the environment.

23.5 Data Interpretation

It is clear that environmental endotoxin measurements from different laboratories are not necessarily comparable because the LAL assay is a comparative rather than an analytical method and because of the lot-to-lot variation in LAL reagent sensitivity described in Section 23.4.2.4. Interlaboratory comparisons found wide-ranging results (>1000-fold variation) in a comparison where laboratories used different LAL preparations (S. Reynolds, University of Iowa, Iowa City, Iowa, and D.T.W. Chun, U.S. Department of Agriculture, Clemson, S.C., personal communications). However, results were in much closer agreement (less than 10-fold variation, but still significantly different) in a second round of the interlaboratory comparison study, in which all laboratories used the same extraction and dilution protocol. Similarly, in a separate interlaboratory comparison, including some of the same laboratories, the variation among laboratories was less than 10 fold (but significant) when each laboratory used its own extraction and dilution method but all laboratories used LAL from a single lot (i.e., the comparison controlled the kit used by the laboratories). Thus, there are at least two sources of variation in endotoxin estimates from LAL assays, and these sources of variation appear to have a multiplicative interaction. Once methods and reagents are better standardized, investigators will be able to determine whether a given LAL result for samples from different environments can be considered to have the same meaning. At present, it appears that only data from one laboratory assaying samples from one environment while using one LAL lot can be considered entirely comparable. However, within one laboratory, the results of endotoxin analyses may be internally consistent and interpretable.

Experimental studies of human exposure to cotton dust and the majority of recent field studies suggest an endotoxin threshold for acute airflow obstruction in the range of 45 to 330 EU/m^3 [see 23.2.1 and 23.2.2]. Thus, most current data support a threshold for acute airflow obstruction within a 10-fold range. These values are generally 10 to 100 times the background endotoxin levels reported in the studies. Unfortunately, some laboratories report background levels as high as 35 ng/m^3 (presumably equivalent to 350 EU/m^3) (Teeuw et al., 1994). Other researchers have reported backgrounds of 18 pg/m^3 with acute health effects observed at 64 pg/m^3 (presumably equivalent to 0.18 and 0.64 EU/m^3) (Kateman et al., 1990).

For the time being, the imprecision of the LAL assay over time and among laboratories makes it impossible to establish a TLV at a given endotoxin level. Some recent guidelines proposed for endotoxin exposure limits assume that the measurement difficulties will be overcome by limiting the measurements to "the techniques employed in laboratories with extensive experience" (Rylander, 1997). However, it is far from clear that this limitation will produce consistent results. Therefore, ACGIH proposes a practical alternative approach for limiting endotoxin exposure using relative limit values (RLVs). This approach is based on an essential comparison of endotoxin activity levels in the environment in question with simultaneously determined background levels. Investigators must ensure that background endotoxin levels are determined under appropriate and representative conditions. Outdoor endotoxin sources may be important and outdoor air may not represent an appropriate background, particularly in certain agricultural environments. In urban and suburban areas, outdoor endotoxin levels during the growing season are the most reasonable basis for comparison. When simultaneously collected background samples cannot come from such environments, the laboratory should ensure that it has previously analyzed appropriate background samples using the same methods and LAL lots as for the present samples so that current background and exposure-zone measurements may be put in the proper context. Because outdoor endotoxin levels are very low in winter, it may be possible to detect lower indoor-derived endotoxin levels at that time of year. Even then, investigators should exercise caution in interpreting results if indoor endotoxin exposures do not exceed appropriate outdoor background levels observed during the growing season.

23.6 Remediation and Prevention
23.6.1 Recommendations — Relative Limit Values

RLVs are proposed because of the observation that endotoxin levels between 10- and 100-fold higher than background are frequently associated with adverse health effects. Therefore, it should be assumed that endotoxin plays a role and action should be taken to reduce exposure when (a) there are health effects consistent with endotoxin exposure (e.g., fatigue, malaise, cough, chesttightness, and acute airflow obstruction), and (b) endotoxin exposures exceed 10 times simultaneously determined, appropriate background levels. Thus, 10 times background is proposed as an RLV action level in the presence of respiratory symptoms.

In environments with a potential for endotoxin exposure but no current complaints, endotoxin levels should not exceed 30 times the appropriate background. Thus, 30 times background is a maximum RLV in the absence of symptoms. The factor of 30 was chosen because (a) the log scale is probably the appropriate one for endotoxin exposure–response relationships (Castellan et al., 1987; Vogelzang et al., 1998), and (b) 30 provides a safety factor that is approximately halfway on the log scale between the action level of 10 times background and 100 times background where health effects have been observed fairly consistently. When exposures exceed the RLV action level or maximum RLV, appropriate remedial actions should be taken [see 23.6.2].

It is important that the appropriate background samples be considered carefully [see 23.5] in determining the relative endotoxin level for any particular circumstance. Other than as described here, it is premature to make recommendations regarding low-level endotoxin exposures such as may occur in homes and offices.

The imprecision of the LAL assay over time and among laboratories makes it impossible to establish a TLV at a given endotoxin level. Therefore, the ACGIH proposes a practical alternative approach for limiting endotoxin exposure using relative limit values (RLVs).

In the presence of respiratory symptoms, ACGIH proposes an RLV action level of 10 times background. In the absence of symptoms, a maximum RLV of 30 times background is proposed. Consider carefully what constitutes an appropriate background sample when determining the relative endotoxin level for a particular circumstance.

23.6.2 Remedial Actions

The source of endotoxin exposure is obvious in many industrial and agricultural environments. Biocides are generally ineffective at controlling GNB, may be important sensitizers or adjuvants themselves (Preller et al., 1995), and when they do work, may permit growth of fungi and GPB with equally or more serious health effects (Burge and Muilenberg, 1995) [see 16.1 and 16.2.1]. The best ways to minimize exposures to endotoxin are to (a) eliminate contamination with GNB, (b) limit aerosolization of materials contaminated with GNB, (c) contain operations that may generate aerosols containing endotoxin, and (d) as a last resort, provide personal respiratory protection for workers who may be exposed to endotoxin. Cool-mist and ultrasonic humidifiers as well as recirculating spray humidifiers sometimes found in building ventilation systems should be considered potential sources of high-level endotoxin exposure [see 10.4.4.4]. Otherwise, little is known about endotoxin sources in office buildings and homes. It is too early to say whether—aside from removing contaminated humidifiers—there is any cause to take remedial actions in these settings, much less what those actions would be.

23.7 Bacterial Cell-Wall Components
23.7.1 Characteristics
Bacterial cell walls contain a number of compounds that have important biological activities in mammals. How-

ever at present, little is known about the contribution of these compounds to possible health effects that may result from inhalation of bacteria. Cell walls of both GNB and GPB contain peptidoglycan (PG), accounting for ~50% of the cell-wall mass. Teichoic acids account for another ~40% of the cell wall. While GNB have a thin cell wall containing limited amounts of these polymers, GPB contain significantly larger amounts. PG generally consists of repeating units composed of N-acetylglucosamine and N-acetylmuramic acid in β-(1-4) linkages. These units are attached to tripeptide side chains containing both D and L amino acids and, in some cases, diaminopimelic acid. The peptide side chains are the sites of cross links between polymer chains. The teichoic acids are polymers of ribitol or glycerol phosphates with side groups consisting of various sugars, D-alanine, and choline. Lipoteichoic acids (LTAs) are additional components of GPB walls. LTAs are composed of unbranched 1,3-linked polyglycerophosphate chains covalently bound to a membrane diacylglyceroglycolipid (Greenberg et al., 1996).

23.7.2 Health Effects
PGs adsorbed from the gut (e.g., after intestinal by-pass) may be important in causing arthritis in humans (Ely, 1980; Wells et al., 1989). Animal models suggest that the large size of PG polymers is an important factor in their ability to produce persistent arthritis (Janusz et al., 1986). Infections with large amounts of GPB in the blood stream (sepsis) can produce life-threatening toxic reactions. Experiments with purified PG and LTA suggest that, by themselves, neither of these compounds is capable of reproducing the effects of Gram-positive sepsis. PG appears to be ~10,000-fold less potent than LPS at eliciting cytokine responses from macrophages (Redl et al., 1989). However, PG and LTA appear to act synergistically and together can reproduce the effects of Gram-positive sepsis (De Kimpe et al., 1995). There is also mounting evidence that LTA or associated lipids activate macrophage cytokine production through the same receptor responsible for the endotoxic effects of LPS (Kusunoki et al., 1995; Suda et al., 1995; Cleveland et al., 1996).

It is known that a small polymer derived from the PG of *Bordetella pertussis*, a GNB, is the essential toxic agent causing death of respiratory epithelial cells resulting in pertussis (whooping cough) (Luker et al., 1995). However, little is known about the effects of inhaling nonpathogenic GPB. A study of volunteers exposed to high concentrations of organic dust in a swine barn demonstrated that PG and associated compounds, as well as LPS, may play an important role in certain responses to organic dust. Zhiping et al. (1996) found PG was associated with increased serum levels of the cytokine IL-6, increased circulating levels of granulocytes, total leukocyte concentration, and fever, but not lung-function changes. Endotoxin and LPS, as measured by an LAL assay and GC–MS, were associated with serum IL-6 concentration, symptoms, and lung-function change, but not with fever or peripheral leukocyte counts. The predominant microorganisms in indoor air are usually GPB from humans [see 18.1.4.2]. Therefore, it is tempting to speculate that PG and associated cell-wall compounds are responsible for symptoms that have been associated with very low indoor ventilation rates. However, at this time, there is no evidence to support this hypothesis.

> The predominant microorganisms in indoor air are usually GPB from humans. Therefore, PG and associated cell-wall compounds could be partly responsible for symptoms that have been associated with very low indoor ventilation rates.

23.7.3 Sampling

PG and associated compounds can be detected in environmental samples by measuring either muramic or diaminopimelic acids as markers. These compounds are not present in nature except as components of PG. Several GC–MS-based methods have been described (Sonesson et al., 1988; Black et al., 1994; Fox and Rosario, 1994; Mielniczuk et al., 1995). Zhiping et al. (1996) demonstrated that at high concentrations (mean inhalable dust, 21 mg/m^3; PG, 6.5 µg/m^3), muramic acid could be detected and used as a marker of PG in air. These investigators collected samples on 0.4-µm pore PC filters using open-face cassettes. Samples were extracted by shaking for 1 hour in water. The extracts were dried and heated to 100°C overnight in 4-M HCl. After extraction with hexane, the aqueous phase was evaporated, subjected to trimethylsylic derivitization, and analyzed for muramic acid (Mielniczuk et al., 1995).

23.7.4 Data Interpretation

Sampling and analysis for PG is an interesting area worthy of further research. However, there currently is no significant body of data relating airborne PG exposure to health effects. Therefore, guidelines for interpreting sampling results and recommendations regarding the significance of observed PG concentrations cannot be made.

23.8 References

Andrews, J.H.; Hirano, S.S., Eds.: Microbial Ecology of Leaves. Springer-Verlag, New York, NY (1992).

Baker, P.J.; Hraba, T.; Taylor, C.E.; et al.: Structural Features that Influence the Ability of Lipid A and its Analogs to Abolish Expression of Suppressor T Cell Activity. Infect. Immun. 60(7):2694-701 (1992).

Black, G.E;, Fox, A.; Fox, K.; et al.: Electrospray Tandem Mass Spectrometry for Analysis of Native Muramic Acid in Whole Bacterial Cell Hydrolysates. Anal Chem. 66(23):4171-6 (1994).

Burge, H.A.; Muilenberg, M.L.: Microbiology of Metalworking Fluids. The Industrial Metalworking Environment — Assessment and Control. American Automobile Manufacturers Association, Dearborn, MI (1995).

Castellan, R.M.; Olenchock, S.A.; Kinsley, K.B.; et al.: Inhaled Endot-

oxin and Decreased Spirometric Values, An Exposure–Response Relation for Cotton Dust. N. Engl. J. Med. 317:605-10 (1987).

Cavagna, G.; Foa, V.; Vigliani, E.C.: Effects in Man and Rabbits of Inhalation of Cotton Dust or Extracts and Purified Endotoxins. Brit. J. Ind. Med. 26:314-21 (1969).

Christiani, D.C.; Ye, T.T.; Wegman, D.H.; et al.: Cotton Dust Exposure, Across-Shift Drop in FEV1, and Five-Year Change in Lung Function. Am. J. Respir. Crit. Care Med. 150(5 Pt 1):1250-5 (1994).

Clapp, W.D.; Becker, S.; Quay, J.; et al.: Grain Dust-Induced Airflow Obstruction and Inflammation of the Lower Respiratory Tract. Am. J. Respir. Crit. Care Med. 150(3):611-7 (1994).

Cleveland, M.G., Gorham, J.D.; Murphy, T.L.; et al.: Lipoteichoic Acid Preparations of Gram-Positive Bacteria Induce Interleukin-12 through a CD14-Dependent Pathway. Infect. Immun. 64(6):1906-12 (1996).

Coughlin, R.T.; Peterson, A.A.; Haug, A.; et al.: A pH Titration Study on the Ionic Bridging within lipopolysaccharide Aggregates. Biochim. Biophys. Acta. 821(3):404-12 (1985).

De Kimpe, S.J.; Kengatharan, M.; Thiemermann, C.; et al.: The Cell Wall Components peptidoglycan and lipoteichoic Acid from Staphylococcus aureus Act in Synergy to Cause Shock and Multiple Organ Failure. Proc. Natl. Acad. Sci. U.S.A. 92(22):10359-63 (1995).

Donham, K.; Haglind, P.; Peterson, Y.; et al.: Environmental and Health Studies of Farm Workers in Swedish Swine Confinement Buildings. Br. J. Ind. Med. 46(1):31-7 (1989).

Douwes, J.; Versloot, P.; Hollander, A.; et al.: Influence of Various Dust Sampling and Extraction Methods on the Measurement of Airborne Endotoxin. Appl. Environ. Microbiol. 61:1763-1769 (1995).

Ely, P.H.: The Bowel Bypass Syndrome: A Response to Bacterial Peptidoglycans. J. Am. Acad. Dermatol. 2(6):473-87 (1980).

Engelhardt, R.; Mackensen, A.; Galanos, C.: Phase I Trial of Intravenously Administered Endotoxin (Salmonella abortus equi) in Cancer Patients. Can. Res. 51(10):2524-30 (1991).

Enterline, P.E.; Sykora, J.L.; Keleti, G.; et al.: Endotoxins, Cotton Dust, and Cancer. Lancet. 2(8461):934-5 (1985).

Finney, D.J.: Statistical Method in Biological Assay. Charles Griffin and Company, Ltd., London, England (1978)

Fogelmark, B.; Lacey, J.; Rylander, R.: Experimental Allergic Alveolitis after Exposure to Different Microorganisms. Int. J. Exp. Pathol. 72(4):387-95 (1991).

Fox, A.; Rosario, R.M.T.: Quantitation of Muramic Acid, a Marker for Bacterial Peptidoglycan, in Dust from Hospital and Home Air-Conditioning Filters using Gas Chromatography-Mass Spectrometry. Indoor Air. 4:239-247 (1994).

Gail, L.; Jensch, U.E.: Measurement and Calculation of Endotoxin Inactivation by Dry Heat. Detection of bacterial endotoxins with the Limulus amebocyte lysate test. S.W. Watson, J. Levin and T.J. Novitsky. New York, Alan R. Liss:273-281 (1987).

Galanos, C.; Freudenberg, M.A.; Matsuura, M.: Mechanisms of the Lethal Action of Endotoxin and Endotoxin Hypersensitivity. Adv Exp Med Biol. 256:603-19 (1990).

Galanos, C.; Luderitz, O.; Westphal, O.: Preparation and Properties of a Standardized lipopolysaccharide from Salmonella abortus equi (Novo-Pyrexal). Zentralbl Bakteriol [Orig A]. 243(2-3):226-244 (1979).

Gordon, T.: Dose-Dependent Pulmonary Effects of Inhaled Endotoxin in Guinea Pigs. Env Res. 59:416-426 (1992).

Gordon, T.; Galdanes, K.; Brosseau, L.: Comparison of Sampling Media for Endotoxin-Containing Aerosols. Appl. Occup. Environ. Hyg. 7(7):472-7 (1992).

Greenberg, J.W.; Fischer, W.; Joiner, K.A.: Influence of :Lipoteichoic Acid Structure on Recognition by the Macrophage Scavenger Receptor. Infect. Immun. 64(8):3318-25 (1996).

Gyntelberg, F.; Suadicani, P.; Nielsen, J.W.; et al.: Dust and the Sick Building Syndrome. Indoor Air 4:223-38 (1994).

Heederik, D.; Smid, T.; Houba, R.; et al.: Dust-Related Decline in Lung Function among Animal Feed Workers. Am. J. Ind. Med. 25(1):117-9 (1994).

Hines, C. J.; Milton, D.K.; Larsson, L.; et al.: Spatial and Temporal Variability of Endotoxin Exposures in an Office Building. In: 6th Annual NIOSH Interdivisional Aerosol Symposium. Ohio State University, Columbus, OH. National Institute for Occupational Safety and Health (1997)

Higgins, K.; Chew, G.; Davidian, M.; et al.: Documenting Variability in Indoor Allergen ELISAs. J. Allergy Clin. Immunol. 97(1 (Part 3)):216 (1996).

Hollander, A.; Heederik, D.; Versloot, P.; et al.: Inhibition and Enhancement in the Analysis of Airborne Endotoxin Levels in Various Occupational Environments. Am. Ind. Hyg. Assoc. J. 54(11):647-53 (1993).

Hudson, A.R.; Kilburn, K.H.; Halprin, G.M.; et al.: Granulocyte Recruitment to Airways Exposed to Endotoxin Aerosols. Am. Rev. Resp. Dis. 115:89-95 (1977).

Jacobs, R.R.: Airborne Endotoxins: An Association with Occupational Lung Disease. Appl. Ind. Hyg. 4:50-6 (1989).

Jacobs, R. R.: Endotoxins in the Environment. Int. J. Occup. Environ. Heal. 3(1): S3-S5 (1997).

Janusz, M.J.; Esser, R.E.; Schwab, J.H.: In Vivo Degradation of Bacterial Cell Wall by the Muralytic Enzyme mutanolysin. Infect. Immun. 52(2):459-67 (1986).

Kateman, E.; Heederik, D.; Pal, T.M.; et al.: Relationship of Airborne Microorganisms with the Lung Function and Leucocyte Levels of Workers with a History of Humidifier Fever. Scand. J. Work. Environ. Health 16:428-33 (1990).

Kennedy, S.M.; Christiani, D.C.; Eisen, E.A.; et al.: Cotton Dust and Endotoxin Exposure–Response Relationships in Cotton Textile Workers. Am Rev Respir Dis. 135:194-200 (1987).

Kotani, S.; Takada, H.: Structural Requirements of Lipid A for Endotoxicity and Other Biological Activities — An Overview. Adv. Exp. Med. Biol. 256(13):13-43 (1990).

Kusunoki, T.; Hailman, E.; Juan, T.S.; et al.: Molecules from Staphylococcus aureus that Bind CD14 and Stimulate Innate Immune Responses. J. Exp. Med. 182(6):1673-82 (1995).

Levin, J.: The Limulus amebocyte lysate Test: Perspectives and Problems. Prog. Clin. Biol. Res. 231:1-23 (1987).

Loppnow, H.; Dürrbaum, I.; Brade, H.; et al.: Lipid A, the Immunostimulatory Principle of lipopolysaccharides Adv. Exp. Med. Biol. 256(561):561-6 (1990).

Luker, K.E.; Tyler, A.N.; Marshall, G.R.; et al.: Tracheal cytotoxin Structural Requirements for Respiratory Epithelial Damage in Pertussis. Mol. Microbiol. 16(4):733-43 (1995).

Maitra, S.K.; Schotz, M.C.; Yoshikawa, T.T.; et al.: Determination of Lipid A and Endotoxin in Serum by Mass Spectroscopy. Proc. Natl. Acad. Sci. 75:3993-3997 (1978).

Malmberg, P.; Larsson, K.: Acute Exposure to Swine Dust Causes Bronchial Hyperresponsiveness in Healthy Subjects. Eur. Respir. J. 6(3):400-4 (1993).

Michel, O.; Ginanni, R.; Duchateau, J.; et al.: Domestic Endotoxin Exposure and Clinical Severity of Asthma. Clin. Exp. Allergy 21(4):441-8 (1991).

Michel, O.; Ginanni, R.; Sergysels, R.: Relation between the Bronchial Obstructive Response to Inhaled lipopolysaccharide and Bronchial Responsiveness to Histamine. Thorax 47(4):288-91 (1992).

Michel, O.; Kips, J.; Duchateau, J.; et al.: Severity of Asthma is Related to Endotoxin in House Dust. Am. J. Respir. Crit. Care Med. 154:1641-1646 (1996).

Mielniczuk, Z.; Mielniczuk, E.; Larsson, L.: Gas Chromatography-Mass Spectrometry Methods for Analysis of 2- and 3-hydroxylated Fatty Acids: Application for Endotoxin Measurement. J. Microbiol. Methods. 17:91-102 (1993).

Mielniczuk, Z.; Mielniczuk, E.; Larsson, L.: Determination of Muramic Acid in Organic Dust by Gas Chromatography-Mass Spectrometry. J. Chromatogr. B. Biomed. Appl. 670(1):167-72 (1995).

Milton, D.K.: Bacterial Endotoxins: A Review of Health Effects and Potential Impact in the Indoor Environment. In: Indoor Air and Human Health, pp. 179-195. R.B. Gammage and B.A. Berven, Eds. Lewis Publishers, Boca Raton, FL (1996).

Milton, D.K.; Amsel, J.; Reed, C.E.; et al.: Cross-Sectional Follow-up of a Flu-like Respiratory Illness among Fiberglass Manufacturing Employees: Endotoxin Exposure Associated with Two Distinct Sequelae. Am. J. Ind. Med. 28:469-488 (1995).

Milton, D.K.; Feldman, H.A.; Neuberg, D.S.; et al.: Environmental Endotoxin Measurement: The Kinetic Limulus Assay with Resistant-Parallel-Line Estimation. Environ. Res. 57:212-30 (1992).

Milton, D.K.; Gere, R.J.; Feldman, H.A.; et al.: Endotoxin Measurement: Aerosol Sampling and Application of a New Limulus Method. Am. Ind. Hyg. Assoc. J. 51(6):331-7 (1990).

Milton, D.K.; Johnson, D.K.: Endotoxin Exposure Assessment in Machining Operations. In: The Industrial Metalworking Environment: Assessment and Control. American Automobile Manufacturers Association, Dearborn, MI (1995).

Milton, D.K.; Johnson, D.K.; Park, J.H.: Environmental Endotoxin Measurement: Interference and Sources of Variation in the Limulus Assay of House Dust. Am. Ind. Hyg. Assoc. J. (1997).

Milton, D.K.; Wypij, D.; Kriebel, D.; et al.: Endotoxin Exposure–Response in a Fiberglass Manufacturing Plant. Am. J. Ind. Med. 29:3-13 (1996).

Morris, N.M.; Catalano, E.A.; Berni, R.J.:3-Hydroxymyristic Acid as a Measure of Endotoxin in Cotton Lint and Dust. Am. Ind. Hyg. Assoc. J. 49:81-88 (1988).

Morrison, D.C.; Vukajlovich, S.W.; Ryan, J.L.; et al.: Structural Requirements for Gelation of the Limulus Amebocyte Lysate by Endotoxin. Prog. Clin. Biol. Res. 231:55-73 (1987).

Novitsky, T.J.; Schmidt-Gengenbach, J.; Remillard, J.F.: Factors Affecting Recovery of Endotoxin Adsorbed to Container Surfaces. J. Parenter. Sci. Tech. 40:284-286 (1986).

Obayashi, T.; Tamura, H.; Tanaka, S.; et al.: A New Chromogenic Endotoxin-Specific Assay Using Recombined Limulus Coagulation Enzymes and its Clinical Applications. Clin. Chim. Acta. 149(1):55-65 (1985).

Olenchock, S.A.: Health Effects of Biological Agents: The Role of Endotoxins. Appl. Occup. Environ. Hyg. 9:62-64 (1994).

Olenchock, S.A.; Christiani, D.C.; Mull, J.C.; et al.: Airborne Endotoxin Concentrations in Various Work Areas within Two Cotton Textile Mills in the People's Republic of China. Biomed. Environ. Sci. 3:443-451 (1990).

Olenchock, S.A.; Lewis, D.M.; Mull, J.C.: Effects of Different Extraction Protocols on Endotoxin Analysis of Airborne Grain Dusts. Scand. J. Work Environ. Health. 15:430-5 (1989).

Olenchock, S.A.; Mull, J.C.; Jones, W.G.: Endotoxins in Cotton: Washing Effects and Size Distribution. Am. J. Ind. Med. 4:515-21 (1983).

Parker, J.H.; Smith, G.A.; Fredrickson, H.L.; et al.: Sensitive Assay, Based on Hydroxy Fatty Acids from Lipopolysaccharide Lipid A, for Gram-Negative Bacteria in Sediments. Appl. and Environ. Microbiology. 44(5):1170-1177 (1982).

Pearson, F.C.; Weary, M.E.; Bohon, J.; et al.: Relative Potency of "Environmental" Endotoxin as Measured by the Limulus Amebocyte Lysate Test and the USP Rabbit Pyrogen Test. Endotoxins and Their Detection with the Limulus Amebocyte Lysate Test, pp. 65-77. Alan R. Liss, Inc., New York (1982).

Pernis, B.; Vigliani, E.C.; Cavagna, C.; et al.: The Role of Bacterial Endotoxins in Occupational Diseases Caused by Inhaling Vegetable Dusts. Br. J. Ind. Med. 18:120-9 (1961).

Preller, L.; Heederik, D.; Boleij, J.S.; et al.:Lung Function and Chronic Respiratory Symptoms of Pig Farmers: Focus on Exposure to Endotoxins and Ammonia and use of Disinfectants. Occup. Environ. Med. 52(10):654-60 (1995).

Redl, H.; Schlag, G.; Thurnher, M.; et al.: Cardiovascular Reaction Pattern during Endotoxin or Peptidoglycan Application in Awake Sheep. Circ. Shock. 28(2):101-8 (1989).

Rietschel, E.T.; Brade, H.: Bacterial Endotoxins. Scientific American. 267(2):55-61 (1992).

Roslansky, P.F.; Novitsky, T.J.: Sensitivity of Limulus Amebocyte Lysate (LAL) to LAL-Reactive Glucans. J. Clin. Microbiol. 29(11):2477-2483 (1991).

Rylander, R.: Environmental Exposures with Decreased Risks for Lung Cancer. Int. J. Epidemiol. 19(1):S67-72 (1990).

Rylander, R.; Haglind, P.; Lundholm, M.: Endotoxin in Cotton Dust and Respiratory Function Decrement among Cotton Workers in an Experimental Cardroom. Am. Rev. Respir. Dis. 131:209-13 (1985).

Rylander, R.; Morey, P.: Airborne Endotoxin in Industries processing Vegetable Fibers. Am. Ind. Hyg. Assoc. J. 43(11):811-2 (1982).

Rylander, R.; Bake, B.; Fischer, J.J.; et al.: Pulmonary Function and Symptoms after Inhalation of Endotoxin. Am. Rev. Respir. Dis. 140(4):981-6 (1989a).

Rylander, R.; Sörensen, S.; Goto, H.; et al.:The Importance of Endotoxin and Glucan for Symptoms in Sick Buildings. In: Present and Future of Indoor Air Quality — Proceedings of the Brussels Conference. Excerpta Medica, New York, NY (1989b).

Rylander, R.: Evaluation of the Risks of Endotoxin Exposures. Int. J. Occup. Env. Heal. 3(1):S32-S36 (1997).

Sandström, T.; Bjermer, L.; Rylander, R.: Lipopolysaccharide (LPS) Inhalation in Healthy Subjects Increases Neutrophils, Lymphocytes and Fibronectin Levels in Bronchoalveolar Lavage Fluid. Eur. Respir. J. 5(8):992-6 (1992).

Saraf, A.; Larsson, L; Burge, H.A.; et al.:Quantification of 3-hydroxy Fatty Acids and Ergosterol in Settled House Dust by Gas Chromatography-mass Spectrometry: Relation to Limulus Assay for Endotoxin and to Culture for Fungi. Appl. Environ. Microbiol. 63:2554-2559 (1997).

Schwartz, D.A.; Thorne, P.S.; Jagielo, P.J.; et al.: Endotoxin Responsiveness and Grain Dust-induced Inflammation in the Lower Respiratory Tract. Am. J. Physiol. 267(5 Pt 1):L609-17 (1994).

Schwartz, D.A.; Donham, K.J.; Olenchock, S.A.; et al.: Determinants of Longitudinal Changes in Spirometric Function among Swine Confinement Operators and Farmers. Am. J. Respir. Crit. Care Med. 151(1):47-53 (1995a).

Schwartz, D.A.; Thorne, P.S.; Yagla, S.J.; et al.: The Role of Endotoxin in Grain Dust-induced Lung Disease. Am. J. Respir. Crit. Care Med. 152(2):603-8 (1995b).

Smid, T.; Heederik, D.; Houba, R.; et al.: Dust- and Endotoxin-related Respiratory Effects in the Animal Feed Industry. Am. Rev. Respir. Dis. 146(6):1474-9 (1992a).

Smid, T.; Heederik, D.; Mensink, G.; et al.: Exposure to Dust, Endotoxins, and Fungi in the Animal Feed Industry. Am. Ind. Hyg. Assoc. J. 53(6):362-8 (1992b).

Smid, T.; Heederik, D.; Houba, R.; et al.: Dust- and Endotoxin-related Acute Lung Function Changes and Work-related Symptoms in Workers in the Animal Feed Industry. Am. J. Ind. Med. 25:877-88 (1994).

Sonesson, A.; Larsson, L.; Fox, A.; et al.: Determination of Environmental Levels of Peptidoglycan and Lipopolysaccharide using Gas Chromatography with Negative-ion Chemical-ionization Mass Spectrometry Utilizing Bacterial Amino Acids and Hydroxy Fatty Acids as Biomarkers. J. Chromatogr. 431(1):1-15 (1988).

Sonesson, A.; Larsson, L.; Schutz, A.; et al.: Comparison of the Limulus Amebocyte Lysate Test and Gas Chromatography-mass Spectrometry for Measuring Lipopolysaccharides (Endotoxins) in Airborne Dust from Poultry Processing Industries. Appl. Environ. Microbiol. 56:1271-8 (1990).

Sonesson, H.R.A.; Zähringer, U.; Grimmecke, H.D.; et al.: Bacterial Endotoxin: Chemical Structure and Biological Activity. In: Endotoxin and the Lungs, 77:1-20. K. Brigham, Ed. Marcel Dekker, Inc., New York, NY (1994).

Suda, Y.; Tochio, H.; Kawano, K.; et al.: Cytokine-inducing Glycolipids in the Lipoteichoic Acid Fraction from Enterococcus hirae ATCC 9790. FEMS Immunol. Med. Microbiol. 12(2):97-112 (1995).

Teeuw, K.B.; Vandenbroucke-Grauls, C.M.; Verhoef, J.: Airborne Gram-negative Bacteria and Endotoxin in Sick Building Syndrome. A Study in Dutch Governmental Office Buildings. Arch. Intern. Med. 154(20):2339-45 (1994).

Thelin, A.; Tegler, Ö.; Rylander, R.: Lung Reactions during Poultry Handling Related to Dust and Bacterial Endotoxin Levels. Eur. J. Respir. Dis. 65:266-71 (1984).

Tsuji, K.; Harrison, S.J.: Dry-heat Destruction of Lipopolysaccharide: Dry-heat Destruction Kinetics. Appl. Environ. Microbiol. 36:710-714 (1978).

Ulevitch, R.J.: Recognition of Bacterial Endotoxins by Receptor-dependent Mechanisms. Adv. Immunol. 53:267-89 (1993).

Vogelzang, P. F. J.; Vandergulden, J.W.J.; Folgering, H.; et al.: Endotoxin Exposure as a Major Determinant of Lung Function Decline in Pig Farmers. Am. J. Respir. Crit. Care Med. 157(1): 15-18 (1998).

Walters, M.; Milton, D.K.; Larsson, L.; et al.: Airborne Environmental Endotoxin: A Cross-validation of Sampling and Analysis Techniques. Appl. Environ. Microbiol. 60(3):996-1005 (1994).

Wells, A.F.; Hightower, J.A.; Parks, C.; et al.: Systemic Injection of Group A streptococcal Peptidoglycan-polysaccharide Complexes Elicits Persistent Neutrophilia and Monocytosis Associated with Polyarthritis in rats. Infect Immun. 57(2):351-8 (1989).

Zhiping, W.; Malmberg, P.; Larsson, B.M.; et al.: Exposure to Bacteria in Swine-house Dust and Acute Inflammatory Reactions in Humans. Am. J. Respir. Crit. Care Med. 154:1261-1266 (1996).

Chapter 24

Fungal Toxins And β-(1→3)-*D*-Glucans

Harriet A. Burge with Harriet M. Ammann

24.1 Characteristics
    24.1.1 *Introduction*
    24.1.2 *Mycotoxins in History*
    24.1.3 *Mycotoxin Structure and Function*
    24.1.4 *Toxigenic Fungi and Mycotoxin Production*
24.2 Health Effects
    24.2.1 *Evidence for Mycotoxin Effects*
    24.2.2 *Effects of Mycotoxins*
        24.2.2.1 Macrophage Toxicity and Other Immune System Effects
        24.2.2.2 Inflammation
        24.2.2.3 Carcinogenicity
        24.2.2.4 Cardiovascular Effects
        24.2.2.5 Neurologic Effects
        24.2.2.6 Synergistic, Additive, or Antagonistic Effects
    24.2.3 *Inhalation or Other Respiratory Tract Exposures in Animals*
    24.2.4 *Human Inhalation Exposure to Mycotoxins*
        24.2.4.1 Agricultural and Manufacturing Exposures
        24.2.4.2 Indoor Air Exposures
    24.2.5 Risk Extrapolation
24.3 Sample Collection
24.4 Sample Analysis
    24.4.1 *Assays for Fungi or Fungal Spores*
        24.4.1.1 Culture of Toxigenic Fungi
        24.4.1.2 Fungal Spore Identification (Microscopy)
    24.4.2 *Assays for Mycotoxins*
    24.4.3 *Bioassays for Relative Toxicity*
    24.4.4 *Biological Exposure Indices*
24.5 Data Interpretation
    24.5.1 *Observation of Visible Fungal Growth*
    24.5.2 *Identification of Toxigenic Fungi*
    24.5.3 *Identification of Specific Mycotoxins or Toxicity*
24.6 Remediation and Prevention
24.7 β-(1→3)-*D*-Glucans
    24.7.1 *Characteristics*
    24.7.2 *Health Effects*
    24.7.3 *Sampling*
    24.7.4 *Data Interpretation*
    24.7.5 *Remediation and Prevention*
24.8 References

24.1 Characteristics

24.1.1 Introduction

The fungi produce many agents that can be toxic with sufficient exposure. In general, these agents fall into two classes: (a) secondary products of metabolism (e.g., mycotoxins, antibiotics, and VOCs), and (b) structural components [e.g., β-$(1{\rightarrow}3)$-D-glucans]. Mycotoxins are nonvolatile, relatively low-molecular-weight secondary metabolic products that may affect exposed persons in a variety of ways, the best known of which are deleterious. Antibiotics are usually large molecules that are recognized primarily by their effects on other microorganisms, although antibiotics may also be toxic to human cells. Some fungal toxins are recognized primarily because they accumulate in large fungal fruiting bodies (i.e., the "mushroom toxins"). Because toxicity is associated exclusively with ingestion, these compounds are not discussed here. Rather, this chapter focuses on mycotoxins (non-volatile, low-molecular-weight secondary metabolites of mycelial fungi) and the β-$(1{\rightarrow}3)$-D-glucans, while Chapter 26 covers fungal and other MVOCs.

Fungal taxonomy is a dynamic science, and new data result in changing views on how fungi should be named. *Stachybotrys chartarum* is an example of this kind of change. Many early publications used the name *Stachybotrys atra* for this fungus; later authors used *Stachybotrys chartarum*. In other cases, new studies may reveal characteristics of a fungus that lead to its reclassification. These kinds of changes have occurred for several important toxigenic species in the genera *Penicillium* and *Fusarium*. Analytical laboratories that provide specific fungal identifications should be aware of these changes and can be consulted when such questions arise. For this chapter, the names the authors of the cited publications used have been maintained.

24.1.2 Mycotoxins in History

Throughout history, mycotoxins have played an important role in human and animal health. Ergotism, because of its distinctive and dramatic symptoms (e.g., gangrene, convulsions, and death), was documented as early as 430 BCE when it afflicted the Spartans (Kendrick, 1992). This disease, also known as Saint Anthony's Fire, is caused by ingestion of rye products contaminated with fungi containing ergot alkaloids. These toxins affect smooth muscles (causing blood vessel constriction) and the central nervous system. However, the pivotal event in the recognition of mycotoxins as a serious cause of animal and human disease occurred in 1960 in the U.K. when over 100,000 turkeys died as a result of an outbreak of Turkey X disease. The common factor in these deaths was that affected livestock had been fed Brazilian peanut meal that was highly contaminated with a fungus identified as *Aspergillus flavus*. Scientists isolated a series of toxins from the meal, which were subsequently named afla-

toxins. This discovery provided a strong stimulus for mycotoxin research, and hundreds of mycotoxins have been identified since then. Most of this research has focused on agricultural settings and contamination of livestock feed.

In the past decade, concern regarding mycotoxin exposure in non-industrial or agricultural indoor environments has been steadily increasing (Jarvis, 1990; Miller, 1992; Hendry and Cole, 1993; Johanning et al., 1993, 1996). The concern over mycotoxins has been fueled by the potent health effects elicited in laboratory animals, clear associations between respiratory disease and fungal exposure in agricultural settings, concern about military uses, and conclusions derived from anecdotal case studies.

24.1.3 Mycotoxin Structure and Function

Mycotoxins are by-products of fungal metabolic processes (Betina, 1989; Jarvis, 1989; Miller, 1994). These compounds are considered secondary metabolites because they are natural products not necessary for fungal growth and are derived from a few precursors formed during primary metabolism (Kendrick, 1992). Mycotoxins do not constitute a chemical category because they have no molecular features in common (Búlock, 1980). Rather, the chemical structures of mycotoxins are quite diverse and include polyketides, terpenes, and indoles. The function of fungal toxins has not been clearly established. However, they are considered to play a role in regulating competition with other microorganisms (Janzen, 1977; Wicklow, 1981; Demain, 1989; Williams et al., 1989), and mycotoxins probably help parasitic fungi invade host tissues. Table 24.1 lists some well-known fungal secondary metabolites and some of their health effects. Note that the amounts of toxin needed to elicit effects varies widely (by orders of magnitude) among toxins. For example, T-2 toxin is far more potent than epicladosporic acid.

24.1.4 Toxigenic Fungi and Mycotoxin Production

Fungi that have been shown to produce mycotoxins will be described here as toxigenic fungi. The most frequently studied mycotoxins are produced by species of *Aspergillus*, *Fusarium*, *Penicillium*, *Stachybotrys*, and *Myrothecium*. However, toxins have been detected from many other fungi under some growth conditions. More than one fungal species or even genus may produce a single mycotoxin. For example, *Aspergillus versicolor*, *Emericella nidulans* (the teleomorph of *Aspergillus nidulans*), and *Cochliobolus sativus*, among others, can make sterigmatocystin. Conversely, a single fungal species may produce more than one mycotoxin. For example, *S. chartarum* produces many toxic substances including satratoxin F, G, and H, roridin E, and verrucarin J (Jarvis et al., 1986, 1995).

The kinds and amounts of toxin a fungus produces depend on the fungal strain, the substrate it is metabolizing, and, possibly, the presence or absence of other organisms. Not all strains of potentially toxigenic fungal species produce mycotoxins under laboratory conditions, even when grown on "natural" or building-related substrates. Whether a toxigenic fungus produces mycotoxins in a given situation appears to depend on environmental conditions, especially growth substrate, temperature, pH, and other factors (Wicklow, 1981; Land et al., 1993; Nikulin et al., 1994; Aziz and Moussa 1997; Rao et al., 1997).

Mycotoxins accumulate in fungal spores, mycelia, and growth substrates in ratios dependent on fungal species and strain (Wicklow and Shotwell, 1982; Palmgren and Lee, 1986; Sorenson et al., 1987). Exposure to mycotoxins occurs when (a) colonized substrate material is ingested (i.e.,

TABLE 24.1. Some Common Fungi, Mycotoxins, and Health Effects from Ingestion, Dermal, or Inhalation Exposure (derived from Kendrick, 1992; Arafat and Musa 1995; Osborne et al., 1996).

CAVEAT: The entries in this table provide examples of the toxins and health effects that have been associated with some fungi. The listed fungi may also produce other toxins, fungi other than those listed may produce some of these toxins, and health effects other than those listed may also be associated with these toxins.

Fungus	Mycotoxin	Possible Health Effect
Acremonium spp.	Cephalosporin	Antibiotic
Alternaria alternata, Phoma sorghina	Tenuazoic acid	Nephrotoxic, hepatotoxic, hemorrhagic
Aspergillus clavatus	Cytochalasin E Patulin	Affects cell division, Inhibits protein synthesis, nephrotoxic, carcinogenic
Aspergillus flavus, Aspergillus parasiticus	Aflatoxins	Mutagenic, carcinogenic, hepatotoxic
Aspergillus fumigatus	Fumitremorgens Gliotoxin	Tremorgenic Cytotoxic
Aspergillus nidulans, Aspergillus versicolor, Cochliobolus sativus	Sterigmatocystin	Hepatotoxic, carcinogenic
Aspergillus ochraceus, Penicillium verrucosum, Penicillium viridicatum	Ochratoxin A	Nephrotoxic, hepatotoxic, carcinogenic
Cladosporium spp.	Epicladosporic acid	Immunosuppressive
Cladosporium cladosporioides	Cladosporin, Emodin	Antibiotics
Claviceps purpurea	Ergot alkaloids	Vasoactive (cause smooth muscles to constrict), hallucinogenic
Fusarium graminearum	Deoxynivalenol Zearalenone	Emetic Estrogenic
Fusarium moniliforme	Fumonisins	Neurotoxic, hepatotoxic, nephrotoxic, carcinogenic
Fusarium poae, Fusarium sporotrichoides	T-2 toxin	Hemorrhagic, immunosuppressive, causes nausea and vomiting
Penicillium chrysogenum	Penicillin	Antibiotic
Penicillium crustosum	Penitrem A, Roquefortine C	Tremorgenic Neutrotoxic (tremorgenic)
Penicillium expansum	Citrinin, Patulin, Roquefortine C	Nephrotoxic, carcinogenic Inhibits protien synthesis, nephrotoxic, carcinogenic Neurotoxic (tremorgenic)
Penicillium griseofulvum, Penicillium viridicatum	Griseofulvin	Tumorigenic, teratogenic, hepatotoxic
Pithomyces chartarum	Sporidesmin Phylloerythrin	Hepatotoxic Causes photosensitization and eczema
Stachybotrys chartarum (atra)	Satratoxins, Verrucarins, Roridins Stachybocins	Inflammatory agents, Immunosuppressive, cause dermatitis, hemotoxic, hemorrhagic
Tolypocladium inflatum	Cyclosporin	Immunosuppressive

when moldy food is eaten), (b) fungal and toxin-containing materials are handled so that skin contact occurs, or (c) spores, mycelial fragments, or colonized substrate materials are aerosolized and inhaled. Spores are considered the most common vehicle for mycotoxin inhalation and can contain significant toxin concentrations (Table 24.2).

Mycotoxin inhalation exposure occurs in particulate form. Note that exposure to potentially toxic fungal VOCs may occur, but these VOCs are not considered mycotoxins (Hendry and Cole, 1993) [see also 26.1.2]. Fungi that produce potent mycotoxins are seldom abundant in outdoor ambient air. Materials from which most exposure to toxigenic fungi occurs include agricultural and industrial substrates (e.g., stored grain, wood chips, and municipal compost) and delignified cellulose (e.g., paper). The fungi common on these substrates differ, with *Aspergillus* and *Penicillium* spp. being common on agricultural materials and *S. chartarum* common on paper products (Burg et al., 1981; Sorenson et al., 1981; Pasanen et al., 1994; Andersson et al., 1997).

24.2 Health Effects

24.2.1 Evidence for Mycotoxin Effects

Initial evidence for the existence of mycotoxins was provided by epidemics of ingestion-related disease in humans (e.g., ergotism) and animals (e.g., Turkey-X disease). For many years, primary interest in mycotoxins was focused in the human and animal food industries, and most of our knowledge of the health effects of mycotoxins is derived from the agricultural literature.

Efforts to elucidate the mechanisms of mycotoxin effects led to the use of *in-vitro* and *in-vivo* toxicity assays designed to test specific kinds of effects. Among these are (a) assays that assess non-specific toxicity by measuring either death rates or changes in metabolic rates (e.g., the brine shrimp, tetrahymena, and yeast metabolic assays) (Binder et al., 1997; Hlywka et al., 1997; Martin-Martin-Gonzalez et al., 1997), (b) assays that evaluate mutagenicity or carcinogenicity (e.g., the Ames assay) (Maron and Ames, 1983; Canas and Aranda, 1996; Cariello and Piegorsch,

1996), and (c) assays that allow measurement of specific chemical mediators released in response to mycotoxin challenge (e.g., cell culture and animal instillation assays) (Hanelt et al., 1994; Richard et al., 1994; Nikulin et al., 1997; Norred et al., 1997). In addition, whole animals may be exposed to toxins by ingestion, injection, or inhalation and the overall health of an animal, development of specific diseases, or specific changes in organs monitored.

Data on human health effects come primarily from epidemiological studies, where populations are studied for exposure and specific health effects (Hayes et al., 1984; Autrup et al., 1993). In such studies, biological indicators or environmental measurements are used as exposure measures. In addition, a limited amount of case-study data support an association between mycotoxin exposure and disease. For case studies, relatively intensive exposure assessments are conducted to determine the source for disease in one or a few persons (Croft et al., 1986; DiPaolo et al., 1994; Lougheed et al., 1995).

24.2.2 Effects of Mycotoxins

Mycotoxins are usually potent cytotoxins that cause cell disruption and interfere with essential cellular processes. Some mycotoxins are potent carcinogens, some are vasoactive, and some penetrate the blood-brain barrier to cause CNS effects (Table 24.1). Often, a single mycotoxin can elicit more than one type of toxic effect. There are hundreds of different mycotoxins, and a review of all of them is beyond the scope of this document. In addition, new toxins and their health effects are continually being described. Only effects considered most relevant for indoor exposures are discussed here. An investigator especially interested in mycotoxins should monitor relevant current literature.

24.2.2.1 Macrophage Toxicity and Other Immune System Effects
The pulmonary macrophage is an important part of the defense system in the lung, where these cells actively ingest and remove foreign particles. Many mycotoxins (including T-2 toxin, patulin, penicillic acid, aflatoxin, and satratoxins) have been shown to interfere with

TABLE 24.2. Toxins from Spores of Several Fungal Species Cultivated on Laboratory Media; Quantification by High-Pressure Liquid Chromatography or Thin-Layer Chromatography (Palmgren and Lee, 1986; Land et al., 1993)

Fungus	Toxin	ng/g Spores	ng/10^6 Spores
Aspergillus fumigatus	Fumitremorgen B, Verruculogen	—	6–80
	F migaclavine C	930,000	9.890
Aspergillus niger	Aurasperone C	460,000	0.114
Aspergillus parasiticus	Aflatoxin B$_1$	16,600	0.976
Penicillium oxalicum	Secalonic Acid D	1,890	0.025
Aspergillus parasiticus (mutant Nor-1)	Norsolorinic Acid	280	0.023

macrophage functioning or to selectively kill macrophages (Richards and Thurston, 1975; Gerberick et al., 1984; Sorenson et al., 1985, 1986; Sorenson and Simpson, 1986; Jakab et al., 1994; Nikulin et al., 1996, 1997).

Aspergillus fumigatus carries a toxin in the spore wall that diffuses rapidly into water and inhibits macrophage function (Mitchell et al., 1997). This and other toxins may facilitate colonization of the airways of asthmatics leading to ABPA (Amitani et al., 1995; Murayama et al., 1996) [see 25.2.2.2]. Gliotoxin is produced during mycelial growth of *A. fumigatus*. This toxin causes fragmentation of DNA, especially in thymocytes, making gliotoxin, a potent immunosuppressive agent, and possibly facilitating invasive aspergillosis in immunosuppressed persons (Sutton et al., 1996; Waring and Beaver, 1996; Waring et al., 1997).

24.2.2.2 Inflammation Fungi are well-known causes of allergic disease, stimulating the production of IgE-mediated conditions as well as HP. However, in addition to the allergen-induced release of inflammatory agents that mediate these conditions, some kinds of fungal spores directly stimulate the release of mediators of inflammation. Agents that pulmonary macrophages may release in response to fungal spore exposure include cytokines, reactive oxygen metabolites, and chemotactic factors (Milanowski et al., 1995; Ruotsalainen et al., 1995; Shahan et al., 1998).

24.2.2.3 Carcinogenicity Aflatoxin B1 is recognized as the most potent known natural carcinogen and is used as a measure against which the carcinogenicity of other chemicals is evaluated. The aflatoxins are produced by several members of the genus *Aspergillus* (*A. flavus*, *A. parasiticus*) and teleomorphic ascomycetes. Aflatoxin B1 is actually a "pro-carcinogen" that must be transformed to the carcinogenic state within the body. Following ingestion exposure, this transformation occurs in the liver, and the result is liver cancer. However, mammalian airway epithelium is also efficient in activating aflatoxin B1 to the carcinogenic form (Wilson et al., 1990; Putt et al., 1995; Kelly et al., 1997). In addition, aflatoxin B1 administered by airway instillation results in binding both in the lung and in the liver, indicating translocation of the active form of the toxin (Biswas et al., 1993).

A number of other mycotoxins are known to be carcinogenic (Table 24.1). Of these, sterigmatocystin has been of most interest as an indoor airborne agent. Sterigmatocystin is produced by *A. versicolor*, *E. nidulans*, and *C. sativus* and is a precursor of aflatoxin.

24.2.2.4 Cardiovascular Effects As discussed above, the ergot alkaloids cause constriction of smooth muscle, which includes blood vessels and the uterine wall. The gangrene that results from ingestion of ergot alkaloids is caused by restriction of blood flow. The disease alimentary toxic aleukia (ATA) is associated with ingestion of trichothecene toxins (e.g., T-2 toxin) and is characterized by tachycardia (rapid heart beat). The trichothecene chemotherapy agent anguidine (diacetoxyscirpenol) causes hypotension (low blood pressure). Fumonisin ingestion produces significant changes in a number of cardiovascular parameters (Bukowski et al., 1982; Smith et al., 1996).

24.2.2.5 Neurologic Effects Several of the diseases related to ingestion of mycotoxins include neurologic symptoms. ATA is associated with headaches and convulsions. Exposure to *A. fumigatus* has been associated with dementia and tremors (Gordon et al., 1993). Treatment of cancer patients with diacetoxyscirpenol may result in abnormal electro-encephalograms, hallucination, headaches, and transient coma (Goodwin et al., 1978; Bukowski et al., 1982).

24.2.2.6 Synergistic, Additive, or Antagonistic Effects Exposures to fungi and mycotoxins (except under controlled conditions) are likely to be associated with exposure to other agents as well. Andersson et al. (1997) recovered satratoxin, glucans, and endotoxin from water-damaged gypsum board, as well as a variety of GNB, GPB, mycobacteria, *S. chartarum*, several species of *Penicillium* and *Mucor*, *A. versicolor*, *Chaetomium globosum*, *Cladosporium herbarum*, *C. sphaerospermum*, and *Paecilomyces varioti*. The release of mediators of inflammation, which play a role in organic dust toxic syndrome (ODTS), are enhanced by mixed exposures to bacteria, endotoxins, and fungal spores in human airway epithelial cells obtained by bronchoalveolar lavage (Norn, 1994). The release of superoxide anions stimulated by fungal spores and bacteria is enhanced by pretreatment with LPS (Shahan et al., 1994).

Koshinsky and Khachatourians (1992) used a yeast metabolic assay to document complex changes in the effects of several trichothecene mycotoxins when administered alone or in mixtures. At very low-effect levels, interactions appeared antagonistic, while at higher-effect levels, relationships appeared to be synergistic, with all relationships toxin specific. Association of mycotoxins with particles seems to amplify adverse effects in the lungs, possibly due to increased toxin retention in the respiratory tract (Coulombe et al., 1991).

24.2.3 Inhalation or Other Respiratory Tract Exposures in Animals

Most of the data on health effects of mycotoxins comes from the literature on ingestion exposure. However, increasing interest in mycotoxins as airborne disease agents have stimulated efforts to document effects that may occur with respiratory tract exposure. Studies focusing on animal models have used cells washed from the lungs, instillation or aerosol exposures to purified toxins, or instillation of whole spores. Creasia et al. (1990) found that T-2 toxin delivered by inhalation was 2 to 20 times more toxic for mice, rats, swine, and guinea pigs than intravenously administered toxin.

A number of recent studies have reported the toxicity of *S. chartarum* toxins in laboratory animals. Nikulin et al. (1996, 1997) used intranasal instillation of whole spores of two *S. atra (chartarum)* strains in mice. One strain contained spirolactones and spirolactams, whereas the other more toxic strains also contained satratoxins. The instillation of 10^6 spores (one instillation) produced bronchiolar and interstitial inflammation with evidence of hemorrhage.

Aflatoxin B1, administered by inhalation and instillation, has also been shown to suppress alveolar macrophage phagocytosis and the primary splenic antibody response, indicating both a local and a systemic effect on immunity (Jakab et al., 1994). Animal instillation of *A. fumigatus* resulted in influx of neutrophils, lymphocytes, and red blood cells to the lung (Milanowski, 1998). This inflammatory response was also demonstrated for the GNB *Erwinia herbicola*. Ochratoxin was found to be rapidly absorbed following intratracheal instillation in rats, with a toxin half life of 127 hours (Breitholtz-Emanuelsson et al., 1995).

24.2.4 Human Inhalation Exposure to Mycotoxins

24.2.4.1 Agricultural and Manufacturing Exposures

Organic Dust Toxic Syndrome ODTS is a flu-like illness that may follow exposure to organic dusts. Cases resembling the disease have been reported from as early as 1927 (May et al., 1986). ODTS symptoms are considered to result from exposure to complex mixtures that may include endotoxin, glucans, antigens, and mycotoxins. Even relatively low-level exposures to complex mixtures result in release of mediators of inflammation (Wintermeyer et al., 1997). As mentioned in Section 24.2.3, laboratory animal experiments indicate a synergistic effect between endotoxin and fungal spore components (Shahan et al., 1994). Some evidence exists that a family history of atopy is a risk factor for ODTS development, possibly indicating a role for antigens in the disease (Sisgaard et al., 1994). At least in some cases, symptoms are indistinguishable from early stages of HP (Weber et al., 1993). Although ODTS has been called pulmonary mycotoxicosis, the role of mycotoxins remains to be elucidated.

Aflatoxin In agricultural and food processing settings, epidemiological studies have linked aflatoxin inhalation exposure to various forms of cancer (Van Nieuwenhuize et al., 1973; Baxter et al., 1981; Hayes et al., 1984; Olsen et al., 1988; Autrup et al., 1991, 1993). Exposure to aflatoxin aerosols has also been blamed for interstitial pneumonitis in textile workers (Lougheed et al., 1995).

Other Mycotoxins Exposure to *S. chartarum*, *Trichoderma* spp., and *Acremonium* spp. was documented in women handling paper flower pots in a large horticultural operation (Dill et al., 1997). Airborne exposure was assessed using culture and spore traps. No culturable spores were isolated, but up to 7500 fungal spores/m³ of air were recovered. Symptoms reported were skin inflammation and scaling. The potential for exposure to verruculogen, fumitremorgen, and ochratoxin has been shown in agricultural and industrial environments (Northrup and Kilburn, 1978; Land et al., 1987; DiPaolo et al., 1993). Gordon et al. (1993) reported a case of dementia and tremors in a person exposed to *A. fumigatus* during silo unloading. The authors attributed the symptoms to a tremorgenic mycotoxin. Exposure of farm workers to aerosols from straw that was supporting the growth of *S. chartarum* has also resulted in toxicosis (Drobotka et al., 1945).

24.2.4.2 Indoor Air Exposures-Aflatoxin

For environments such as offices, schools, hospitals, and residences, there is little evidence to link inhalation exposure to aflatoxins with any form of cancer. One epidemiological study found that persons who lived in residences from which culture of dust produced colonies of *A. parasiticus* (known to produce aflatoxin) had a relatively high incidence of leukemia (Wray, 1975; Wray et al., 1982). There was a significant difference in prevalence of precipitins to *A. flavus* in cases compared to controls. However, a definitive causal relationship has yet to be established between environmental exposure and disease.

Other Mycotoxins One of the relatively well-defined case studies of human mycotoxicosis from indoor exposure involved a house with heavy growth of *S. atra (chartarum)* resulting from inadequate home maintenance (Croft et al., 1986). Water damage had occurred over a period of several years, and extensive fungal growth was evident on the ceiling of an upstairs bedroom and in the air ducts. *S. atra* spores were identified in indoor air samples, and a series of highly toxic trichothecene mycotoxins were isolated from both ceiling material and debris found in the air ducts. The occupants' complaints (headaches, sore throats, hair loss, flu symptoms, diarrhea, fatigue, dermatitis, and generalized malaise) were consistent with chronic trichothecene intoxication. These symptoms disappeared after the house was thoroughly cleaned.

Recently, pulmonary hemorrhage in infants has been reported in home environments from most of which toxin-producing strains of *S. chartarum (atra)* were recovered. An abundance of other fungi were recovered as well (CDC, 1995, Jarvis et al., 1996; Sorenson et al., 1996; Montana et al., 1997). When matched from recoveries from control homes, airborne levels of both total fungi and *S. chartarum* were higher in case homes (Etzel et al., 1998). While the overall pattern of toxins in strains isolated from case and control homes differed, there were isolates of similar toxicity in both groups (Jarvis et al., 1998). These data suggest that living in water-damaged homes that have become extensively colonized with fungi increases the risk of this kind of illness, although the actual role for *S. chartarum* exposure remains to be confirmed.

Other case studies implicating inhalation of toxin-containing fungal spores in human disease have been reported (Auger et al., 1990; Johanning et al., 1996). However, a causal relationship between exposure to specific spore-borne mycotoxins produced by fungi growing in buildings and symptoms of ill-health among the occupants of the buildings has not been unequivocally demonstrated.

24.2.5 Risk Extrapolation

Toxicity (i.e., relative potency on a mass basis) varies greatly among different kinds of mycotoxins. The mycotoxin dose required to cause a specific effect varies with (a) the toxin, (b) the experimental animal species used to assess toxicity, and (c) the administration route. The widely differing sensitivities among animal species make extrapolation of laboratory data to human health effects problematic. In addition, much of the toxicological data on mycotoxins is related to ingestion exposure, and extrapolation to inhalation exposures introduce further uncertainties into assessment of the risks associated with indoor air exposure.

Extrapolating human risk from laboratory animal data involves many assumptions. Two of these are (a) that effects on human cells will parallel those in animal models, and (b) that natural human exposures are equivalent to those used for laboratory animals. In fact, animals clearly differ in their responses to environmental agents. Further, exposure of experimental animals is usually at much higher doses than would be expected for humans in order to achieve a response in a small exposure population over a relatively short time period. In addition, environmental conditions for the test animals are carefully controlled and the animals are adults in excellent condition. To account for these factors, safety or uncertainty factors of ≥100 are usually applied to animal work when it is extrapolated to humans.

To be useful to investigators, laboratory animal data must be related to the kinds of human exposures that are expected in terms of particle size, nature of particulate exposure, time course for the exposures, levels, and so forth. At this time, inadequate data exist to accurately predict the risk associated with human inhalation exposure to mycotoxins in non-agricultural indoor environments.

24.3 Sample Collection

Methods commonly used to estimate exposure to mycotoxins in indoor air include (a) sampling air or sources for propagules of fungi known to be toxigenic, with identification either by culture or microscopic analysis [see 24.4.1], and (b) collecting samples for actual mycotoxin analysis [see 24.4.2]. Mycotoxin exposure occurs in particulate form, so air sampling protocols must consider

the likely particle size range, which for the spores of most toxigenic fungi is ~2 to 10 μm in smallest diameter. Most of the commonly used bioaerosol samplers are relatively efficient across this particle size range [see 11.3.2.3 and Tables 11.1 and 11.2]. Therefore, selection of an air sampling device depends primarily on compatibility with the analytical method to be used (Pasenen et al., 1993, 1994).

24.4 Sample Analysis

Establishing possible exposure to mycotoxins is based on (a) analysis of environmental samples for potentially toxigenic fungi [see 24.4.1], (b) analysis of environmental samples for mycotoxins themselves [see 24.4.2 and 24.4.3], or (c) biological monitoring of potentially exposed persons [see 24.4.4].

24.4.1 Assays for Fungi or Fungal Spores

24.4.1.1 Culture of Toxigenic Fungi Most methods for detecting fungi involve culturing viable microorganisms from either air or source samples. Many *Aspergillus* and *Penicillium* spp. grow well in culture, but species identification is necessary to assess the potential for toxin production [see 24.1.4]. However, mycotoxin presence is independent of fungal viability, and mycotoxins can be recovered from both living and dead fungal materials as well as from growth substrates. In addition, some toxigenic fungi cannot compete well with common fungi on rich culture media, such as MEA [see 19.4]. For example, if common *Aspergillus* or *Penicillium* spp. are abundant in a sample, *S. chartarum* may be obscured due to the rapid growth of these competing fungi. It has also been estimated that the half life of *S. chartarum* spores may be as short as one month and that ≥80% of spores in an environmental sample may not be culturable (Miller, 1992).

It bears emphasizing that culture of a potentially toxigenic fungus from an indoor environment is not the same as detecting mycotoxin in that environment [see 24.5]. Also, an isolate's potential for toxin production cannot be definitively determined in the laboratory. Toxin production depends strongly on environmental growth conditions (e.g., growth substrate, temperature, and pH), and conditions in a laboratory may not duplicate those in the building from which an isolate was collected. Therefore, toxin production or lack thereof in the laboratory does not necessarily reflect what occurred in the original building.

24.4.1.2 Fungal Spore Identification (Microscopy) Complementary sampling methods that include direct spore counts or toxin analysis could overcome the disadvantages of fungal culture [see 24.4.1.1]. *S. chartarum* spores are readily identified microscopically, and spore counts from air samples are likely to provide a more accurate estimation of the potential for toxin exposure than would attempts to culture the relatively short-lived spores of this slow-growing fungus. However, many (perhaps most) toxigenic fungi produce spores of a type common to other, apparently non-toxigenic species. It may be possible to identify fungi to species level from adhesive-tape or source samples if spore-bearing structures remain intact. However, it is usually not possible to differentiate species of *Aspergillus* and *Penicillium* by examination of individual spore characters or, in fact, even to distinguish the two genera on spore-trap air samples. However, the use of culture and microscopy together can provide reasonably accurate measures of exposure to toxigenic fungi.

24.4.2 Assays for Mycotoxins

Ideally, assessment of environmental mycotoxins would focus on tests for actual toxins in air or source samples. However, all of the currently available analytical methods for mycotoxin detection were designed for agricultural products and the crossover to use on samples from the indoor environment may be problematic in some instances (e.g., air samples with low mycotoxin concentrations).

Samples for chemical analysis of mycotoxins are usually collected directly from sources (e.g., settled dust and materials on which fungi are growing) (AIHA, 1996). Filter air samples have also been analyzed for mycotoxin content (Pasanen et al., 1993, 1994). Thin layer chromatography (TLC) has been a popular technique to measure mycotoxins, but it is relatively insensitive and susceptible to interference. TLC may be better used to screen samples or as a cleanup procedure prior to more sensitive analyses, which may be more expensive and time intensive. Other TLC methods are being developed (e.g., two-dimensional, high-performance, and bi-directional methods), which are more sensitive for mycotoxin analysis (Scott, 1995). Sophisticated analytical techniques, such as HPLC and GC–MS, can be used to quantify specific mycotoxins. Samples must undergo extensive organic solvent cleanup to minimize interfering compounds. Other than in research laboratories, use of these techniques is limited because the necessary reference standards, expensive analytical equipment, and technical expertise generally are not available.

Immunoaffinity chromatography is a relatively new technology that uses highly specific antibodies to separate mycotoxins from a sample. Commercially available assay kits are sensitive, rapid, and easy to use. However, commercial assays exist for only a few agriculturally significant mycotoxins (e.g., aflatoxin, ochratoxin, zearalone, T-2 toxin, deoxynivalenol, and fumonisins). The accuracy of some of the kits is comparable to HPLC analysis (Dorner and Cole, 1989). Confirmation of mycotoxin identity should be the last step of an analytical method. However, simple comparison of sample retention time and fluorescence value to those of reference standards is not adequate to definitively identify a mycotoxin (Scott, 1995).

24.4.3 Bioassays for Relative Toxicity

Another approach to evaluating potential mycotoxin exposure is to subject environmental samples to assays that quantitatively detect a potential to produce toxic effects. When used for environmental samples, bioassays detect effects rather than identify the specific type or types of toxins present. These assays were discussed in Section 24.2.1 with respect to their use in elucidating the mechanisms of toxicity of specific mycotoxins. For environmental samples, these assays have the powerful advantages of sensitivity and independence from prior knowledge of the kinds of toxins likely to be present. The most feasible use of bioassays is as a qualitative screening method, although there have been attempts to increase their quantitative capability.

24.4.4 Biological Exposure Indices

Biomarkers generally fall into two categories, that is, markers of exposure and markers of response. Both approaches have been used to study mycotoxins. Exposure to aflatoxins M1 and P1 has been documented using affinity chromatography to detect the toxins in the urine of humans exposed through their diet (Groopman et al., 1985). Aflatoxins were detected in breast milk from 88% of 113 lactating mothers in Sierra Leone, and ochratoxin A was detected in 35% of these same mothers, indicating exposure to the toxins (Jonsyn et al., 1995). Aflatoxin B1 has been detected in urine and in samples of lung tissue collected at autopsy from three patients who were apparently occupationally exposed to toxin-containing aerosols (Dvoráckova and Pichová, 1986). Groopman et al. (1985, 1992) also studied aflatoxin-DNA adducts in the urine of their subjects. These data indicate not only exposure, but also a specific response, and emphasize the potential carcinogenicity of these compounds. Skin tests and serological measures of specific immunoglobulins are biomarkers that are commonly used to assess human response to antigen exposure. These same measures have been used to infer exposure to both fungi and their toxins (Johanning et al., 1993).

24.5 Data Interpretation

Investigators may encounter five indicators of potential exposure to mycotoxins:

1. Observation of fungal growth in an indoor environment.
2. In source samples, identification of fungi known to produce mycotoxins.
3. In air samples, identification of fungi known to produce mycotoxins.
4. In source samples, identification of mycotoxins.
5. In air samples, identification of mycotoxins.
6. Detection of biomarkers for mycotoxins or fungi in biological specimens (e.g., urine, breast milk, or blood).

The first of these provides the weakest evidence of possible mycotoxin exposure; the next five provide progressively stronger evidence. Chapter 14 also discusses how exposure–response relationships are established using various types of information.

24.5.1 Observation of Visible Fungal Growth

The visual observation of indoor fungal growth generally should lead to recommendations to remediate the underlying cause. However, the presence of fungi in indoor sources, although presenting a potential for human exposure to both glucans and mycotoxins, does not allow the assumption of a relationship between existing disease among occupants and possible mycotoxin exposure. Such a relationship must be established through a logical series of steps, including documentation of actual exposure and establishment of a clear connection between exposure and the type of disease present. The fact that fungi are growing indoors does not necessarily imply the presence of mycotoxins nor the exposure of occupants to any mycotoxins that may be present. Toxigenic species of fungi apparently do not always produce mycotoxins. Even when toxins are present, the fungi may not be releasing sufficient toxin-containing particles to cause disease.

24.5.2 Identification of Toxigenic Fungi

When a fungal species known to be toxigenic is identified in an environmental sample, it is often assumed that the fungus is producing mycotoxins and releasing toxin-containing particles. Indeed, the identification of toxigenic fungi in source or air samples implies a greater risk for toxin exposure than the mere presence of unidentified fungal growth. However, an association between toxin exposure and existing disease must still be established.

While the presence of a toxigenic fungus is not proof of human mycotoxin exposure, neither is the converse true. The majority of fungi commonly isolated indoors have not been tested for toxin production. Some fungi (e.g., *Cladosporium cladosporioides*) produce toxins of limited potency. However, exposure to a mixture of fungal products may lead to toxic symptoms even if no single mycotoxin is present in an amount sufficient to cause disease. Therefore, failure to identify fungal species known to be toxigenic should not automatically lead an investigator to presume that mycotoxins are not present.

24.5.3 Identification of Specific Mycotoxins or Toxicity

Isolation of fungi known to produce toxins or identification of mycotoxins in source samples are reasons to suspect mycotoxin exposures [see 24.5.2]. However, only detection of specific mycotoxins or toxicity in air samples provides convincing environmental evidence of exposure [see 24.4.2]. Once possible exposure is established, it remains to be determined whether the level of toxin or toxicity detected is sufficient to cause disease. Unfortunately, inhalational dose–response data are available for very few mycotoxins and for no toxin mixtures. Given the relative insensitivity of most assays for specific toxins, the presence of measurable toxin in a standard air sample (e.g., an 8-hour personal sample collected at 2 L/min) probably indicates that potentially hazardous exposures are occurring. However, when searching for causes of existing disease, it is essential to make a clear connection between an identified biological agent and symptoms known to be associated with exposure to that agent. Investigators must always consider possible causes of disease other than exposure to fungi and fungal products.

24.6 Remediation and Prevention

Readers are directed to Chapter 10 for (a) discussions of measures to control indoor fungal growth [see 10.2, 10.3, 10.4, and 10.5], and (b) Chapter 15 for a discussion of cleaning and remediation of existing contamination [see 15.2, 15.3, and 15.6].

Where toxigenic fungi have been identified, it is especially important to recognize the potential for hazardous exposures during remediation. Information and appropriate protection should be provided for all persons who may potentially be exposed during the clean-up process (e.g., remediation workers and building occupants). Appropriate protection for remediation workers would involve skin covering and respiratory protection where, respectively, direct contact with contaminated material and inhalation exposure may occur. Protection for building occupants may require relocation during remediation and the use of appropriate containment to prevent dissemination of toxin-containing particles.

24.7 β-(1→3)-D-Glucans
24.7.1 Characteristics

Glucans are glucose polymers that are structural components of most fungal cell walls (Table 24.3). β-(1→3)-D-glucans may exist as unbranched or branched chains and form single-stranded or stable, triple-strand helices (Williams et al., 1996). The glucans may be chemically bound to chitin (an acetyl glucosamine polymer that forms the basic fibrillar structure of fungal cell walls) or may form a soluble matrix in which the chitin microfibrils are embedded (Sietsma and Wessels, 1981).

24.7.2 Health Effects

The β-(1→3)-D-glucans act as potent T-cell adjuvants and have been studied intensively for their anti-tumor activities (Kiho et al., 1991; Kraus and Franz, 1991; Kitamura et al., 1994). Glucans stimulate macrophages and neutrophils and modulate the endotoxin-mediated release of tumor necrosis factor α from macrophages and thus increase resistance to infections with GNB (Saito et al., 1992; Adachi et al., 1994a,b; Brattgjerd et al., 1994; Sakurai et al., 1994; Zhang and Petty, 1994; Williams et al., 1996).

Glucans may play a role in both virulent and opportunistic fungal infections (San-Blas, 1982) and may also be involved in the development of HP by affecting the inflammation-regulating capacity of airway macrophages. In insoluble form, glucans cause a gradual decrease in the number of macrophages and lymphocytes in the lung wall and appear to have an effect on the lung similar to that of endotoxin (Fogelmark et al., 1994).

Glucans are a part of the complex mixture of exposures that result in ODTS [see 24.2.4.1]. In Guinea pigs, exposure to glucans was found to cause increased breathing rate and increases in neutrophils, lymphocytes, and red blood cells in the lung (Milanowski, 1997).

24.7.3 Sampling
Documenting possible exposure to glucans is accomplished using bioassays, although exposure to most fungi implies glucan exposure as well. The β-$(1\rightarrow3)$-D-glucans act similarly to endotoxin in some LAL assays (Rylander et al., 1992; Obayashi et al., 1995), and these specific tests have been used for glucan measurement. Antibodies have also been produced that are specific for some glucans, and immunoassay methods have been developed for glucan analysis (Adachi et al., 1994a; Hirata et al., 1994; Douwes et al., 1996). The LAL assays are the most sensitive, but do not differentiate between β-$(1\rightarrow3)$- and β-$(1\rightarrow6)$-D-glucans, whereas immunoassays are more specific for the former (Douwes et al., 1996). These assays can be performed on extracts of materials or filter air samples.

24.7.4 Data Interpretation
Exposure to some concentration of β-$(1\rightarrow3)$-D-glucans can be assumed whenever fungi are isolated from an environment. β-$(1\rightarrow3)$-D-glucans have been proposed as agents possibly responsible for some cases of BRSs (Rylander et al., 1992). However, it is difficult to imagine how BRSs could routinely be associated with such ubiquitous agents unless other factors also play a role. Even in problem environments, fungal concentrations [and presumably, although not studied, β-$(1\rightarrow3)$-D-glucan concentrations] are likely lower than concentrations frequently encountered outdoors. Workers exposed to high concentrations of β-$(1\rightarrow3)$-D-glucan (during drybagging of the compound produced for use as a food additive in processed cheese) showed no irritant symptoms (Donald Milton, personal observation during a walkthrough visit at a biotechnology company). It is possible that the glucan measurements in the Rylander study were surrogates for exposure to other fungal products. It is also possible that exposure to a mixture of fungal products (including glucans) such as may occur in some indoor environments plays an important role in development of BRSs.

24.7.5 Remediation and Prevention
Recommendations for remediation of fungal contamination and prevention of exposure to β-$(1\rightarrow3)$-D-glucan are the same as those outlined in Section 24.6 for fungal toxins and Section 19.6 for fungi.

24.8 References
Adachim, Y; Ohno, N.; Yadomae, T.: Preparation and Antigen Specificity of an Anti-β-$(1\rightarrow3)$-D-glucan Antibody. Biol. Pharm. Bull. 1508-1512 (1994a).

Adachi, Y.; Okazaki, M.; Ohno, N.; et al.: Enhancement of Cytokine Production by Macrophages Stimulated with 1-3-β-D-glucan, Grifolan (GRN) Isolated from *Grifola frondosa*. Biol. Pharm. Bull. 17:1554-1560 (1994b).

AIHA: Total (Viable and Nonviable) Fungi and Substances Derived from Fungi in Air, Bulk, and Surface Samples. In: Field Guide for the Determination of Biological Contaminants in Environmental Samples, pp. 119-130. H.K. Dillon, P.A. Heinsohn, and J.D. Miller, Eds. American Industrial Hygiene Association, Fairfax, VA (1996).

Amitani, R.; Taylor, G.; Elezis, E.N.; et al.: Purification and Characterization of Factors Produced by *Aspergillus fumigatus* which Affect Human Ciliated Respiratory Epithelium. Infection Immunity 63(9):3266-71 (1995).

Andersson, M.A.; Nikulin, M.; Köljalg, U.; et al.: Bacteria, Molds, and Toxins in Water-damaged Building Materials. Appl. Environ. Microbiol. 63(2):387-393 (1997).

Arafat, W.; Musa, M.N.: Patulin-induced Inhibition of Protein Synthesis I Hepatoma Tissue Culture. Res. Comm. in Molecular Pathol. Pharmacol. 87(2):177-86 (1995).

Auger, P.L.; St. Onge, M.; Aberman, A.; et al.: Pseudo Chronic Fatigue Syndrome in Members of one Family which was Cured

TABLE 24.3. Principal Cell-Wall Polysaccharides of Some Common Environmental Fungi

Fungal Group	Common Genera	Cell-Wall Polysaccharide
Zygomycetes	*Rhizopus*	Chitin-chitosan
Ascomycetes	*Alternaria* *Aspergillus* *Cladosporium* *Penicillium*	Chitin-glucan
Saccharomycetaceae	*Candida* *Saccharomyces*	Mannan-glucan
Sporobolomycetaceae	*Sporobolomyces*	Chitin-mannan

by Eliminating the *Penicillium brevicompactum* Moulds Found in their Home. In: Proceedings of the Fifth International Conference on Indoor Air Quality and Climate. Toronto, Canada, July 1990. Vol 1:181-185 (1990).

Autrup, J.L.; Schmidt, J.; Seremet, T.; et al.: Determination of Exposure to Aflatoxins among Danish Workers in Animal Feed Production through the Analysis of Aflatoxin B1 Adducts to Serum Albumin. Scand. J. Work Environ. Health 17:436-440 (1991).

Autrup, J.L.; Schmidt, J.; Seremet, T.; et al.: Exposure to Aflatoxin B in Animal Feed Production Plant Workers. Scand. Environ. Health Perspectives 99:195-197 (1993).

Aziz, N.H.; Moussa, L.A.: Influence of White Light, Near-UV Irradiation and Other Environmental Conditions on Production of Aflatoxin B1 by *Aspergillus flavus* and Ochratoxin A by *Aspergillus ochraceus*. Nahrung 41(3):150-4 (1997).

Baxter, C.S.; Wey, H.E.; Burg, W.R.: A Prospective Analysis of the Potential Risk Associated with Inhalation of Aflatoxin-contaminated Grain Dusts. Food Cosmet. Toxicol. 19:765-679 (1981).

Betina, V.: Mycotoxins: Chemical, Biological, and Environmental Aspects. Elsevier, Amsterdam (1989).

Binder, J.; Horvath, E.M.; Heidegger, J.; et al.: A Bioassay for Comparison of the Toxicity of Trichothecenes and their Microbial Metabolites. Cereal Research Communications 25 (3 Part 1):489-491 (1997).

Biswas, G.; Raj, H.G.; Allameh, A.; et al.: Comparative Kinetic Studies on Aflatoxin B1 Binding to Pulmonary and Hepatic DNA of Rat and Hamster receiving the Carcinogen Intratracheally. Teratogenesis, Carcinogenesis, Mutagenesis 13(6):259-268 (1993).

Brattgjerd, S.; Evensen, O.; Lauve, A.: Effect of Injected Yeast Glucan on the Activity of Macrophages in Atlantic Salmon, *Salmo salar* L., as Evaluated by in vitro Hydrogen Peroxide Production and Phagocytic Capacity. Immunology 83:288-294 (1994).

Breitholtz-Emanuelsson, A.; Fuchs, R.; Hult, K.: Toxicokinetics of Ochratoxin A following Intratracheal Administration. Natural Toxins 3:101-103 (1995).

Bukowski, R.; Vaughn, C.; Bottomley, R.; et al.: Phase II Study of Anguidine in Gastrointestinal Malignancies: A Southwest Oncology Group Study. Cancer Treat. Rep. 66:381-383 (1982).

Búlock, J.D.: Mycotoxins as Secondary Metabolites. In: The Biosynthesis of Mycotoxins. P.S. Steyn, Ed. (1980).

Burg, W.R.; Shotwell, O.L.; Saltzman, B.E.: Measurements of Airborne Aflatoxins during the Handling of Contaminated Corn. Am. Ind. Hyg. Assoc. J. 42(1):1-11 (1981).

Canas, P.; Aranda, M.: Decontamination and Inhibition of Patulin-induced Cytotoxicity. Environ. Toxicology Water Quality 11(3):249-253 (1996).

Cariello, N.F.; Piegorsch, W.W.: The Ames Test — The Two-fold Rule Revisited. Mutation Research: Genetic Toxicology. 369(1-2):23-31 (1996).

Centers for Disease Control: Acute Pulmonary Hemorrhage/Hemosiderosis among Infants — Cleveland, January 1993-November 1994. MMWR 44:4 (1995).

Coulombe, Jr., R.A.; Huie, J.M.; Ball, R.W.; et al.: Pharmacokinetics of Intratracheally Administered Aflatoxin B1. Toxicol. Appl. Pharm. 109:196-206 (1991).

Creasia, D.A.; Thurman J.D.; Wannemacher, R.W.; et al.: Acute Inhalation Toxicity of T-2 Mycotoxin in the Rat and Guinea Pig. Fund. Appl. Toxicol. 14:54 (1990).

Croft, W.A.; Jarvis, B.B.; Yatawara, C.S.: Airborne Outbreak of Trichothecene toxicosis. Atmos. Environ. 20:549-552 (1986).

Demain, A.L.: Functions of Secondary Metabolites. In: Genetics and Molecular Biology of Industrial Microorganisms. S.W. Queener and G. Hageman, Eds. ASM Washington, DC (1989).

DiPaolo, N.; Guarnieri, A.; Garosi, G.; et al.: Inhaled Mycotoxins Lead to Acute Renal Failure. Nephrology, Dialysis, Transplantation. 9 Suppl 4:116-20 (1994).

Di Paolo, N.; Guarnieri, A.; Loi, F.; et al.: Acute Renal Failure from Inhalation of Mycotoxins. Nephron 64:621-625 (1993).

Dill, I.; Trautmann, C.; Szewzyk, R.: Massenentwicklung von Stachybotrys chartarum auf kompostierbaren Pflanztopfen aus Altpapier. Mycoses 40 Suppl 1:110-4 (1997).

Dorner, J.W.; Cole, R.J.: Comparison of Two ELISA Screening Tests with Liquid Chromatography for Determining Aflatoxins in Raw Peanuts. J. Assoc. Off. Anal. Chem. 72:962-964 (1989).

Douwes, J.; Doekes, G.; Montijn, R.; et al.: Measurement of $(1-3)Glucans in the Occupational and Home Environment with an Inhibition Enzyme Immunoassay. Appl. Environ. Microbiol. 62:3176-3182 (1996).

Drobotka, V.G.; Maruschenko, P.E.; Aizeman, B.E.; et al.: Stachybotryotoxicosis, A New Disease of Horses and Humans. Am. Rev. Sov. Med. 2:238-242 (1945).

Dvorácková, I.; Pichová, V.: Pulmonary Interstitial Fibrosis with Evidence of Aflatoxin B$_1$ in Lung Tissue. J. Tox. Environ. Health 18:153-157 (1986).

Etzel, R.A.; Montana, E.; Sorenson, W.G.; et al.: Acute Pulmonary Hemmorhage in Infants Associated with Exposure to Stachybotrys atra and Other Fungi. Arch. Pediatr. Adolsc. Med. 152L757-762 (1998).

Fogelmark, B.; Sjostrand, M.; Rylander, R.: Pulmonary Inflammation Induced by Repeated Inhalations of Beta(1,3)-D-glucan and Endotoxin. Int. J. Exp. Path. 75:85-90 (1994).

Gerberick, G.F.; Sorenson, W.G.; Lewis, D.M.: The Effects of T-2 Toxin on Alveolar Macrophage Function in vitro. Environmental Res. 33(1):246-60 (1984).

Goodwin, W.; Haas, C.D.; Fabian, C.; et al.: Phase I Evaluation of Anguidine (diacetoxyscirpenol, NSC-141537). Cancer 42:23-26 (1978).

Gordon, K.E.; Masotti, R.E.; Waddell, W.R.: Tremorgenic Encephalopathy — A Role of Mycotoxins in the Production of CNS Disease in Humans. Can. J. Neurological Sciences 30(3):237-239 (1993).

Groopman, J.D.; Donahue, P.R.; Zhu, J.; et al.: Aflatoxin Metabolism in Humans: Detection of Metabolites and Nucleic Acid Adducts in Urine by Affinity Chromatography. Proc. Natl. Acad. Sci. 82:6492-6496 (1985).

Groopman, J.D.; Donahue, P.R.; Pikul, A.; et al.: Molecular Dosimetry of Urinary Aflatoxin DNA-Adducts in People Living in Guanxi Autonomous Region, People's Republic of China. Cancer Res. 52:45-52 (1992).

Hanelt, M.; Gareis, M.; Kollarczik, B.: Cytotoxicity of Mycotoxins Evaluated by the MTT-Cell Culture Assay. Mycopathologia 128(3):167-174 (1994).

Hayes, R.B.; Van Nieuwenhuize, J.P.; Raatgever, J.W.; et al.: Aflatoxin Exposures in the Industrial Setting: An Epidemiological Study of Mortality. Food Chem. Toxicol. 22:39-43 (1984).

Hendry, K.M.; Cole, E.C.: A Review of Mycotoxins in Indoor Air. J. Toxicol. Environ. Health 38:183-198 (1993).

Hirata, A.; Komoda, M.; Itoh, W.; et al.: An Improved Sandwich ELISA Method for the Determination of Immunoreactive Schizophyllan. Biol. Pharm. Bull. 17:1437-40 (1994).

Hlywka, J.J.; Beck, M.M.; Bullerman, L.B.: The Use of the Chicken Embryo Screening Test and Brine Shrimp (*Artemia salina*) Bioassays to Assess the Toxicity of Fumonisin B-1 Mycotoxin. Food Chemical Toxicology 35(10-11):991-999 (1997).

Jakab, G.J.; Hmieleski, R.R.; Zarba, A.; et al.: Respiratory Aflatoxicosis — Suppression of Pulmonary and Systemic Host Defenses in Rats and Mice. Toxicol. Appl. Pharmacol. 125(2):198-205 (1994).

Janzen, D.: Why Fruits Rot, Seeds Mold, and Meats Spoil. Am. Nat. 11:691 (1977).

Jarvis, B.B.; Lee, Y.W.; Comezoglu, S.N.; et al.: Trichothecenes Produced by *Stachybotrys atra* from Eastern Europe. Appl. Environ. Microbiol. 51(5):915-918 (1986).

Jarvis, B.B.; Mycotoxins — An Overview. In: Natural Toxins: Characterization, Pharmacology and Therapeutics. C.L. Ownby and G.V. Ocell, Eds. Pergamon Press, New York, NY (1989).

Jarvis, B.B.: Mycotoxins and Indoor Air Quality. Biological Contaminants in Indoor Environments, ASTM STP 1071, P.R. Morey, J.C. Feeley, J.A. Otten, Eds. American Society for Testing Materials, Philadelphia, PA (1990).

Jarvis, B.B.; Salemme, J.; Morais, A.: *Stachybotrys* Toxins 1. Natural Toxins 3(1):10-16 (1995).

Jarvis, B.B.; Zhou, Y.; Jiang, J.; et al.: Toxigenic Molds in Water-Damaged Buildings: Dechlorogriseofulvins from *Memnoniella echinata*. J. Natural Prod. 59(6):553-4 (1996).

Jarvis, B.B.; Sorenson, W.G.; Hintikka, E.L.; et al.: Study of Toxin Production by Isolates of Stachybotrys chartarum and Memnoniella echinata Isolated During a Study of Pulmonary hermosiderosis inInfants. Appl. Environ. Microbiol. 64(10): 3620-3625 (1998).

Johanning, E.; Biagnin, R.; Hull, D.L.; et al.: Health and Immunology Study Following Exposure to Toxigenic Fungi (*Stachybotrys chartarum*) in a Water-Damaged Office Environment. Int. Arch. Occup. Environ. Health 68:207-218 (1996).

Johanning, E.; Morey, P.R.; Jarvis, B.B.: Clinical-Epidemiological Investigation of Health Effects Caused by *Stachybotrys atra* Building Contamination. Proc. Indoor Air '93 Helsinki, Vol 1, 225:230 (1993).

Jonsyn, F.E.; Maxwell, S.M.; Hendrickse, R.G.: Ochratoxin A and Aflatoxins in Breast Milk Samples from Sierra Leone. Mycopathologia 131(2):121-6 (1995).

Kelly, J.D.; Eaton, D.L.; Guengerich, F.P.; et al.: Aflatoxin B-1 Activation in Human Lung. Toxicol. Appl. Pharmacol. 144(1)88-95 (1997).

Kendrick, B.: The Fifth Kingdom. Mycologue Publications, Newburyport, MA (1992).

Kiho, T.; Sakushima, M.; Wang, S.R.; et al.: Polysaccharides in Fungi. XXVI. Two Branched (163)-β-d-glucans from Hot Water Extract of Yu er (*Auricularia sp*). Chem. Pharm. Bull. 39(3):798-800 (1991).

Kitamura, S.; Hori, T.; Kurita, K.; et al.: An Antitumor, Branched (163)-β-D-glucan from a Water Extract of Fruiting Bodies of *Cryptoporus volvatus*. Carbohydrate Res. 263(1):111-21 (1994).

Koshinsky, H.A.; Khachatourians, G.G.: Trichothecene Synergism, Additivity, and Antagonism: The Significance of the Maximally Quiescent Ratio. Nat. Toxins 1:38-47 (1992).

Kraus, J.; Franz, G.: β(1-3)Glucans: Anti-Tumor Activity and Immunostimulation. In: Fungal Cell Wall and Immune Response. J.P. Latge, D. Boucias, Eds. NATO ASI Series, H53:431-444, SpringerVerlag, Berlin, Heidelberg. (1991)

Land, C.J.; Lundstrom, H.; Werner, S.: Production of Tremorgenic Mycotoxins by Isolates of *Aspergillus fumigatus* from Sawmills in Sweden. Mycopathologia 124(2):87-93 (1993).

Land, K.J.; Hult, K.; Fuchs, R.; et al.: Tremorgenic Mycotoxins from *Aspergillus fumigatus* as a Possible Occupational Health Problem in Sawmills. Appl. Environ. Microbiol. 53(4):787-790 (1987).

Lougheed, M.D.; Roos, J.O.; Waddell, W.R.; et al.: Desquamative Interstitial Pneumonitis and Diffuse Alveolar Damage in Textile Workers — SpringerVerlag, Berlin, Heidelberg.Potential Role of Mycotoxins. Chest 108(5):1196-1200 (1995).

Maron, D.M.; Ames, B.N.: Revised Methods for the *Salmonella* Mutagenicity Test. Mutation Res. 113:173-215 (1983).

Martin-Gonzalez, A.; Benitez, L.; Soto, T.; et al.: A Rapid Bioassay to Detect Mycotoxins Using a Melanin Precursor Overproducer Mutant of the Ciliate *Tetrahymena thermophila*. Cell. Biol. Internat. 21(4):213-6 (1997).

May, J.J.; Stallones, L.; Darrow, D.; et al.: Organic Dust Toxicity (Pulmonary mycotoxicosis) Associated with Silo Unloading. Thorax 41:919-923 (1986).

Milanowski, J.; Sorenson, W.G.; Dutkiewicz, J.; et al.: Chemotaxis of Alveolar Macrophages and Neutrophils in Response to Microbial Products Derived from Organic Dust. Environ. Res. 69(1):59-66 (1995).

Milanowski, J.: Experimental Studies on the Effects of Organic Dust-derived Agents on Respiratory System — Comparison between Endotoxins and Glucans. Inhalation Toxicol. 9(4):369-388 (1997).

Milanowski, J.: Effect of Inhalation of Organic Dust-derived Micro-

bial Agents on the Pulmonary Phagocytic Oxidative Metabolism of Guinea Pigs. J. Toxicol. Environ. Health 53(1):5-18 (1998).

Miller, J.D.: Fungi as Contaminants in Indoor Air. Atmos. Environ. 26(A):2163-2172 (1992).

Miller, J.D.: Mycotoxins. In: Organic Dusts: Exposure, Effects, and Prevention, pp. 87-92. R. Rylander and R.R. Jacobs, Eds. Lewis Publishers, Boca Raton, FL (1994).

Mitchell, C.G.; Slight, J.; Donaldson, K.: Diffusible Component from the Spore Surface of the Fungus *Aspergillus fumigatus* which Inhibits the Macrophage Oxidative Burst is Distinct from Gliotoxin and other Hyphal Toxins. Thorax 52(9):796-801 (1997).

Montaña, E.; Etzel, R.A.; Allan, T.; et al.: Environmental Risk Factors Associated with Pediatric Idiopathic Pulmonary Hemorrhage and Hemosiderosis in a Cleveland Community. Pediatrics 99(1 Suppl S):E51-E58 (1997).

Murayama, T.; Amitani, R.; Ikegami, Y.; et al.: Suppressive Effects of *Aspergillus fumigatus* Culture Filtrates on Human Alveolar Macrophages and Polymorphonuclear Leucocytes. European Respir. J 9(2):293-300 (1996).

Nikulin, M.; Pasanen, A.L.; Berg, S.; et al.: *Stachybotrys atra* Growth and Toxin Production in Some Building Materials and Fodder under Different Relative Humidities. Appl. Environ. Microbiol. 60(9):3421-3424 (1994).

Nikulin, M.; Reijula, K.; Jarvis, B.B.; et al.: Experimental Lung Mycotoxicosis in Mice Induced by *Stachybotrys atra*. Intern. J. Experimental Pathol. 77(5):213-8 (1996).

Nikulin, M.; Reijula, K.; Jarvis, B.B.; et al.: Effects of Intranasal Exposure to Spores of *Stachybotrys atra* in Mice. Fund. Appl. Toxicol. 35(2):182-8 (1997).

Norn, S.: Microorganism-induced or Enhanced Mediator Release: A Possible Mechanism in Organic Dust Related Diseases. Amer. J. Industrial Med. 25:91-95 (1994).

Norred, W.P.; Plattner, R.D.; Dombrinkkurtzman, M.A.; et al.: Mycotoxin-induced Elevation of Free Sphingoid Bases in Precision-cut Rat Liver Slices — Specificity of the Response and Structure-activity Relationships. Toxicology Appl. Pharmacol. 147(1):63-70 (1997).

Northrup, S.C.; Kilburn, K.H.: The Role of Mycotoxins in Pulmonary Disease. In: Mycotoxic Fungi, Mycotoxins, Mycotoxicoses, An Encyclopedic Handbook, Vol 3, pp. 91-108. T. Wyllie, L. Morehouse, Eds. Marcel Dekker, New York, NY (1978).

Obayashi, T.; Yoshida, M.; Mori, T.; et al.: Plasma (1-3) β–d Glucan Measurement in Diagnosis of Invasive Deep Mycosis and Fungal Febrile Episodes. Lancet 345:17-20 (1995)

Olsen, J.H.; Dragsted, L.; Autrup, H.: Cancer Risk and Occupational Exposure to Aflatoxins in Denmark. Brit. J. Cancer 58:392-396 (1988).

Osborne, B.G.; Ibe, F.; Brown, G.L.; et al.: The Effects of Milling and Processing on Wheat Contaminated with *Ochratoxin A*. Food Additives Contaminants 13(2):141-53 (1996).

Palmgren, M.S.; Lee, L.S.: Separation of Mycotoxin-containing Sources in Grain Dust and Determination of their Mycotoxin Potential. Environ. Health Persp. 66:105-108 (1986).

Pasanen, A.L.; Nikulin, M.; Tuomainen, M.; et al.: Laboratory Experiments on Membrane Filter Sampling of Airborne Mycotoxins Produced by *Stachybotrys atra Corda*. Atmos. Environ. 27A(1):9-13 (1993).

Pasanen, A.L.; Tuomainen, M.; Hintikka, E.L.; et al.: Collection of Airborne Mycotoxins on Membrane Filters. In: Health Implications of Fungi in Indoor Environments, pp 317-324. R.A. Samson, B. Flannigan, M.E. Flannigan, et al., Eds. Elsevier, New York, NY (1994).

Putt, D.A.; Ding, X.X.; Coon, M.J.; et al.: Metabolism of Aflatoxin B1 by Rabbit and Rat Nasal Mucosa Microsomes and Purified Cytochrome P450, Including Isoforms 2A10 and 2A11. Carcinogenesis 16(6):1411-1417 (1995).

Rao, C.Y.; Fink, R.C.; Wolfe, L.B.; et al.: A Study of Aflatoxin Production by *Aspergillus flavus* Growing on Wallboard. J. Am. Biol. Safety Assoc. 2:36-42 (1997).

Richard, J.L.; Peden, W.M.; Williams, P.P.: Gliotoxin Inhibits Transformation and its Cytotoxic to Turkey Peripheral Blood Lymphocytes. Mycopathologia 126(2):109-114 (1994).

Richards, J.L.; Thurston, J.R.: Effect of Aflatoxin on Phagocytosis of *Aspergillus fumigatus* Spores by Rabbit Alveolar Macrophages. Appl. Micobiol. 30:44 (1975).

Ruotsalainen, M.; Hyvarinen, A.; Nevalainen, A.; et al.: Production of Reactive Oxygen Metabolites by Opsonized Fungi and Bacteria Isolated from Indoor Air, and their Interactions with Soluble Stimuli, fMLP or PMA. Environ. Res. 69(2):122-31 (1995).

Rylander, R.; Persson, K.; Goto, H.; et al.: Airborne β-1,3-Glucan May be Related to Symptoms in Sick Buildings. Indoor Environ. 1:263-267 (1992).

Saito, K.; Nishijima, M.; Ohno, N.; et al.: Activation of Complement and Limulus coagulation System by an Alkali-soluble Glucan Isolated from *Omphalia lapidescens* and its Less Branched Derivates. Chem. Pharm. Bull. 40:1227-1230 (1992).

Sakurai, T.; Ohno, N.; Yadomae, T.: Changes in Immune Mediators in Mouse Lung Produced by Administration of Soluble (1-3) β-*D*-glucan. Biol. Pharm. Bull. 17:617-622 (1994).

San-Blas, G.: The Cell Wall of Fungal Human Pathogens: Its Possible Role in Host-Parasite Relationships. Mycopathologia 79:159-184 (1982).

Scott, P.M.: Mycotoxin Methodology. Food. Add. Contamin. 12(3):395-403 (1995).

Shahan, T.A.; Sorenson, W.G.; Lewis, D.M.: Superoxide Anion Production in Response to Bacterial Lipopolysaccharide and Fungal Spores Implicated in Organic Dust Toxic Syndrome. Environ. Res. 67(1):98-107 (1994).

Shahan, T.A.; Sorenson, W.G.; Paulauskis, J.D.; et al.: Concentration- and Time-dependent Upregulation and Release of the Cytokines MIP-2, KC, TNF, and MIP-1a in Rat Alveolar Macrophages by Fungal Spores Implicated in Airway Inflammation. Amer. J. Respir. Cell. Mol. Biol. 18:435-440 (1998).

Sietsma, J.H.; Wessels, J.G.H.: Solubility of (163)-β-*D*/(166)-β-*D*-Glucan in Fungal Walls: Importance of Presumed Linkage between Glucan and Chitin. J. Gen. Microbiol. 125:209-212 (1981).

Sisgaard, T.; Malmros, P.; Nersting, L.; et al.: Respiratory Disorders and Atopy in Danish Refuse Workers. Amer. J. Resp. Crit. Care Med. 149(6):1407-1412 (1994).

Smith, G.W.; Constable, P.D.; Haschek, W.M.: Cardiovascular Responses to Short-term Fumonisin Exposure in Swine. Fund. Appl. Toxicol. 33(1):140-148 (1996).

Sorenson, W.G.; Simpson, J.; Peach 3d, M.J.; et al.: Aflatoxin in Respirable Corn Dust Particles. J. Toxicol. Environ. Health 7(3-4):669-672 (1981).

Sorenson, W.G.; Simpson, J.; Castranova, V.: Toxicity of the Mycotoxin Patulin for Rat Alveolar Macrophages. Environ. Res. 38(2):407-416 (1985).

Sorenson, W.G.; Gerberick, G.F.; Lewis, D.M.; et al.: Toxicity of Mycotoxins for the Rat Pulmonary Macrophage in vitro. Environ. Health Persp. 66:45-53 (1986).

Sorenson, W.G.; Simpson, J.: Toxicity of Penicillic Acid for Rat Alveolar Macrophages in vitro. Environ. Res. 41(2):505-513 (1986).

Sorenson, W.G.; Frazer, D.G.; Jarvis, B.B.; et al.: Trichothecene Mycotoxins in Aerosolized Conidia of *Stachybotrys atra*. Appl. Environ. Microbiol. 53(6):1370-1375 (1987).

Sorenson, B.; Kullman, G.; Hintz, P.: NIOSH Health Hazard Evaluation Report. HETA 95-0160-2571. Centers for Disease Control and Prevention, National Center for Environmental Health (1996).

Sorenson, W.G.; Jarvis, B.B.; Zhou, Y.; et al.: Toxine im Zusammenhalt mit *Stachybotrys* and *Memnoniella* in Hausern mit Wasserschaden — 18th Mycotoxin Workshop. M.Gareis and R. Scheuer, Eds. Institut fur Mikrobiologie und Toxikologie der Bundesanstalt fur Fleischforschung, Kulbach, Germany (1996).

Sutton, P.; Waring, O.; Mullbacher, A.: Exacerbation of Invasive Aspergillosis by the Immunosuppressive Fungal Metabolite, Gliotoxin. Immunology Cell Biology 74(4):318-22 (1996).

Van Nieuwenhuize, J.P.; Herber, R.F.M.; DeBruin, A.; et al.: Aflatoxins — Epidemiological Study on the Carcinogenicity of Prolonged Exposure to Low Levels among Workers of a Plant. Tijdschr. Soc. Geneesk. 51:754-769 (1973).

Waring, P.; Beaver, J.: Gliotoxin and Related Epipolythiodioxopiperazines. General Pharmacology 27(8):1311-6 (1996).

Waring, P.; Khan, T.; Sjaarda, A.: Apoptosis Induced by Gliotoxin is Preceded by Phosphorylation of Histone H3 and Enhanced Sensitivity of Chromatin to Nuclease Digestion. J. Biological Chemistry 272(29):17929-36 (1997).

Weber, S.; Kullman, G.; Petsonk, E.; et al.: Organic Dust Exposures from Compost Handling — Case Presentation and Respiratory Exposure Assessment. Amer. J. Industr. Med. 24(4):365-374 (1993).

Wicklow, D.T.; Shotwell, O.L.: Intrafungal Distribution of Aflatoxins among Conidia and Sclerotia of *Aspergillus flavus* and *Aspergillus parasiticus*. Can. J. Microbiol. 29:1-5 (1982).

Wicklow, D.T.: Interference Competition. In: The Fungal Community. D.T. Wicklow and G.C. Carroll, Eds. Marcel Dekker, New York, NY (1981).

Williams, D.H.; Stone, M.J.; Hauck, P.R.; et al.: Why Are Secondary Metabolites (Natural Products) Biosynthesized? J. Nat. Prod. 52:1189 (1989).

Williams, D.L.; Mueller, A.; Browder, W.: Glucan-based Macrophage Stimulators: A Review of their Anti-infective Potential. Clin. Immunother. 5(5):392-399 (1996).

Wilson, D.W.; Ball, R.W.; Coulombe Jr., R.A.: Comparative Action of Aflatoxin B1 in Mammalian Airway Epithelium. Cancer Res. 50:2493-2498 (1990).

Wintermeyer, S.F.; Kuschner, W.G.; Wong, H.; et al.: Pulmonary Responses after Wood Chip Mulch Exposure. J. Occupational Environ. Med. 39(4):308-314 (1997).

Wray, B.B.; Harmon, C.A.; Rushing, E.J.; et al.: Precipitins to an Aflatoxin-producing Strain of *Aspergillus flavus* in Patients with Malignancy. J. Cancer Res. Clin. Oncol. 103:181-185 (1982).

Wray, B.B.: Mycotoxin-producing Fungi from House Associated with Leukemia. Arch. Environ. Health 30(12):571-3 (1975).

Zhang, K.; Petty, H.R.: Influence of Polysaccharides on Neutrophil Function: Specific Antagonists Suggest a Model for Cooperative Saccharide-associated Inhibition of Immune Complex-triggered Superoxide Production. J. Cell Biochem. 56:225-235 (1994).

Chapter 25

Antigens

Cecile S. Rose

25.1 Characteristics
    25.1.1 *Pollen Grains and Fungal Spores*
    25.1.2 *Bacteria, Protozoa, and Fungi*
    25.1.3 *House Dust Mites*
    25.1.4 *Cockroaches*
    25.1.5 *Birds*
    25.1.6 *Mammals*
25.2 Health Effects
    25.2.1 *Immune System Responses*
    25.2.2 *Antigen-Mediated Diseases*
        25.2.2.1 Allergic Asthma and Rhinitis
        25.2.2.2 Allergic Bronchopulmonary Mycoses and Allergic Sinusitis
        25.2.2.3 Hypersensitivity Pneumonitis
        25.2.2.4 Antigens as Causes of BRI and BRS
    25.2.3 *Medical Testing*
        25.2.3.1 Immunological Tests
        25.2.3.2 Pulmonary Function Tests
        25.2.3.3 Exposure Challenge Tests
25.3 Sample Collection
    25.3.1 *Source Samples*
    25.3.2 *Air Samples*
    25.3.3 *Sample Handling*
25.4 Sample Analysis
25.5 Data Interpretation
    25.5.1 *Fungal Allergens and Bacterial Antigens*
    25.5.2 *Dust-Mite Allergens*
    25.5.3 *Cockroach Allergens*
    25.5.4 *Cat and Dog Allergens*
25.6 Remediation and Prevention
25.7 References

25.1 Characteristics

Antigens are substances that can induce a detectable hypersensitivity or immune response. Inhalation of airborne antigens—many of which are of biological origin—can cause a spectrum of immune illnesses broadly referred to as hypersensitivity diseases. An immune response consists of specific antigen recognition and the recruitment of an arsenal of sensitized cells and antibodies. An allergen (Gr. altered action) is a biological or chemical substance that causes a specific immune response called an allergic reaction. Thus an allergy is the immune-mediated state of hypersensitivity (i.e., an exaggerated or inappropriate immune response) that results from exposure to an allergen

(IOM, 1993). The body must recognize a substance as "foreign" for it to be immunogenic. This is one reason why people rarely react to human proteins (e.g., shed human skin scales in the indoor environment). Table 25.1 lists some of the terms used in the following discussion of antigen-mediated diseases. This chapter addresses airborne antigens and allergens (aeroallergens) but not contact or food allergens.

Although most antigens are relatively large (>10 kilodaltons, kd), some smaller compounds can also be antigenic. Many substances, including some metals and low molecular weight chemicals, may function as antigens if attached to larger molecules. Plant and animal proteins are usually the most potent antigens, although polysaccharides and some other compounds can also induce a strong immune response under certain conditions. The ability of a substance to cause an immune response is also related to the complexity of the molecule. Proteins with repeating units of a single amino acid are generally not antigenic nor are pure, single-sugar polysaccharides, such as cellulose (Goodman, 1980).

Antigens may become aerosolized as part of intact microorganisms or spores, as fragments or excretions from organisms, or attached to inorganic particles. Table 25.2 lists some of the most common airborne antigens and their sources. Antigen exposures have increased due to people spending more time indoors and recent changes in homes and offices (e.g., higher mean indoor temperatures, reduced ventilation, laundering with cool-wash detergents that may not remove allergens effectively, and widespread use of carpeting) (Platts-Mills, 1994; Platts-Mills et al., 1995).

25.1.1 Pollen Grains and Fungal Spores
Pollen grains and fungal spores are major outdoor allergens, but indoor pollen concentrations are generally lower than outdoor concentrations. Fungal air concentrations also follow this pattern unless fungi are growing indoors [see 7.4.2.1].

25.1.2 Bacteria, Protozoa, and Fungi
Reservoir water in devices that aerosolize water droplets (e.g., humidifiers, cool-mist vaporizers, evaporative condensors, and hot tubs) is never sterile and can support the growth of antigenic bacteria, protozoa, and fungi. Such devices are of concern as sources of airborne, respirable, particulate antigens as well as infectious agents such as *Legionella* spp., amebae, and opportunistic fungal pathogens (Tyndall et al., 1995). HP outbreaks have been reported in connection with contaminated humidification and air-wash systems in office buildings (Arnow et al., 1978; Morey et al., 1984).

25.1.3 House Dust Mites
Several studies have documented a dose–response relationship between cumulative exposure to dust-mite allergen and subsequent sensitization of exposed persons (Peat et al., 1993; Ollier and Davies, 1994; Marks et al., 1995; Verhoeff et al., 1995). House dust mites proliferate and their droppings accumulate in bedding, upholstery, carpeting, and similar sites, where RH and temperature conditions are favorable [see 22.1.3]. Besides dwellings, house dust mites and related allergens have been found in office buildings and schools (Lundblad, 1991; Hung et al., 1993; Menzies et al., 1993; Janko et al., 1995) and even in clothing (Tovey et al., 1995). Other mites may also be allergenic (Powell, 1994).

25.1.4 Cockroaches
Many insects have been identified as sources of airborne allergens (e.g., moths, crickets, locusts, beetles, and flies) (IOM, 1993; Powell, 1994; Squillace, 1995). However, the insect repeatedly recognized as a common source of indoor al-

TABLE 25.1. Terms Used with Antigen-Mediated Diseases (adapted from IOM, 1993)

Term	Definition
Allergen	Biological or chemical substance that causes a specific immune response called an allergic reaction, characterized by hypersensitivity.
Allergy	Immune-mediated state of hypersensitivity that results from exposure to an allergen.
Antibody	A protein molecule formed by the immune system in response to the body's contact with an antigen; antibodies to infectious agents play a role in preventing re-infection; certain types of antibodies to allergens (e.g., IgE) can cause adverse hypersensitivity reactions (see Immunoglobulin).
Antigen	Substance that, when introduced into the body, stimulates production of a specific immune response, which may include antibody production and hypersensitivity.
Asthma	An inflammatory disease of the airways often associated with variable airflow obstruction and with allergy to inhaled substances.
Atopy	Hereditary or familial predisposition to produce IgE antibodies to environmental allergens and develop allergic disease.
Hypersensitivity disease	Disease for which a subsequent exposure to an antigen produces a greater effect than that produced on initial exposure.
Hypersensitivity pneumonitis (HP) (extrinsic allergic alveolitis)	An inflammatory disease of the lung parenchyma (alveoli and terminal airways) caused by inhalation of and immune response to foreign substances such as particles of biological origin and some industrial chemicals.
Immunoglobulin	A protein (or family of proteins) that participates in an immune reaction as an antibody; an antibody belonging to one of five classes represented by letters (e.g., IgA, IgD, IgE, IgG, and IgM).
Immune response	Specific antigen recognition and recruitment of an arsenal of sensitized cells and antibodies.

lergens is the cockroach, for example, *Blattella germanica* (the German cockroach), *Periplaneta americana* (the american cockroach), and *Blattella orientalis* (the oriental cockroach). The first two species are the most common in the U.S. Cockroaches can become abundant in any environment where sanitary practices are inadequate. Therefore, environments where food is routinely handled and stored are prone to cockroach contamination. Cockroaches are able to survive low ambient humidity much better than dust mites and, unlike mites, are able to actively search for water (Squillace, 1995). The requirement for moisture implies that cockroaches will be found in the highest concentrations in kitchens, bathrooms, and other similiar areas. Cockroaches are a perennial problem in animal-care facilities, where the insects are often re-introduced with shipments of animal food (Kang and Chang, 1985). Cockroach allergen may be associated with cast skins, whole bodies, egg shells, fecal particles, and saliva (Squillace, 1995). Proteins derived from cockroaches are associated with particles >5 µm diameter and become airborne only when a room is disturbed (Platts-Mills and Carter, 1997).

25.1.5 Birds

Avian antigens are highly immunogenic and have been found in the feathers, droppings, and dried eggs of turkeys, chickens, geese, ducks, parakeets, parrots, budgerigars, pigeons, doves, lovebirds, and canaries (Squillace, 1995). The highest exposures to avian proteins are associated with cleaning bird lofts, cages, and coops. Bird-breeder's lung is the most common form of HP in Great Britain, where ~12% of the general population keep budgerigars as pets.

25.1.6 Mammals

The most common animals kept as pets are cats and dogs, and 33% to 50% of U.S. homes have a mammalian pet (Ledford, 1994). Small-particle allergen aerosols from cats, dogs, and other mammals accumulate in environments where such animals live (Luczynska, 1995). Dog allergens are found in saliva, skin dander, and urine, but not hair. Thus, short-haired and long-haired dog breeds may be equally allergenic. Cat allergens derive both from saliva (from the animal's grooming practices) and skin and are potent antigens. Cat allergens, being much smaller and lighter than dust-mite allergens, are readily found in the air of households with cats (Ingram et al., 1995). In homes with low rates of outdoor air ventilation, cat allergens have been found on particles below 2 µm in diameter (Anderson and Baer, 1981). In environments with high air-change rates (e.g., laboratory animal-care facilities), particles carrying cat allergen were larger. Cat allergens have been detected in office buildings, schools, new homes, allergists's offices, hospitals, and shopping malls (Lundblad, 1991; Enberg et al., 1993; Hung et al., 1993; Janko et al., 1995). Antigens from rodents are found primarily in the animals' urine, which can be aerosolized as liquid droplets or in dust. Rodent allergy is common among laboratory and animal-care personnel who work with mice, rats, guinea pigs, rabbits, and hamsters (Powell, 1994; Sjöstedt et al., 1995).

25.2 Health Effects

25.2.1 Immune System Responses

The upper and lower airways—the primary targets for inhaled antigens—respond to antigenic material in a variety of ways. The body's response depends on the source material, host factors (e.g., genetic factors and prior exposure), and the duration and intensity of exposure. All organisms probably contain proteins, glycoproteins (protein-polysaccharide combinations), or polysaccharides that can be antigenic. Therefore, many bioaerosols induce health effects via the actions of antigens on the immune system. The

TABLE 25.2. Biological Sources of Some Common Airborne Allergens

Agent	Antigenic Product	Common Sources
Pollen	Soluble antigens	Outdoor air, settled dust
Bacteria	Organisms, soluble antigens	Outdoor air, settled dust, indoor growth, humidifiers
Amebae	Soluble antigens	Humidifiers, wash waters
Fungi	Organisms, spores, fragments, soluble antigens	Outdoor air, settled dust, indoor growth, humidifiers
Arthropods		
Dust mites	Fecal pellets, dried body fragments	Settled dust
Cockroaches	Fecal particles, saliva, dried body fragments	Settled dust
Birds	Dried eggs, droppings, serum	Settled dust
Mammals		
Cats	Skin dander, saliva	Settled dust
Dogs	Skin dander, saliva, urine	Settled dust
Rodents	Urine	Settled dust, bedding

reason some persons develop allergies and others do not is unclear, but genetic factors have a major influence on peoples' responses to antigen exposures.

Antibodies (produced in response to specific antigens) belong to a group of molecules called "immunoglobulins" (Ig), of which there are several classes represented by letters (e.g., IgM, IgG, IgE, and IgA). IgM is produced shortly following antigen exposure but plays a transitory role in the immune process, generally disappearing after a few weeks (Greenberg and Stave, 1994). IgG follows IgM (Figure 25.1) and is the largest (70% to 75% of the total immunoglobulin pool) and longest-lasting of the antibody classes. IgG (along with IgM and IgA) is responsible for immunity to infectious agents, and specific IgG can be detected in some individuals with HP. IgE adheres to certain cells and activates them to release products, including histamine, causing many of the typical clinical allergy manifestations. IgA is the predominant immunoglobulin in seromucous secretions, such as saliva, and is found in the gastrointestinal, respiratory, and urogenital tracts.

Figure 25.1 shows primary and secondary antibody responses for IgM and IgG. The primary response is characterized by (a) a lag phase before antibody is detected, (b) appearance of IgM before IgG, (c) a log phase when the antibody titers rise rapidly, (c) a plateau phase during which the antibody titers stabilize, and (d) a decline phase during which the antibodies decrease. The actual time course and titers depend on the nature of the antigenic challenge and characteristics of the host. In comparison with a primary antibody response, the response to a second antigenic exposure (a) appears more quickly and lasts longer, (b) reaches a higher titer, and (c) consists primarily of IgG.

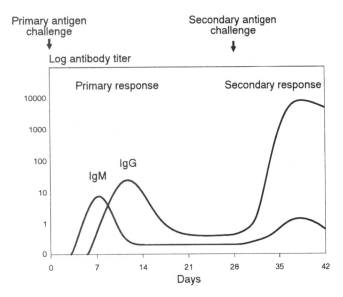

FIGURE 25.1 Primary and secondary antibody responses.

A primary antigen exposure could occur in the workplace, or a previously exposed and sensitized person may be re-exposed on the job. Exposure to infectious agents may occur naturally (e.g., by contact with an infected person or an infectious agent in the environment) or through immunization. For many agents, infection results in immunity from re-infection because the body forms protective antibodies and a re-exposure is met with a more rapid and intense response. Unfortunately, some immunological responses can be harmful, even life threatening.

When presented with an allergen, the immune system's reactions fall into four broad hypersensitivity response categories outlined in Table 25.3. These responses represent normal immune reactions that act inappropriately. These four reactions do not necessarily occur in isolation from each other, and they sometimes cause inflammatory reactions and tissue damage. The first three of these are humoral (blood and lymph) responses, that is, they require the elaboration of specific antibodies. A Type IV response requires no antibody, only the activation of thymus-derived T-lymphocytes and other immune effector cells.

TABLE 25.3. Hypersensitivity Responses to Antigen Exposure

- Type I, IgE-dependent allergic response, also called anaphylactic or immediate-type allergic reactions [e.g., allergic asthma, rhinitis, conjunctivitis, and urticaria (skin rash or hives) and less common anaphylactic shock (a severe and sometimes fatal reaction)].
- Type II, cytotoxic reactions, in which IgG and IgM are directed against host tissues.
- Type III, immune complex reaction, in which soluble antigens and antibodies (IgG or IgM) form immune complexes that deposit in target tissues (e.g., Arthus reaction and serum-sickness).
- Type IV, cell-mediated (or lymphocyte-mediated) immunity, also called delayed type hypersensitivity [e.g., contact hypersensitivity (poison oak or ivy), positive tuberculin skin test reaction, and HP].

25.2.2 Antigen-Mediated Diseases

The pathogenesis of all hypersensitivity diseases involves (a) repeated antigen exposure, (b) immunological sensitization of a host to an antigen, and (c) immune-mediated damage to the host [see 8.2.1]. Time is required for the body to develop an immunological sensitization. Therefore, latency (absence of response or disease on first contact) is characteristic of all hypersensitivity diseases. The spectrum of diseases associated with airborne antigen exposure may involve the upper or lower airways and includes allergic asthma, allergic rhinitis, allergic sinusitis, atopic dermatitis (skin inflammation with itching), allergic mycosis (fungosis, most commonly allergic aspergillosis), and HP. The first three of these are allergic diseases that depend on the antigen-stimulated production of specific IgE antibodies [see 25.2.2.1]. Allergic mycoses result from immune reactions to fungi colonizing the airways without tissue invasion [see 25.2.2.2]. HP results from activation of specific T-lymphocytes (cell-mediated immunity) and is associated with the production of specific immunoglobulins (IgG and IgM) [see 25.2.2.3]. These conditions are also described in Chapters 3 and 8.

Healthy airways have a great capacity to eliminate microbial cells and fungal spores that may deposit there [see also 9.3.4]. Therefore, inhaled bacteria and fungi are rarely pathogenic under normal conditions. The lung's first line of defense against inhaled microorganisms consists of opsonization (i.e., processing that makes foreign cells more susceptible to the action of phagocytes). Next, alveolar macrophages phagocytize (ingest) and kill the microorganisms. The mucus-epithelial cell barrier, the major mechanism for clearing the airways, removes the engulfed material (Kauffman et al., 1995). In addition to macrophages, other cells (e.g., neutrophils) are recruited into the lungs to deal with the foreign material. The neutrophils release enzymes, which may be protective if the particle is an infectious agent, but repeated excessive release of these materials can damage lung tissue (Robinson, 1994).

25.2.2.1 Allergic Asthma and Rhinitis *Allergic asthma* is a condition marked by labored breathing and accompanied by wheezing, a sense of chest constriction, and often coughing or gasping [see also 3.3.2 and 8.2.1.2]. An estimated 8% to 12% of the U.S. population has asthma (IOM, 1993), which ranges from intermittent to mild, moderate, or severe persistent asthma (Lemanske and Busse, 1997). In persons with fungal asthma, exposure to fungal spores or hyphal fragments results in the formation of IgE antibodies. Sensitization to fungi is high in childhood and declines rapidly with age, suggesting that younger children may be less proficient in clearing fungal particles from the airways (Niemeyer and de Monchy, 1992).

Allergic rhinitis ("hay fever") is typically manifested by a runny or congested nose, irritated and inflamed throat and eyes (allergic conjunctivitis), and sneezing due to IgE-mediated inflammation and histamine release [see also 3.3.1 and 8.2.1.1]. The prevalence rate for allergic rhinitis has been estimated at 10% to 20% and varies with skin test reactivity (i.e., the rate is higher in skin-test-positive people) (IOM, 1993; Naclerio and Solomon, 1997).

Allergens produced by dust mites (Chapter 22), fur-bearing pets (mainly cats and dogs), indoor insects (primarily cockroaches), and fungi (Chapter 19) are known to produce the majority of indoor hypersensitivity diseases. Allergic asthma and rhinitis may occur in sensitized individuals from exposure to such antigens. In a survey of dust-mite prevalence, allergen load was suggested as a risk factor for asthma (Arlian et al., 1992). Only 13 of the homes of 252 mite-sensitive asthmatics had <100 mites/g in at least one sampling site, and 83% of

the homes had an average mite density of >100 mites/g of dust. Improvement in asthma symptoms following reduction in exposure to house dust mites was reported as early as 1970. More recent evidence supports a role for dust-mite allergen inhalation in some sensitized individuals with atopic dermatitis (Chapman et al., 1983; Wickman et al., 1993).

25.2.2.2 Allergic Bronchopulmonary Mycoses and Allergic Sinusitis Fungal colonization (i.e., growth without tissue invasion) of the central airways can occur in individuals with long-standing, severe asthma and leads to a syndrome known as allergic bronchopulmonary mycosis (ABPM) (Rose, 1994; Lemanske and Busse, 1997). *Aspergillus fumigatus* and other *Aspergillus* spp. are the fungi most commonly associated with this condition, which then is termed allergic bronchopulmonary aspergillosis (ABPA). The clinical picture for ABPA is characterized by recurrent episodes of bronchial obstruction. These episodes may be associated with fever; malaise; brownish mucus plugs, containing fungal hyphae, in sputum; peripheral blood eosinophilia, an abnormal increase of eosinophils, a type of white blood cell; and hemoptysis, expectoration of blood from the respiratory tract. A chest radiograph typically shows fleeting shadows and areas of subsegmental atelectasis (collapse) due to mucus plugging. A chest CT scan usually shows bronchiectasis, a chronic inflammatory or degenerative condition with dilation of one or more bronchi or bronchioles. Diagnosis of ABPM relies on finding a positive reaction to a specific fungus on a skin-prick test and detecting fungal-specific IgE in a patient's serum. Another form of fungal colonization (i.e., localized growth in the nasal sinuses) can cause allergic sinusitis (DeShazo et al., 1997). This condition is characterized by chronic sinus symptoms and increased fungal-specific serum IgE, most commonly to *A. fumigatus*.

25.2.2.3 Hypersensitivity Pneumonitis HP is an inflammatory granulomatous lung disease that occurs in persons sensitized to respirable bioaerosols (Rose, 1996) [see also 3.3.3 and 8.2.1.3]. Diagnosis of HP is often difficult because it requires a high index of suspicion on the part of physicians and careful collection of patient histories (IOM, 1993). HP is caused by continuous or repeated exposure to various antigenic substances (e.g., organic dusts containing bacterial, fungal, or avian proteins) and some chemicals (e.g., isocyanates). Once sensitized, individuals respond to very low exposures to environmental antigens and may not tolerate indoor environments that are of no apparent concern for nonsensitized occupants.

HP is often named for the occupation or environment involved, for example, wood trimmer's disease, farmer's lung, pigeon-breeder's disease, and humidifier lung (Richerson, 1994). However, association of HP with particular work environments may contribute to failure to recognize the disease in populations with less obvious workplace exposures. HP outbreaks have also occurred in various office, residential, and recreational environments. Antigenic bioaerosols have been traced to contaminated ventilation systems (Acierno et al., 1985; Hodgson et al., 1987), microbial contamination following cafeteria flooding (Hodgson et al., 1985), a water spray system (Arnow et al., 1978), microbially contaminated hot tubs (Jacobs, 1986), and decorative fountains.

25.2.2.4 Antigens as Causes of BRI and BRS Diseases that do not involve stimulation of the immune system (e.g., toxic and irritant syndromes, such as reactions to formaldehyde, environmental tobacco smoke, or scents) should not be called "allergies" or "hypersensitivity" diseases, even if some building occupants respond with symptoms similar to allergic reactions or appear more sensitive than others. Physician-diagnosed allergy, allergic asthma, allergic rhinosinusitus, ABPM, or HP that can be associated with exposure in a particular indoor environment would be considered a BRI [see 3.1].

25.2.3 Medical Testing

A range of medical examination methods are used to diagnose antigen-mediated conditions. These include collecting a good medical history along with a physical exam, immunological testing, pulmonary function tests, chest radiographic studies, sometimes lung biopsy, and exposure challenge in the laboratory or workplace [see also 3.3 and 8.2.1].

25.2.3.1 Immunological Tests The primary immunological assays used to detect specific antibodies in humans include direct skin tests, the Ouchterlony double diffusion assay, the RAST (Radio-AllergoSorbent Test), and the ELISA (Enzyme-Linked ImmunoSorbent Assay). As with tests for antigens in environmental samples [see 6.5.2], immunoassays for medical testing depend on very specific antigen–antibody interactions. For example, anti-cockroach antibodies in a patient's serum will not recognize mite allergen. Skin testing takes place directly on a human subject and may entail a small risk of adverse reaction. The other tests are carried out on serum from blood samples and present no risk for a subject.

Skin Prick Testing. Skin prick testing is commonly used to confirm immediate-type IgE-mediated hypersensitivity reactions and to identify the offending antigens. Thus, skin tests may provide evidence of immunological sensitization (e.g., a person may be shown to react to dust-mite allergen), but are not diagnostic of allergic disease (e.g., the person's current symptoms may or may not be due to exposure to dust-mite allergens). Prick testing involves placing a drop of antigen extract onto the skin of an individual's forearm or back and introducing anti-gen into the epidermis with a needle prick (Norman and Peebles, 1997). An immediate reaction, a wheal (a lump or swelling) and erythema (reddening), is read at 15 to 20 min. Some crude antigen extracts can produce false-positive results because of direct skin irritation or endotoxin contamination (Michel, 1991). Appropriate controls should be placed to assess false-positive and false-negative reactions. Dust-mite and cat allergen extracts are available that contain standardized amounts of major antigens, but other skin test reagents (e.g., fungal extracts) are fairly crude preparations of a limited number of species (Horner et al., 1995). Cross reactivity has also been observed, that is, the interaction of an antigen with an antibody formed against a different antigen with which the first antigen shares closely related or common antigenic determinants (IOM, 1993; Horner et al., 1994, 1995). This can cause confusion if a person tests positive to a certain biological agent but that agent is not found in a problematic environment, whereas another related agent is present but its significance is not apparent.

Double-Immunodiffusion Assay. Precipitating antibodies (precipitins) to a growing list of antigens demonstrate exposure to a variety of causative agents of HP. In the Ouchterlony double-immunodiffusion assay for precipitins, two or more small wells are cut in an agar gel (Kurup and Fink, 1997). Sample material is placed in one well and serum in another. The sample material may be known antigen extracts or extracts from environmental samples containing an unidentified, potentially antigenic material. The serum may be from an affected individual or a person known to be sensitive to a particular antigen. Serum from a single patient serum can be tested simultaneously with multiple antigen extracts, or multiple patient sera can be tested against a single antigen extract.

If present, the antigens and serum antibodies (IgG) diffuse radially from their original positions in the agar, meeting between the wells. A precipitate forms as a visible band if an antigen is present that matches antibodies in a test serum or a known-positive one (Crowle, 1980). Precipitin testing provides supporting evidence when positive but is inconclusive when negative. This is due to (a) the high rate of positive precipitins in asymptomatic, exposed individuals, (b) the temporal variation of precipitins in symptomatic individuals with HP, and (c) the lack of an exhaustive battery of test antigens with which to identify relevant antibodies (Newman et al., 1987; Kurup and Fink, 1997). Therefore, precipitin testing never independently proves or disproves a diagnosis of HP.

RAST and ELISA Tests. RAST is probably the most commonly used test for measuring specific IgE antibodies directed against a particular antigen and uses a radioactive label. ELISA uses an enzymatic colorimetric label and is a very sensitive tool to detect antigen-specific IgE as well as antibodies of other classes (e.g., IgG, IgA, or IgM) (Carpenter, 1997). For the RAST and ELISA assays, a known antigen is attached to a solid phase (usually cellulose beads or plastic microtiter wells) and then incubated first with patient serum and then with radio- or enzyme-labeled antihuman immunoglobulin. Section 6.5.2 describes the complementary use of the ELISA method to detect antigens in environmental samples. The number of labeled anti-human immunoglobulin molecules that bind to a test allergen reflects the amount of specific IgE in a patient's serum (IOM, 1993).

25.2.3.2 Pulmonary Function Tests Asthma and HP are characterized by changes in lung function over time (IOM, 1993). Pulmonary function tests, which measure such changes in lung function, have many applications in clinical medicine and research related to indoor allergens. These tests include spirometry, lung-volume measurement, diffusing-capacity testing, exercise studies, and rhinomanometry (measurement of nasal airflow). Pulmonary function tests must be administered by medical personnel or trained technicians. However, serial peak-flow measurements and serial spirometry using microprocessor-controlled pneumotachographs can be performed by patients independently. Valid test results require cooperation and maximum effort on a subject's part. Peak-flow meters are portable, and a patient can take measurements throughout the day and over several days, making these meters useful in asthma diagnosis and management. Pulmonary function tests can help a physician determine if a patient has asthma or HP, if the patient's condition improves with treatment, and where and when the patient becomes ill.

25.2.3.3 Exposure Challenge Tests Identifying the antigen responsible for a person's hypersensitivity disease and its source in a problem building can be difficult. However, sensitized occupants may be excellent detectors of antigen presence. Therefore, an alternative diagnostic method is direct patient challenge, that is, sending a person into a suspect environment after a period of absence, exposing a patient to suspect materials from an environment (Greenberg and Stave, 1994), or conducting controlled exposures in a clinic and monitoring lung function (Norman and Peebles, 1997). There is a significant danger for a test subject in this method, and a physician should only consider such a trial when prior illness has been mild and the symptoms have been reversible. Often, exposure challenge is considered when a subject likely will return to an environment in any case, which provides a physician an opportunity to observe any adverse reactions that may occur in response to antigen re-exposure.

25.3 Sample Collection

Assessing indoor antigen exposure and measuring antigen concentrations require careful observation and, occasionally, collection of source or air samples and measurement of their specific antigen content. This chapter does not cover sampling for culturable microorganisms, microbial spores, chemical markers of microorganisms, whole arthropods, guanine, or pollen grains that may serve as indicators of the presence of biological antigens (IOM, 1993) [see 18.3, 19.3, and 22.3].

25.3.1 Source Samples
Source samples are usually dust samples collected with vacuum devices (AIHA, 1996; Luczynska, 1997) [see 12.1.2 and 12.2.2]. For a number of reasons, recommended threshold concentrations for the major indoor antigens (i.e., dust mite, cat, cockroach, and dog allergens) are based on concentrations in settled dust rather than in air (IOM, 1993). The amount of antigen in dust or other source material is much higher than that typically found in indoor air (e.g., $\mu g/g$ versus ng/m^3). In addition, dust samples are easy to collect with inexpensive adaptors that fit onto most standard household vacuum cleaners, and collection time is generally short (AIHA, 1996). Some antigens (e.g., dust-mite allergen) are carried on fairly large particles that settle quickly and whose air concentrations depend on local disturbances, which may lead to inaccurate exposure measurements.

25.3.2 Air Samples
Sampling for airborne antigens is far less common than source sampling. The presence of known allergenic pollen or microorganisms is often taken as presumptive evidence that antigens from these biological agents are also present. Air samples for antigen assay are generally collected using high-volume filtration devices, although liquid impingers have also been used (Reed et al., 1993; Burge, 1995; Luczynska, 1997). Chapter 11 provides more information on air sampling instruments. Air sample volume must be only a fraction of the total room volume to obtain representative estimates of airborne antigen concentration. However, this is seldom a problem except in confined spaces or for unusually high sampling rates (e.g., >1000 L/min). In small areas, it may be difficult to balance the requirement to collect representative samples with the need for large-volume air samples—required by the typically low antigen concentrations in indoor air and the relative insensitivity of available assays.

25.3.3 Sample Handling
Dry dust, bulk, and air samples generally should be transported to an analytical laboratory at ambient temperature. Refrigeration may be appropriate for impinger liquids, dust samples for live mite identification [see 22.3.3], and samples for isolating culturable microorganisms [see 6.3 and the chapters on the individual microbial agents]. Freezing (–20°C) is recommended for unextracted samples that will

be stored for long periods before analysis by immunoassay (AIHA, 1996; Yunginger and Swanson, 1997). Liquid samples and filter extracts may be freeze dried.

25.4 Sample Analysis

Samples collected for antigen analysis should be processed as soon as possible after collection because sample stability during storage is unknown for some materials. Dust samples are sieved to remove large particles (e.g., >300 μm), and fine dust fractions and filter air samples are extracted in appropriate buffer solutions [see 22.4]. Extracts and liquid samples (e.g., impinger fluids) typically are evaluated using a panel of monoclonal antibodies in immunoenzymetric assays to measure indicator molecules for the primary indoor aeroallergens (Table 25.4).

TABLE 25.4. Primary Indoor Aeroallergens (molecule size given in parentheses)

Allergen	Source
Der f 1 (25 kd)	Dermatophagoides farinae, dust mite
Der p 1 (25 kd)	Dermatophagoides pteronyssinus, dust mite
Fel d 1 (35 kd)	Felis domesticus, domestic cat
Can f 1 (27 kd)	Canis familiaris, domestic dog
Bla g 1/Bla g 2 (25 to 36 kd)	Blattella germanica, German cockroach
Per a 1 (20,000 to 25,000 MW)	Periplanetta americana, American cockroach

25.5 Data Interpretation

Beyond the allergens mentioned in Table 25.4, a lack of well-characterized, standardized reagents poses problems in the testing of environmental and clinical samples, respectively, for the presence of antigens and immunological responses to them (IOM, 1993; Horner et al., 1995; Yunginger and Swanson, 1997). The World Health Organization (Geneva, Switzerland)/International Union of Immunological Societies and the U.S. Food and Drug Administration (Office of Biologics Research and Review, Center for Drugs and Biologics, Bethesda, Md.) provide standardized extracts or reference proteins for some specific allergens. The AIHA Biosafety Committee (1996) has written that research laboratories have developed immunoassays for latex-derived proteins and some fungal and bacterial products, but that such specialized research assays generally are not available for routine IH investigations. Further, the precision of immunoassays for indoor antigens is rather poor and the error from variation in replicate sample collection and analysis can result in over-

all coefficients of variation exceeding 20%. In addition, people have varying sensitivities to antigens. Therefore, investigators should not interpret too strictly comparisons of recommended allergen thresholds and actual environmental measurements.

25.5.1 Fungal Allergens and Bacterial Antigens

Chapters 7, 6, 18, and 19 discuss the interpretation of bacterial and fungal sampling data. Over 60 species of fungi are known to produce allergens that cause allergic rhinitis (hay fever) and asthma, and many more are probably also involved. Purified and characterized allergens have been prepared for *Alternaria alternata, Aspergillus fumigatus,* and *Cladosporium herbarum.* However, none of these allergens has been adequately tested for the analysis of environmental samples. To date, the limited analysis of fungal allergens in indoor environments precludes estimation of a risk level for symptom exacerbation or even determination of what is a high level (Horner et al., 1994, 1995). Researchers have also attempted to determine threshold concentrations of fungal spores associated with immunological sensitization or symptom development (Levetin, 1995). While there may be threshold concentrations below which no symptoms are experienced (Dhillon, 1991), this is probably not an absolute value but a gradient based on individual sensitivities and the kind of fungus (Levetin, 1995).

A variety of bacterial and fungal products, as well as whole organisms and fragments, have been recognized as important etiological agents of occupational allergies and respiratory disorders in workers in biotechnology industries (Biagini et al., 1996). Exposure to bacterial antigens has been associated with work-related asthma, HP, humidifier fever, and a disease resembling ABPM. Like fungi, bacteria secrete enzymes that can act as allergens. *Bacillus* spp. are used to produce proteases that are added to laundry detergents for stain removal (IOM, 1993). ACGIH has adopted a ceiling limit for subtilisins (100% pure crystalline enzyme) of 0.00006 mg/m^3 (ACGIH, 1998). However, data on bacterial antigens is limited, many bacterial antigens are poorly characterized, and exposure to bacterial antigens in non-manufacturing indoor environments is interpreted on a case-by-case basis.

If HP cases can be connected to a specific work environment, it may be worthwhile to collect samples of fungi and bacteria from potential sources in the workplace. It then may be possible to test sera from affected individuals for precipitating antibodies—antibodies that form insoluble precipitates with specific antigens—to identified microorganisms. Reed et al. (1983) used this approach to provide evidence that a humidification system was the source of a building-related HP epidemic.

25.5.2 Dust-Mite Allergens

Concentrations of *Der p* 1 or *Der f* 1 allergen above 10 μg/

g in settled dust have been associated with an increased risk of dust-mite sensitization. Concentrations above 2 µg/g have been associated with an increased risk of allergic symptoms in sensitized individuals (Hamilton and Adkinson, 1996) [see also 22.5]. Moderate to high concentrations (≥2 µg/g dust) of dust-mite allergen have been detected in office carpets and upholstered chairs (Hung et al., 1993; Janko et al., 1995).

25.5.3 Cockroach Allergens

Cockroaches are the only insects repeatedly recognized as a common source of indoor allergens (IOM, 1993). However, significant levels or thresholds of cockroach allergen exposure have not been determined (Pollart et al., 1991; Ledford, 1994). Any cockroach allergen detected in an environment (i.e., any amount above the LDL of the assay used) identifies an indoor area that places cockroach-allergic persons at some risk for symptoms and places allergy-prone (susceptible) persons at some risk for sensitization and symptom development.

25.5.4 Cat and Dog Allergens

Cat and dog allergens have been detected in schools and offices where these animals were not kept, suggesting that persons with animal contact may carry allergens on their clothing (Enberg et al., 1993; Hung et al., 1993; IOM, 1993). Hung et al. (1993) most often found moderate to high concentrations of cat allergen (≥1 µg Fel d 1/g dust) in the offices of persons who had cats at home. Fel d 1 concentrations above 8 µg/g have been proposed as a threshold for sensitization. Fewer people appear to be sensitive to dog than to cat allergens, perhaps because more dogs are housed outdoors and owners may bathe dogs fairly frequently. Concentrations of dog allergen (Can f 1) sufficient to result in the development of symptoms have not been established (Ledford, 1994). In one study, Can f 1 concentrations ranged from <0.3 to 23 µg/g in dust from houses without dogs and from 10 to 10^4 µg/g in those with dogs (Schou et al., 1991).

25.6 Remediation and Prevention

Preventing or minimizing exposure to airborne antigens is the best means of avoiding health problems and providing people a safe and comfortable indoor environment. Control methods for limiting exposure to house dust mites and microbial contaminants are discussed in Chapters 10 and 22. Cockroaches can be a major problem in some parts of the country, but good housekeeping practices and moisture control will go far to keep cockroach populations under control. Chemical insecticide use is discouraged for pest problems that can be controlled by other means. However, regular pesticide application may be necessary in environments where large quantities of high-protein organic material are stored. Some insecticides can be used in the form of bait (Platts-Mills and Carter, 1997).

Nonhuman mammalian danders are best controlled by keeping animals out of indoor environments. Recent data suggest that weekly washing of cats, when combined with other avoidance measures, greatly reduces the allergen load in a house and possibly also in a cat owner's work environment. In animal-care facilities, high ventilation rates and proper facility design (e.g., downdraft rooms) are required and can serve to maintain relatively low air concentrations of small-particle allergens (Reed et al., 1993). Section 10.5.2.2 discusses ventilation for control of airborne antigens. Remediation of avian allergens in indoor environments where someone has developed HP is difficult. Remediation may require removal of carpets and extensive cleaning of furnishings and possibly ventilation systems.

25.7 References

Acierno, L.J.; Lytle, J.S.; Sweeney, M.J.: Acute Hypersensitivity Pneumonitis Related to Forced Air Systems — A Review of Selected Literature and a Commentary on Recognition and Prevention. J. Environ. Health 48:138-141 (1985).

ACGIH: TLVs® and BEIs®. Threshold Limit Values for Chemical Substances and Physical Agents. Biological Exposure Indices. American Conference of Governmental Industrial Hygienists, Cincinnati, OH (1998).

AIHA: Allergens. Method for Collecting Bulk Dust Samples to Assess Environmental Allergen Exposure. In: Field Guide for the Determination of Biological Contaminants in Environmental Samples, pp. 149-160. American Industrial Hygiene Association, Fairfax, VA (1996).

Anderson, M.C.; Baer, H.: Allergenically Active Components of Cat Allergen Extracts. J. Immunol. 127:972 (1981).

Arlian, L.G.; Bernstein, D.; Bernstein, I.L.; et al. Prevalence of Dust Mites in the Homes of People with Asthma Living in Eight Different Geographic Areas of the United States. J. Allergy Clin. Immunol. 90:292-300 (1992).

Arnow, P.M.; Fink, J.N.; Schlueter, D.P.; et al.: Early Detection of Hypersensitivity Pneumonitis in Office Workers. Am. J. Med. 64:236-242 (1978).

Biagini, R.E.; Driscoll, R.J.; Bernstein, D.I.; et al.: Hypersensitivity Reactions and Specific Antibodies in Workers Exposed to Industrial Enzymes at a Biotechnology Plant. J. Appl. Toxicol. 16:139-145 (1996).

Burge, H.A.: Aerobiology of the Indoor Environment. Occup. Med. 10:27-40 (1995).

Carpenter, A.B.: Enzyme-linked Immunoassays. In: Manual of Clinical Laboratory Immunology, pp. 20-29. N.R. Rose, E.C. de Macario, J.D. Folds, et al., Eds. ASM Press, Washington, DC (1997).

Chapman, M.D.; Rowntree, S.; Mitchell, E.B.; et al.: Quantitative Assessment of IgG and IgE Antibodies to Inhalant Allergens in Patients with Atopic Dermatitis. J. Allergy Clin. Immunol. 72:27-33 (1983).

Crowle, A.J.: Precipitin and Microprecipitin Reactions in Fluid Medicine and in Gels. In: Manual of Clinical Immunology. N.R. Rose, H. Friedman, Eds. American Society Microbiologists, Washington, D.C. (1980).

DeShazo, R.D.; Chapin, K.; Swain, R.E.: Fungal Sinusitis. N. Eng. J. Med. 337:254-259 (1997).

Dhillon, M.: Current Status of Mold Immunotherapy. Ann. Allergy 66:385 (1991).

Enberg, R.N.; Shamie, S.M.; McCullough, J.; et al.: Ubiquitous Presence of Cat Allergen in Cat-free Buildings: Probable Dispersal from Human Clothing. Ann. Allergy. 70:471-474 (1993).

Goodman, J.W.: Immunogenicity and Antigenic Specificity. In: Basic and Clinical Immunology, pp. 44-52. H.H. Fudenberg, D.P. Stites, J.L.

Caldwell, et al., Eds. Lang Medical Publications, Los Altos, CA (1980).

Greenberg, G.N.; Stave, G.M.: Clinical Recognition of Occupational Exposure and Health Consequences. Physical and Biological Hazards of the Workplace, pp. 243-256. P.H. Wald, G.M. Stave, Eds. Van Nostrand Reinhold, New York NY (1994).

Hamilton, R.G.; Adkinson Jr., N.F.: Assessment of Human Allergic Diseases. In: Clinical Immunology Principles and Practice, pp. 2169-86. R.R. Rich, T.A. Fleisher, B.D. Schwarz, Eds. (1996).

Hodgson, M.J.; Morey, P.R.; Attfield, M.; et al.: Pulmonary Disease Associated with Cafeteria Flooding. Arch. Environ. Health 40:96-101 (1985).

Hodgson, M.J.; Morey, P.R.; Simon, J.S.; et al.: An Outbreak of Recurrent Acute and Chronic Hypersensitivity Pneumonitis in Office Workers. Am. J. Epidemiol. 125:631-638 (1987).

Horner, W.E.; Lehrer, S.B.; Salvaggio, J.E.: Fungi. In: Indoor Air Pollution. Immunology and Allergy Clinics of North America 14:551-566 (1994).

Horner, W.E.; Hebling, A.; Salvaggio, J.E.; et al.: Fungal Allergens. Clin. Microbiol. Rev. 8:161-179 (1995).

Hung, L.L.; Lewis, F.A.; Yang, C.S.; et al.: Dust Mite and Cat Dander Allergens in Office Buildings in the Mid-Atlantic Region. In: Environments for People. Proceedings of IAQ92 Conference, San Francisco, California, pp. 163-170. American Society of Heating, Refrigerating and Air-Conditioning Engineers, Inc., Atlanta, GA (1993).

Ingram, J.M.; Sporik, R.; Rose, G.; et al.: Quantitative Assessment of Exposure to Dog (Can f 1) and Cat (Fel d 1) Allergens: Relation to Sensitization and Asthma among Children Living in Los Alamos, New Mexico. J. Allergy Clin. Immunol. 96(4):449-56 (1995).

Institute of Medicine (IOM): Indoor Allergens: Assessing and Controlling Adverse Health Effects. A.M. Pope, R. Patterson, and H. Burge, Eds. National Academy Press, Washington, D.C. (1993).

Jacobs, R.L.; Throner, R.E.; Holcomb, J.R.; et al.: Hypersensitivity Pneumonitis Caused by *Cladosporium* in an Enclosed Hot-tub Area. Ann. Int. Med. 105:204-206 (1986).

Janko, M.; Gould, D.C.; Vance, L.; et al.: Dust Mite Allergens in the Office Environment. Am. Ind. Hyg. Assoc. J. 56:1133-1140 (1995).

Kang, B.; Chang, J.C.: Allergic Impact of Inhaled Arthropod Material. Clin. Rev. Allergy 3:363 (1985).

Kauffman, H.F.; Tomee, J.F.; van der Werf, T.S.; et al.: Review of Fungus-induced Asthmatic Reactions. Am. J. Respir. Crit. Care Med. 151(6):2109-15; Discussion 2116 (1995).

Kurup, V.P.; Fink, J.N.: Immunological Tests for Evaluation of Hypersensitivity Pneumonitis and Allergic Bronchopulmonary *aspergillosis*. In: Manual of Clinical Laboratory Immunology, pp. 908-915. N.R. Rose, E.C. de Macario, J.D. Folds, Eds. ASM Press, Washington, D.C.(1997).

Ledford, D.K.: Indoor Allergens. J. Allergy Clin. Immunol. 94:327-334 (1994).

Lemanske, R.F.; Busse, W.W.: Asthma. JAMA. 278:1855-1873.

Levetin, E (1995). Fungi. In: Bioaerosols, pp. 87-120. H. Burge, Ed. Lewis Publishers, Boca Raton, FL (1994).

Lundblad, F.P.: House Dust Mite Allergy in an Office Building. Appl. Occup. Environ. Hyg. 6:94-96 (1991).

Luczynska, C.M.: Mammalian Aeroallergens. In: Bioaerosols, pp. 149-161. H.A. Burge, Ed. Lewis Publishers, Boca Raton, FL (1995).

Luczynska, C.M.: Sampling and Assay of Indoor Allergens. J. Aerosol Sci. 28:393-399 (1997).

Marks, G.B.; Tovey, E.R.; Green, W.; et al.: The Effect of Changes in House Dust Mite Allergen Exposure on the Severity of Asthma. Clin. Exp. Allergy 25(2):114-8 (1995).

Menzies, R.; Tamblyn, R.; Comtois, P.; et al.: Case-control Study of Microenvironmental Exposures to Aero-allergens as a Cause of Respiratory Symptoms — Part of the Sick Building Syndrome (SBS) Symptom Complex. In: Environments for People, pp. 201-210. Proceedings of IAQ'92 Conference, San Francisco, California. American Society of Heating, Refrigerating and Air-Conditioning Engi-

neers, Inc., Atlanta, GA (1993).

Michel, O.; Ginanni, R.; Le, B.B.; et al.: Effect of Endotoxin Contamination on the Antigenic Skin Test Response. Ann. Allergy. 66:39-42 (1991).

Morey, P.R.; Hodgson, M.J.; Sorenson, W.J.; et al. Environmental Studies in Moldy Office Buildings: Biological Agents, Sources, Preventive Measures. Ann. Am. Conf. Govt. Ind. Hyg. 10:21 (1984).

Naclerio, R.; Solomon, W.: Rhinitis and Inhalant Allergens. JAMA, 278:1842-1848 (1997).

Newman, L.S.; Storey, E.; Kreiss, K.: Immunologic Evaluation of Occupational Lung Disease. In: Occupational Medicine: State of the Art Reviews, pp. 345-71 (1987).

Niemeyer, N.R.; de Monchy, JGR. Age-dependency of Sensitization to Aero-allergens in Asthmatics. Allergy 47:431-435 (1992).

Norman, P.S.; Peebles, R.S.: In vivo Diagnostic Allergy Testing Methods. In: Manual of Clinical Laboratory Immunology, pp. 875-880. N.R. Rose, E.C. de Macario, J.D. Folds, et al., Eds. ASM Press, Washington, DC (1997).

Ollier, S.; Davies, R.J.: A Report on the Relationship between Allergen Load, Primary Immunologic Sensitization and the Expression of Allergic Disease. Respir. Med. 88(6):407-15 (1994).

Peat, J.K.; Tovey, E.; Mellis, C.M.; et al.: Importance of House Dust Mite and Alternaria Allergens in Childhood Asthma: An Epidemiological Study in two Climatic Regions of Australia. Clin. Exp. Allergy 23:812-820 (1993).

Platts-Mills, T.A.: How Environment Affects Patients with Allergic Disease: Indoor Allergens and Asthma. Ann. Allergy 72(4):381-4 (1994).

Platts-Mills, T.A.E.; Sporik, R.B.; Wheatley, L.M.; et al.: Is There a Dose–response Relationship between Exposure to Indoor Allergens and Symptoms of Asthma? J. Allergy Clin. Immunol. 96:435-440 (1995).

Platts-Mills, T.A.; Carter, M.C.: Asthma and Indoor Exposure to Allergens. N. Engl. J. Med. 336:1382-1384 (1997).

Pollart, S.M.; Smith, T.F.; Morris, E.C.; et al.: Environmental Exposure to Cockroach Allergens: Analysis with Monoclonal Antibody-based Enzyme Immunoassays. J. Allergy Clin. Immunol. 87:505-510 (1991).

Powell, G.S.: Allergens. In: Physical and Biological Hazards of the Workplace, pp. 458-475. P.H. Wald, G.M. Stave, Eds. Wiley, New York, NY (1994).

Reed, C.E.; Swanson, M.C.; Lopez, M.; et al.: Measurement of IgG Antibody and Airborne Antigen to Control an Industrial Outbreak of Hypersensitivity Pneumonitis. J. Occup. Med. 25(3):207 (1983).

Reed, C.E.; Swanson, M.C.; Li, J.T.C.: Environmental Monitoring of Protein Aeroallergens. In: Asthma in the Workplace, pp.249-275. Marcel Dekker, Inc., New York, NY (1993).

Richerson, H.B.: Hypersensitivity pneumonitis. In: Organic Dusts: Exposure, Effects, and Prevention, pp. 139-160. R.R. Rylander, and R.R. Jacobs, Eds. Lewis Publishers, Boca Raton, FL (1994).

Robinson, B.W.: Cell reactions. In: Organic Dusts: Exposure, Effects, and Prevention, pp. 95-107. R.R. Rylander and R.R. Jacobs, Eds. Lewis Publishers, Boca Raton, FL (1994).

Rose, C.S.: Fungi. In: Physical and Biological Hazards of the Workplace, pp. 392-416. P.H. Wald, G.M. Stave, Eds. Wiley, New York, NY (1994).

Rose, C.S.: Hypersensitivity Pneumonitis. In: Occupational and Environmental Respiratory Disease, pp. 201-215. P. Harber, M.B. Schenker, and J.R. Balmes, Eds. Mosby Yearbook, Inc. (1996).

Schou, C.; Hanseb, G.N.; Lintner, T.; et al.: Assay for Major Dog Allergen Can f I: Investigation of House Dust Samples and Commercial Dog Extracts. J. Allergy Clin. Immunol. 88:847-853 (1991).

Sjöstedt, L.; Willers, S.; Ørbaek, P.: Laboratory Animal Allergy — A Review. Indoor Environ. 4:67-79 (1995).

Squillace, S.P.: Allergens of Arthropods and Birds. In: Bioaerosols, pp. 133-147. H.A. Burge, Ed. Lewis Publishers, Boca Raton, FL (1995).

Verhoeff, A.P.; van Strien, R.T.; van Wijnen, J.H.; et al.: Damp Housing and Childhood Respiratory Symptoms: The Role of Sensitization to Dust Mites and Molds. Am. J. Epidemiol. 141(2):103-10 (1995).

Tovey, E.R.; Mahmic, A.; McDonald, L.G.: Clothing — An Important Source of Mite Allergen Exposure. J. Allergy Clin. Immunol. 96:999-1001 (1995).

Tyndall, R.L.; Leman, E.S.; Bowman, E.K.; et al.: Home Humidifiers as a Potential Source of Exposure to Microbial Pathogens, Endotoxins, and Allergens. Indoor Air. 5:171-178 (1995).

Wickman, M.; Lennarat Nordvall, S.; Pershagen, G.; et al.: Sensitization to Domestic Mites in a Cold Temperate Region. Am. Rev. Respir. Dis. 148:58-62 (1993).

Yunginger, J.W.; Swanson, M.C.: Quantitation and standardization of allergens. In: Manual of Clinical Laboratory Immunology, pp. 868-874. N.R. Rose, E.C. de Macario, J.D. Folds, et al. Eds. ASM Press, Washington, DC (1997).

Chapter 26

Microbial Volatile Organic Compounds

Harriet M. Ammann

26.1   Volatile Organic Compounds Emitted by Microorganisms
    26.1.1   *Questions about MVOCs*
    26.1.2   *Microbial Ecology and MVOC Production*
    26.1.3   *Status of MVOC Sampling and Analysis*
26.2   Characteristics
    26.2.1   *Primary and Secondary Microbial Metabolism*
        26.2.1.1   Primary Metabolism
        26.2.1.2   Secondary Metabolism
    26.2.2   *Odorous MVOCs*
        26.2.2.1   Fungal VOCs
        26.2.2.2   Bacterial VOCs
        26.2.2.3   Odor Mixtures
    26.2.3   *Effects of Growth Substrate and Environmental Conditions*
26.3   Health Effects
    26.3.1   *Indoor VOC Sources*
    26.3.2   *Health Effects of VOCs*
    26.3.3   *Mechanisms for VOC Effects*
    26.3.4   *In-Vitro Studies*
    26.3.5   *Migration of MVOCs*
26.4   Sample Collection
    26.4.1   *Sorbent Sampling*
    26.4.2   *Whole Air Sampling*
26.5   Sample Analysis
26.6   Types of MVOC Studies
    26.6.1   *Laboratory Studies of VOC Emissions from Stock Fungi*
    26.6.2   *Laboratory Studies of VOC Emissions from Fungi Isolated from Contaminated Materials*
    26.6.3   *Putative MVOCs from Microbially Contaminated Buildings*
26.7   Data Interpretation and Applying Study Information
    26.7.1   *Use of MVOC Detection to Indicate Microbial Contamination*
    26.7.2   *Health Consequences of Exposure to MVOCs*
26.8   References

26.1 Volatile Organic Compounds Emitted by Microorganisms

26.1.1 Questions about MVOCs

Some investigators are beginning to use VOC analysis to search for microbial VOCs (MVOCs). Several compounds (e.g., 1-octen-3-ol, geosmin, 3-methylfuran, 3-methyl-2-butanol, 2-pentanol, 2-hexanone, 3-octanone, 2-octen-1-ol, 2-methyl-isoborneol, and 2-isopropyl-3-methoxypyrazine, 3-methyl-1-butanol, 2-heptanone, and 3-octanol) have been studied as indicators of microbial contamination in buildings, both for research purposes and as investigative tools. However, only a few laboratories currently can apply such analysis effectively in building investigations. The uncertainties associated with attributing VOCs to microbial sources preclude using this approach for routine investigations.

There are two primary research reasons for addressing MVOC production in buildings (a) to determine whether these compounds have specific toxic properties and could be produced in sufficient amounts to affect human health, and (b) to determine whether there are "signature," "fingerprint," or "marker" compounds that investigators could use to determine if microorganisms are growing in a building. At first glance, it seems that these fundamental questions would have straightforward answers. Laboratory approaches to answering these questions might involve identifying the VOCs that microorganisms produce, determining if any of the compounds are known to cause adverse health effects, and measuring the amounts of these compounds that microorganisms produce. Field measurement approaches to answering these questions might involve measuring the VOCs in indoor air and comparing these with known MVOC emissions from microorganisms in culture, especially if the bacteria or fungi studied in the laboratory were isolated from the building under study. In practice, the answers are more difficult to formulate.

The dynamics of microbial growth in buildings almost always involves a mixture of microorganisms and other life forms, which comprise a microecosystem. Further, many VOCs that microorganisms produce also have non-microbial origins or the original biological producer is not directly involved in the compound's presence in a building (e.g., natural materials, such as limonene and pinene, used in cleaning agents). Thus, indoor air contains VOCs from a combination of microbial and other sources [see 26.3.1]. Standard laboratory methods for culturing organisms cannot recreate the conditions and interactions found in nature. Therefore, researchers cannot be certain what MVOCs microorganisms may produce in nature. Transfer of contaminated materials from a building to a test chamber for VOC emission measurements may provide a better picture. However, in such tests, it is seldom possible to identify the individual microorganisms that produced the compounds detected.

The answer to the first question posed above—whether MVOCs could contribute to the health effects reported by the occupants of contaminated buildings—is a qualified yes. This answer depends on the nature and concentration of the MVOCs produced and the sensitivity of the people exposed. The second question—whether there are signature, fingerprint, or marker VOCs related to fungal or bacterial growth in buildings that can be quickly and easily determined by field measurements such as those IHs routinely perform—cannot be answered at this time because sufficient information is not yet available. MVOC sampling is useful for research purposes to determine what VOCs microorganisms produce under defined laboratory culture conditions. Sampling is also useful to learn whether field measurements can identify MVOCs that characterize indoor microbial growth. However, VOC sampling is not particularly useful as a means to establish whether microbial growth is impacting the health of building occupants and requires remediation.

26.1.2 Microbial Ecology and MVOC Production
Fungi and bacteria can grow on many different substrates and exploit a wide spectrum of environmental niches. These niches encompass broad ranges of temperature, pH, moisture, and nutrients. Some microorganisms may

be commensal or symbiotic with others [see 17.5]. Microorganisms must compete for available ecological niches and to do so have developed toxic and antibiotic products to inhibit or kill their competitors. The underlying commonality among toxic microbial products is cytotoxicity, through such mechanisms as inhibition of DNA, RNA, or protein synthesis as well as interference with membrane or enzyme functions. Fungal and bacterial toxins are harmful to other microorganisms but also—perhaps only incidentally—to higher plants and animals, including humans.

Few mycotoxins are known to be readily volatilized. In the fungal genera that have been studied (primarily *Aspergillus* and *Penicillium* spp.), certain MVOCs are produced during mycotoxin production and at the time of spore formation. Some of these compounds are volatile (e.g., the terpenes), have strong odors (e.g., pinene, limonene, and other essential oils), and are biologically active. Terpenes are isomeric, unsaturated hydrocarbons made of 5-carbon isoprene units, so that the number of carbon atoms is a multiple of 5 (e.g., 10, 15, 20, 30, or 40). Fifteen-carbon terpenes are called sesquiterpenes ($C_{15}H_{24}$). Higher plants (e.g., certain shrubs in the California chaparral ecosystem) also produce terpenes that serve a competitive function. These aromatic compounds, produced in the plants' leaves, enter the soil where the leaves fall. In the soil, these terpenes inhibit other shrubs from growing too closely and competing for water (Muller, 1969). Some mycotoxins, such as the trichothecenes, are sesquiterpenes. Some sesquiterpenes (but not the trichothecenes) are volatile and can be recovered from fungal cultures through VOC sampling. Therefore, it is possible that some substances that could be classified as mycotoxins, because of their biological activity, may be volatile and considered to be MVOCs [see also 26.2.1.2].

The competitive advantage conveyed by antibiotics, mycotoxins, and other potentially toxic products, such as terpenes, change the composition and succession of organisms in microbial and plant communities (Odum, 1971). Buildings themselves are dynamic in their microenvironments. Among the factors that shift the availability of suitable conditions for microbial growth are geographical location, season, weather conditions, ventilation system design and operation, building materials, moisture intrusion, pest colonization, and human and animal activities. Microorganisms growing in water-damaged or moist areas of buildings also play a role in ecological dynamics (Price et al., 1995). The interactions of microorganisms with each other and with building factors result in a succession of microbial genera, species, and strains over time. Nutrients available to support microbial growth also change as successions of organisms process the compounds and perhaps themselves leave metabolites that other microorganisms can utilize. As a result, the microbial growth observed in a building at a given time reflects only that moment in the progression of microbial succession (Ström et al., 1990; Yang, 1995). Likewise, MVOC air samples represent snapshots of such moments, even though the time period may span several hours.

26.1.3 Status of MVOC Sampling and Analysis

Observing sampling snapshots may obscure the actual picture of what is happening in a building. VOC profiles can change with time of day, day of the week, and season of the year as occupant activities and the interplay of environmental sources and sinks alter the VOC mixture in a building (Weschler and Hodgson, 1993). For instance, sampling may occur when a potentially toxigenic fungus is in its early vegetative growth phase, during which it does not produce mycotoxins and associated MVOCs. MVOC sampling at this time would not capture those products of secondary metabolism that have been associated with mycotoxin production. Therefore, such samples would not indicate the presence of this potentially important fungus. The same fungus grown in a laboratory, where it has abundant, rich nutrients available and no competition from other organisms, may never produce mycotoxins or associated MVOCs (Jarvis, 1995). Again, the toxigenic potential of a fungus isolated from a problem building would be undetected if sampling relied on MVOC production in laboratory culture.

Important organisms may also be present in a building but inhibited at the time samples are collected (Yang, 1995). Collected microorganisms may be overgrown in culture and not detected, or an organism may be slow to grow and not appear during the observation period. Nonetheless, an organism comprising a minor part of a microbial community could be a significant contributor to health problems if environmental conditions in the building were different in the past or changed in the future.

Other complications of microbial sampling can also affect an investigator's ability to use MVOC measurements to determine the nature and extent of microbial growth within a building. Capture and analysis methods may yield a non-representative assessment of MVOC production, in either the complex ecosystem of a building or in the more controlled conditions of a laboratory. The inability to capture and detect all organisms present or to assess the MVOC-producing capabilities of those isolated may result in an inaccurate picture of a microbial community.

26.2 Characteristics

A discussion of what is known about microbial metabolism that leads to VOC production may provide a context for interpreting available studies. Fungi and bacteria found contaminating indoor spaces produce MVOCs as products of primary and secondary metabolism. While the exact mechanisms by which MVOC production occurs are not well understood, some distinctions can be made between these metabolic processes.

26.2.1 Primary and Secondary Microbial Metabolism

26.2.1.1 Primary Metabolism Primary metabolism extracts energy from nutrient molecules and produces building blocks for essential cell components. Most fungi use aerobic respiration, releasing carbon dioxide and water as end products. Some fungal fermentation ends with the production of ethanol or lactic acid. Both fungi and bacteria produce extracellular enzymes that metabolize hydrocarbons, forming a variety of compounds, including MVOCs.

26.2.1.2 Secondary Metabolism Secondary metabolism is not essential to maintaining the life of an organism and requires extra energy. The function of secondary metabolites seems to be to give fungi or bacteria a competitive advantage over other microorganisms seeking to occupy the same ecological niche. Therefore, toxic metabolites may give a fungus or bacterium a competitive edge in a nutrient-poor medium (e.g., on building materials, insulation, or wallpaper) (Jarvis, 1995). Secondary metabolites include mycotoxins, antibiotics (e.g., penicillin and streptomycin), and some VOCs (e.g., terpenes) [see 18.2.5, 19.2.5, and 24.1.3]. Many products of secondary bacterial or fungal metabolism are known to be biologically active, and even antibiotics and mycotoxins may have some degree of volatility. Therefore, the distinction among these products (i.e., identification as an antibiotic, mycotoxin, or MVOC) may be only semantic and not functionally valid.

If competition for energy and water are among the factors that drive the production of secondary metabolites, then information about such products derived strictly from laboratory cultures may be incomplete. Laboratory conditions may not reproduce the complex, stressful, and competitive situations microorganisms find when struggling to survive in nature. This is one reason why the MVOCs microorganisms produce in laboratory culture may differ from those they produce in buildings. Some toxigenic strains may also lose their ability to produce toxins after several generations in pure culture. Therefore, MVOCs associated with toxin production could more likely be found in buildings—where competition occurs—than in laboratory culture [see also 26.6].

The highest MVOC production associated with secondary fungal metabolism (specifically terpenes and sesquiterpenes) seems to occur prior to and during sporulation and mycotoxin production (Abramson et al., 1983; Zeringue et al., 1993). Zeringue et al. (1993) investigated the relationship between the production of aflatoxins and that of odorous sesquiterpenes by *Aspergillus flavus* in culture. These investigators demonstrated that of eight strains tested, the four aflatoxic ones produced sesquiterpenes that peaked during aflatoxin production and, otherwise, were absent. Nonaflatoxic strains produced no sesquiterpenes. Each of the aflatoxic strains produced a distinctive fingerprint of sesquiterpene types and relative amounts. Detection of such MVOCs could be useful in determining whether a building is contaminated with toxin-producing fungi if the observed relationship between the production of mycotoxins and that of sesquiterpenes or odorous secondary metabolites holds for other toxigenic fungi.

26.2.2 Odorous MVOCs

Because of their volatility, many microbial metabolites enter the air where microorganisms grow. Microorganisms may release fermentation products (e.g., ethanol) in sufficient quantity that the human olfactory sense and VOC sampling can detect them. Volatile products may be aliphatic or aromatic compounds (containing sulfur, amines, and other substituted groups) which are detectable by smell because their odor thresholds are low. It is useful to know that, in general, MVOCs and their odors are indicators of microbial growth and may reflect microorganism presence even when growth is hidden. More particularly, the finding that specific (and odorous) volatile secondary metabolites are associated with mycotoxin production and spore formation may be important to indicate the presence of toxigenic fungi, for which remediation may be especially important to protect occupant health. Although the odor thresholds for some compounds have been determined, what is not known is the minimal amount of microbial growth needed to generate detectable concentrations, nor is it possible to use odor intensity to indicate the degree of indoor microbial growth.

26.2.2.1 Fungal VOCs Some primary fungal metabolites are pungent (e.g., those identified with certain cheeses) and others evoke unpleasant associations such as "dirty socks" or "locker rooms" or are described as "moldy," "musty," or "mildewed." Still other fungi release odors that are considered pleasant and are described with such terms as "fruity," "earthy," or "snow-pea-pod-like." For example, the fungal metabolite 1-octen-3-ol has a strong resinous, fungal odor in its pure state, but at low concentrations—close to the human odor threshold—it resembles the odor of mushrooms. This compound has also been described as having an old paper, mildew-like mustiness. A number of fungi, including *Aspergillus niger*, *A. flavus*, and *Penicillium roqueforti* have been found to produce 1-octen-3-ol (Harris et al., 1986). On the other hand, 2-octen-1-ol has a strong musty, oily odor. While some other fungi produce these two alcohols, they are responsible for the characteristic odor of *A. flavus* (Kaminski et al., 1972). Benzyl cyanide may be responsible for the grassy odor that *Botrytis cinerea* produces (Gravesen et al., 1994).

Ström et al. (1990) found that odors characteristic of microbial growth in buildings indicated the production of secondary metabolites. In addition to terpenes, other compounds may also be products of secondary metabolism, such as 2-methyl-isoborneol (1,2,7,7-tetramethyl-exo-bicyclo [2,2,1] heptan-2-ol—a terpene-derived alcohol) and geosmin (1-10-dimethyl-*trans*-9-decalol). The genus *Chaetomium* is a known producer of geosmin which, together with 2-methyl-isoborneol, has been found to contribute to musty, earthy odors (Harris et al., 1986). *Penicillium expansum* is also a producer of geosmin (Mattheis and Roberts, 1992). *Penicillium casiecolum*, *B. cineria*, and *P. roqueforti* produce 2-methyl-isoborneol (Karahadian et al., 1985a,b; Harris et al., 1986).

26.2.2.2 Bacterial VOCs Many bacteria, especially anaerobic ones, produce odorous MVOCs that microbiologists use to identify bacterial isolates. However, information on the MVOCs bacteria produce in building environments is even more limited than for fungi. Harris et al. (1986) studied the odorous MVOCs produced by two actinomycetes, *Streptomyces griseus* and *Streptomyces odorifer* (Table 26.1). In this case, the investigators identified both the bacterial genus and the species studied but, in other studies, organism identification is often limited to the genus—and even that is not always known. *Streptomyces* spp. also produce geosmin, which was mentioned in Section 26.2.2.1. McJilton and colleagues (1990) reported field studies of illness-related odors in two office buildings and an eating establishment. Odors characterized as "old cheese" and "locker room or dirty socks" were attributed to 2-methyl propionic acid and 1-butoxy-2-propanol. The latter compound was also found in the headspace of a pure culture of a bacterium (identified only as "red, Gram-negative rods") isolated from the problem areas where the 1-butoxy-2-propanol was detected.

26.2.2.3 Odor Mixtures Odors considered characteristic of certain microorganisms can change or be obscured when the organisms producing them are growing in association with other organisms, on varied substrates, or under different moisture or temperature conditions. Odor perception may change with VOC concentration [see 26.2.2.1] but may also depend on what other compounds are present. Seifert and King (1982) found that odors associated with the growth of different strains of *Aspergillus clavatus* in culture changed from a pleasant ester-like aroma to a variety of unpleasant odors. Depending on fungal strain, the odors were described as "unpleasant, fetid-decaying fish," "weak, moldy, rubbery," or "weak, rotten, and fetid." However, no specific compound isolated from the cultures had the odor associated with the mixture of compounds obtained from the parent strain.

26.2.3 Effects of Growth Substrate and Environmental Conditions

Microbial species and strain as well as laboratory culture methods affect the nature and amount of VOCs microorganisms produce. VOCs measured in microbially contaminated buildings may differ from the VOCs that fungi and bacteria from such buildings produce in the laboratory because the identity of all of the environmental contaminants may not be known (Bayer and Crow, 1993a,b; Bayer et al., 1995; Wessén et al., 1995). Ström et al. (1990) noted that the kinds of organisms found in problem buildings depended on the nature of the building materials present. Also, moisture, pH, and temperature along with the availability of nutrients and oxygen determined fungal and bacterial succession (Ezeonu et al., 1994b; Price et al., 1995).

Researchers studying VOC emissions from microorganisms in culture can choose the composition of the growth medium they use. However, they also must determine other important parameters; for example: select a solid or liquid medium, agitate liquid cultures or leave them still, choose one or more incubation temperatures, and decide when within the growth phase or age of a culture to test for VOC emissions. Of these variables, growth substrate may play the greatest role (Börjesson et al., 1989, 1990; Whillans and Lamont, 1995). This is especially noticeable when comparing MVOC production from microbial growth on laboratory media and natural substrates, for example, grains (Börjesson et al., 1990) (Table 26.1) or building materials (Bayer and Crow, 1993 a,b; Ezeonu et al., 1994a; Larsen and Frisvad, 1994b) (Tables 26.1, 26.2, and 26.3)].

> The tables presented in this chapter should only be used to compare (a) the VOCs a given microorganism may produce when grown on different substrates, or (b) what VOCs different organisms may produce when grown on a particular substrate. The information should not be construed to represent either the types or amounts of VOCs microorganisms may produce. The tables do not provide sufficient details to allow readers to identify the organisms that may be sources of specific MVOCs measured during field investigations.

One study comparing volatile compounds produced by a mixture of microorganisms grown in culture and on various building materials demonstrated variability in both MVOC kind and concentration (Wessén et al., 1995). Larsen and Frisvad (1994b) used standard laboratory media along with a medium containing wallpaper paste and noted differences in fungal VOC production on the two substrates (Table 26.1). Batterman (1995) also found that fungi emitted different compounds on different media, and Batterman et al. (1991) suggested that MVOC composition and quantity change throughout the phases of fungal growth. Other studies of MVOCs emitted from fungal cultures of varying ages support this concept (Börjesson et al., 1990; Whillans and Lamont, 1995).

It has also been suggested that metabolic changes occur in microorganisms as preferred substrates are exhausted or different substrates become available (Rivers et al., 1992). The bacteria in this study were postulated to use first simple sugars, emitting ethanol as a product of fermentation. Following sugar depletion, the researchers suggested that the bacteria moved to amino acid degradation, producing methyl mercaptan and other odorous sulfur-containing compounds. Variations in MVOC production may also be expected with environmental changes in temperature, humidity, moisture availability, light, and the presence of other fungi and bacteria (Ström et al., 1990; Yang, 1995).

Halogenated or metal-containing VOCs might also be released as microorganisms metabolize substrates containing these elements. For instance, fungi growing on arsenic-containing media (Challenger et al., 1933) and wallpaper (Gravesen et al., 1994) have released the highly toxic gas arsine. The common fungus *Scopulariopsis brevicaulis* (frequently isolated from soil, house dust, old carpets, and wallpaper) has the ability to convert inorganic nitrogen to organic nitrogen compounds in laboratory culture. This organism may be able to decompose arsenic compounds, which at one time were applied to wall-to-wall carpets as antistatic agents. Gravesen et al. (1994) reported that *B. cinerea* may produce benzyl cyanide (phenyl acetonitrile).

26.3 Health Effects

Potential adverse health effects are a reason for interest in MVOCs, in addition to determining whether these compounds could serve as markers of microbial contamination and perhaps even to determine risk levels by identifying the presence of particularly problematic organisms. Characterizing MVOCs is important in trying to establish whether they cause health effects individually or in aggregate and whether they are produced in sufficient quantities to contribute to total VOC burdens associated with specific toxic or other reactions.

26.3.1 Indoor VOC Sources

There are many indoor VOC sources other than microorganisms (e.g., paints, varnishes, cleaning materials, solvents, fabrics, carpets, plastics, building materials, and indoor combustion). Human body odor is also, to a large degree, due to bacteria growing in the mouth and on the skin as well as in the gastrointestinal tract. Personal-care products may also contribute to the indoor VOC load. Some VOCs emanate from both living and non-living sources, and others solely from one or the other. For instance, ethanol, toluene, heptane, octane, nonane, xylene, styrene, acetone, and formaldehyde could originate from either living or non-living sources.

Health effects related to MVOC exposure have not been specifically studied. Nevertheless, studies of asthma

TABLE 26.1. Selected Studies of VOCs from Stock Fungi and Bacteria in Laboratory Culture

CAVEAT: The studies in this and other tables are not comparable except in very general terms. The information is presented only to illustrate that methods of sample collection, organism culture, and VOC analysis can change the composition and quantities of VOCs found. These tables are NOT to be used in building investigations as references of MVOC emissions for the purpose of identifying what microorganisms may be present. Nor are these tables to be used to anticipate possible health outcomes that may result from observed MVOC profiles or the presence of particular microorganisms in the indoor environment.

Microorganism	VOC Recovered	Culture/Substrate		Reference
Aspergillus flavus	1-octen-3-ol	saline wheat meal slurry		Kaminski et al., 1972
Alternaria sp.	1-octanol	saline wheat meal slurry		Kaminski et al., 1974
Cephalosporium sp.				
Fusarium sp.				
P. chrysogenum				
P. citrinum				
P. funiculosum				
P. raistrickii				
P. viridicatum				
A. niger	*cis*-2-octen-1-ol			
A. ochraceus	3-methyl-butanol			
A. parasiticus	3-octanone			
A. clavatus	1-octen-3-ol	yeast extract Czapek's		Seifert and King, 1982
(3 strains)	4-methylbenzaldehyde	agar/steam extract		
	phenylacetaldehyde			
	2-methylphenol			
	+46 other aliphatic aldehydes,			
	ketones, and aromatic compounds			
A. niger	1-octen-3-ol	wholewheat bread		Harris et al., 1986
P. roqueforti	1-octen-3-ol	slurry/headspace		(odorous
	2-methyl-isoborneol			compounds)
Botryris cineria	2-methyl-isoborneol	bread slurry; agar/		
Streptomyces griseum	2-methyl-isoborneol	headspace		
	geosmin			
Streptomyces odorifer	3-methyl-1-butanol			
	3-methyl-butanol			
	geosmin			
Penicillium	3-methyl-1-butanol	Czapek's agar,		Börjesson et al., 1990
aurantiogriseum	ethanol	Norkran's agar,		
	1-propanol	oats, wheat		
	acetic acid	oats		
	3-methylfuran			
	3-octen-2-ol			
	1-octen-3-ol			
	terpenes			
P. clavigerum	geosmin	CYA/headspace		Larsen and
P. vulpinum	2-methoxy-3-isopropyl-pyrazine			Frisvad, 1994a
	2-methoxy-3-*sec*-butyl-pyrazine			(CYA = Czapek
	mono- and sesquiterpenes			yeast autolysate
				agar)

Microorganism	VOC Recovered	SYES	Wallpaper agar	Reference
Aspergillus versicolor	2-methyl-1-propanol	+	+	Larsen and
	3-methyl-1-butanol	+	+	Frisvad, 1994b
	1-pentanol	+	+	(SYES = Sigma
	2-heptanone	+	+	yeast extract
	1-octen-3-ol	+	+	sucrose agar;
	3-octanone	+	+	wallpaper agar/
	1,3-dimethyl benzene	+	+	ether extract)
P. aurantiogriseum	2-methyl-1-propanol	+	+	
	3-methyl-hexane	+	+	
	3-methyl-1-butanol	+	+	
	4-methyl-3-hexanone	+	-	
	1-octen-3-ol	+	+	
	2 sesquiterpenes (no I.D.)	+	+	

TABLE 26.1 Selected Studies of VOCs from Stock Fungi and Bacteria in Laboratory Culture (cont.)

Microorganism	VOC Recovered	Culture/Substrate		Reference
		SYES	Wallpaper Ager	Larsen and Frisvad, 1994b (continued)
Penicillium brevicompactum	2-methyl-1-propanol	+	+	
	3-methyl-1-butanol	+	+	
	1-pentanol	+	+	
P. chrysogenum	2-methyl-1-propanol	+	+	
	1-heptene	+	-	
	3-methyl-3-buten-1-ol	+	-	
	3-methyl-1-butanol	+	+	
	1,3-nonadiene	+	-	
	3-octanone	+	+	
	3-octanol	+	+	
P. commune	2-methyl-1-propanol	+	+	
	3-methyl-1-butanol	+	+	
	1-pentanol	+	+	
	monoterpene 1 (no I.D.)	+	+	
	3-octanone	+	+	
	3-octanol	+	+	
	monoterpene 2 (no I.D.)	+	+	
	2-methyl-isoborneol	+	+	
P. expansum	2-methyl-1-propanol	+	+	
	3-methyl-1-butanol	+	+	
	1-pentanol	+	+	
	β-pinene	+	+	
	m-methyl-anisole	+	+	
	limonene (1 of 3 strains)	+	+	
	sesquiterpenes (tentative I.D.):			
	α-zingiberene	+	+	
	trans-α-bergamotene	+	+	
	bis-abolene	+	+	
	β-farnesene	+	+	
P. glabrum	2-methyl-1-propanol	+	+	
	3-methyl-1-butanol	+	+	
	1-pentanol	+	+	
	3-methyl-3-hexanone	+	-	
	3-methyl-1-butanol acetate	+	-	
	styrene	+	-	
	3-octanone	+	+	
	sesquiterpenes (no I.D.)	+	+	
P. olsonii	2-butanone	+	+	
	2-methyl-1-propanol	+	+	
	3-methyl-1-butanol	+	+	
	1-pentanol	+	+	
	unsaturated monterpenes (no I.D.)	+	+	
A. versicolor	1,3-pentadiene	MEA and dichloran-glycerol agar		Sunesson et al., 1995 (MEA = malt extract agar)
	2,3,5-trimethyl-1,3-hexadiene			
	2-methyl-1-butanol			
	3-methyl-1-butanol			
	1-octen-3-ol			
	3-methyl-2-butanone			
	3-octanone			
	4-methyl-3-hexanone			
	5-ethyl-4-methyl-3-heptanone			
	anisole			
	3-methoxyanisole			
	dimethyl disulfide			
	sabinene			
	β-myrcene			
	1,8-cineol			
	γ-terpenine			
	β-bisabolene			
	β-himachalene			
	cuparene			

TABLE 26.1 Selected Studies of VOCs from Stock Fungi and Bacteria in Laboratory Culture (cont.)

Microorganism	VOC Recovered	Culture/Substrate	Reference
Cladosporidium cladosporoides	3-pentanone 3-methylfuran sesquiterpenes (no I.D.)	MEA and dichloran-glycersol agar	Sunesson et al., 1995 (MEA = malt extract agar)
Paecilomyces variotii	2-methyl-1,3-pentadiene octane xylene trimethyl benzene 2-propanol 2-methyl-1-propanol 2-methyl-1-butanol 3-methyl-1-butanol 2-butanone furan 2,4-dimethylfuran 1-methylpropyl formate 2-methylpropyl formate 1,2,3,4-tetramethyl-4-methylethenyl) benzene α-curcumene sesquiterpenes (no I.D.)		
P. commune	heptane 2-methyl-1-propanol 2-methyl-1-butanol 3-methyl-1-butanol cresol acetone 2-butanone cyclopentanone 3-methylfuran 2,5-dimethylfuran 3-methylanisole methyl acetate ethyl acetate propyl acetate 2-methylpropyl acetate 3-methylbutyl acetate ethyl propanoate ethyl butanoate ethyl-2-methyl butanoate dimethyl sulfide dimethyl disulfide ethanthioic acid-S-(2-methyl) butylester α-pinene β-pinene camphene limonene geosmin methyl-(1-methylethenyl) benzene α-curcumene sesquiterpenes (no I.D.)		
Phialophora fastigata	3-methyl-1-butanol 1-octen-3-ol acetone 2-butanone cyclopentanone methyl-3-methyl butanoate caryophyllene sesquiterpenes (no I.D.)		

and respiratory symptoms have implicated damp buildings (primarily residences), moldy odors, and the presence of visible indoor fungal growth as risk factors for these illnesses (Brunekreef et al., 1989; Jaakola et al., 1993; Spengler et al., 1993; Norbäck et al., 1995; Verhoeff et al., 1995). The logical explanation for the allergic effects reported in these studies might be exposure to antigenic bioaerosols, but mycotoxins and MVOCs may also play a role. However, the complexity of factors involved in the initiation and exacerbation of asthma makes it difficult to determine which aspects of biological growth in damp buildings are responsible for associated health effects. The specific mechanisms by which various risk factors may be involved have not been determined.

26.3.2 Health Effects of VOCs

At the concentrations found in industrial workplaces, it is clear that VOCs are mucous membrane irritants (Ware et al., 1993). However, such concentrations are generally orders of magnitude higher than those found in non-industrial indoor environments. VOCs at such concentrations also affect the central nervous system, producing such symptoms as headache, attention deficit, inability to concentrate, and dizziness (Fidler et al., 1987; Kraut et al., 1988; Cone and Sult, 1992; Fiedler et al., 1992). In an investigation of the effects of industrially emitted VOCs on people living in surrounding communities, Ware et al. (1993) addressed VOC concentrations more in keeping with those found in indoor air. This epidemiological study gave an estimated odds ratio of 1.08 (95% confidence interval: 1.02 to 1.14) for chronic lower respiratory symptoms associated with a 2-mg/m^3 change in a mixture of 15 industrially produced VOCs. Some of these were VOCs commonly found indoors.

It is difficult to determine what portion of indoor VOCs arise from common fungi, bacteria, and other microorganisms, but MVOCs likely are responsible for only a small fraction of total VOCs, measuring in the low μg/m^3 range. However, a severely contaminated building may have a significant MVOC contribution derived from microbial growth. For example, Miller et al. (1988) measured a total putative fungal VOC concentration of ~2 mg/m^3. This could represent a significant contribution to the concentration of total VOCs at which Otto et

TABLE 26.2. VOCs from Fungi Grown in Laboratory Culture and from Mixed Culture on Building Material Substrates

CAVEAT: The studies in this and other tables are not comparable except in very general terms. The information is presented only to illustrate that methods of sample collection, organism culture, and VOC analysis can change the composition and quantities of VOCs found. These tables are NOT to be used in building investigations as references of MVOC emissions for the purpose of identifying what microorganisms may be present. Nor are these tables to be used to anticipate possible health outcomes that may result from observed MVOC profiles or the presence of particular microorganisms in the indoor environment.

Microorganism	VOC Recovered	Culture/Substrate	Reference
Acremonium obclavatum	ethanol acetone 2-butanone methyl benzene cyclohexane 2-ethyl hexanol benzene	enriched agar	Ezeonu et al., 1994a
Aspergillus obclavatum in mixed culture	cylotrisilane limonene pentane arsenous acid benzene	duct liner (the liner was exposed to a laboratory culture of *A. obclavatum* and the fungus was identified microscopically)	
A. versicolor	1,3-dimethoxy benzene methyl benzene cyclotetrasiloxane	enriched agar	
A. versicolor in mixed culture	methyl benzene cyclotrisiloxane ethanol xylene limonene ethylhexanol benzene	duct liner (see above)	

al. (1990) and Hudnell et al. (1992) found trigeminal nerve effects (i.e., 25 mg/m³). The trigeminal nerve is part of the "common chemical sense" that responds to the perception of chemical pungency rather than odor. Perceived pungency produces reflex constriction of the airways and inflammation resulting in nasal stuffiness as well as symptoms of headache, malaise, memory loss, and reduced ability to concentrate (Cometto-Muniz and Cain, 1993). In a controlled exposure study, Koren et al. (1990, 1992) investigated the inflammatory response to a VOC mixture previously used by Mølhave et al. (1986), Otto et al. (1990), and Hudnell et al. (1992). Relative to controls, Koren et al. (1990, 1992) found significant increases in polymorphonuclear neutrophils (PMNs) in nasal lavage fluid both immediately and 18 hours after a 4-hour exposure to the mixture. Because PMN influx into an injured site is a strong indicator of inflammation, this study showed that low concentrations of VOCs common in indoor air can cause acute inflammation of the upper airways.

26.3.3 Mechanisms for VOC Effects

The work of a number of researchers has lead to pos-sible explanations of etiological mechanisms that could account for the associations described in Section 26.3.2 on the health effects of VOCs. Meggs and Cleveland (1993) examined ten patients who reported multiple physical and mental complaints in response to odorous organic chemicals. Rhinitis was a common complaint. Rhinolaryngoscopic findings from examinations of the subjects' nasal passages and throats were abnormal in all cases and included swelling of mucous membranes and excess mucus secretion. Meggs (1994) implicated neurogenic inflammation as being activated by exposure to chemical irritants. This kind of inflammation arises through mediation of substances released by nerve endings rather than from direct damage to respiratory epithelium. Whether VOCs of microbial origin could initiate neurogenic inflammation of the type Meggs and Cleveland (1993) described has not been specifically studied.

The odors that microorganisms produce can also contribute to some of the adverse reactions observed in contaminated buildings (Shusterman, 1992). Warren et al. (1992) used continuous measurement of respiratory behavior and

TABLE 26.3. VOCs Found in Problem Buildings and Fungi and Bacteria Cultured from Air Samples from these Buildings

CAVEAT: The studies in this and other tables are not comparable except in very general terms. The information is presented only to illustrate that methods of sample collection, organism culture, and VOC analysis can change the composition and quantities of VOCs found. These tables are NOT to be used in building investigations as references of MVOC emissions for the purpose of identifying what microorganisms may be present. Nor are these tables to be used to anticipate possible health outcomes that may result from observed MVOC profiles or the presence of particular microorganisms in the indoor environment.

Microorganism Isolated from Test Environment	Mixed VOCs Recovered from Air, Listed Compounds not Associated with Particular Fungi		Reference
42 total fungal isolates *Alternaria* sp. *Cladosporium* sp. *Penicillium* spp.[A]	3-methyl-butanol 2-hexanone 2-heptanone heptane benzene octane toluene hexanol nonane	ethyl benzene 3-pentanone ethenyl benzene 2-butoxy ethanol benzaldehyde naphthalene decanol cyclohexane-1-methyl- 4-1-methyl-ethenyl	Miller et al., 1988
Organism unknown "red Gram-negative bacterium"	1-butoxy propanol		McJilton et al., 1990
Organisms unknown	acetone isopropanol	ethanol	Bayer and Crow, 1993
12 *Penicillium* spp.	acetone 2-pentanone	ethyl acetate	
Organisms unknown greater in complaint houses	Total VOCs (13 putative MVOCs), concentration		Ström et al., 1994
Organisms unknown	acetaldehyde formaldehyde	acetone	Schleibinger et al., 1995

[A] Most common fungus isolated from air samples.

nasal patency (openness) to examine responses to well controlled odor stimulation. These researchers observed rapid changes in respiratory behavior related to stimulant concentration and perceived odor intensity as well as degree of nasal irritation. Alarie (1973) noted changes in respiration, cough, and signs of tissue irritation in animals exposed to gases varying in reactivity.

Shim and Williams (1986) examined the effects of cologne on the pulmonary function of people with asthma, observing symptoms of wheezing and chest tightness in asthmatic patients but not controls. Additionally, 57 of 60 patients queried about asthma and odor responses reported respiratory reactions to odors. Of these, 85% reported worsening of asthma as a response to insecticides, 78% to perfume or cologne, 73% to fresh-paint smells, 60% to automobile exhaust or gas fumes, and 37% to cooking odors. The odors that induced asthma symptoms were not necessarily unpleasant, and some obnoxious odors failed to produce asthma. Fungal and other microbial odors were not on the questionnaire, but at least some of the substances listed had VOC components.

In the occupational literature, cacosmia is defined as nausea, headaches, and subjective distress in individuals exposed to neutral odors (Ryan et al., 1988). In that fraction of the population who are cacosmic, stimulation of the olfactory sense by a number of odorous substances (including "musty" smells) can produce headache, nausea, and other aversive responses. Bell et al. (1993a,b) reported that as many as 30% of the young, non-occupationally exposed adults and older adults they studied were thought to be cacosmic.

26.3.4 In-Vitro Studies

In-vitro assays of cultured swine cells have indicated that exposure to fungal homogenates can cause adverse effects (Gareis, 1995), but little information exists that specifically implicates MVOCs. Smith et al. (1992) exposed monolayers of cultured human lung cells to spore extracts made from fungi isolated from damp homes. These investigators found significant cytotoxicity associated with several *Penicillium* and *Aspergillus* spp. It was unclear whether the cellular toxicity resulted from adsorbed

TABLE 26.4. VOCs Found in Problem Buildings and Fungi and Bacteria Cultured from Materials from these Buildings[A]

CAVEAT: The studies in this and other tables are not comparable except in very general terms. The information is presented only to illustrate that methods of sample collection, organism culture, and VOC analysis can change the composition and quantities of VOCs found. These tables are NOT to be used in building investigations as references of MVOC emissions for the purpose of identifying what microorganisms may be present. Nor are these tables to be used to anticipate possible health outcomes that may result from observed MVOC profiles or the presence of particular microorganisms in the indoor environment.

Microorganism	VOC Recovered	Culture/Substrate	Reference
Unknown bacteria and fungi	VOCs were looked for, and putative MVOCs were recovered in varying amounts	laboratory cultures	Ström et al., 1994
A. versicolor *Pseudomonas* spp. *Trichtoderma viride* *Bacillus* spp.	2-methoxy-1-propanol 1-butanol 3-methyl furan 3-methyl-1-butanol	linoleum from wet basement	
A. versicolor *Penicillium* spp. *Pseudomonas* spp.	3-methyl-2-butanol 2-pentanone 2-hexanone	wood-studding from wet basement	
A. versicolor *Aureobasidium pullulans* *Penicillium* spp. *Phialophora fastigata* *Pseudomonas* sp. *Streptomyces* sp.	2-octen-1-ol 3-heptanone 3-octanol 3-octanone 2-isoproyl-3-methoxy pyrazine 2-methoxy borneol geosmin 1-butanone 2-methyl-1-propanol	water-damaged chipboard from apartment	

[A] The reported VOCs were measured in microbially contaminated houses and were those thought by the authors to be of microbial origin, based on other studies. The specific organisms that produced the compounds are not known, but those bacteria and fungi listed were identified and counted by epifluorescent microscopy after isolation from building materials and growth in the laboratory. Emissions were measured after 5 to 10 days growth on malt extract agar with streptomycin sulfate (for fungi) or tryptone glucose extract agar with cycloheximide (for bacteria).

mycotoxins or adsorbed MVOCs. Joki et al. (1993) exposed tracheal tissue to solutions containing volatile compounds from two actinomycetes and two fungi—a *Penicillium* sp. and *Trichoderma viride*. These researchers compared the effects of MVOCs from these test organisms to that of lipopolysacheride (LPS) from the bacterium *Klebsiella pneumoniae*. LPSs from this bacterium are known to inhibit ciliary beat frequency (CBF), that is, the rate at which the cilia of tracheal tissues move. VOCs produced by the actinomycetes and older fungal cultures increased CBF by statistically significant amounts, analogous to the effects of inflammatory mediators.

How the above *in-vitro* effects relate to the respiratory symptoms reported by individuals in microbially contaminated buildings is unknown. Disturbance of particle clearance by the mucociliary escalator could be part of the total picture of respiratory responses. Volatile compounds do adsorb to the surface of respirable particles (Gebefuegi and Kettrup, 1994), and such particles could carry VOCs (including MVOCs) to the deepest parts of the lungs. It remains to be investigated whether fungal and bacterial VOCs affect airway inflammation or other airway responses, as might be suggested by their similarity to inflammatory mediators with respect to CBF increase.

26.3.5 Migration of MVOCs
Toxic effects have been reported when fungal contamination was hidden and air samples failed to identify a problem. Explanations related to difficulties in capturing and culturing *Stachybotrys chartarum* spores could account for these findings. However, as yet unidentified volatile compounds also may be responsible for some of the symptoms commonly associated with the presence of this fungus. MVOCs from fungi and bacteria commonly found in Swedish homes have been shown to penetrate plastics—used as moisture barriers—as well as PVC-coated wallpapers (Ström et al., 1994). These authors postulated that researchers' inability to correlate MVOC and airborne spores concentrations (e.g., Miller et al., 1988) may be due to encapsulation of fungal growth within structural building components. Although visible microbial growth is hidden, the MVOCs they produce penetrate to spaces where investigators can smell or measure the compounds.

26.4 Sample Collection
There are two major methods for collecting VOCs; sorbent sampling and whole air sampling. Sorbent sampling uses materials (adsorbents) that attract and hold VOCs for later analysis. Sorbent sampling can be further divided into active and passive sorbent sampling. Whole air sampling uses evacuated canisters to collect air samples. The methods described here are for sampling VOCs in general, but they also hold for collecting MVOCs.

At this time, MVOC collection and analysis is done primarily for research purposes. Therefore, VOC sampling is not used to identify indoor bacterial or fungal growth during investigations of possible indoor microbial contamination. Careful use of blanks and standards is necessary in such testing because MVOC concentrations tend to be low relative to VOCs from other indoor sources.

26.4.1 Sorbent Sampling
Active sorbent sampling is a well-developed technique in which sample air is drawn through a small tube filled with one or several sorbents to which VOCs will adhere. This method allows very small amounts of VOCs to be extracted from large air volumes. Depending on ambient VOC concentrations, indoor sampling may require only a few or as much as several hundred liters of air to collect detectable amounts of VOCs. Sorbent-tube sampling is useful for field collection of VOCs because the sampling equipment is small, portable, and relatively inexpensive (McJilton et al., 1990). Passive sorbent sampling is another well-developed methodology. However, this approach has not been applied to MVOC studies because the method is only suitable for high VOC concentrations, such as those found in industries using solvents.

Common sorbents used for both active and passive VOC sampling include activated charcoal and carbon black (e.g., Carbosieve®, Anasorb®, and Carbotrap®), silica gel, polyurethane foam, polymers (e.g., Tenax®, Chromosorb®, and Poropak®), and resins (e.g., XAD-4®). Williams and Sievers (1984) used a sorbent called Eu-sorb that seems to capture aldehydes and ketones and thus may be useful for collecting MVOCs. A VOC's properties (e.g., boiling point and polarity) determine which sorbent will best capture it. VOC stability on sorbents varies and may be limited for resins and polymers.

Sorbent selection must address potential problems of breakthrough and competitive sorption. Breakthrough occurs when a sorbent becomes saturated, in which case not all VOCs are captured, giving erroneously low measurements. Competitive sorption occurs between polar and non-polar compounds (i.e., high concentrations of nonpolar compounds displace polar compounds on a sorbent). Use of charcoal as a primary sorbent, followed by silica gel or Tenax® as a backup, is recommended if high concentrations of nonpolar compounds are expected.

26.4.2 Whole Air Sampling
Whole air sampling, with cryogenic preconcentration, can capture a wide spectrum of VOCs. Such samples are collected in evacuated stainless-steel canisters (e.g., Summa®) that have low background VOC concentrations. Canisters provide enhanced VOC stability over Teflon®, Tedlar®, or Mylar® sample bags, which are difficult to clean and may introduce artifacts. Sample collection is simple with

vacuum canisters, involving only opening previously evacuated canisters in the environments to be sampled and closing them after drawing in air. Alternatively, a pump may be attached to entrain air. Unfortunately, canisters tend to be expensive and bulky (e.g., a typical 6-L canister weighs several kilograms and stands 0.3 m high). However several GC–MS samples can be withdrawn from the large volume of a canister. Direct cryogenic sampling condenses VOCs by passing air over a cold trap or cryotrap, with the trap maintained at cryogenic temperatures (–196° to –90°C) until analysis. Use of a cryotrap is expensive and water accumulation may be a problem. Therefore, this method is less suitable for field sample collection than for sample recovery in the laboratory.

26.5 Sample Analysis

The VOC analyses described here are conducted in research laboratories and not in the field by IHs, EHPs, and IEQ consultants. However, adaptation of research methods to enable investigators to measure specific MVOCs on site are under development in several laboratories and have been published (AIHA, 1996).

VOCs captured on sorbents must be transferred to an analytical instrument without contaminating or decomposing them through improper handling. Either thermal or chemical means can be used to desorb and transfer samples, but VOC concentration measurements by these two methods may differ. For thermal desorption, a collection medium is heated to release VOCs. Thermal desorption may cause a sorbent to degrade, introducing artifacts to the solvent mixture. For chemical desorption, a small quantity of solvent (e.g., ethanol, ether, methanol, or carbon disulfide) is injected into a sorbent container. Both thermal and chemical desorption have been used with polymer sorbents. Chemical desorption is more often used with silica gel or carbon media.

The VOC concentration in a canister used to collect a whole air sample is likely to be low, and some method to concentrate the collected compounds must be used. A cold trap or cryotrap, cooled by liquid nitrogen, is one example. After the flow of sample air into a trap is terminated, a carrier gas (e.g., helium or nitrogen) is fed through. The trap is then rapidly heated above the boiling points of the VOCs to be analyzed, and the sample is introduced into a GC.

Sorbents used in GC analysis can be chosen to capture lighter or heavier compounds. However, lighter compounds may be lost if the chosen sorbent preferentially captures heavier compounds (Whillans and Lamont, 1995). As discussed earlier relative to sorbent sampling, breakthrough can occur if high concentrations of certain VOCs pass through a column. Thus the ability to accurately determine the concentrations of the VOCs that break through may be lost. Depending on the composition of the VOC mixture being adsorbed, competitive sorption between

polar and nonpolar compounds can lead to displacement of polar ones from a sorbent, giving an inaccurate picture of the composition of a VOC mixture (Batterman, 1995).

Differences in GC application (e.g., in choice of carrier gases and columns) may cause variations in VOC separation. Adding to the difficulty of comparing results, researchers have used different detection methods, for example, several applications of MS together with other detectors, such as flame ionization detectors (FIDs), photoionization detectors (PIDs), dual or tandem MS (MS-MS), and MS with liquid chromatography.

High resolution GC is generally used to separate and partially identify the components of a VOC mixture. VOCs are separated in a GC column by their elution time, which depends on interactions between the VOCs and the column contents. Determining factors include such parameters as polarity, molecular size, and other physical/chemical interactions. After a sample has passed through a GC column, a number of different means can be used to identify the separated components or peaks.

MS can refine the identification of separated VOC peaks that were tentatively identified by their elution times. MS gives the molecular weight of molecules or molecular fragments as they are ionized after passing through a GC column. The spectra that result are often characteristic and can be compared to spectra libraries. For example, a library of $>10^5$ spectra is available from the National Institute of Standards and Technology) (NIST, U.S. Department of Commerce, Washington, DC). MS-MS can enhance identification of sample components.

FIDs respond to the carbon content of compounds and can be used to further enhance detection. PIDs respond to compounds based on their ionization potential and are also very sensitive. PIDs are especially good for identifying unsaturated aromatic compounds and sulfur-containing compounds but are less effective in identifying alkenes, ketones, and aldehydes.

26.6 Types of MVOC Studies

There has been no systematic examination of the specific MVOCs produced by the many microorganisms that may grow in indoor spaces. In fact, relatively few microorganisms have been analyzed for MVOC production. Among the available studies, few relate specifically to MVOC production in buildings. Rather, the majority of existing studies concern microorganisms important in foods (e.g., cheeses and grains) and investigate the production of both desirable and undesirable odors and flavors.

Available studies fall into four general categories (a) laboratory measurements of VOC emissions from pure stock cultures of specific microbial species or strains grown on defined media or food stuffs, (b) laboratory measurements of VOCs from wild-type organisms isolated from contaminated building materials when grown on defined media and actual building materials, (c) field measurements of VOCs in problem buildings and laboratory isolation of airborne microorganisms from these buildings, and (d) field measurements of VOCs in problem buildings and laboratory isolation of microorganisms from contaminated materials from these buildings.

Comparison of results among the available studies is difficult because of differences in the conditions under which the MVOCs were collected, the building substrates on which the microorganisms originally grew, and the substrates and laboratory media used to isolate microorganisms. Preparation of material for testing has also varied, and intracellular compounds released by homogenizing cultures and media could differ from those VOCs released into the headspace of a culture plate, a liquid growth medium, or a building environment. Researchers have also used different analytical procedures, further limiting accurate comparison of measurements.

Laboratory studies of MVOCs emitted by fungi in culture typically identify the volatile compounds recovered. Studies have used various methods to quantify VOC concentrations, for example, reporting compound percentages relative to total VOCs or reporting VOC mass per dry weight of fungus or weight of fungus-containing medium. Other researchers have focused on specific VOC attributes such as the odors associated with GC–MS peaks, using subjective odor evaluations. However, for a number of reasons, neither MVOC origin nor amount measured is strictly comparable across studies. Distinction must be made between the VOCs originating from the growth medium itself and those produced by microbial metabolism. Of the VOCs that microorganisms produce, what is captured and what is identified and quantified depend on the method of sample collection employed, the nature of the sorbent, the eluant or desorption techniques used, and the analytical techniques applied (Batterman, 1995).

There are two primary differences between MVOCs captured from laboratory cultures and growth in buildings. First, in laboratories, fungi and bacteria are grown on defined culture media or substrates with known composition; however, in buildings, the substrates are not well defined and may be highly variable, even changing. Second, in laboratories, organisms are grown in pure culture so that MVOCs are attributable to a single species or strain; whereas, in buildings, MVOCs frequently arise from mixtures of organisms whose identities, relative concentrations, and ecological interactions at the time of MVOC sampling are unknown.

Some researchers looking at MVOCs have focused on determining whether MVOCs are produced in sufficient amounts to contribute to the total exposure burden that has been identified with BRSs. Others investigators have searched for specific VOCs that would serve as signatures of microbial growth. Still others have looked for unique toxic MVOCs that could be associated with specific health effects.

26.6.1 Laboratory Studies of VOC Emissions from Stock Fungi

A clear relationship can be made between a fungal or bacterial species growing in the laboratory on defined medium and the MVOCs a microorganism produces at various growth stages [see 26.2.3]. However, relatively few species and strains of fungi have been studied in this manner. The related genera *Aspergillus* and *Penicillium* are frequently recovered in considerable concentrations from contaminated indoor spaces and have been examined most extensively. VOCs from these fungi have been studied because a number of these compounds are odorous and contribute undesirable tastes and smells to foods. These fungi have also been studied because a number of species and strains are toxigenic in laboratory culture and possibly in buildings. In addition to mycotoxins, these strains produce volatile terpenes (Zeringue et al., 1993; Larsen and Frisvad, 1994b). In competitive building environments, more species and strains of these two genera may produce toxins and their associated MVOCs than do so in the laboratory. Larsen and Frisvad (1994b) demonstrated that mycotoxin production in *Aspergillus* and *Penicillium* spp. is associated with production of sesquiterpenes, including odorous compounds such as pinene, limonene, and camphene (Table 26.1). The production of odorous compounds in association with mycotoxin production is a potentially useful feature for determining whether fungal growth in a building is likely to cause significant health effects [see 26.2.1.2 and 26.2.2].

26.6.2 Laboratory Studies of VOC Emissions from Fungi Isolated from Contaminated Materials

The work described in Section 26.6.1 can be carefully controlled but may not accurately reflect the MVOCs that would be found in building environments [see 26.2.3]. It is important to know if fungal growth on building materials produces VOCs different from those the fungi produce on standard laboratory media. Table 26.2 summarizes MVOCs measured from fungi recently isolated from building environments when grown on agar or building substrates in a laboratory.

26.6.3 Putative MVOCs from Microbially Contaminated Buildings

Because there are myriad VOC sources indoors, it is difficult to attribute VOCs captured in buildings to specific microorganisms. This difficulty is not resolved by isolating microorganisms from problem buildings, growing them in pure culture, and measuring their MVOC emissions [see 26.6.2]. Although microorganisms may produce characteristic MVOCs or combinations of MVOCs, no

VOCs have been found that are uniquely produced by any given microbial species or strain. Therefore, the presence of a particular microorganism cannot be predicted from an indoor VOC sample, nor does the isolation of a particular microorganism indicate that particular MVOCs will necessarily be present. Table 26.3 summarizes VOCs detected in microbially contaminated buildings and the fungi and bacteria isolated from the air in these buildings. Table 26.4 summarizes MVOCs detected in problem buildings and the microorganisms recovered from contaminated materials from these buildings.

26.7 Data Interpretation and Applying Study Information

26.7.1 Use of MVOC Detection to Indicate Microbial Contamination

Studies indicate that many microorganisms produce VOCs and that the MVOCs produced overlap with VOCs from other sources. A few MVOCs (e.g., 1-octen-3-ol, 3-methyl-1-butanol, 2-hexanone, and 2-heptanone) may originate specifically from fungi but not solely any single fungus. Other MVOCs (e.g., ethanol, which microorganisms may produce in large amounts) have many additional non-microbial sources. Nevertheless, microbial fermentation deriving from building contamination should be considered when the alcohol is detected by smell or air sampling unless the building is a bar, brewery, or bakery or cleaning products containing ethanol were recently used. Some researchers have proposed using detection of particular VOCs to determine the presence and nature of fungal contamination in indoor air (AIHA, 1996; AQS, 1997). However, the usefulness of sampling for specific compounds to identify fungal contamination remains to be confirmed.

The fraction of the total VOC burden in buildings that may arise from microorganisms is not known, but it can be expected to vary with the nature and extent of indoor microbial growth and the presence of other sources. The few available laboratory measurements indicate that microorganisms may contribute significant amounts of VOCs. However, at present, no simple, rapid, inexpensive, and reliable method is available to identify the nature and extent of indoor microbial growth from measurements of indoor VOC concentrations. Also, no library is available that gives MVOC profiles for combinations of microbial species and substrates (Batterman, 1995). Some degree of fungal and bacterial growth can be found in most indoor environments and humans also contribute VOCs. Researchers need to determine typical MVOC emissions from such background sources.

The presence of moldy, mildewy, musty, spoiled, or rotten odors should be considered a strong indication that a building requires remedial attention. This is especially true if the odors are from terpenes, because these compounds have been identified with mycotoxin biosynthesis

and fungal sporulation, at least in *Aspergillus* spp. However, terpenes are also components of some cleaning products, and the presence of pinene or limonene, for instance, does not necessarily indicate microbial growth. Better markers of microbial growth would be fungi-associated terpene-derived products, such as isoborneol or geosmin. It is not very likely that specific organisms can be identified based on detection of characteristically odorous MVOCs or other more common VOCs. Growth substrate is extremely important in determining which metabolites microorganisms release. Until investigators locate contamination, they do not know the types of materials on which microorganisms are growing. When microbial contamination is obvious, it should be appropriately removed and the underlying moisture source should be controlled. In such cases, VOC sampling is relatively redundant except for research purposes.

Indoor MVOC concentrations and plate counts of culturable microorganisms from environmental air samples have not necessarily been found to be correlated. Microbial growth may be encapsulated within building or structural materials that do not permit spores or microbial fragments to escape, yet which are permeable to VOCs. In such situations, MVOC identification can justify a continued search for and lead to the discovery of hidden microbial growth. Still, odor tracing is probably more useful for locating non-visible or encapsulated microbial contamination than is VOC sampling.

While odor is a useful marker of microbial growth, odor perception may change when various substances are combined, as would be the case when several organisms grew in a building and their MVOCs combined with each other and VOCs from non-microbial sources. VOCs considered characteristic of a fungus or bacterium grown in pure culture on defined medium in a laboratory may not be at all characteristic when that microorganism is part of a microbial mixture growing on an actual building material where the individual contaminants may be at different growth stages. Nevertheless, generic recognition of moldy or musty odors is useful for judging whether there is microbial growth in a building.

26.7.2 Health Consequences of Exposure to MVOCs

At present, the specific contribution of MVOCs to building-related health problems has not been studied. Studies of the health effects of VOCs in general, as well as studies of odor effects, indicate that MVOCs may contribute to the symptoms and complaints that occur among the occupants of microbially contaminated problem buildings. In themselves, odors have been implicated in the exacerbation of asthma and other illnesses. In particular, cacosmics (used here to describe people with a heightened sensitivity to odors but who otherwise are considered healthy) may respond adversely to microbial and other odors.

Other health effects from MVOC exposure may include signs and symptoms of direct and neurogenic inflammation as well as neurotoxic symptoms. The nature and severity of these responses varies with MVOC type and concentration. In addition, the aggregate of a VOC mixture may produce an effect even if the individual components are below known effect thresholds. Therefore, reactions to MVOC exposure may add to the health problems of allergy, infection, and toxicity already known to be related to exposure to biological agents.

26.8 References

Abramson, D.; Sinha, R.M.; Mills, J.T.: Mycotoxins and Odor Formation in Barley Stored at 16 and 20% Moisture in Manitoba. Cereal Chem. 60:350 (1983).

AIHA: Fungal Volatile Organic Compounds in Air. In: Field Guide for the Determination of Biological Contaminants in Environmental Samples, pp. 121-122. American Industrial Hygiene Association, Fairfax, VA (1996).

Alarie, Y.: Sensory Irritation by Airborne Chemicals. CRC Crit. Rev. Toxicol. 2:299-363 (1973).

Air Quality Services: Microbial Volatile Organic Compounds (MVOCs) as an Indicator of Microbial Contaminination in Buildings. Technical Brief No. 97-7A. Air Quality Services, Atlanta, GA (1997).

Batterman, S.; Bartoletta, N.; Burge, H.: Fungal Volatiles of Potential Relevance to Indoor Air Quality, Annual Meeting of the Air and Waste Management Association. Vancouver, Canada, Paper 91-62.9. June (1991).

Batterman, S.A.: Sampling and Analysis of Biological Volatile Organic Compounds. In: Bioaerosols, pp. 249-268. H.A. Burge, Ed. Lewis Publishers, Ann Arbor, MI (1995).

Bayer, C.W.; Crow, S.: Odorous Volatile Emissions from Fungal Contamination — Proceedings of IAQ 93. Philadelphia, PA, November 7-10, 1993, pp. 165-170. Operating and Maintaining Buildings for Health, Comfort, and Productivity. K.Y. Teichman, Ed. American Society of Heating, Refrigerating and Air-conditioning Engineers, Atlanta, GA (1993a).

Bayer, C.W.; Crow S.: Detection and Characterization of Microbially Produced Volatile Organic Compounds — Proceedings of Indoor Air '93, 6th International Conference on Indoor Air Quality and Climate. Helsinki, Finland, July 4-8, 1993. Vol. 2., pp. 33-38 (1993b).

Bayer, C.W.; Crow, S.A.; Noble, J.A.: Production of Volatile Organic Emissions by Fungi — Proceedings of IAQ '95, Practical Engineering for IAQ. Denver, Colorado, October, 1995, pp. 101-105. American Society of Heating, Refrigerating and Air-conditioning Engineers, Atlanta, GA (1995).

Bell, I.R.; Schwartz, G.E.; Petersen, J.M.; et al.: Self-reported Illness from Chemical Odors in Young Adults without Clinical Symptoms or Occupational Exposures. Arch. Environ. Health 48:6-13 (1993a).

Bell, I.R.; Schwartz, G.E.; Petersen, J.M.; et al.: Possible Time-dependent Sensitization to Xenobiotics: Self-Reported Illness from Chemical Odors, Foods and Opiate Drugs in an Older Population. Arch. Environ. Health 48:315-327 (1993b).

Börjesson, T.; Stöllman, U.; Adamek, P.; et al.: Analysis of Volatile Compounds for Detection of Molds in Stored Cereals. Cereal Chem. 60:300-304 (1989).

Börjesson, T.; Stöllman, U.; Schnürer, J.: Volatile Metabolites and other Indicators of *Penicillium aurantiogriseum* Growth on Different Substrates. Applied and Environmental Microbiology, 56(12):3705-3710 (1990).

Brunekreef, B.; Dockery, D.; Speizer, F.E.; et al.: Home Dampness and Respiratory Morbidity in Children. American Review of Respiratory Disease 140:1363-1367 (1989).

Challenger, F.; Higginbotham, C.; Ellis, L.: The Formation of Organo-

metalloidal Compounds by Microorganisms. I. *Trimethylarsine* and *dimethylarsine*. J. Chem. Soc. Trans. 95-101 (1933).

Cometto-Muñiz, J.E.; Cain, W.S.: Efficacy of Volatile Organic Compounds in Evoking Nasal Pungency and Odor. Arch. Environ. Health 48(5):309-314 (1993).

Cone, J.E.; Sult, T.A.: Acquired Intolerance to Solvents following Ppesticide/Solvent Exposure in a Building: A New Group of Workers at Risk for Multiple Chemical Sensitivities? Toxicol. and Ind. Health 8(4):29-39 (1992).

Ezeonu, I.M.; Price, D.L.; Simmons, R.B.; et al.: Fungal Production of Volatiles during Growth on Fiberglass. Appl. Environ. Microbiol. 60:4172-4173 (1994a).

Ezeonu, I.M.; Noble, J.A.; Simmons, R.B.; et al.: Effect of Relative Humidity on Fungal Colonization of Fiberglass Insulation. Appl. Environ. Microbiol. 60:2149-2157 (1994b).

Fidler, A.T.; Baker, E.L.; Letz, R.E.: Neurobehavioural Effects of Occupational Exposure to Organic Solvents among Construction Painters. British Journal of Occup. Med. 44:292-308 (1987).

Fiedler, N.; Maccia, C.; Kipen, H.: Evaluation of Chemically Sensitive Patients. J.O.M. 34 (5):529-538 (1992).

Gareis, M.: Cytotoxicity of Samples Originating from Problem Buildings — Proceedings from the International Conference: Fungi and Bacteria in Indoor Environments: Health Effects, Detection and Remediation. Saratoga Springs, NY, pp. 139-145. E. Johanning and C.S. Yang, Eds. (1995).

Gebefuegi, I.L.; Kettrup, A.: Adsorption of Volatile and Semivolatile Organic Compounds on Particle Surface as Indoor Contaminants. In: Indoor Air an Integrated Approach, pp. 55-58. L. Morawska, N.D. Bofinger, M. Maroni, Eds. Elsevier Science Inc., Tarrytown, New York (1995).

Gravesen, S.; Frisvad, J.C.; Samson, R.A.: Microfungi. Munksgaard Copenhagen, p. 141 (1994).

Harris, N.D.; Karahadian, C.; Lindsay, R.C.: Musty Aroma Compounds Produced by Selected Molds and Actinomycetes on Agar and Whole Wheat Bread. J. Food Protection 49(12):964-970 (1986).

Hudnell, H.K.; Otto, D.A.; House, D.E.; et al.: Exposure of Humans to a Volatile Organic Mixture. In Sensory, II. Archives of Environmental Health 47(1):31-38 (1992).

Jaakola, J.J.K.; Jaakola, N.; Ruotsalainen, R.: Home Dampness and Molds as Determinants of Respiratory Symptoms and Asthma in Preschool Children. J. Exposure Analysis Environ. Epidemiol. 3, Suppl. 1:129-142 (1993).

Jarvis, B.B.: Mycotoxins in the Air: Keep Your Buildings Dry or the Bogeyman Will Get You — Proceedings from the International Conference: Fungi and Bacteria in Indoor Environments: Health Effects, Detection and Remediation. Saratoga Springs, NY, October 67, 1994, pp. 35-44. E. Johanning and C.S. Yang, Eds. (1995).

Joki, S.; Saano, V.; Reponen, T.; et al.: Effect of Indoor Microbial Metabolites on Ciliary Function in Respiratory Airways — Proceedings of Indoor Air '93, 6th International Conference on Indoor Air Quality and Climate. Helsinki, Finland, July 4-8, 1993, Vol. 1., pp. 259-263 (1993).

Kaminski, E.; Libbey, L.M.; Stawicki, S.; et al.: Identification of the Predominant Volatile Compounds Produced by *Aspergillus flavus*. Applied Microbiol. 24(5):721-726 (1972).

Kaminski, E.; Stawicki, S.; Wasowicz, E.: Volatile Flavor Compounds Produced by Molds of *Aspergillus, Penicillium*, and *Fungi imperfecti*. Applied Microbiol. 27(6):1001-1004 (1974).

Karahadian, C.; Josephson, D.B.; Lindsay, R.C.: Volatile Compounds from *Penicillium sp.* Contributing Musty-earthy Notes to Brie and Camembert cheese flavors. J. Agric. Food Chem. 33:339-343 (1985a).

Karadian, C.; Josephson, D.B.; Lindsay, R.C.: Contribution of *Penicillium sp.* to the flavors of Brie and Camembert cheese flavors. J. Dairy Sci. 68:1865-1877 (1985b).

Koren, H.; Devlin, R.; House, D.; et al.: The Inflammatory Response of the Human Upper Airways to Volatile Organic Compounds (VOC) — Proceedings of the 5th International Conference on Indoor Air Quality and Climate, Indoor Air '90. Toronto, Canada, July 29-August 3, 1990, 1:325-329 (1990).

Koren, H.S.; Graham, D.E.; Devlin, R.B.: Exposure of Humans to a Volatile Organic Mixture. Inflammatory response, III. Archives of Environ. Health 47(1):39-44 (1992).

Kraut, A.; Lili, R.; Marcus, M.; et al.: Neurotoxic Effects of Solvent Exposure on Sewage Treatment Workers. Arch. Environ. Health 43(4):263-268 (1988).

Larsen, T.O.; Frisvad, J.C.: A Simple Method for Collection of Volatile Metabolites from Fungi Based on Diffusive Sampling from Petri Dishes. J. Microbial Methods. 19:297-305 (1994a).

Larsen, T.O.; Frisvad, J.C.: Production of Volatiles and Presence of Mycotoxins in Conidia of common indoor *Penicillia* and *Aspergilli*. In: Health Implications of Fungi in Indoor Environments, pp. 251-279. Air Quality Monograph, Vol. 2. R.A. Samson, B. Flannigan, M.E. Flannigan, et al., Eds. Elsevier, New York, NY (1994b).

Mattheis; Roberts: Identification of Geosmin as a Volatile Metabolite of *Penicillium expansum*. Appl. Environ. Microbiol. 58(9):3170-3172 (1992).

McJilton, C.E.; Reynolds, S.J.; Streifel, A.J.; et al.: Bacteria and Indoor Odor Problems — Three Case Studies. Am. Ind. Hyg. Assoc. J. 51(10):545-549 (1990).

Meggs, W.J.: RADS and RUDS — The Toxic Induction of Asthma and Rhinitis. Clinical Toxicology 32(5):487-501 (1994).

Meggs, W.J.; Cleveland, C.H.: Rhinolaryngoscopic Examination of Patients with Multiple Chemical Sensitivity Syndrome. Arch. Environ. Health 48:14-18 (1993).

Miller, J.D.; Laflamme, A.M.; Sobol, Y.; et al.: Fungi and Fungal Products in Some Canadian Houses. Internat. Biodeterior. 24:103-120 (1988).

Mølhave, L.; Bach, B.; Peterson, F.: Human Reaction to Low Concentrations of Volatile Organic Compounds. Environ. Intl. 12:167-175 (1986).

Muller, C.H.: Allelopathy as a Factor in Ecologic Progress. Vegetatio. 18:348-357 (1969).

Norbäck; Wieslander, G.; Ström, G.; et al.: Exposure to Volatile Organic Compounds of Microbial Origin (MVOC) during Indoor Application of Water-based Paints. Indoor Air 5:166-170 (1995).

Odum, E.P.: Fundamentals of Ecology, 3rd edition, p. 227. W.B. Saunders Co., Philadelphia, PA (1971).

Otto, D.; Mølhave, L.; Rose, G.; et al.: Neurobehavioral and Sensory Irritant Effects of Controlled Exposure to a Complex Mixture of Volatile Organic Compounds. Neurotoxicology and Teratology 12:649-652 (1990).

Price, D.L.; Ezeonu, I.M.; Simmons, R.B.; et al.: Fungal Colonization and Water Activity Studies of Heating, Ventilation, and Air Conditioning System Insulation Materials from a Sick Building. In: Indoor Air an Integrated Approach, pp. 321-324. L. Morawska, N.D. Bofinger, M. Maroni, Eds. Elsevier Science Inc., Tarrytown, NY (1995).

Rivers, J.C.; Pleil, J.D.; Wiener, R.W.: Detection and Characterization of Volatile Organic Compounds Produced by Indoor Air Bacteria. J. Exp. Anal. Environ. Epidemiol. Suppl. 1:177 (1992).

Ryan, C.M.; Morrow, L.A.; Hodgson, M.: Casosmia and Neurobehavioral Dysfunction Associated with Occupational Exposure to Mixtures of Organic Solvents. Am. J. Psychiatry 145(11):1442-1445 (1988).

Schleibinger, H.; Böck; Rüden, H.: Occurrence of Microbiologically Produced Aldehydes and Ketones (MVOC) from Filter Materials of HVAC Systems — Field and Laboratory Experiments — Proceedings of IAQ '95: Practical Engineering for IAQ. Denver, CO, October 1995, pp. 91-100. American Society of Heating, Refrigerating and Air-conditioning Engineers, Atlanta, GA (1995).

Shim, C.; Williams, H.: Effect of Odors in Asthma. Amer. J. of Med. 80:18-22 (1986).

Shusterman, D.: Critical Review: The Health Significance of Environmental Odor Pollution. Arch. Environ. Health 47(1):76-91 (1992).

Seifert, R.M.; King, A.D.: Identification of Some Volatile Constituents of *Aspergillus clavatus*. J. Agri. Food Chem. 30:786-790 (1982).

Smith, J.E.; Anderson, J.G.; Lewis, C.W.; et al.: Cytotoxic Fungal Spores in the Atmosphere of the Damp Domestic Environment. FEMS Microbiology Letters 100:337-344 (1992).

Spengler, J.; Neas, L.; Nalai, S.; et al.: Respiratory Symptoms and Housing Characteristics — Proceedings of Indoor Air '93, 6th International Conference on Indoor Air Quality and Climate. Helsinki, Finland, July 4-8, 1993, Vol 1., 165-170 (1993).

Ström, G.; Palmgren, U.; Wessén, B.; et al.: The Sick Building Syndrome — An Effect of Microbial Growth in Building Constructions? — Proceedings of Indoor Air '90, 5th International Conference on Indoor Air Quality and Climate. Toronto, Ontario, Canada, July 29-Aug. 3, 1990, Vol 1., pp. 173-178 (1990).

Ström, G.; West, J.; Wessén; et al.: Quantitative Analysis of Microbial Volatiles in Damp Swedish Houses. In: Health Implications of Fungi in Indoor Environments, pp. 291-305, Air Quality Monograph, Vol. 2. R.A. Samson, B. Flannigan, M.E. Flannigan, et al., Eds. Elsevier, New York, NY (1994).

Sunesson, A.L.; Vaes, W.H.J.; Nilsson, C.A.; et al.: Identification of Volatile Metabolites from Five Fungal Species Cultivated on Two Media. Applied and Environ. Microbiol. 61(8):2911-2918 (1995).

Verhoeff, A.P.; van Strien, R.T.; van Wijnen; et al.: Damp Housing and Childhood Respiratory Symptoms: The Role of Sensitization to Dust Mites and Molds. Am. J. of Epidemiol. 141(2):103-110 (1995).

Ware, J.S.; Spengler, J.D.; Neas, L.M.; et al.: Respiratory and Irritant Effects of Ambient Volatile Organic Compounds: The Kanawha County Health Study. Am. J. Epidemiol. 137 (12):1287-1301 (1993).

Warren, D.W.; Walker, J.C.; Drake, A.F.; et al.: Assessing the Effects of Odorants on Nasal Airway Size and Breathing. Physiology and Behavior 51:425-430 (1992).

Weschler, C.J.; Hodgson, A.T.: Indoor VOCs: Is What You Measure on Tuesday Night the Same as what You Measure on Wednesday afternoon? — IAQ '92 Environments for People. San Francisco, CA, October 19-21, 1992, pp. 213-219. American Society of Heating, Refrigerating, and Air-conditioning Engineers, Atlanta, GA (1993).

Wessén, B.; Ström, G.; Schoeps, K.O.: MVOC Profiles — A Tool for Indoor-Air Quality Assessment. In: Indoor Air an Integrated Approach, pp. 67-70. L. Morawska, N.D. Bofinger, M. Maroni, Eds. Elsevier Science, Inc., Tarrytown, NY (1995).

Whillans, F.D.; Lamont, G.S.: Fungal Volatile Metabolites Released into Indoor Air Environments: Variation with Fungal Species and Growth Media. In: Indoor Air an Integrated Approach, pp. 47-50. L. Morawska, N.D. Bofinger, M. Maroni, Eds., Elsevier Science, Inc., Tarrytown, NY (1995).

Williams, E.J.; Sievers, R.E.: Synthesis and Characterization of a New Sorbent for Use in the Determination of Volatile, Complex-forming Organic Compounds in Air. Anal. Chem. 56:2523 (1984).

Yang, C.S.: Understanding the Biology of Fungi Found Indoors — Proceedings from the International Conference: Fungi and Bacteria in Indoor Environments: Health Effects, Detection and Remediation. Saratoga Springs, NY, October 6-7, 1994, pp. 131-137. E. Johanning, C.S. Yang, Eds. (1995).

Zeringue, H.J., Jr.; Bhatnagar, D.; Cleveland, T.E.: $C_{15}H_{24}$ Volatile Compounds Unique to Aflatoxigenic Strains of *Aspergillus flavus*. Applied and Environmental Microbiology 59(7):2264-2270 (1993).

INDEX

Numbers in this index refer to section numbers, not page numbers. A "T" preceding a number refers to a table number. An "F" preceding a number refers to a figure number.

A

Acanthamoeba spp., 20.1.1.2, 20.1.4, 20.2, 20.2.1.2, 20.4.1, 20.4.4

Acaricides, 22.6.1

Acarus siro (see Storage mites)

Actinomycetes, 8.2.2.1, 18.1.1, 18.5.1, 26.2.2.2

Acute viral diseases, 8.2.3.3, 21.2

Acute hazard/health risk, 5.2.1.2

Adhesive-tape sampling, 12.2.2.1

Aerodynamic diameter (d_a), 11.3.2.4

Aflatoxins, 6.5.2, 24.1.2, 24.2.2.3, 24.4.2, T24.1, T24.2, 26.2.1.2

Agar-contact sampling, 12.2.2.1

Agreement ratios, 13.6.2

Aggressive sampling, (see also Worst-case assessment)

Agriculture/agricultural environments, 3.1, 3.3.3, 3.4.1, 4.2, 7.4.2.1, 8.2.2, 9.1, 15.2, 19.4.1.1, 19.4.3, 23.2.2.2, 23.2.2.3, 23.2.3, 23.5, 23.6.2, 24.1.2, 24.1.4, 24.2.4.1,

Air barriers/air retarders, 10.4.3.1, 10.4.3.2, F10.3, F10.4

Air cleaners, 9.8.6.4, 10.5.6

Air conditioning/conditioners, 3.2.1, 3.4.2, 4.3.1, A4.A, T22.6

Air samplers/air sampling (Ch 11), T11.1
    selection, 5.2.2, T5.7, T11.1, 11.5

Air washers, 4.3.4, 4.3.4.3

Airborne transmission, 9.4.1.2, 9.7

Airflow calibration, 11.4.1, 11.4.2, T11.3

Alimentary toxic aleukia (ATA), 24.2.2.4

Allergen standardization, 25.5

Allergens (Ch. 25, 25.1-26.6), T6.1, 7.4.5
    amebic allergens, antigens, 20.2, 20.2.2, 20.4.5, 25.1.2, T25.2
    bird antigens, 12.1.3.2, 25.1.5, T25.2
    bacterial antigens, T6.1, 6.5.2, 7.4.5.3, 25.1.2, T25.2, 25.5.1
    cat allergen, 4.4.4, 7.4.5.5, 25.1.6, T25.2, 25.5.4
    cockroach allergen, 4.4.4, 6.5.2, 7.4.5.4, 12.1.3.2, 25.1.4, T25.2, 25.5.3
    defined, 25.1, T25.1
    dog allergen, 4.4.4, 7.4.5.5, 25.1.6, T25.2, 25.5.4
    fungal allergens/allergic fungi, T6.1, 6.5.2, 7.4.5.2, 19.2.2, 19.5.1.4, 25.1.1, 25.1.2, T25.2, 25.5.1
    house dust and storage mite allergens, 4.4.4, T6.1, 6.5.2, 7.4.5.1, 12.1.3.2, 22.2, 22.4, 22.5, 25.1.3, T25.2
    insect allergens, 25.1.4
    rodent antigens, 4.4.4, 25.1.6

Allergic alveolitis (see Hypersensitivity pneumonitis)

Allergic bronchopulmonary aspergillosis (ABPA), T3.2, 24.2.2.1, 25.2.2.2

Allergic bronchopulmonary mycosis (ABPM), T3.2, 25.2.2.2, 25.5.1

Allergic responses (see Immune responses)

Allergies, defined, 25.1, T25.1

Alternaria spp., 7.4.5.2, 19.4.4, 25.5.1

Amebae (Ch 20), T6.1, 7.4.3, 8.2.2.1, 9.4.1.1, 20.1.2

American Industrial Hygiene Association (AIHA), 5.2.3, 6.9.2, 19.5.1.3

Analytical methods
    broad methods, 6.1.1.1,
    focused methods, 6.1.1.3
    indicator methods, 6.1.1.2

Andersen samplers (see Multiple-hole impactors)

Animals droppings (see also Bats, Birds, Cats, Cockroaches, Dogs, and House dust mites) 4.2, 4.3.2.2, 15.6, 19.5.3

Anthrax, 9.4.1.2, 18.2.1

Antibodies, defined, T25.1

Antigens (see also Allergens), defined, T25.1

Antimicrobial agents (Ch 16), 16.2, T16.1
    precautions, 16.3.4
    use, 10.2.4

Ascomycetes, 19.1.3.2, T19.1

Aspergillosis, 24.2.2.1

Aspergillus spp., 7.4.5.2, 8.4, 24.1.4, 24.2.2.3, 26.1.2, 26.2.2.1, 26.3.4, 26.6.1
    A. flavus 24.1.2, 26.2.1.2
    A. fumigatus, 19.4.4, 19.5.1.2, T24.1, T24.2, 24.2.2.1, 24.2.2.5, 24.2.3, 24.2.4.1, 25.5.1

Aspiration, 3.5.1, 9.4.2

Assayable biological contaminants, 1.2.5, 2.1, 6.1.1.3, 6.1.2.3

Assays (see Analytical methods)

Asthma (allergic), T3.2, 3.3, 3.3.2, 3.4.3, 8.2.1.2, 18.2.2, 8.5.3, 19.2.2.1, 25.5.1, 26.3.1, 26.7.2
    defined, T3.2, T25.1, 25.2.2.1
    diagnosis and treatment, 3.3.2, 8.2.1.2, 8.2.1.4

Atopy/atopic persons, defined, T25.1

Average/baseline/typical exposure assessment, 5.2.1, T5.4, T5.6, 5.2.1.2, 14.2.3.1

B

Bacteria (Ch 18), 4.2, T6.1, 7.4.1
    in indoor air, 7.4.1.2, 18.1.4.2
    in outdoor air, 7.4.1.1, 4.2, 18.1.4.1

Bacteria/amebae interactions, 20.1.3, 20.1.4

Bacterial cell-wall components (Ch 23 23.7.1-23.7.4), 6.7, 18.2.3

Balamuthia spp., 20.2, 20.2.1.2

Basidiomycetes, 19.1.3.2, T19.1

Bats/bat droppings, 4.2, 4.3.2.2, 4.6.1, 9.2.1, 19.2.1, 19.3.2, 19.5.1.2, 19.5.3.3
Beta-glucans (see Glucans) (Ch 24)
Bioaerosols, 1.1, 17.1
Bioassays, 6.5
Biocide (Ch 16), 3.3.2, 16.2, T16.1
    for sample preservation, 18.3.4, 19.3.4, 23.3.2
    precautions, 16.2.3, T16.5, 16.2.6
    selection, 16.2.2.3, T16.3
    use, 10.2.4, 10.6, 15.3, 15.4
Biofilm, 4.3.2.2, 4.3.4.2, 18.1.4.2, 18.3.2
Biological agent, 1.1
    culturable and countable biological agents, 1.2.2, 1.2.3, 2.1
Biological contamination, 1.1
Biologically derived airborne contaminants, 1.1, 17.1
Biomarkers, 14.1.4.4, 24.4.4
Birds/bird droppings/nesting material, 4.2, 4.3.2.2, 4.6.1, 9.2.1, 19.2.1, 19.3.2, 19.5.1.2, 19.5.3.3
Blastomyces dermatitidis/blastomycosis, 9.4.1.2, 19.2.1, 19.5.3.3
Blastomycetes, 19.1.3.3
Blattella spp., 25.1.4
    Bla g, T25.4
Blomia spp., 22.1.2, T22.1, 22.1.2.2, 22.2
Bounce (particle), 11.3.2.5
Bronchitis, T3.2, 9.3.6
Brucellosis, 9.4.1.2, 18.2.1
Building Assessment Survey and Evaluation (BASE) studies, 7.3.2
Building envelope, 10.4.2
Building inspections/walkthrough inspections (Ch 4)
Building-related illnesses (BRIs), 3.1, T3.1, T3.2, 3.8, 7.3.1, 8.1, 8.6, 25.2.2.4
Building-related symptoms (BRSs), 3.1, T3.1, T3.2, 3.2.1, 3.2.2, 3.6, 3.7, 3.8, 7.3.1, 8.1, 8.3, 19.2.3.2, 19.2.4, 23.2.2.3, 24.7.4, 25.2.2.4, 26.6
    diagnosis and treatment, 8.3.1, 8.3.2
Bulk samples/sampling (12.1-12.1.4.2), 6.3.1.3

C

Cacosmia/Cacosmic persons, 26.3.3, 26.7.2
Cancer, 3.4.3/carcinogens/carcinogenicity, 24.2.2, 24.2.2.3
Canis familiaris/*Can f* I, 7.4.5.5, T25.4, 25.5.4
Carpets/carpeting, T4.1, 4.4.4, 18.3.2, 22.1.3.1, T22.6
Case definitions, 2.2.1, 8.6.1
Cause-effect relationships, 1.2.3, 14.1.1, T14.1
Cats (see Allergens, cat allergens; *Canis familiaris*)
Cell walls/cell-wall components (Ch 23), T6.1
Central nervous system (CNS) effects, 26.3.2
Centrifugal impaction/impactors, 11.3.1, F11.5, 11.5.1.3, 18.3.3
Chain of custody, 5.3
Chemical assays, 6.7

Chi-square test, 13.6.3, 13.6.6
Chicken pox (see Varicella)
Chlamydia/chlamydiae/chlamydial diseases, 18.1.3
Chlorine dioxide, 16.2.4.1
Chronic hazard/health risk, 5.2.1.2
Cladosporium herbarum, 7.4.5.2, 19.4.4, 25.5.1
Classification of biological agents, 6.2, 17.2, 18.1.2, 19.1.3, 20.1.2, 21.1.2, 22.1.2
Cleaning programs/management, 10.7.2
Cluster analysis, 13.6.3, 13.6.4
Coccidioides immitis/coccidioidomycosis, 9.2.1, 9.4.1.2, 19.2.1, 19.5.3.3
Cockroaches, 25.1.4
Coelomycetes, 19.1.3.3
Cold climates, 10.4.3.1
Collection fluids/liquids, 11.3.3.3
Colony-forming units (CFU), 18.4.1
Common cold, T3.2, 8.2.3.3, 9.1, 9.2.2, 21.1.4
Communication, 2.7, 8.6.1
Competition/competitors, microbial, 26.1.2
Compost, 18.1.3, 19.5.3.3, 24.1.4
Concentration calculations, 13.2.1, 13.2.2, 13.2.3, 13.2.4,
Condensers (evaporative), 3.4.2, 3.5.1, 4.3.2.2, 10.5.4, 10.6
Condesate pans (see Drain pans)
Condensation/control of, 10.4.3, 10.4.3.1, 10.4.3.2
Confidentiality, 8.6.1
Confirmed presence, 19.5.1.3
Conidia, 19.1.2
Conjunctivitis, T3.2, 3.3.1, 3.4.3
Contact transmission, 9.4.1.1
Contagious disease (see Infectious disease)
Containment, 9.8.2, 15.2.1, 15.2.2, 15.2.3, T15.1,
Contamination
    fecal/sewage contamination, 6.1.1.2, 10.2.4, 18.5.3
    microbial, 4.6.1
Cooling coils, F4.1, 4.3.4.1, 10.4.4.2
Cooling towers, 3.5.1, 4.2, 4.3.2.2, 10.5.4, 10.6, 15.7, 18.3.2
Cryptococcus neoformans/cryptococcosis, 3.5.2, 4.3.2.2, 9.2.1, 9.4.1.2, 15.6, 19.5.1.2, 19.5.3.3
Culture/culture-based analyses, 6.1.1.1, 6.1.2.1, 6.3, 12.1.3.1, 12.2.3.1
Culture media (see also Media), 6.3.2
    for bacteria
        nutrient agar, 6.3.2, 18.4.1.1, T18.4
        R2A agar, 6.3.2, 18.4.1.1, T18.4
        soybean casein digest agar (SCDA), 6.3.2, 18.4.1.1, T18.4
        tryptic soy agar/trypticase soy agar (TSA) (see soybean casein digest agar)
    for fungi
        cellulose agar, 19.4.1.1, T19.5
        dichloran glycerol (DG18), 6.3.2, 19.4.1.1, T019.5

malt extract agar (MEA), 19.4.1.1, T19.5
rose bengal, 6.3.2, 19.4.1.1
Cut-off/cutpoint diameter (d_{50}), 11.3.2.3, T11.2, F11.9
Cyclone samplers, 11.3.1, F11.5b, 11.3.3.2, 11.5.1.4
Cysts (amebic), 20.1.1.1, 20.1.1.2, 20.2.1.1, 20.3.2, 20.4.4, 20.5
Cytotoxic effects/assays, 6.5.3, 24.2.2, 26.1.2, 26.3.4

D

Data, 7.2.1
analysis (Ch 13), 7.2.3
distributions, 13.3.1, 13.3.2
evaluation (Ch 14)
interpretation (Ch 7), 7.2.4
quality, 7.2.2
Dehumidifiers/dehumidification, 4.3.1
Deposition (particle), 9.8.6.2
Dermatitis, T3.2
Dermatophagoides farinae, T22.1, 22.1.2.2, 22.2, 22.4.3
Der f 1, 22.4.3, T25.4, 25.5.2
Dermatophagoides microceras, T22.1, 22.1.2.2
Dermatophagoides pteronyssinus, T22.1, 22.1.2.2, 22.2, 22.4.3
Der p 1, 22.4.3, T25.4, 25.5.2
Dermatophagoides spp., 20.1.2
Detection limits, 5.2.3, T5.6, 14.2.2
Dose-response relationships, 1.2.4, 7.3.2, 8.5.1, 14.1.1, 14.1.4.1
Drain pans, F4.1, 4.3.4, 4.3.4.2, , 10.4.4.3, F10.7, 18.3.2
Drift eliminators, 10.6
Droplet nuclei, 9.4.1, 18.1.4.2
Droplet transmission, 9.4.1, 9.4.1.1
Ducts/ductwork, F4.1, T4.1, 4.3.5.1, 10.4.4.5
Duct cleaning, 15.3, 15.5.1, 16.2.2.2
Duct linings, 4.3.4.1, 4.3.5.1,
Dust mites (see House dust mites) (Ch. 22)
Dust samples/sampling, 12.1.1, 12.1.2, 22.3.1.1, 25.3

E

Echinamoeba spp., 20.1.4
Ehrliciae, 18.1.3
Endospores, 18.1.1
Endotoxin (Ch. 23, 23.1-23.7.4), 3.2.2, 3.3.2, 3.4.1, 3.4.3, T6.1, 6.7, 7.4.6, 8.2.2.1, 12.1.3.2, 18.2.2, 18.2.3, T18.3, 19.2.3.2, box 23-2, 24.2.2.6
indoor concentrations, 23.1.2, 23.2.2.3
outdoor concentrations, 23.1.2
acute effects, 23.2.2.1
sample analysis
chemical assay, 23.1.2, 23.4.1, 23.4.3
chronic effects, 23.2.2.2
endotoxin units (EU), 23.1.2
extraction, 23.3.1, 23.4.1
dilution-dependent (DDI), 23.4.2.3
dilution-independent (DII), 23.4.2.3

Limulus amebocyte lysate (LAL) assay, 6.5.3, 23.2.2.3, 23.4.2, 24.7.3
relative limit value (RLV), 7.4.6, 23.5, 23.6.1, 23.6.2
Enzyme-linked immunosorbent assays (ELISAs), 6.5.2, F6.1, 25.2.3.1
Enzymes
bacterial proteases, 6.7, 7.4.5.3, 18.2.2, 25.5.1
fungal enzymes, 19.1.4
Epidemiology/epidemiological investigations, 2.2.1, 3.2.1, 3.7, T3.3
Equilibrium relative humidity (ERH), 10.3.3, T10.1
Ergosterol, 6.1.1.2, 6.7, 12.1.3.2, 19.1.4, 19.3.2, 19.4, 19.4.3
Ergotism, 24.1.2, 24.2.1
Ethanol, 16.2.6, 19.1.4, 26.2.1.1, 26.2.2, 26.2.3, 26.3.1, 26.7.1
Ethylene oxide, 16.2.5
Euroglyphus spp., 22.1.2, T22.1, 22.1.2.2, 22.2
Evaporative coolers, 4.3.1
Exotoxins, 18.2.3, T18.3
Exposure challenge tests, 8.2.1.2, 25.2.3.3
Exposure (far-field/near-field exposure), 9.7.1, 9.7.2
Exposure-response relationships, 1.2.2, 14.1.4.2
Extensive contamination, 4.6.2, 7.3.1, 14.1.2, 15.2, T15.1, 15.2.3, 15.2.3.6, 15.3, 15.5.2, 16.1.1, 16.2.4

F

Factor analysis, 13.6.3
Faenia spp., 18.1.1, 18.2.2
Fancoil units, T4.1, 4.4.2
Farmer's lung, 3.3.3, 18.2.2, 25.2.2.3
Fatty acids, 6.7, 23.4.3
Felis domesticus/   *Fel d* I, 7.4.5.5, T25.4, 25.5.4
Filters/filtration
air cleaning, F4.1, T4.1, 4.3.3, 9.8.6.4, 10.4.4.1, 10.5.3
high-efficiency filters, 9.7.2
air sampling, 11.3.1, F11.7, 11.3.4, 11.5.1.5
arrestance, 10.5.3
dust spot efficiency, 10.5.3
Flooding/floods, 10.4.1
Formaldehyde, 16.2.5
Fumonisins, T24.1, 24.2.2.4, 24.4.2
Fungal toxins (Ch 24), 3.2.2, 3.4.3, T6.1, 6.7, 7.4.8, 19.2.3.1, 26.1.2, 26.2.1.2, 26.6.1
Fungus/fungi (Ch 19), T6.1, 7.4.2
in outdoor air, 4.2
Fusarium spp., 24.1.4

G

Geosmin, 26.2.2.1, 26.2.2.2, 26.7.1
Glucans (Ch. 24, 24.7.1-24.7.5), 3.2.2, 6.1.1.2, 6.5.2, 6.7, 7.4.9, 19.1.4, 19.2.3.2, 19.3.2, 19.4, 19.4.3, 24.2.2.6
Grain/grain dust, 8.2.2, T8.1, , 23.1.2, 23.3.2, 24.1.4
Gram-negative bacteria (GNB), 7.4.1.3, 18.1.2, 18.2.2, 18.4.1.3

Gram-positive bacteria (GPB), 8.2.2.1, 18.1.2
Gram stain/Gram stain reaction, 18.1.2
Granulomatous amebic encephalitis (GAE), T3.2, 20.2.1.2
Gravitational deposition (see Settle plates/slides)
Growth media (see Culture media)
Guanine, 6.1.1.2, T6.1, 6.7, 7.4.5.1, 22.4.4
Guidelines, 2.1, 7.3.2, 14.2.6, 19.5.1.1, 19.5.1.2, 19.1.5.3, 19.5.1.4

H

Hantavirus pulmonary syndrome (HPS), 3.5.2, T3.2, 4.6.2, 9.4.1.2, 21.1.4
Hartmannella spp., 20.1.4, 20.2
Hay fever (see Rhinitis)
Hazard categories (infectious agents), 9.3.3
Health-care facilities/workers, 1.2.4, 2.6.3, 3.1, 8.2.3.2, 9.1, 9.2.2, 9.8.4, 9.8.5, 10.5.3, 16.2.3
Health effects (of bioaerosols) (Ch 3)
Heat exchangers, T4.1, 4.3.4
Hidden microbial contamination/negative samples, 14.2.3.3, 14.2.5, 26.3.5
Histoplasma capsulatum/histoplasmosis, T3.2, 3.5.2, 4.3.2.2, 9.2.1, 9.4.1.2, 15.6, 19.2.1, 19.5.1.2, 19.5.3.3
Hot, humid climates, 10.4.3.2
House dust mites (Ch 22), 23.2.2.3
   moisture/humidity requirements, 22.1.4.1, 22.1.4.2, 22.1.4.3, T22.6
Human-source/human-commensal organisms, 4.4.4, 7.4.1.2, 9.2.2, 18.1.4.2, 18.5.2
Humidifier fever, 3.4.1, 8.2.2, 18.2.3, 23.2.2.1
Humidifiers, 3.3.2, F4.1, T4.1, 4.3.1, 4.3.4.3, 4.3.5.1, 8.2.2, 8.2.2.1, 10.4.4.4, 18.3.2, 23.2.2.1, 23.2.3, 23.6.2
HVAC systems, 4.1.3, 4.3, 4.3.1, F4.1, T4.1, access, 10.5.1
Hydrogen peroxide, 16.2.5, 20.6
Hydrophilic fungi, 19.1.5.2
Hypersensitivity diseases (see also Allergic responses), 3.3, 3.7, 8.2.1, 25.1
   defined, T3.2, T25.1
   treatment/medical management, 8.2.1.4
Hypersensitivity pneumonitis (HP), 3.3, 3.3.3, 4.3.2.2, 19.2.2.2, 24.2.2.2, 25.2.2.3, 25.5.1
   defined, T3.2, T25.1
   diagnosis and treatment, 3.3.3, 8.2.1.3, 8.2.1.4
Hypersensitivity responses (see Immune responses)
Hypochlorites, 16.2.4.2, 20.6
Hyphomycetes, 19.1.3.3
Hypotheses, 2.3, 5.1.1

I

Ig (immunoglobulin), 3.3, T25.1, 25.2.1
Immune system responses, T25.1, 25.2.1, T25.3
Immunization/immunity, 1.2.4, 8.2.3.3, 8.2.3.4, 8.4, 9.8.1, 21.6

Immunoassays, 6.1.1.3, 6.5.2, 19.4.4, 20.2.5, 24.7.3
Immunocompromised/immunodeficient/immuno-suppressed persons, 8.4, 9.2.1, 19.2.1
Impactors/impaction, 5.2.3.1, 5.2.3.2, inertial, 11.3.1, F11.1, 11.3.2, F11.8
Impingers/impingement, liquid, 11.3.1, F11.6, 11.3.3.1, 11.5.1.4
Incubation conditions, 6.3.3
Incubation period (infectious disease), 9.3
Index cases (see Sentinel cases)
Indicator organisms/species, 1.2.3, 7.4.2.2, 19.5.1.1, 19.5.3.2
Indoor/outdoor comparisons, 5.2.1.4, 7.4.2.1, 13.6.3, 14.2.3.2, 19.5.3.1
Induction units, 4.4.2
Infections/infectious diseases, 1.2.4, 2.1, T3.2, 3.5, 20.2.1
Infectious dose, 9.3.1
Inflammation/inflammatory agents, 24.2.2.2
Influenza, T3.2, 8.2.3.4, 9.2.2, 9.4.1.1, 9.4.1.2, 21.1.4, 21.2
Inhalation fevers, T3.2, 3.4, 8.2.2
   humidifier fever, 8.2.2.1, 20.2.2, 20.5, 25.5.1
   Pontiac fever, 8.2.2.2
Insects/insect fragments, 4.2, 4.4.4, T4.1, A4.A, 10.1, 10.5.3, 12.1.1, 25.1.4, 25.2.2.1, 25.5.5.3
Insulation, 4.3.5, 4.4.4, T4.1, 10.4.3.1, 10.4.3.2, 12.1.3, 12.2.4, 26.2.1.2
Irritants/irritation, 3.4.3, 8.2.1.1, 26.3.2
Isoaxial/isokinetic air sampling, 10.2.1

K

Kingdoms, 17.2, F17.1

L

Latent/latency period, 8.2.1
Legal decisions/legal disputes/litigation, 14.2.7
Legionella spp., 8.2.2.2, 18.3.2, T18.5
   L. pneumophila, 8.2.2.2, 18.4.2
   sources, 8.2.2.2, 8.2.3.1, 18.1.4.1, 18.1.4.2
Legionnaires' disease, T3.2, 4.6.1, 4.6.2, 3.5.1, 4.3.2.2, 8.2.3.1, 9.4.1.2, 15.7
   diagnosis and treatment, 8.2.3.1
   control/prevention, 10.2.4, 10.6
Lepidoglyphus destructor (see Storage mites)
Limonene, 26.1.1, 26.1.2, 26.6.1, 26.7.1
Limulus amebocyte lysate (LAL) assay (see Endotoxin)
Lipid A, 23.1.1
Lipopolysaccharides (LPSs), 23.1.1, box23-2
Liquid impingers (see Impingers, impingement/liquid)

M

Macrophages/macrophage function, 24.2.2.1
Mass psychogenic illnesses, 3.6
Measles, 3.5.2, 8.2.3.3, 8.2.3.4, 9.4.1.2, 21.2
Mesophilic microorganisms, 18.1.3, 19.1.5.3
Metabolism/metabolites, 24.1.3, 26.2.1, 26.2.2

Metal-working fluid (MWF), 19.4.1.1, 23.1.2, 23.2.3
Microbial volatile organic compounds (MVOCs) (Ch. 26, 26.1-26.7), 3.2.2, T6.1, 6.7, 7.4.10, 18.2.4, 19.2.4
Micropolyspora faeni (see *Faenia rectivigula*)
Microscopes/microscopy, 6.1.1.1, 6.1.2.2, 6.4, 12.1.3.2
Mildew, 19.1.3.1
Mites (see House dust mites, Storage mites)
Moisture
        carryover, 10.4.4.2
        content (MC), 10.3.2, T10.1
        indices, 10.3
        meters, 4.1.4
Molds (moulds), 19.1.1, 19.1.3.1
Multiple-chemical sensitivity, 3.6
Multiple-hole impactors, 11.3.1, F11.2, 18.3.3, 1.5.1.1
Mumps, 8.2.3.4, 9.4.1.1
Muramic acid, 12.1.3.2, 23.7.3
Mycobacterium spp., 20.1.4
Mycobacterium tuberculosis, 8.2.3.2, 18.3.2, 18.4.1.2
Mycotoxins (see Fungal toxins) (Ch 24)
Myrothecium spp., 24.1.4

N

Naegleria spp., 8.2.2.1, 20.1.1.1, 20.1.4, 20.2, 20.4.1, 20.4.4
        N. fowleri, 20.2.1.1, 20.4.3
Nocardia spp., 18.1.1
Number of samples to collect, 5.2.4, T5.9

O

Ochratoxin, T24.1, 24.2.3, 24.2.4.1, 24.4.2, 24.4.4
Odor thresholds, 26.2.2
Odors, 4.3.4.2, 18.3.2, 19.2.4, 19.5.3, 26.2.2, 26.2.2.1, 26.2.2.2, 26.2.2.3, 26.3.1, 26.7.1, 26.7.2
Opportunistic infections/opportunistic pathogens, T3.2, 3.5.2, 7.4.2.3, 8.4, 9.2.1, 15.6
Organic dust, 4.6.1, 8.2.2, 23.1.2, 23.7.2, 25.2.2.3
Organic dust toxic syndrome (ODTS), T3.2, 3.4.1, 24.2.2.6, 24.2.4.1, 24.7.2
Outdoor air (see also Ventilation rates), F4.1, T4.1, 4.2, 4.3.2, 5.2.1.4,
Outdoor air intakes (OAIs), F4.1, T4.1, 4.3.2.2, 10.5.4
Ozone/ozone generators, 10.5.6, 16.2.4.4, 16.2.5

P

Paper mites, 22.1.2
Particle
        diameter, 4.6.2, 9.4.1, 11.5.2
        removal, 9.8.6
Pathogens/pathogenic microorganisms, 7.4.2.3, 8.4, 9.2.1, 19.5.1.2, 19.5.1.3, 19.5.3.3
Patulin, T24.1, 24.2.2.1
Peak flow measurement/meters, 8.2.1.2, 25.2.3.2
Penicillium spp., 24.1.4, 26.1.2, 26.2.2.1, 26.3.4, 26.6.1
Peptidoglycans (PG), 7.4.7, 18.2.3, 23.7.1, 23.7.2
Performance criteria, air samplers, 11.5.3

Periplaneta americana, 25.1.4
        Per a, T25.4
Personal protective equipment (PPE), 1.2.4, 4.6.2, 8.4, 9.8.5, 15.2, T15.1, 15.2.3.6, 15.6, 16.2.6, 16.3.4, 24.6
Physical examinations, 8.6.1
Pigeons/pigeon droppings/pigeon-breeder's disease, 3.3.3, 19.2.1, 25.2.2.3
Pinene, 26.1.1, 26.1.2, 26.6.1, 26.7.1
Plants/plant litter/microorganisms on plant surfaces, 4.2, T4.1, 4.3.2.2, 8.4, 10.1
Pneumonitis, hypersensitivity (see Hypersensitivity pneumonitis)
Pneumonia, T3.2, 9.3.6
Pollen, 4.2, 25.1.1
Polymerase chain reaction (PCR), 6.1.1.3, 6.6.1, 6.6.2, T6.2
Pontiac fever, T3.2, 4.3.2.2, 8.2.2.2
        diagnosis and treatment, 3.4.2, 8.2.2.2
Porous materials, 4.1.4, 4.3.4.1, 4.4.4, A4.A, 10.2.1, 10.4.1, 10.4.2, 10.4.4.2, 10.4.4.5, 12.2.4, 15.1, 15.2, 15.2.3.4, 15.3, 15.4, 15.5, 15.5.2, 16.2.4, 16.3.3
Positive-hole correction, 11.5.1, A11.B(i)-11.B(iv)
Precipitating antibodies/precipitins, 3.3.3, 25.2.3.1, 25.5.1
Pressure relationships, 9.7.2, 10.4.3.2, F10.5, 15.2.3.2, T15.1
Prevention of microbial contamination (Ch. 10)
        long-term, 10.7.1, 15.8
Primary amebic meningoencephalitis (PAM), T3.2, 20.2.1.1
Principal-component analysis, 13.6.3
Professional conduct, 2.1.4
Proficiency testing
Environmental Microbiology Proficiency Testing (EMPAT), 6.9.2
Protection (see Personal protective equipment)
Psittacosis/*Chlamydia psittaci*, 9.4.1.2, , 15.6, 18.2.1
Pulmonary function/pulmonary function tests, 8.2.1.2, 25.2.3.2
Pulmonary hemorrhage/hemosiderosis, 24.2.4.2

Q

Q-fever, 3.5.2, 9.4.1.2, 18.2.1
Qualifications/proficiency of laboratories and laboratory personnel, 2.1.4, 6.9
Quality control, 6.9.3
Quantum, 9.3.1
Questionnaires, 3.7, 8.5.2

R

Radioallergosorbent test (RAST), 25.2.3.1
Randomization/random sampling, 5.2.1, A5.A, 5.2.1.3
Rank-order comparison, 13.6.5, 13.6.6, T13.5, T13.6, T13.7, T13.8, T13.9
Rapid detection/identification, 11.3.5

Records/recordkeeping, 5.3, T5.10, T5.11
Referral criteria (medical referral), 8.5.2, 8.6.1
Relative humidity (RH), 4.3.4.1, 4.4.3, 10.3.1, T10.1, 10.3.4, 10.4.3
Remediation of microbial contamination (Ch 15) judging effectiveness, 15.5
Replicate samples/tests, 1.2.3
Reportable diseases, 8.5.3
Reservoirs of biological/infectious agents, 2.1.2, 9.2
Respirators/respiratory protection (see Personal protective equipment)
Respiratory infections (Ch 9)
Respiratory tract, 9.3.4, 9.3.5, T9.1, T9.2
Return air, F4.1, 4.5
Rhinitis (allergic), T3.2, 3.3, 3.3.1, 8.2.1.1, 25.2.2.1
rickettsia/rickettsiae/rickettsial diseases, 18.1.3
Risk assessment/communication, 2.7, 4.6.1
Risk factors, 3.7, 8.5
Risks
        acceptable risks, 2.7, 7.3.2, 14.1.4.5
        risk extrapolation, 24.2.5
        significant or substantial risks, 7.4.1.4, 18.5.4
Rodents/rodent droppings/infestation, 4.2, 4.6.1, 9.2.1, 19.3.2
Role of medical professionals, 2.2.1, 3.8, Ch8
Rubella, 8.2.3.3, 8.2.3.4, 9.4.1.2

S
Sample analysis (Ch 6)
Sample handling, 5.1.2, 18.3.4, 19.3.4, 20.3.3, 21.3.2, 22.3.3, 23.3.2, 25.3.3
Sample volume, 5.2.3
Sampling efficiency, 11.2
Sampling time, 5.2.3, T5.5
Satratoxin, T24.1, 24.1.4, 24.2.2.1, 24.2.2.6, 24.2.3
Sentinel cases/health events, 3.3.3, 8.1, 8.5, 8.5.3
Sesquiterpenes, 26.1.2, 26.2.1.2, 26.6.1
Settle plates/slide, 2.4.2.2, 11.3.1
Sick building syndrome (SBS) (see Building-related illnesses, 3.1, T3.1, T3.2
Similarity index, 13.6.4, T13.4
Sinusitis, T3.2, 3.3.1, 8.2.1.1
Skin prick testing, 8.2.1.2, 25.2.3.1
Slit samplers, 11.3.1, F11.3, F11.4, 11.5.1.2, 18.3.3
Soil/soil microorganisms, 4.3.2.2, 8.4, 10.1, 19.2.1, 19.3.2, 19.5.1.4, 19.5.3.3
Sorbents/sorbent sampling, 26.4.1
Sorption isotherms, 10.3.4, F10.1
Source sampling (see bulk and surface sampling) (Ch 12)
Sources, of biological agents, 2.1.2
Sporangiospores, 19.1.2
Spores/bacterial and fungal spores, 18.1.1, 19.1.2
Stachybotrys atra (see *Stachybotrys chartarum*)

Stachybotrys chartarum (atra), 19.4.1.1, 19.5.1.2, 24.1.1, 24.1.4, 24.2.3, 24.2.4.2, 26.3.5
Standards (see Guidelines)
Statistics/statistical tests, 13.5, 13.6,
Statistical significance, 5.2.4, 7.2.3, 13.6.3, 13.6.5, 13.6.6
Sterigmatocystin, 24.2.2.3
Storage mites, 22.1.2.1, 22.1.3.1, T22.2, 22.2
Streptomyces spp., 18.1.1, 26.2.2.2
Supply air, 4.3.5, 4.3.5.2, 4.4.1
Surface samples/sampling (Ch 12, 12.2.1-12.2.4), 6.3.1.3, 12.2.2.2
Surveillance (prospective), 8.6.3
Susceptible persons/susceptibility, 1.2.2, 1.2.6, 9.3.2, 9.8.4.1

T
T-2 toxin, T24.1, 24.1.3, 24.2.2.1, 24.2.2.4, 24.2.3, 24.4.2
TLVs (see Threshold limit values)
Teichoic acids, 23.7.1, 23.7.2
Temperature
        control, 10.4.3, 10.4.3.2
        preferences
                bacteria, 6.3.3.1, 18.1.3
                fungi, 6.3.3.1, 19.1.5.3
Terpenes, 26.1.2, 26.2.1.2, 26.2.2.1, 26.6.1, 26.7.1
Thermoactinomyces spp., 18.1.1
Thermophilic microorganisms, 18.1.3, 18.4.1.2, 19.1.5.3
Threshold Limit Values (TLVs), 1.2, 1.2.6, 23.2.3
Total counts, 1.2.2, 2.1, 18.4.2
Toxigenic fungi, 19.5.1.1, 19.5.1.2, 19.5.1.4, 26.2.2
Toxins (see endotoxin, fungal toxins, and mycotoxins)
Trichothecenes, 24.2.2.4, 24.2.2.6, 24.2.4.2, 26.1.2
Trophozoites (amebic), 20.1.1.1, 20.3.2, 20.4.3, 20.5
Tuberculin skin test (TST), 8.2.3.2
Tuberculosis (TB), T3.2, 3.5.2, 8.2.3.2, 8.5.3
        diagnosis and treatment, 8.2.3.2
        transmission, 8.2.3.2, 9.4.1.2
        multiple-drug resistant (MDR-TB), 8.2.3.2
Tularemia, 9.4.1.2, 18.2.1
Turkey X disease, 24.1.2, 24.2.1

U
Ultraviolet germicidal irradiation (UVGI), 9.8.6.4
Upholstered furniture, 4.4.3, T4.1, 10.2.1, 12.1.1, 12.2.4, 22.1.3.1, T22.6

V
Vaccination (see Immunization)
Vacuum sampling (bulk/surface samples), 4.4.4, 12.1.2, 12.2.2.2, 22.3.1.1, 25.3.1
Valkamfia spp., 20.1.4
Vapor barriers/vapor retarders, 10.4.3.1, 10.4.3.2, F10.3, F10.4
Varicella (chicken pox), 3.5.2, 8.2.3.3, 8.2.3.4, 9.4.1.2, 21.2

Ventilation/ventilation rates, 3.2.1, 4.3.2.1, 9.8.6.3, 10.5.2
    dilution ventilation, 9.8.6.3, 10.4.3.1
    source ventilation, 10.4.3.1
    to control airborne antigens, 10.5.2.2
    to control airborne infectious agents, 10.5.2.1
Viruses (Ch 21), T6.1, 7.4.4
Visible contamination/microbial growth, 7.3.1, 19.3.1, 19.3.2, 19.5.3, 24.5.1, 26.3.1, 26.7.1

W

Walkthrough building inspection (Ch 4)
    checklist, A4.A
Waste-handling/waste-handling environments, 15.1, 15.2, 15.6, T15.1, 19.4.1.1
Water
    activity (a_w), 4.1.4, 10.3.3, T10.1, 19.1.5.2, 19.1.5.3

damage/entry/leaks/intrusion, T4.1, 4.4.3, 10.4.1, F10.2, 10.5.5
Whole air sampling, 26.4.2
Wood/wood chips, 3.3.3, 24.1.4, 25.2.2.3
Worst-case/high exposure assessment, 5.2.1, T5.4, 5.2.1.2, T5.5, T5.6, 5.2.1.2, 14.2.3.1, 14.2.4

X

Xerotolerant microorganisms, 19.1.5.2

Y

Yeasts, T6.1, 19.1.1, 19.1.3.1

Z

Zoonoses/zoonotic infections, 9.2.1
Zygomycetes, 19.1.3.2, T19.1